Biochemistry and Biotechnology

Biochemistry and Biotechnology

Edited by **Oliver Stone**

SYRAWOOD
PUBLISHING HOUSE
New York

Published by Syrawood Publishing House,
750 Third Avenue, 9th Floor,
New York, NY 10017, USA
www.syrawoodpublishinghouse.com

Biochemistry and Biotechnology
Edited by Oliver Stone

International Standard Book Number: 978-1-68286-331-2 (Hardback)

Printed in China.

Contents

	Preface	**VII**
Chapter 1	**Economics and environmental impact of bioethanol production technologies: an appraisal** Anuj Kumar Chandel, Chan ES, Ravinder Rudravaram, M. Lakshmi Narasu, L. Venkateswar Rao and Pogaku Ravindra	**1**
Chapter 2	**Phytochemical antioxidants for health and medicine – A move towards nature** Duduku Krishnaiah, Rosalam Sarbatly and Awang Bono	**20**
Chapter 3	**Telomerase: Roles in aging, cancer and hereditary disease** K Sreenivasulu and M Vijaya Lakshmi	**28**
Chapter 4	**Plant protease inhibitors: a defense strategy in plants** Huma Habib and Khalid Majid Fazili	**36**
Chapter 5	**Biodegradation alternative in the cleanup of petroleum hydrocarbon pollutants** Anthony I Okoh	**54**
Chapter 6	**Cationic lipids used in non-viral gene delivery systems** Sarder Nasir Uddin	**67**
Chapter 7	**Status of biotechnology in Eastern and Central Africa** Godliving Y. S. Mtui	**77**
Chapter 8	**Lignocellulosic ethanol production: Current practices and recent developments** Bishnu Joshi, Megh Raj Bhatt, Dinita Sharma, Jarina Joshi, Rajani Malla and Lakshmaiah Sreerama	**93**
Chapter 9	**RNAi in medicine: current and future perspectives** L. Sudarsana Reddy, V. Sarojamma[$] and V. Ramakrishna	**104**
Chapter 10	**Phenolic compounds and related enzymes as determinants of sorghum for food use** Mamoudou H. Dicko, Harry Gruppen, Alfred S . Traoré, Alphons G. J. Voragen and Willem J. H. van Berkel	**116**
Chapter 11	**Bioconversion of pentose sugars into ethanol: A review and future directions** Anuj Kumar Chandel, Gajula Chandrasekhar, Konakalla Radhika, Rudravaram Ravinder and Pogaku Ravindra	**134**
Chapter 12	**Immunomodulatory effects of carbohydrates and advanced glycation end products** Kankia, H. I.	**147**
Chapter 13	**RNAi as a novel therapeutic platform technology for oncological solutions** Gowhar Shafi, Kaiser Jamil, Atya Kapley, Hemant J. Purohit and Mohana Ch Vamsy	**158**

Chapter 14 **The large-scale use of feruloyl esterases in industry** 174
Ahmed E Fazary and Yi-Hsu Ju

Chapter 15 **Emerging infections and bioterrorism emergencies: Where do we go from here?** 190
I. O. Okonko, M. O. Ojezele, E. T. Babalola, J. C. Nwanze, O. K. Mejeha and T. A. Amusan

Chapter 16 **Non-rhizobial nodulation in legumes** 200
D. Balachandar, P. Raja, K. Kumar and SP. Sundaram

Chapter 17 **Transgenic plants: Successes and controversies** 209
Alfred O. Ubalua

Chapter 18 **Transition metal complexes as potential therapeutic agents** 219
Shazia Rafique, Muhammad Idrees, Anwar Nasim, Haji Akbar and Amin Athar

Chapter 19 **Starch phosphorylase: Biochemical and biotechnological perspectives** 227
Rachana Tiwari and Anil Kumar

Chapter 20 **Biotechnology and food security in developing countries** 242
Nyerhovwo J. Tonukari and Douglason G. Omotor

Chapter 21 **Computational aspects of thematics: application to protein tyrosine phosphatase role in diabetes mellitus** 253
Allam Appa Rao

Permissions

List of Contributors

Preface

The main aim of this book is to educate learners and enhance their research focus by presenting diverse topics covering this vast field. This is an advanced book which compiles significant studies by distinguished experts. This book addresses successive solutions to the challenges arising in the area of application, along with it; the book provides scope for future developments.

Biotechnology and biochemistry are transforming the face of science. These rapidly advancing fields, with their interdisciplinary approach have applications in all major sectors such as agriculture and pharmaceutics. This book integrates the principles of biochemistry and biotechnology. It brings forth the recent concepts and researches in these disciplines and highlights their applications for industrial use. The tools and techniques employed to study microorganisms have also been presented in this all-inclusive book. Some of the most significant aspects of these fields such as enzymes and proteins, cell physiology, genetics etc. have also been discussed. This book will prove to be an advantage text for anyone associated with these fields. Molecular biologists, bioengineers, scientists, researchers and scholars in these fields will find this book helpful.

It was a great honour to edit this book, though there were challenges, as it involved a lot of communication and networking between me and the editorial team. However, the end result was this all-inclusive book covering diverse themes in the field.

Finally, it is important to acknowledge the efforts of the contributors for their excellent chapters, through which a wide variety of issues have been addressed. I would also like to thank my colleagues for their valuable feedback during the making of this book.

Editor

1

Economics and environmental impact of bioethanol production technologies: an appraisal

Anuj Kumar Chandel[1,2], Chan ES[3], Ravinder Rudravaram[1], M. Lakshmi Narasu[2], L. Venkateswar Rao[1] and Pogaku Ravindra[3*]

[1]Department of Microbiology, Osmania University, Hyderabad-500 007, India.
[2]Department of Biotechnology, Jawaharlal Nehru Technological University, Hyderabad-500 007, India.
[3]School of Engineering and Information Technology, Universiti Malaysia Sabah,Kotakinabalu, 88999, Malaysia.

Contemporary industrial developments and rapid pace of urbanization have called for an environmentally sustainable energy sources. Ethanol made from biomass provides unique environmental, economic strategic benefits and can be considered as a safe and cleanest liquid fuel alternative to fossil fuels. There is a copious amount of lignocellulosic biomass worldwide that can be exploited for fuel ethanol production. Significant advances have been made at bench scale towards the fuel ethanol generation from lignocellulosics. However there are still technical and economical hurdles, which make the bioethanol program unsuccessful at commercial scale. This review provides a broad overview on current status of bioethanol production technologies in terms of their economic and environmental viability. These technologies include pretreatment of biomass, the use of cellulolytic enzymes for depolymerisation of carbohydrate polymers into fermentable constituents and the use of robust fermentative microorganisms for ethanol production. Among all the available technologies, dilute acid hydrolysis followed by enzymatic hydrolysis by less expensive and more efficient cellulases has been found more promising towards the potential economics and environmental impact.

Key words: Bioethanol, lignocellulosic feedstock, ethanol fermentation, economics, environmental impact.

Table of contents

1. Introduction
2. Lignocellulosic biomass
2.1. Composition
3. Ethanol Production Technologies
3.1. Pretreatment
3.1.1. Hydrolysis
3.2.1. Acid hydrolysis
3.2.1.1 Dilute acid hydrolysis
3.2.1.1.1 First-stage dilute acid hydrolysis
3.2.1.1.2 Two-stage dilute acid hydrolysis
3.2.1.2. Concentrated acid hydrolysis
3.2.2 Enzymatic hydrolysis
3.2.2.1 Separate hydrolysis and fermentation (SHF)
3.2.2.2 Simultaneous saccharification and fermentation (SSF)
3.3. Direct microbial conversion (DMC)
3.4 Fermentation
3.4.1. Batch fermentation
3.4.2 Fed batch fermentation
3.4.3. Continuous fermentation
3.4.3.1. Immobilized cells
3.4.3.2 Recycling of process stream
4. Economics of ethanol production technologies
5.1. Ethanol and Environment
5.2. Environmental impact of bioethanol production technologies and their Life-cycle assessment (LCA)

6. Conclusion
7. References

INTRODUCTION

In 1925, Henry Ford had quoted ethyl alcohol, ethanol, as "the fuel of the future." He furthermore stated, "The fuel of the future is going to come from apples, weeds, sawdust – almost anything. There is fuel in every bit of vegetable matter that can be fermented." Today Henry Ford's futuristic vision significance can be easily understood.

In the current time, the importance of alternative energy source has become even more necessary not only due to the continuous depletion of limited fossil fuel stock but also for the safe and better environment, with an inevitable depletion of the world's energy supply, there has been an increasing worldwide interest in alternative sources of energy (Wyman, 1999; Lynd, 2004; Herrera, 2004; Herrera, 2006; Lin and Tanaka, 2006, Schubert, 2006; Chandel et al., 2006a; Vertes et al., 2006, Dien et al., 2006). Keeping in view all the above said advantages, biomass based fuel development technologies should rapidly gain momentum and the barriers imposed earlier should be removed for successfully attempting the production of bioethanol at the commercial level.

It is welcome to understand that the use of bioethanol as a source of energy would be more than just complementting for solar, wind and other intermittent renewable energy sources in the long run (Lin and Tanaka, 2006). During the last two decades, advances in technology for ethanol production from biomass have been developed to the point that large-scale production will be a reality in next few years (Yu and Zhang, 2004; Moiser et al., 2006). Ethanol production from biomass can be summarized briefly into following steps: depolymerization of holocellulose polymer into monomeric fermentable substrate, fermentation of depolymerized substrates, and the distillation of the fermentation broth to obtain dehydrated ethanol.

The ethanol yields and processes economics along with the technical maturity and environmental benefits of using ethanol blend fuel are the key parameters that determine the feasibility of bioethanol production (Nguyen and Saddler, 1991).

The burning fossil fuel at the current rate is likely to create an environmental crisis globally. Use of fossil fuel generates carbon dioxide, methane and a significant quantity of nitrous oxide. Most of these harmful gases are formed due to incomplete combustion of fossil fuel; since ethanol contains 35% oxygen that may result in a more complete combustion of fuel and thus reduces tailpipe emissions.

*Corresponding author. E-mail: dr_ravindra@hotmail.com.

Moreover, biomass energy can play an important role in reducing green house gas emissions. Ethanol production process only uses energy from renewable energy sources. Hence no net carbon dioxide is added to the atmosphere, making ethanol an environmentally bene-ficial energy source (Bull et al., 1992; Kheshgi et al., 2000). Furthermore, fuel ethanol from lignocelluloses may also open new employment opportunities in rural areas, and thus make a positive socio-economic impact (Wyman, 2003; Bevan and Franssen, 2006). Developing ethanol as fuel, beyond its current role as fuel oxygenates will require developing lignocellulosic biomass as a feedstock because of its abundantly available and low cost.

The world ethanol production in 2004 was estimated to be 40 giga litres (GL) (Berg, 2004; Kim and Dale, 2004). Brazil and the US are the world leaders, which together accounted for about 60% of the world ethanol production exploiting sugarcane and corn respectively. In India, lignocellulosic biomass (crop residues, forestry and fruit and vegetable waste and weeds) is available in plenty. Renewable fuels particularly ethanol should get more and more attention all over the world.

The important issue that we wish to address affirmatively here is that the bioethanol production, without doubt, needs an economical approach to address the global fuel needs. Research efforts are needed to design and improve the process, which would produce sustainable and economically feasible transportation fuel. Improvement in process economics using new designed cellulases enzyme cocktail are important factors in establishing a cost effective technology, besides the low cost of feedstock (Mojovic et al., 2006; Gray et al., 2006).

For the long haul, it is very important to understand bioethanol production technologies in terms of their economic viability, environmental feasibility and empowering employment opportunities before implementing a fuel ethanol policy.

The choice of the best technology for lignocellulose to bioethanol conversion should be decided on the basis of overall economics (lowest cost), environmental (lowest pollutants) and energy (higher efficiencies)that is, comprehensive process development and optimization are still required to make the process economically viable.

In reality, environmental considerations, energy and tax policies will determine the extent of fuel ethanol utilization in the future (Keim and Venkatasubramanian, 1989) and therefore the role of one and all is very crucial to identify the gravity of the situation associated with bioethanol production and use of it as an alternative fuel.

The focus of this review is on the current status of available ethanol production technologies in terms of their practical cost economics and the desirable environmental impact they have for a whole generation of commuters across the globe.

Lignocellulosic biomass

Composition

Basically, the lignocellulosic biomass comprises of cellulose, hemicellulose and lignin (Hayn et al., 1993). Cellulose is a linear, crystalline homopolymer with a repeating unit of glucose strung together beta-glucosidic linkages. The structure is rigid and harsh treatment is required to break it down (Gray et al., 2006). Hemi-cellulose consists of short, linear and highly branched chains of sugars. In contrast to cellulose, which is a polymer of only glucose, a hemicellulose is a hetero-polymer of D-xylose, D-glucose, D-galactose, D-man-nose and L-arabinose (Saha et al., 2003). The composi-tion of holocellulose (cellulose + hemicellulose) varies with the origin of the lignocellulosic material. Table 1 shows the composition of the selected crop residues, woody materials, vegetable and fruit waste and municipal solid waste and their simulated ethanol production.

Ethanol production has been taken into estimation depending upon the ratio of hexosans (glucan, galactan and mannan) and pentosans (xylan, arabinan) in each biomass source. The ethanol production from each bio-mass sources was calculated / ascertained from US Department of Energy website, which provide "Theoretical ethanol yield calculator" at http://www.eere.ene-rgy.gov/biomass/ethanol_yield_calculator.html.

Ethanol production technologies

Bioconversion of lignocellulosics to ethanol consists of four major unit operations: pretreatment, hydrolysis, fermentation and product separation/ distillation.

Pretreatment

Pretreatment is required to alter the biomass macroscopic and microscopic size and structure as well as its submicroscopic chemical composition and structure so that hydrolysis of carbohydrate fraction to monomeric sugars can be achieved more rapidly and with greater yields (Sun and Cheng, 2002; Moiser et al., 2005). Pretreatment affects the structure of biomass by solubilizing hemicellulose, reducing crystallinity and increase the available surface area and pore volume of the substrate. Pretreatment has been considered as one of the most expensive processing step in biomass to fermentable sugar conversion with cost as high as 30 cents/gallon ethanol produced (Moiser et al., 2005). To asses the cost and performance of pretreatment methods, technoeconomic analysis have been made recently (Eggerman and Elander, 2005). There is huge scope in lowering the cost of pretreatment process through extensive R&D approaches. Pretreatment of cellulosic biomass in cost effective manner is a major challenge of cellulose to ethanol technology research and development.

Native lignocellulosic biomass is extremely recalcitrant to enzymatic digestion. Therefore, a number of thermochemical pretreatment methods have been developed to improve digestibility (Wyman et al., 2005). Recent studies have clearly proved that there is a direct correlation between the removal of lignin and hemi-cellulose on cellulose digestibility (Kim and Holtzapple, 2006). Thermochemical processing options appear more promising than biological options for the conversion of lignin fraction of cellulosic biomass, which can have a detrimental effect on enzyme hydrolysis. It can also serve as a source of process energy and potential co-products that have important benefits in a life cycle context (Sheehan et al., 2003). Pretreatment can be carried out in different ways such as mechanical combination (Cadoche and Lopez, 1989), steam explosion (Gregg and Saddler, 1996), ammonia fiber explosion (Kim et al., 2003), acid or alkaline pretreatment (Damaso et al., 2004; Kuhad et al., 1997) and biological treatment (Keller, et al., 2003).

Hydrolysis

After pretreatment there are two types of processes to hydrolyze the feed stocks into monomeric sugar constituents required for fermentation into ethanol. The hydrolysis methods most commonly used are acid (dilute and concentrated) and enzymatic. To improve the enzymatic hydrolytic efficiency, the lignin-hemicellulose net work has to be loosened for the better amenability of cellulases to residual carbohydrate fraction for sugar recovery. Dilute acid treatment is employed for the degradation of hemicellulose leaving lignin and cellulose network in the substrate. Other treatments are alkaline hydrolysis or microbial pretreatment with white-rot fungi (*Phaenerochate chrysosporium*, *Cyathus stercoreus*, *Cythus bulleri* and *Pycnoporous cinnabarinus* etc.) preferably act upon lignin leaving cellulose and hemicellulose network in the residual portion. However during both treatment processes, a considerable amount of carbohydrates are also degraded, hence the carbohydrate recovery is not satisfactory for ethanol production.

Acid hydrolysis

There are two types of acid hydrolysis process commonly used - dilute and concentrated acid hydrolysis. The dilute acid process is conducted under high temperature and pressure and has reaction time in the range of seconds or minutes. The concentrated acid process uses relatively mild temperatures, but at high concentration of sulfuric acid and a minimum pressure involved, which only creates by pumping the materials from vessel to vessel. Reaction times are typically much longer than for dilute acid

Table 1. Chemical composition of raw materials and simulated ethanol production

Raw material	Cellulose/ Hexosans (H)	Hemicellulose / Pentosans (P)	Lignin	Ethanol yield /kg dry mass	Reference
Sugarcane baggase	33 (H)	30 (P)	29	0.279	Kuhad and Singh, 1993
Wheat straw	30 (H)	24 (P)	18	0.239	Kuhad and Singh, 1993
Sorghum straw	33 (H)	18 (P)	15	0.240	Kuhad and Singh, 1993
Rice straw	32 (H)	24 (P)	13	0.248	Kuhad and Singh, 1993
Oat straw	41 (H)	16 (P)	11	0.252	Kuhad and Singh, 1993
Corn cob	42 (H)	39 (P)	14	0.358	Kuhad and Singh, 1993
Corn stalks	35 (H)	15 (P)	19	0.221	Kuhad and Singh, 1993
Barley straw	40 (H)	20 (P)	15	0.265	Kuhad and Singh, 1993
Ground nut shell	38 (H)	36 (P)	16	0.327	Kuhad and Singh, 1993
Alfalfa stalks	48.5	6.5	16.6	0.209	Shleser, 1994
Rice hulls	36 (H)	15 (P)	19	0.265	Kuhad and Singh, 1993
Eucalyptus grandis	38	13	37	0.225	Shleser, 1994
Eucalyptus saligna	45	12.0	25.0	0.252	Shleser, 1994
Pine	44.0	26.0	29.0	0.310	Olsson and Hagerdal, 1996
Poplar	47.6	27.4	19.2	0.332	Olsson and Hagerdal, 1996
Saw dust	55.0	14.0	21.0	0.305	Olsson and Hagerdal, 1996
Willow	37.0	23.0	21.0	0.265	Olsson and Hagerdal, 1996
Aspen	51	29.0	16.0	0.354	Olsson and Hagerdal, 1996
Spruce	43.0	26.0	29.0		Olsson and Hagerdal, 1996
Birch	40.0	23.0	21.0	0.305	Olsson and Hagerdal, 1996
Lantana camara	42.50	22.70	22.88	0.288	Chandel (Unpublished work)
Prosopis juliflora	45.5	20.38	24.65	0.291	Chandel (Unpublished work)
Saccharum spontaneum	45.10	22.70	24.56	0.300	Gupta, 2006
Eicchornia crassipis	18.2	48.7	3.50	0.296	Nigam, 2002
Paja brava	32.2	28.1	24.0	0.267	Sanchez et al., 2004
News Paper	61	16	21	0.341	Olsson and Hagerdal, 1996
Processed Paper	47	25	12	0.318	Olsson and Hagerdal, 1996
Paper-based municipal solid waste	43	13	6	0.248	Olsson and Hagerdal, 1996

process

Dilute acid hydrolysis

In dilute acid hydrolysis, the hemicellulose fraction is depolymerized at lower temperature than the cellulosic fraction. Dilute sulfuric acid is mixed with biomass to hydrolyse hemicellulose to xylose and other sugars. Dilute acid is interacted with the biomass and the slurry is held at temperature ranging from120 - 220°C for a short period of time. Thus hemicellulosic fraction of plant cell wall is depolymerised and will lead to the enhancement of cellulose digestibility in the residual solids (Nigam, 2002; Sun and Cheng, 2002; Dien et al., 2006; Saha et al., 2005). Dilute acid hydrolysis has some limitations. If higher temperatures (or longer residence time) are applied, the hemicelluosic derived monosaccharides will degrade and give rise to fermentation inhibitors like furan compounds, weak carboxylic acids and phenolic compounds (Olsson and Hahn- Hagerdal, 1996; Klinke et al., 2004; Larsson et al., 1999). These fermentation inhibitors are known to affect the ethanol production performance of fermenting microorganisms (Chandel et al., 2006b). In order to remove the inhibitors and increase the hydrolysate fermentability, several chemicals and biological methods have been used. These methods include overliming (Martinez et al., 2000), charcoal adsorption (Chandel et al., 2006b), ion exchange (Nilvebrant, 2001), detoxification with laccase (Martin et al., 2002; Chandel et al., 2006b), and biological detoxification (Lopez et al., 2004). The detoxification of acid hydrolysates has been shown to improve their fermentability; however, the cost is often higher than the benefits achieved (Palmqvist and Hahn- Hagerdal, 2000; von Sivers and Zacchi, 1996). Dilute acid hydrolysis is carried out in two stages- First-stage and two-stage.

First-stage dilute acid hydrolysis

The lignocellulosic material is first contacted with dilute

sulfuric acid (0.75%) and heated to approximately 50°C followed by transferring to the first stage acid impregnator where the temperature is raised to 190°C. Approximately, 80% of the hemicellulose and 29% of cellulose are hydrolyzed in the first reactor. The hydrolysate is further incubated at a lower temperature for a residence time of 2 h to hydrolyse most of the oligosaccharides into monosaccharides followed by the separation of solid and liquid fractions. The solid material again washed with plentiful of water to maximize sugar recovery. The separated solid material is sent to second stage acid hydrolysis reactor (Figure 1).

Two-stage dilute acid hydrolysis

In two-stage dilute acid hydrolysis process, first, biomass is treated with dilute acid at relatively mild conditions during which the hemicelluose fraction is hydrolyzed and the second stage is normally carried out at higher temperature for depolymerisation of cellulose into glucose. The liquid phase, containing the monomeric sugars is removed between the treatments, thereby avoiding degradation of monosaccharides formed (Figure 1). It is very important to avoid monosaccharide degradation products for improving the ethanol yield. Sanchez et al. (2004) carried out the two-stage dilute acid hydrolysis using Bolivian straw material, *Paja brava*. In first stage, *P. brava* material was pretreated with steam followed by dilute sulfuric acid (0.5 or 1.0% by wt) hydrolysis at temperatures between 170 and 230°C for a residence time between 3 and 10 min. The highest yield of hemicellulose derived sugars were found at a temperature of 190°C, and a reaction time of 5 – 10 min, whereas in second stage hydrolysis considerably higher temperature (230 °C) was found for hydrolysis of remaining fraction of cellulose.

Concentrated acid hydrolysis

This method uses concentrated sulfuric acid followed by a dilution with water to dissolve and hydrolyse the substrate into sugar constituents. This process provides complete and rapid conversion of celluose to glucose and hemicellulose to xylose with a little degradation. The concentrated acid process uses 70% sulfuric acid at 40 - 50°C for 2 to 4 h in a reactor. The low temperatures and pressure will lead to minimize the sugar degradation. The hydrolyzed material is then washed to recover the sugars.

In the next step, the cellulosic fraction has to be deploymerized. The solid residue from first stage is de-watered and soaked in 30 - 40% sulfuric acid for 50 min. at 100°C for further cellulose hydrolysis. The resulting slurry mixture is pressed to obtain second acid–sugar stream (approximately 18% sugar and 30% acid). Both the sugar steams from two hydrolysis steps are combined and may be used for subsequent ethanol production. Iranmahboob et al. (2002) performed the concentrated acid hydrolysis of mixed wood chips and found that maximum sugar recovery (78 - 82% of theoretical yields) was achieved at sulfuric acid concentration (26%) for 2 h of residence time.

The primary advantage of the concentrated acid process is the potential for high sugar recovery efficiency, about 90% of both hemicellulose and cellulose fraction gets depolymerized into their monomeric fractions. The acid and sugar syrup are separated via ion exchange and then acid is reconcentrated through multiple effect evaporators. The remaining lignin rich solids are collected and optionally palletized for fuel generation (Figure 2).

Enzymatic hydrolysis

The acid, alkaline or fungal pretreated lignocellulosics can be saccharified enzymatically to get fermentable sugars (Ghose and Bisaria, 1979; Kuhad et al., 1997; Itoh et al., 2003; Tucker et al., 2003). Bacteria and fungi are the good sources of cellulases, hemicellulases that could be used for the hydrolysis of pretreated lignocellulosics. The enzymatic cocktails are usually mixtures of several hydrolytic enzymes comprising of cellulases, xylanases, hemicellulases and mannanases. In the last decade, new cellulases and hemicellulases from bacterial and fungal sources have continued been isolated and regular efforts have been made for the improved production of enzymetic titers (Aro et al., 2005; Foreman et al., 2003). However, the cellulases were produced at a concentration too low to be useful. There is a group of microorganisms (*Clostridium*, *Cellulomonas, Tricho-derma*, *Penicillium*, *Neurospora*, *Fusarium*, *Aspergillus* etc.) showing a high cellulolytic and hemicellulolytic activity, which are also highly capable of fermenting monosaccharides. Genetic engineering is used to produce super strains, which are capable of hydrolysing cellulose and xylan along with fermentation of glucose and xylose to ethanol (Aristidou and Penttila, 2000; Lin and Tanaka, 2006). The utilization of cellulose by microorganisms involves a substantial set of fundamental phenomena beyond those associated with enzymatic hydrolysis of cellulose (Lynd et al., 2002).

Separate hydrolysis and fermentation (SHF)

Enzymatic hydrolysis performed separately from fermentation step is known as separate hydrolysis and fermentation (SHF) (Sreenath et al., 2001; Wingren et al., 2003). The separation of hydrolysis and fermentation offers various processing advantage and opportunities. It enables enzymes to operate at higher temperature for increased performance and fermentation organisms to operate at moderate temperatures, optimizing the utilization of sugars (Figure 3).

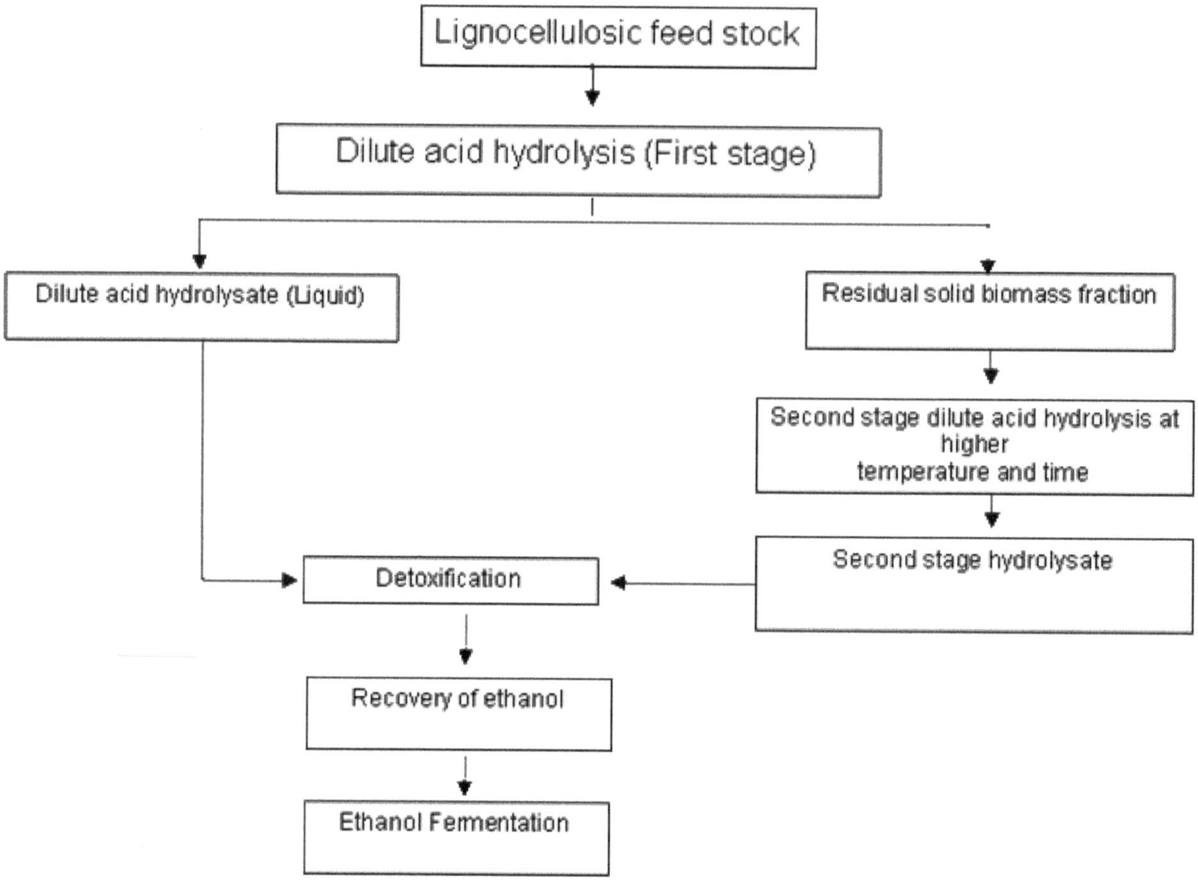

Figure 1. Dilute acid hydrolysis (First-stage and Two-stages) and separate fermentation of pentose and hexose sugars.

Simultaneous saccharification and fermentation (SSF)

The most important process improvement made for the enzymatic hydrolysis of biomass is the introduction of simultaneous saccharification and fermentation (SSF), which has been improved to include the co-fermentation of multiple sugar substrates (Figure 4) (Sreenath et al., 2001; Wingren et al., 2003). This approach combined the cellulase enzymes and fermenting microbes in one vessel. This enabled a one-step process of sugar production and fermentation into ethanol. Simultaneous saccharification of both carbon polymer, cellulose to glucose; and hemicellulose to xylose and arabinose; and, fermentation will be carried out by recombinant yeast or the organism which has the ability to utilize both C_5 and C_6 sugars. According to Alkasrawi et al. (2006) the mode of preparation of yeast must be carefully considered in SSF designing. A more robust strain will give substantial process advantages in terms of higher solid loading and possibility to recirculate the process stream, which results in increased energy demand and reduced fresh water utilization demand in process. Adaptation of yeast to the inhibitors present in the medium is an important factor for consideration in the design of SSF process. More recently, Kroumov et al. (2006) demonstrated an unstructured model of SSF of starch to ethanol by genetically modified strain *Saccharomyces cereviseae* YPB-G, using two hierarchic levels of concept. In first concept, a mechanism of enzymatic hydrolysis of starch to glucose by combined action of two enzymes (alpha-amylase and glucoamylase) secreted by recombinant yeast and the second concept was the enzymatic degradation of starch to glucose and simultaneous utilization of glucose to ethanol by microorganisms. SSF combines enzymatic hydrolysis with ethanol fermentation to keep the concentration of glucose low. The accumulation of ethanol in the fermenter does not inhibit cellulase action as much as high concentration of glucose; so, SSF is good strategy for increasing the overall rate of cellulose to ethanol conversion (Lin and Tanaka, 2006). SSF gives higher ethanol yield while requiring lower amounts of enzyme because end-product inhibition from cellobiose and glucose formed during enzymatic hydrolysis is relieved by the yeast fermentation (Banat et al., 1998). However, it is not feasible for SSF to

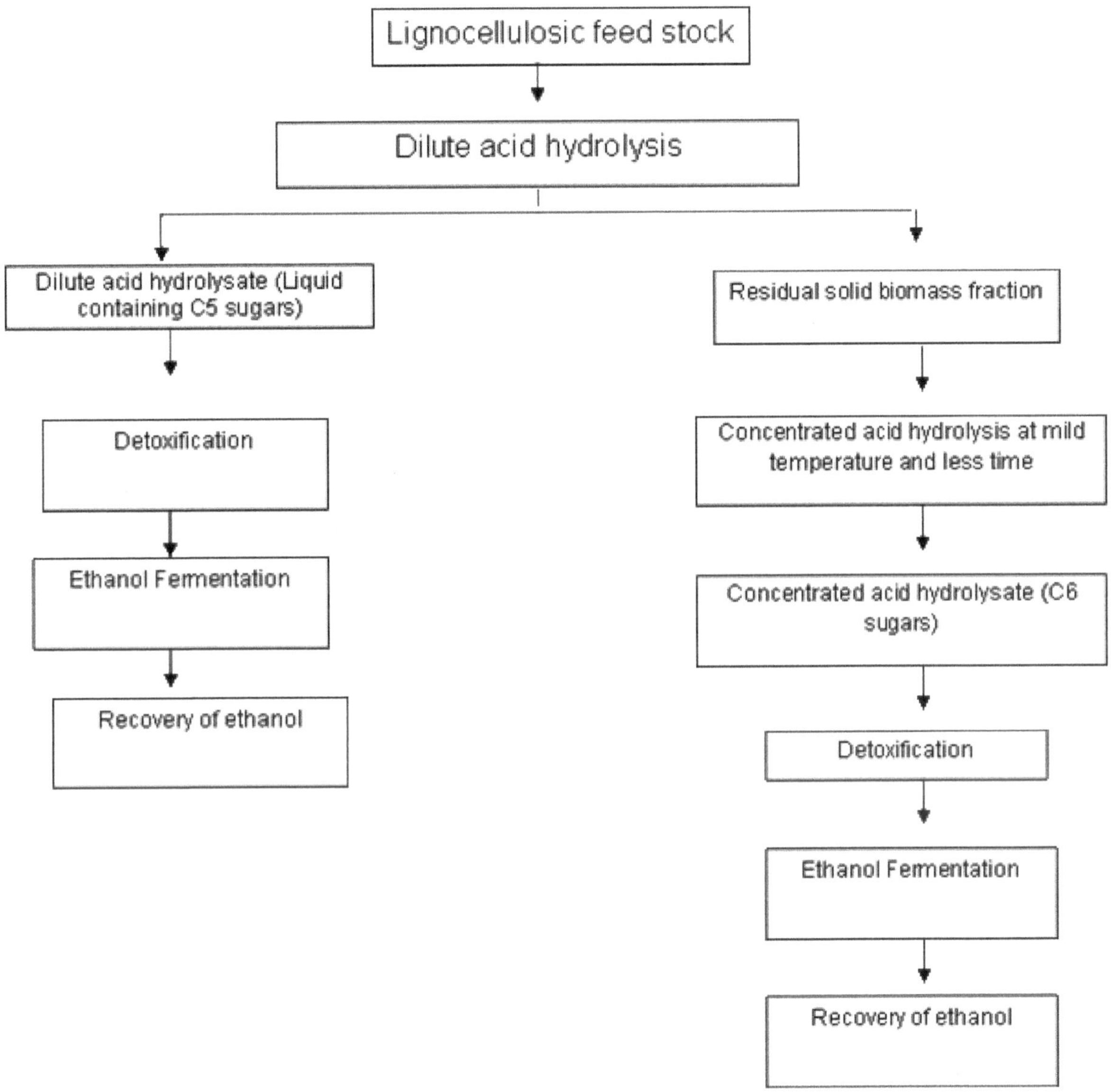

Figure 2. Concentrated acid hydrolysis and separate pentose and hexose sugars fermentation.

meet all the challenges at industrial level due to its low rate of cellulose hydrolysis and most microorganisms employed for ethanol fermentation can not utilize all sugars derived after hydrolysis. To overcome of this problem, the cellulolytic enzyme cocktail should be more stable in wide range of pH and temperature. Also the fermenting microorganisms (yeasts or bacteria) should be able to ferment a wide range of C_5 and C_6 sugars. Recently Matthew et al. (2005) has found some promising ethanol producing bacteria viz. recombinant *E. coli* K011, *Klebsiella oxytoca*, and *Zymomonas mobilis* for industrial exploitation. SSF process has now improved after including the co-fermentation of multiple sugar substrates present in the hydrolysate. This new variant of SSF is known as simultaneous saccharification and co- fermentation (SSCF) (Wilke et al., 1976; Patel et al., 2005; Wyman et al., 2005). SSF and SSCF are preferred over SHF, since both operations can be performed in the same tank resulting in lower cost, higher ethanol yield and shorter processing time (Wright et al., 1988). The most upgraded form of biomass to ethanol conversion is consolidated bioprocessing (CBP) - featuring cellulose production, cellulose hydrolysis and fermentation in one step-is a highly integrated approach with outstanding potential (Lynd et al., 2005). It has the potential to provide the lowest cost route for biological conversion of cellulo-

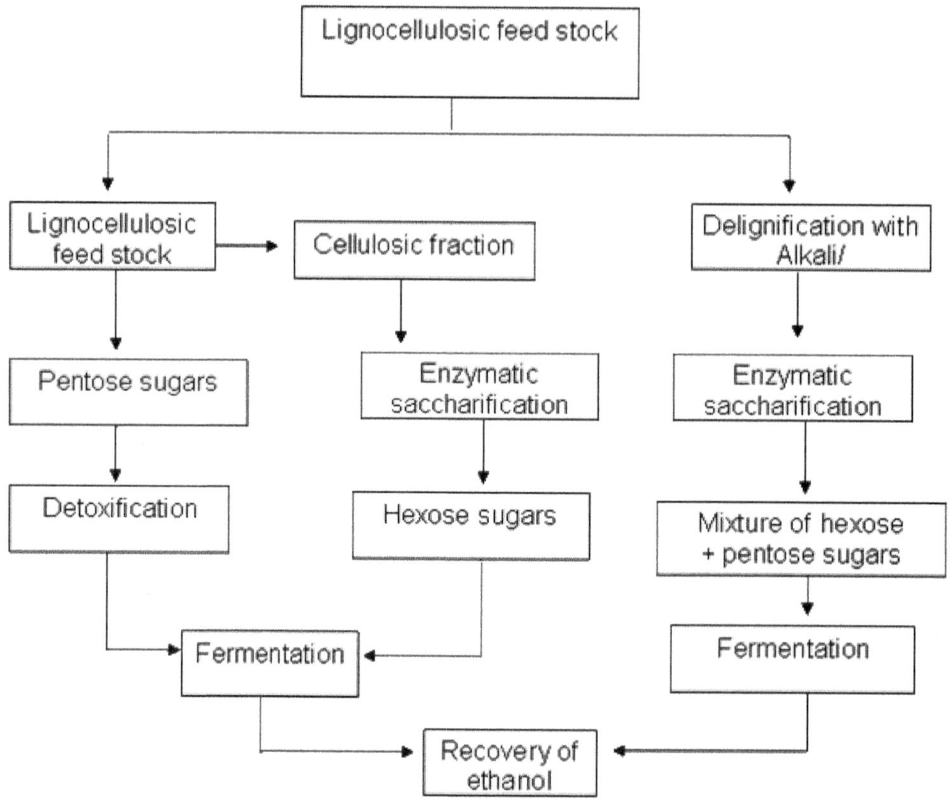

Figure 3. SHF with separate pentose and hexose sugars and combined sugar fermentation.

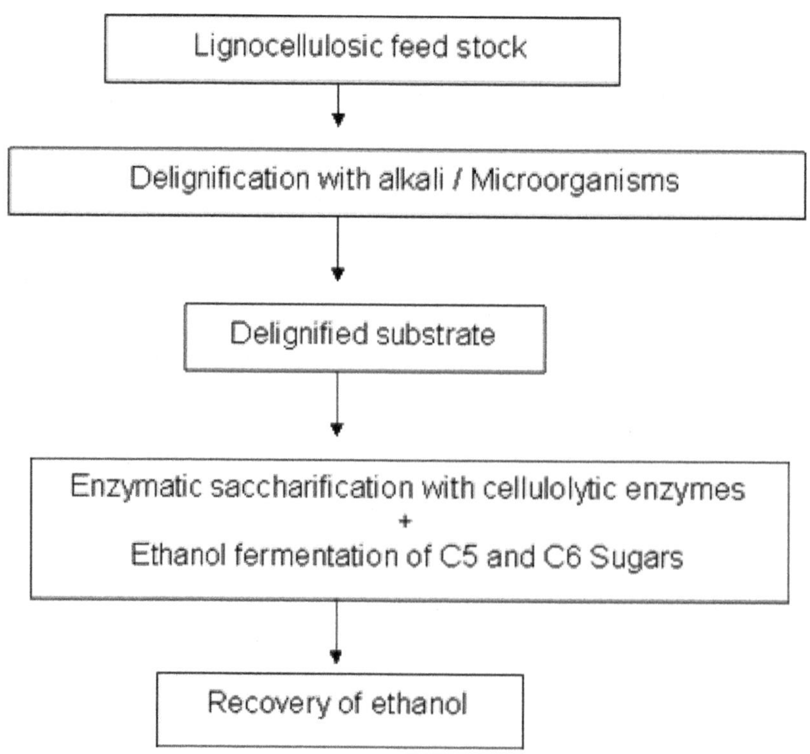

Figure 4. SSF with combined sugars (pentoses and hexoses) fermentation.

losic biomass to ethanol with high rate and desired yields.

Direct microbial conversion (DMC)

DMC is a method of converting cellulosic biomass to ethanol in which both ethanol and all required enzymes are produced by a single microorganism. The potential advantage of DMC is that a dedicated process step for the production of cellulase enzyme is not necessary. Cellulase enzyme production (or procurement) contributes significantly to the cost involved in enzymatic hydrolysis process. However, DMC is not considered the leading process alternative. This is because there is no robust organism available that can produce cellulases or other cell wall degrading enzymes in conjunction with ethanol with a high yield. Singh and Kumar (1991) found that several strains of *Fusarium oxysporum* have the potential for converting not only D-xylose, but also cellulose to ethanol in a one-step process. Distinguishing features of *F. oxysporum* for ethanol production in comparison to other organisms are identified. These include the advantage of *in situ* cellulase production and cellulose fermentation, pentose fermentation, and the tolerance of sugars and ethanol. The main disadvantage of *F. oxysporum* is its slow conversion rate of sugars to ethanol as compared to yeast.

Fermentation

Bioconversion of various biomass sources into ethanol by different microorganisms has been summarized in Table 2. The sugar syrup obtained after cellulosic hydrolysis is used for ethanol fermentation. The ability to ferment pentoses along with hexoses is not widespread among microorganisms (Toivolla et al., 1984), *S. cereviseae* is capable of converting only hexose sugars to ethanol. The most promising yeasts that have the ability to use both C_5 and C_6 sugars are *Pichia stipitis*, *Candida shehatae* and *Pachysolan tannophilus*. However, ethanol production from sugars derived from starch and sucrose has been commercially dominated by the yeast *S. cereviseae* (Lin and Tanaka, 2006). Thermotolerant yeast could be more suitable for ethanol production at industrial level. In high temperature process energy savings can be achieved through a reduction in cooling costs. Considering this approach, Sree et al. (1999) developed solid state fermentation system for ethanol production from sweet sorghum and potato employing a thermotolerant *S. cereviseae* strain (VS3).

Researches are now focusing on developing recombinant yeast, which can greatly improve the ethanol production yield by metabolizing all form of sugars, and reduce the cost of operation. In this contention the researchers have made efforts by following two approaches. The first approach has been to genetically modify the yeast and other natural ethanologens additional pentose metabolic pathways. The second approach is to improve ethanol yields by genetic engineering in microorganisms that have the ability to ferment both hexoses and pentoses (Jeffries and Jin, 2000; Dien et al., 2003; Katharia et al., 2006). Jeffries and Jin (2004) compiled the recent developments happened towards the genetic engineering of yeast metabolism and concluded that strain selection through mutagenesis, adaptive evolution using quantitative metabolism models may help to further improve their ethanol production rates with increased productivities. Piskur et al., (2006) showed the recent developments in comparative genomics and bioinformatics to elucidate the high ethanol production mechanism from *Saccharomyces* sp.

Though new technologies have greatly improved bioethanol production yet there are still a lot of problems that have to be solved. The major problems include maintaining a stable performance of genetically engineered yeast in commercial scale fermentation operation (Ho et al., 1998, 1999), developing more efficient pre-treatment technologies for lignocellulosic biomass, and integrating optimal component into economic ethanol production system (Dien et al., 2000). Sridhar and co-workers (2002) made an effort to improve the thermo tolerance of yeast isolates by treating them with UV radiation.

Fermentation can be performed as a batch, fed batch or continuous process. The choice of most suitable process will depend upon the kinetic properties of microorganisms and type of lignocellulosic hydrolysate in addition to process economics aspects.

Batch fermentation

Traditionally, ethanol has been produced batch wise. At present, nearly, all of the fermentation ethanol industry uses the batch mode. In batch fermentation, the microorganism works in high substrate concentration initially and a high product concentration finally (Olsson and Han-Hagerdal, 1996). The batch process is a multi-vessel process, allows flexible operation and easy control over the process. Generally batch fermentation is characterized by low productivity with an intensive labour (Shama, 1988). For batch fermentation, elaborate preparatory procedures are needed; and because of the discontinuous start up and shut down operations, high labour costs are incurred. This inherent disadvantage and the low productivity offered by the batch process have led many commercial operators to consider the other fermentation methods.

Fed batch fermentation

In fed batch fermentation the microorganism works at low substrate concentration with an increasing ethanol concentration during the course of fermentation process. Fed batch cultures often provide better yield and productivities than batch cultures for the production of microbial metabolites. For practical reasons, therefore, some continuous operations have been replaced by fed batch process

Table 2. Various raw materials for ethanol production.

Raw Material	Pretreatment and Saccharification	Fermentation conditions	Microorganism	Reference
Sugarcane baggase	Dilute acid hydrolysis	Batch	*C. shehatae* NCIM3501	Chandel et al., 2006b
Wheat straw	Dilute acid, Enzymatic hydrolysis	SSF, SHF	*E. coli FBR5*	Saha et al., 2005
Rice straw	Auto hydrolysis	Batch	*C. shehatae* NCIM3501	Abbi et al., 1996
Sorghum straw	Steam explosion, enzymatic	SSF	*Kluyveromyces marxianus* CECT10875	Ballesteros et al., 2004
Corn stover	Steam, enzymatic	Fed- batch	*S. cereviseae* TMB3400	Ohgren et al., 2006
Barley husk	Steam, enzymatic	SSF	*S. cereviseae*	Palmarola et al, 2005
Sun flower stalk	Steam, enzymatic	Batch	*S. cereviseae var ellipsoideus*	Sharma et al., 2002
Sugarcane leaves	Alkaline H2O2	SSF	*S. cereviseae* NRRL-Y-132	Krishna et al., 2001
Wheat bran	Dilute acid, Enzymatic hydrolysis	Batch	*S. cereviseae*	Palmarola et al, 2005
Ground nut shell	Acid hydrolysis	Batch	*S. cereviseae*	Akpan et al., 2005
Alfalfa fibers	Liquid hot water	SSF, SHF	*C. shehatae* FPL-702	Sreenath et al., 2001
Aspen	Acid hydrolysis	Continuous, Immobilized cells	*P. stipitis* R	Parekh et al., 1987
Saw dust	Acid hydrolysis	Batch, Continuous up-flow reactors	*Clostridium thermosaccharolyticum* ATCC 31925	Liu et al., 1988
Pine	Acid hydrolysis	Continuous stirred tank reactor, Immobilized cells	*P. stipitis* NRRL-1724	Qureshi et al., 1991
Poplar	Steam explosion, Enzymatic	SSF, SHF	*S. cerevisiae*	Cantarella et al., 2004
Birch	Acid hydrolysate	Batch	*S. cerevisiae*	Johanssen et al., 2001
Spruce	Dilute acid hydrolysis	Fed batch	*S. cerevisiae*	Taherzadeh., 1999
Willow	Steam	Batch	*E. coli* K011	Olsson et al., 1995
Paja brava	Dilute acid hydrolysis, Two stage	Batch	*C. shehatae, P. stipitis, Pachysolen tannophilus*	Sanchez et al., 2004
Eicchornia crassipes	Dilute acid hydrolysis	Batch	*P. stipitis*	Nigam, 2002
Saccharum spontaneum	Dilute acid hydrolysis, Enzymatic	Batch, Fed Batch	*P. stipitis* NCIM 3498	Gupta, 2006
Cassava starch	Starch liquifaction	Batch, Continuous Co-immobilized cells	*S. diastaticus Zymomonas mobilis*	Amutha and Gunashekhran, 2001
Apple pomace		SSF	*S. cerevisiae* ATCC 24702	Ngadi and Correial, 1992
Pine apple canary waste	Juice Extraction	Continuous, Immobilized cells	*S. cereviseae* ATCC 24553	Nigam, 2000
Banana pulp waste	Juice Extraction	Continuous, Cell recycles.	*S. uvarum* NCIM culture 3528	Joshi et al., 2001
Finger Millet (*Eleusine corcana* flour)		High gravity fermentation	*S. cerevisiae*	Reddy and Reddy, 2006
Municipal solid waste (MSW)	Acid pretraetment	Batch	*S. cerevisiae*	Mtui and Nakamura, 2005
News print	Acid hydrolysate	Batch	*E. coli* B (pLOI297)	Lawford and Rousseau, 1993
Industrial waste		SSF	*K. marxianus, S. cereviseae*	Kadar et al., 2004

(Schugerl, 1987). Keeping the low feed rate of substrate solution containing high concentration of fermentation inhibitors such as furfural, hydroxymethyl furfural and phenolics, the inhibitory effect of these compounds to yeast can be reduced. Complete fermentation of an acid hydrolysate of spruce, which was strongly inhibiting in batch fermentation, has been achieved without any detoxification treatment (Taherzadeh, 1999). The productivity in fed batch fermentation is limited by the feed rate which, in turn, is limited by the cell mass concentration. The specific ethanol productivity has also been reported to decrease with increasing cell mass concentration (Lee and Chang, 1987; Palmqvist et al., 1996). Ideally, the cell density should be kept at a level providing maximum ethanol productivity and yield.

Continuous fermentation

Continuous fermentation can be performed in different kind of bioreactors – stirred tank reactors (single or series) or plug flow reactors. Continuous fermentation often gives a higher productivity than batch fermentation, but at low dilution rates which offers the highest productivities. Alexander et al., (1989) studied the effect of shift in temperature and aeration in steady state continuous culture of *C. shehatae* to determine the effects of ethanol on xylose metabolism. The accu-mulation of ethanol exerted a delayed inhibitory effect on the specific rate of substrate utilization. Continuous operation offers ease of control and is less labor intensive than batch operation. However contamination is more serious in this operation. Since the process must be interrupted, all the equipments must be cleaned, and the operation started again with the growth of new inoculum. The continuous process eliminates much of the unproductive time associated with cleaning, recharging, adjustment of media and sterilization. A high cell density of microbes in the continuous fermenter is locked in the exponential phase, which allows high productivity and overall short processsing of 4 - 6 h as compared to the conventional batch fermentation (24 - 60 h). This results in substantial savings in labour and minimizes investment costs by achieving a given production level with a much smaller plant.

Immobilized cells

A limitation to continuous fermentation is the difficulty of maintaining high cell concentration in the fermenter. The use of immobilized cells circumvents this difficulty. Immobilization by adhesion to a surface (electrostatic or covalent), entrapment in polymeric matrices or retention by membranes has been successful for ethanol production from hexoses (Godia et al., 1987). The applications of immobilized cells have made a significant advance in fuel ethanol production technology. Immobilized cells offer rapid fermentation rates with high productivity - that is, large fermenter volumes of mash put through per day, without risk of cell washout. In continuous fermentation, the direct immobilization of intact cells helps to retain cells during transfer of broth into collecting vessel. Moreover, the loss of intracellular enzyme activity can be kept to a minimum level by avoiding the removal of cells from downstream products (Najafpour, 1990). Immobilization of microbial cells for fermentation has been developed to eliminate inhibition caused by high concentration of substrate and product and also to enhance ethanol productivity and yield. Abbi et al., (1996) observed that the rate of sugar consumption by immobilized cells of *C. shehatae* NCL-3501 was slightly lower than that of free cells, thus leading to higher ethanol production. When microorganisms are attached to solid supports, fluid viscosity is lower which contributes to better mixing and mass transfer in the system. The work on ethanol production in an immobilized cell reactor (ICR) showed that ethanol production using *Z. mobilis* was doubled. Amutha and Gunasekaran (2001) reported ethanol production, 46.7 g/l from 150 g/l liquefied cassava starch from co-immobilized cells of *Saccharomyces diastaticus* and *Z. mobilis*. Yamada et al. (2002) successfully used recombinant *Z. mobilis* with high sugar concentration (12-15%) and further observed the significant role of increased biomass concentration in bioreactor performance for the improved ethanol production. A repeated batch fermentation system was used to produce ethanol using an immobilized osmotolerant *S. cereviseae*, in which ethanol concentration as high as 93 g/l was recorded at 200 g/l glucose concentration (Sree et al., 2000). Nigam (2000) has reported that the ethanol production rate as high as 42.8 g/l/h was achieved from the fermentation of pineapple canary derived sugars by *S. cereviseae* ATCC 24553.

Recycling of process stream

In an environmentally sustainable process, the use of fresh water, the amount of wastewater and the energy consumption must be minimized. The water consumption is decreased by recirculating process streams for use in the washing and hydrolysis steps (Palmqvist and Hahn-Hagerdal, 2000). Recirculating part of the dilute ethanol stream from the fermenter can increase the ethanol concentration in the feed to the distillation stage. However, computer simulations have shown that recirculation of streams leads to the accumulation of non-volatile inhibitory compounds (Galbe and Zacchi, 1992; Palmqvist et al., 1996).

To increase the ethanol productivity, cell recycling has been employed by several workers (Fein et al., 1984; Maleszka et al., 1981), while retaining the simplicity of the batch process. Cell recycling generally does not increase the sugar consumption or ethanol production but the time required for the fermentation can be reduced by 60 - 70%. Schneider (1989) observed a reduction in ethanol production after third cell cycle and suggested the decrease in ethanol production was due to the limitations of oxygen and sugar as a result of an increase in cell density.

Economics of ethanol production technologies

To be competitive, and find economic acceptance, the cost for bioconversion of biomass to liquid fuel must be lower than the current gasoline prices (Wayman and Parekh, 1990; Subramanian et al., 2005). It seems how-ever; now much more attainable because of increasing efforts of researchers working towards improvisation in the efficiency of biomass conversion technologies (Vertes et al., 2006).

However there is still huge scope to bring down the cost of biomass-to-ethanol conversion. The cost of feedstock and cellulolytic enzymes are two important parameters for low cost ethanol production. Biomass feedstock cost represents around 40% of the ethanol production cost (Hamelinck et al., 2005). An analysis of the potential of bioethanol in short and long term (2030) in terms of performance, key technologies and economic aspects such as cost per kilometer driven has been conducted recently by Hamelinck and Faaij (2006). In this analysis, the production cost of bioethanol was found to be within the range of 16-22 €/GJ_{HHV} (Euro/Giga Joules High Heating Value) at present and down to 13 €/GJ_{HHV} in future (2030). The feedstock cost is major parameter influencing the ethanol production cost at a rate of 2-3 €/GJ_{fuel}. Employing integrated approaches using the larger industrial facilities by integrated action plan along with cheap feedstock and potent cellulases could make the process more economically viable (Sun and Cheng, 2002; Dien et al., 2006)

Securing long supply of energy sources requires not only that existing fuel resources be utilized as economically as possible along with their diversification uses. Keeping this point of view, bioethanol is identified and portrayed as the entity for ensuring energy security in future fuel interests and global requirements.

The choice of feedstock for ethanol production depends upon its availability and the ongoing uses. For example, agroresidues such as wheat straw, sorghum and barley straw are not preferable for bioethanol production due to their use as animal fodder. On the contrary, agroresidues like sugarcane bagasse, rice straw, rice bran, groundnut shell, corn stover, *Brassica carniata* stalks and soyabean stalks etc can be used directly because these sources are not preferably used as fodder for livestock. Some dedicated energy sources like damaged rice and sorghum grains (Suresh et al., 1999), sunflower stalks and hulls (Sharma et al., 2000), *Eicchornia crassipies* (Nigam, 2003), *P. brava* (Sanchez et al., 2004), alfalfa fibers (Sreenath et al., 2001), residual starch and crushed wheat grains (Davis et al., 2006), agro waste (Campo et al., 2006) and *Saccharum spontaneum* (Gupta, 2006) are more feasible sources for bioethanol production. Also, organic waste and municipal solid waste (MSW), which contain significant amount of cellulose could be cost economic friendly and solve the problem of solid waste storage and management.

In India, Department of Biotechnology, Govt. of India funded a nationwide research project towards the use of selective weeds like *Lantana camara, Prosopis juliflora* and fruit or vegetable waste into ethanol conversion due to their vast abundance, low cost and rich in fermentable carbohydrates.

An important factor for reducing the cost of bioethanol production is to use larger industrial facilities rather than smaller ones. Ward and Singh (2002) also suggested the integrated approach (Process engineering, fermentation and enzyme and metabolic engineering) could improve the ethanol production economics. By increasing the plant size, the investment per unit output of product falls off, a ten-fold increase in size reducing the unit cost to less than one-half and thereby reducing unit capital cost charges and conversion cost reducing profitability (Wayman and Parekh, 1990; Henke et al., 2006).

To further improve the economy of ethanol production, energy integration of the ethanol production, to already existing plants such as pulp and paper plants is necessary. O' O'Boyle et al. (1991) reported that the cost of producing ethanol from pine with a diluted acid hydrolysate process, was estimated to be 3.22 SEK L^{-1} in a stand alone plant in comparison to 2.54 SEK L^{-1} with an integrated plant. They projected that the cost of bioethanol can be reduced from US$ 1.22 per liter to about US $ 0.31 per liter on the basis of continuous improvement in pretreatment of biomass, enzyme application and fermentation.

Aristidou and Penttila (2000) reported that the total cost of ethanol will be dropped from more than $1.0 per litre to ~ $0.3-0.5 per litre, with a projected cost of less than $0.25 per litre in the near future.

Wilke et al. (1981) has made the first effort to analyse the cost of the conversion of biomass to ethanol process based on a SHF operation and concluded that neat ethanol could compete with gasoline at the oil prices at $20 to $30 per barrel. Foody (1988) has emphasized the importance of improvement in cost effective cellulases production and outlined the potential for bringing down the price of bioethanol from 25-55 cents to 10-28 cents per liter. Subsequently, Wright (1988) and Hinman et al. (1992) evaluated the benefits of SSF process over to SHF technology and reported the cost based on economic evaluation of bioethanol keeping view the SSF mode of operation. Wooly et al. (1999) explained the further economic analysis of bioethanol ($ 0.78 per gallon) and suggested a projected cost of as low as $ 0.20 per liter by 2015 if enzymatic processing and biomass improvement targets are met. The projected cost of ethanol production from cellulosic biomass as per the earlier estimates ($4.63/gallon in 1980) has been reduced by almost a factor of four ($1.22/gallon) over the last 20 years (Wyman, 1999; Wyman, 2001).

However, Kadam et al. (2000) reported that ethanol production cost can be achieved at $ 1.20 per liter by using two-stage dilute acid hydrolysis process. Krishnan et al. (2000) worked out on economics of ethanol production from dry-milled corn starch in fluidized–bed reactors using immobilized *Z. mobilis* cells. They analysed the cost of ethanol for 15 million gal/year production plant using Aspen Plus (Aspen Technology, Cambridge; MA) process simulation software and concluded that the operating cost savings of ethanol production in the range of 1.1 - 3.1 cents/gallon. Kwaiatkowski and Co-workers (2006) developed a model for cost evaluation of ethanol production from 40 million gal/year ethanol producing facility using corn dry-grind process technology. The model was developed using Super Pro Designer R software and data collected from ethanol producers, technology suppliers, equipment manufactures and engineers working in the different industries. The cost of ethanol was found to be increased from US$ 0.235/L to US$ 0.365/L as the price of corn increased from US$ 0.071 to US$ 0.125 /kg.

The bioconversion of pentosans and hexosans from lignocellulosic biomass to ethanol need to be achieved with the maximum efficiencies in terms of higher yield and improved productivity to make the biomass-to-ethanol process economical. The economics of the ethanol process is determined by the cost of sugar. The average biomass cost amounts to ~$0.06 per kg of sugar, or a contribution to the feedstock costs for ethanol production of as low as $0.10 per liter (Aristidou and Penttila, 2000).

Wingren et al. (2003) evaluated the SHF and SSF economic using cellulase enzymes in both configurations with SSF being less expensive by about 10%; and estimated the ethanol production cost of 0.56 – 0.67 $/L. Later, Wingren et al., 2004 studied the effect of reduction in yeast and enzyme concentration in SSF process and concluded final ethanol production cost 4.80 SEK/L (2.34 US$/gal). According to the National Renewable Energy Laboratory (NREL, Colarado, USA) estimations, ethanol production cost of 20 cents per litre is possible in another 15 years from lignocellulose biomass employing designer cellulases and SSCF (simultaneous saccharification and co-fermentation) process (Ghose and Ghose, 2003). However, in both the process, the use of cellulase makes the process cost effective (Mielenz, 2001; Alzate et al., 2005; Gray et al., 2006). According to the analysis of US DOE (US Department of energy), if the enzyme cost can be brought for less than 10 cents per gallon of ethanol the cost of making ethanol could drop as low as 75 cents / gallon (Griffith and Atlas, 2005). Apart from focusing the economics of cellulase production cost, several studies are being carried out to improve the ethanol production by improving acid hydrolysis process. Luong and Tseng (1984) evaluated the technoeconomics of ethanol production under continuous culture using immobilized cells of *Z. mobilis* using plug-flow reactor. They found this technology economically attractive and finally concluded that at least 4 cents/gal of ethanol could be saved using immobilized cells rather than the conventional batch system. They further suggested that by switching from batch to immobilized processing; the fixed capital investment is substantially reduced, thus increasing the profitability of ethanol production by fermentation. Von Sivers et al. (1994) analysed the cost economics of ethanol production (48 cents/gallon) from detoxified willow hemicellulosic hydrolysate using recombinant *E. coli* K011. Zacchi and Axelsson (1989) suggested preconcentrating dilute sugar solution using reverse osmosis is economically feasible to getting high ethanol concentration in fermented broth. Cysewki and Wilke (1976) described cell recycle and vacuum fermentation processes for continuous ethanol production on a production capacity of 78,000 gal ethanol / day employing molasses as the fermentation substrate and estimated ethanol production cost 82.3 and 80.6 cent/gal, for the cell recycle and vacuum processes, respectively.

More recently Zhang (2006) analyzed the cost economics of ethanol production from a small-size lignocellulose refinery with a capacity of 100 tones per day producing approximately 3 million gallons of ethanol plus co-products. He estimated the cost of ethanol production ~$1.00-1.20 per gallon.

The distillation cost of per unit amount of ethanol produced is substantially higher at low ethanol concentrations, the researchers have dealt with the idea of concentrating sugar solutions prior to fermentation (Cyweski et al., 1976; Oh et al., 2000; Iraj et al., 2002). Ethanol distillation cost can be further improved by using membrane distillation process. It has the lowest operational cost, flexible, simple to use and is easy to maintain. It is most efficient and cost effective option among the available distillation processes (Gonsalves, 2006).

Ethanol and Environment

Ethanol represents closed carbon dioxide cycle because after burning of ethanol, the released carbon dioxide is recycled back into plant material because plants use CO_2 to synthesize cellulose during photosynthesis cycle (Wyman, 1999; Chandel et al., 2006a). Ethanol production process only uses energy from renewable energy sources; no net carbon dioxide is added to the atmosphere, making ethanol an environmentally beneficial energy source (Figure 5). In addition, the toxicity of the exhaust emissions from ethanol is lower than that of petroleum sources (Wyman and Hinman, 1990). Ethanol derived from biomass is the only liquid transportation fuel that does not contribute to the green house gas effect (Foody, 1988).

As energy demand increases the global supply of fossil fuels cause harm to human health and contributes to the green house gas (GHG) emission. Hahn-Hagerdal (2006) alarmed to the society by seeing the security of oil supply and the negative impact of the fossil fuel on the environment, particularly on GHG emissions. The reduction of GHG pollution is the main advantage of utilizing biomass conversion into ethanol (Demirbas, 2007). Ethanol contains 35% oxygen that helps complete combustion of fuel and thus reduces particulate emission that pose health hazard to living beings. A study conducted by Bang-Quan et al. (2003) on the ethanol blended diesel (E10 and E30)

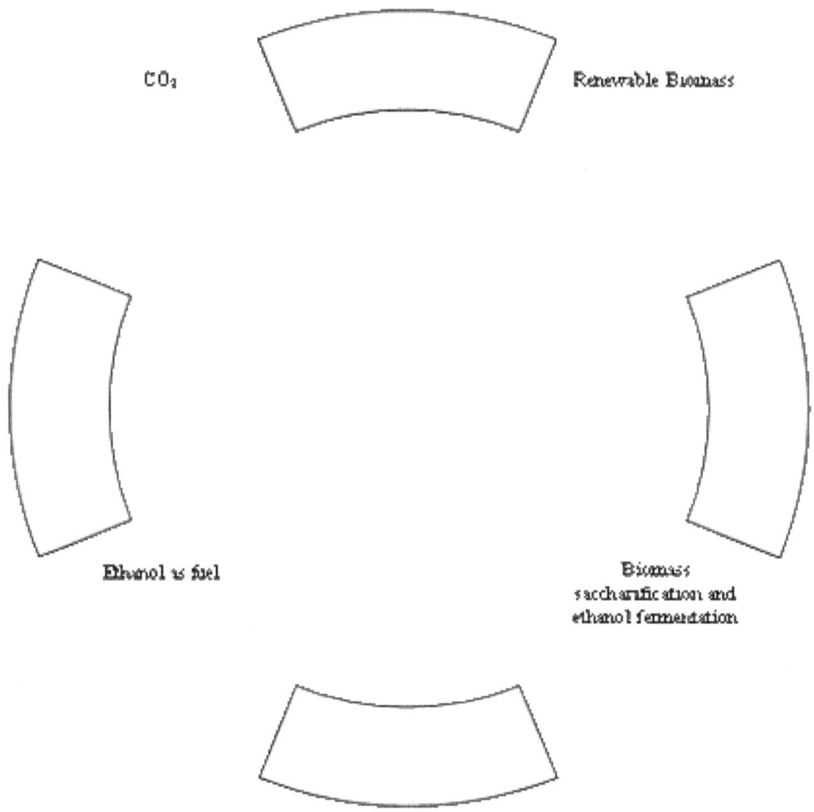

Figure 5. Ethanol represents closed CO_2 cycle.

combustion at different loads found that addition of ethanol to diesel fuel simultaneously decreases cetane number, high heating value, aromatics fractions and kinematic viscosity of ethanol blended diesel fuels and changes distillation temperatures. These factors lead to the complete burning of ethanol and less emissions. With its ability to reduce ozone precursors by 20 - 30%, bio-ethanol can play a significant role in reducing the harmful gasses in metro cities worldwide. Ethanol blended diesel (E-15) causes the 41% reduction in particulate matter and 5% NO_x emission (Subramanian et al., 2005; Chandel et al., 2006a). One of the disadvantage in using ethanol as fuel is that aldehyde predominantly acetaldehydes emissions are higher than those of gasoline. However acetaldehydes emissions generate less adverse health effects in comparison to formaldehydes emitted from gasoline engines (Gonsalves, 2006).

Environmental impact of bioethanol production technologies and their life-cycle assessment (LCA)

Life-cycle assessment (LCA) is a conceptual framework and methodology for the assessment of environmental impacts of product systems on a cradle-to-grave basis (Graedel, 1999; Tan et al., 2002). Analysis of a system under LCA encompasses the extraction of raw materials and energy resources from the environment, the conversion of these resources into the desired products, the utilization of the product by the consumer, and finally the disposal, reuse, or recycle of the product after its service life (Tan et al., 2002). The LCA approach is an effective way to introduce environmental considerations in process and product design or selection (Azapagic, 1999). Based on life cycle assessment (LCA) studies, ethanol production technologies can be compared. Energy production and utilization cycles based on cellulosic biomass have near-zero green house gas emissions on a life cycle basis (Lynd et al., 1991). Biomass utilization into ethanol production offer environmental benefits in terms of nonrenewable energy consumption and global warming impact. Kim and Dale (2005) studied LCA emphasizing corn and soyabean production and their utilization into bioethanol and biodiesel production and concluded that both the biofuels have environmental benefits in terms of nonrenewable energy consumption and global warming impact. However biomass utilization into ethanol also tends to increase acidification and eutrophication, primarily because large nitrogen, phosphorus are released after cultivation of crops. Lechon et al. (2005) studied the LCA of ethanol production from wheat and barley grain and found that barley is a better option than wheat in terms of fulfillment of the green house gas emissions.

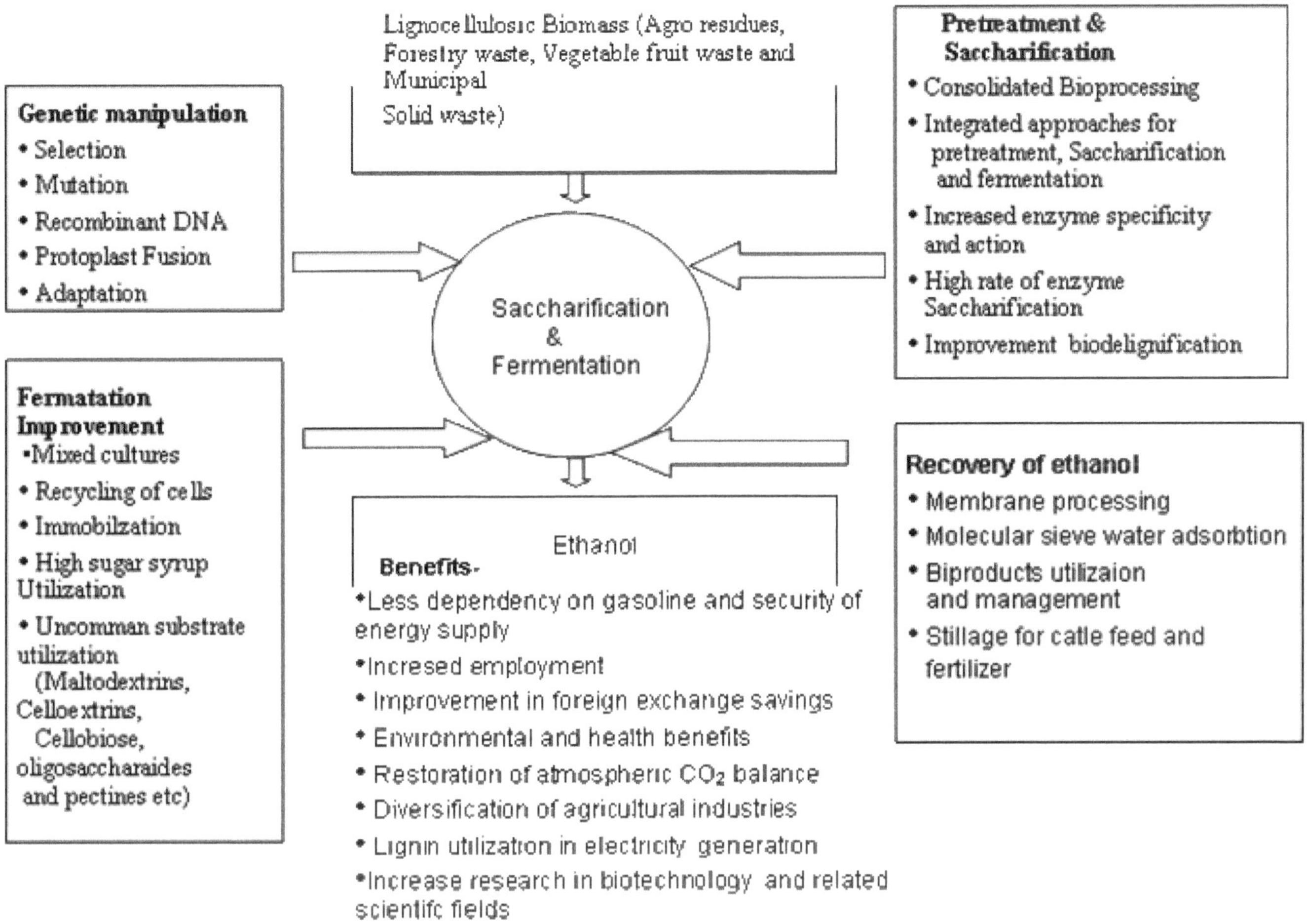

Figure 6. Coordinated action for improvement of biomass to ethanol and its long term benefits.

In an extensive study by Kadam (2000), LCA of acid and enzymatic hydrolysis was compared. All environmental flows were examined from the product life cycle, its production and extraction from raw materials through intermediate conversion process, transportation, distribution and use. Dilute acid process was found better than the enzyme process in terms of greenhouse gas potential, natural resource depletion, acidification potential and eutrophication potential (Kadam, 2002). The reason is dilute acid process sends a much higher proportion of biomass to the boiler for electricity production, which in turn offsets large amounts of emissions (Kadam, 2000). Conversely the concentrated acid process is a net consumer of energy in terms of high acid load and reaction temperature, results in very high values for green house gas (GHG) effect and other impact parameters (Kadam, 1999). Kemppainen and Shonnard (2005) compared the energy consumption and environmental impact for ethanol production using timber wood and recycled news print. The news print conversion into ethanol has a slightly lower overall composite environmental index compared to the timber process. However ethanol production from timber takes less energy, electricity and produces fewer emissions.

A study conducted by Hu et al. (2004) revealed that the E-85 fueled vehicle is better vehicle than the gasoline fueled car by balancing of all the 3E"s the energy, environmental and economic aspects. E-85 fuelled FFV (fossil fuelled vehicle) is about 15% higher efficient when compared to gasoline fuelled car. It also lowers the pollutant emission viz. particulate matter, CO_2, CO, NOx emission than gasoline fuelled car. E-85 fuelled vehicle is higher in total energy consumption and a good combined energy indicator. This was also in agreement with the recent life cycle-based (well to wheel) studies of fuel / propulsion alternatives for light–duty vehicles with a focus on lignocelluloses derived fuel ethanol by reducing 86% lower life cycle green house gas emissions as compared to the gasoline (Fleming et al., 2006).

Tonan et al. (2006) has discussed the integrated assessment of energy conversion processes by evaluating the thermodynamic, economic and environmental parameters and found that water and air emissions of the plant producing ethanol are relatively low. The tansformity of ethanol (1.32 x 10^5 seJ/J) is quite high if compared to that of fossil fuel (5.4 x 10^4 seJ/J). This is even when bioethanol production is driven by a large amount of non-renewable inputs (fertilizer, fuel, machinery).

Conclusion

In spite of laboratory based bioethanol success stories, the production of fuel ethanol at plant scale still remains a challenging issue. A positive solution to this issue could bring economic advantage not only for fuel and power industry, but also benefit the environmental rehabilitation and balance issues and cause.

Worldwide, there is only one company, Iogen Corporation, Canada (http://www.iogen.ca), produce bio-ethanol at commercial scale using wheat straw and corn stover. In India, despite plentiful availability of biomass, there is no commercial ethanol production plant from lingocelluloses. The key to the establishment of a commercial bioethanol production facility and the reduction in capital thereof, resulting in lessening of operating costs from each of the units of operations will be an achievement par standards of excellence and utility! Industry attention, not just the accolades is required for searching the answers to the fast paced fuel drain phenomena threatening to takeover into as a major crisis or even worse an economic depression by the end of 21st century!

For a flourishing bioethanol industry, government support is critical in correcting tax anomalies, exemption from excise and sales tax, deregulation of feedstock and its pricing and encouraging pilot projects and R&D work on bioethanol. Advances in pretreatment by acid catalyzed hemicellulose hydrolysis or employing an integrated approach in the form of consolidated bioprocessing with application of novel, tailored cocktails of enzymes for the cellulose breakdown coupled with the recent development of genetically engineered microorganism those ferment all possible sugars in biomass to ethanol at high productivity are the major key factors to make bioethanol program successful at commercial scale (Figure 6).

The other important aspect by deploying the bioethanol option is its benefit to the environment. Ethanol is one of the best tools to fight vehicular pollution; its clean burning reduces the harmful gasses and particulate emissions that pose health hazard. The implementation of bioethanol policy can be helpful in improving in envi-ronment and rural economic development with sustainable agricultural practices and enhancement of biomass feedstock conscious usage towards the bioethanol industry will bring up the new age farmer into the limelight and horizon of activities and threshold of business to become renewed with options to deal better in life! A better farmer will ultimately usher in a better livelihood for one and all!

ACKNOWLEDGEMENT

We gratefully acknowledge the financial support from Universiti Malaysia Sabah,Kotakinabalu, Malaysia.

REFERENCE

Abbi M, Kuhad RC, Singh A (1996a). Fermentation of xylose and rice straw hydrolysate to ethanol by *Candida shehatae* NCL-3501. J. Ind. Microbiol. 17: 20-23.

Abbi M, Kuhad RC,Singh A (1996b.) Bioconversion of pentose sugars to ethanol by free and immobilized cells of *Candida shehatae* NCL-3501: Fermentation behaviour. Proc. Biochem.31: 55-560.

Alexender MA, Chapman TW, Jeffries TW (1989). ontinuos- culture responses of *Candida shehatae* to shits in temperature and aeration: Implications for ethanol inhibition. Appl. Environ. Microbiol. 55: 2152-2154.

Amutha R, Gunasekaran P (2001). Production of ethanol from liquefied cassava starch using co-immobilized cells of *Zymomonas mobilis* and *Saccharomyces diastaticus* J. Biosci. Bioeng. 92: 560-564.

Alkasrawi A, Rudolf A, Lid´en G, Zacchi G (2006). Influence of strain and cultivation procedure on the performance of simultaneous saccharification and fermentation of steam pretreated spruce. Enzyme Microb. Technol. 38: 279–286.

Alzate CAC, Sánchez Toro OJ (2005). Energy consumption analyses of integrated flow sheets for production of fuel ethanol from lignocellulosic biomass. Energy (In Press).

Aristidou A, Penttilä M (2000). Metabolic engineering applications to renewable resource utilization. Curr. Opin. Biotechnol. 11: 187–198.

Aro N, Pakula T, Penttiillae M (2005). Transcriptional regulation of plant cell wall degradation by filamentous fungi. FEMS Microbiol. Rev. 29: 719-739.

Banat IM, Nigam P, Singh D, Marchant P and McHale AP (1998) Ethanol production at elevated temperatures and alcohol concentrations. Part I: Yeasts in general. World J Microbiol Biotechnol. 14: 809–821.

Berg C (2004). World Ethanol Production. The Distillery and Bioethanol Network. Available at http://www.distill.com/ world ethanol production.htm.

Bull SR, Riley CJ, Tyson KS and Costella R (1992). Total fuel cycle and emission analysis of biomass to ethanol . In: Energy from Biomass and wastes vol . XVI (Klass DL ed.) Institute of gas technology , Chicago , IL. pp. 1-14.

Bustos G, Ramirez JA, Garrote G, Vazquez M (2003). Modeling of the hydrolysis of sugarcane bagasse with hydrochloric acid. Appl. Biochem. Biotechnol. 104: 51-67.

Cadoche L, Lopez GD (1989). Assessment of size reduction as a preliminary step in the production of ethanol from lignocellulosic wastes. Biol. Wastes 30: 153–157.

Chaudhuri BK, Sahai V (1993). Production of cellulose enzyme from lactose in batch and continuous cultures by a partially constitutive strain of *Trichoderma reesei.* Enzyme Microb. Technol. 15: 513–518.

Cysewski GR, Wilke CR (1976). Utilization of cellulosic materials through enzymatic hydrolysis. I. Fermentation of hydrolysate to ethanol and single cell protein. Biotechnol. Bioeng.18: 1297– 1313.

Damaso MCT, Castro Mde, Castro RM, Andrade MC, Pereira N (2004). Application of xylanase from *Thermomyces lanuginosus* IOC-4145 for enzymatic hydrolysis of corncob and sugarcane bagasse. Appl. Biochem. Biotechnol. 115: 1003-1012.

Dien BS, Nichols NN, O'Bryan PJ, Bothast RJ (2000). Development of new ethanologenic *Escherichia coli* strains for fermentation of lignocellulosic biomass. Appl. Biochem. Biotechnol. 84/86: 181-196.

Dien BS, Cotta MA, Jeffries TW (2003). Bacteria engineered for fuel ethanol production current status. Appl. Microbiol. Biotechnol. 63: 258–266.

Dien BS, Jung HJG, Vogel KP, Casler MD, Lamb JAFS, Iten L, Mitchell RB, Sarath G (2006). Chemical composition and response to dilute-acid pretreatment and enzymatic saccharification of alfalfa, reed canarygrass and switchgrass. Biomass and Bioenergy In Press

Draude KM, Kurniawan CB, Duff S JB (2001). Effect of oxygen delignification on the rate and extent of enzymatic hydrolysis of lignocellulosic material. Biores. Technol. 79(2): 113-120.

Eggeman T, Elander TR (2005). Process and economic analysis of pretreatment technologies Biores. Technol. 8: 2019-2025.

Fan LT, Gharpuray MM, Lee YH (1987). In: Cellulose Hydrolysis Biotechnology Monographs. Springer, Berlin. p. 57.

Fein JE, Tallim SR, Lawford GR (1984). Evaluation of D-xylose fermenting yeasts for utilization of a wood-derived hemicellulose hydrolysate. Can. J. Microbiol. 30: 682–690.

Fleming JS, Habibi S, MacLean HL (2006). Investigating the sustainability of lignocellulose-derived fuels for light-duty vehicles. Trans. Res. Part D 11: 146–159.

Foody B (1988). "Ethanol from Biomass: The Factors Affecting It's Commercial Feasibility." Iogen Corporation, Ottawa, Ontario,Canada.

Foreman PK , Brown D , Dankmeyer L, Dean R , Diener S , Dunn-Coleman NS, Goedegebuur F, Houfek TD, England GJ, Kelley AS, Meerman HJ, Mitchell T, Mitchinson C, Olivares HA, Teunissen PJM, Yao J, Ward M (2003). Transcriptional regulation of biomass-degrading enzymes in the filamentous fungus *Trichoderma*. J. Biol. Chem. 278: 31988-31997.

Galbe M, Zacchi G (1992). Simulation of ethanol production processes based on enzymatic hydrolysis of lignocellulosic materials using Aspen. Appl. Biochem. Biotechnol. 34-35: 93-104.

Ghose TK, Bisaria VS (1979). Studies on mechanism of enzymatic hydrolysis of cellulosic substances. Biotechnol Bioeng. 21:131–146.

Goel AK (2002). A sugar industry perspective. Ethanol: Winrock International India, New Delhi, August 2002. p. 5.

Godia F, Casas C, Sola C (1987). A survey of continuous ethanol fermentation systems using immobilized cells. Process Biochem. 22-22: 43–48.

Gray KA, Zhao L and Emptage M (2006). Bioethanol. Curr Opin Chem Biol 10:141-146.

Gregg DJ, Saddler JN (1996) Factors affecting cellulose hydrolysis and the potential of enzyme recycle to enhance the efficiency of an integrated wood to ethanol process. Biotechnol Bioeng. 51: 375 – 383.

Griffith M, Atlas RM (2005). Preemptive bioremediation: applying biotechnology for clean industrial products and processes. In: R.M. Atlas and J. Philp (eds.). Bioremediation: Applied microbial solutions for real-world environmental cleanup. ASM press, Washington, D.C. pp. 318-356.

Gupta P (2006). Bioconversion of *Saccharum spontaneum* into fuel ethanol by *Pichia stipitis* NCIM3498. M.Sc.Thesis. Department of Microbiology, University of Delhi, South Campus, New Delhi, India.

Hayn M, Klinger R, Esterbauer H (1993). Isolation and partial characterization of a low molecular weight endoglucanase from *Trichoderma reesei. Suominen, P.* and *Reinikainen, T.* Proceedings of the second TRICEL symposium on *Trichoderma reesei* cellulases and other hydrolases. Helsinki, Foundation for Biotechnical and Industrial Fermentation Research. pp. 147-151.

Herrera S (2004). Industrial biotechnology- a chance at redemption. Nature Biotechnol. 22:671-675

Hinman ND, Schell DJ, Riley CJ, Bergeron PW, Walter PJ (1992). Preliminary estimate of the cost of ethanol production for SSF technology. Appl. Biochem. Biotechnol. 34/35: 639-649.

Ho NWY, Chen Z, Brainard AP (1998). Genetically engineered *Saccharomyces* yeast capable of effective co-fermentation of glucose and xylose. Appl. Environ. Microbiol. 64: 1852–1859.

Ho NWY, Chen Z, Brainard A, Sedlak M (1999). Successful design and development of genetically engineered *Saccharomyces* yeasts for effective cofermentation of glucose and xylose from cellulosic biomass to fuel ethanol. Adv. Biochem. Eng. Biotechnol. 65: 164-192.

Holtzapple MT, Lundeen JE, Sturgis R (1992). Pretreatment of lignocellulosic municipal solid waste by ammonia by explosion (AFEX). Appl. Biochem. Biotechnol. 34/35: 5–21.

Hu Z, Pu G, Fang F, Wang C (2004). Economics, environment, and energy life cycle assessment of automobiles fueled by bio-ethanol blends in China. Renewable Energy 29: 2183-2192.

International Energy Agency (2001). World Energy outlook 2000. Second edition, France, International energy agency, February 2001, 75775 Paris cedex 16, France.

Iraj N, Giti E, Lila A (2002). Isolation of flocculating Saccharomyces cerevisiae and investigation of its performance in the fermentation of beet molasses to ethanol. Biomass Bioenergy 23: 481–486.

Itoh H, Wada M, Honda Y, Kuwahara M & Watanabe T (2003). Bioorganosolve pretreatments for simultaneous saccharification and fermentation of beech wood by ethanolysis and white-rot fungi. J. Biotechnol. 103 (3): 273-280.

Jeffries TW, Jin YS (2000). Ethanol and thermotolerance in the bioconversion of xylose by yeasts. Adv Appl Microbiol 47: 221–268.

Kadam KL, Camobreco Glazebrook, Forrest LH, Jacobson WA, Simeroth DC, Blackburn W, NKC (1999). Environmental Life cycle implications of fuel oxygenate production from California biomass. NREL Report no. NREL/TP-580-25688, National Renewable Energy Laboratory Golden, Colorado.

Kadam KL, McMillan JD (2003). Availability of corn stover as a sustainable feedstock for bioethanol production. Bioresour Technol 18: 17–25

Kapoor RK, Chandel AK, Kuhar S, Gupta R, Kuhad RC (2006). Bioethanol from crop residues, production forecasting and economics: An Indian perspective. In: Lignocellulose biotechnology: current and future prospects (Kuhad RC and Singh A eds). I.K.International, New Delhi, India. (In Press).

Keim CR, Venkatasubramanian K (1989). Economics of current biotechnological methods of producing ethanol. Trends Biotechnol. 7: 22-29.

Keller FA, Hamilton JE, Nguyen QA (2003). Microbial pretreatment of biomass: potential for reducing severity of thermochemical biomass pretreatment. Appl. Biochem. Biotechnol. 27-41: 105-108.

Kheshgi HS, Prince RC, Marland G (2000). The potential of biomass fuels in the context of global climate change; Focus on transportation fuels. Annual Rev. Energy Environ. 25: 199-244

Kim HT, Kim JS, Sunwoo C, Lee YY (2003). Pretreatment of corn stover by aqueous ammonia. Biores Technol. 90: 39-47.

Kim S, Dale EB (2004). Global potential bioethanol production from wasted crops and crop residues. Biomass Bioenergy 26: 361–375.

Kim S, Holtzapple MT (2006). Lime pretreatment and enzymatic hydrolysis of corn stover. Bioresour Technol. 96: 1994-2006.

Klinke HB,Thomsen AB, Ahring BK (2004). Inhibition of ethanol producing yeast and bacteria by degradation products produced during pretreatment of biomass. Appl. Microbiol. Biotechnol. 66:10-26.

Krishna SH, Reddy TJ, Chowdary GV (2001). Simultaneous saccharification and fermentation of lignocellulosic wastes to ethanol using thermotolerant yeast. Biores. Technol. 77: 193-196.

Kroumov AD, M´odenes AN, de Araujo Tait MC (2006). Development of new unstructured model for simultaneous saccharification and fermentation of starch to ethanol by recombinant strain. Biochem. Eng. J. 28: 243–255.

Kuhad RC, Singh A (1993). Lignocellulose biotechnology: current and future prospects. Crit Rev Biotechnol 13: 151–172.

Kuhad RC, Singh A, Ericksson KE (1997). Microorganisms and enzymes involved in the degradation of plant fiber cell walls. Adv Biochem Eng Biotechnol 57: 45-125

Larsson S, Palmqvist E, Hahn Hagerdal B, Tengborg C, Stenberg K, Zacchi G, Nilvevrant NO (1999). The generation of fermentation inhibitors during dilute acid hydrolysis of soft wood. Enzyme MicrobTechnol 24: 151-159.

Lee CW, Chang HN (1987). Kinetics of ethanol fermentations in membrane cell recycle fermentors. Biotechnol. Bioeng. 29: 1105 – 1112.

Lin Tanaka (2006). Ethanol fermentation from biomass resources: Current state and prospects. Appl. Microbiol. Biotechnol. 69: 627-642.

Lopez MJ, Nichols NN, Dien BS, Moreno J,Bothast RJ (2004). Isolation of microorganisms for biological detoxification of lignocellulosic hydrolysates. Appl. Microb. Biotechnol. 64: 125-131.

Lynd LR, Weimer PJ, van Zyl WH, Pretorius IS (2002). Microbial cellulose utilization: fundamentals and biotechnology. Microbiol. Mol. Biol Rev. 66: 506-577.

Lynd LR and Wang MQ (2004). A product-nonspecific framework for evaluating the potential of biomass-based products to displace fossil fuels. J. Ind. Ecol. 7: 17-32.

Lynd LR, van Zyl WH, McBride JE, Laser M (2005). Consolidated bioprocessing of cellulosic biomass: an update. Curr. Opin. Biotechnol. 16: 577–583.

Maleszka R, Veliky IA, Schneider H (1981). Enhanced rate of ethanol production from D-xylose using recycled or immobilized cells of *Pachysolen tannophilus.* Bioctehnol. Lett. 3: 415-420.

Mamma D, Christakopoulos P, Koullas D, Kekos D, Macris BJ, Koukios

E (1995). An alternative approach to the bioconversion of sweet sorghum carbohydrates to ethanol. Biomass Bioenergy 8: 99-103.

Martín C, Galbe M, Wahlbom CF, Hägerdal BH, Jönsson LJ (2002). Ethanol production from enzymatic hydrolysates of sugarcane bagasse using recombinant xylose-utilising *Saccharomyces cerevisiae*. Enz. Microb. Technol. 31: 274-282.

Martinez A, Rodriguez ME, York SW, Preston JF, Ingram LO (2000). Effects of $Ca(OH)_2$ treatments ("overliming") on the composition and toxicity of bagasse hemicellulose hydrolysates. Biotechnol Bioeng. 69: 526-536.

Matthew H, Ashley O, Brian K, Alisa E, Benjamin JS (2005). Wine making 101.

Mohagheghi AW, Dowe NW, Schell DW, Chou YCW, Eddy C (2004). Performance of a newly developed integrant of *Zymomonas mobilis* for ethanol production on corn stover hydrolysate. Biotechnol. Lett. 26: 321-325.

Moiser N, Wyman C, Dale B, Elander R, Lee YY, Holtzapple M, Ladisch M (2005). Features of promising technologies for pretreatment of lignocellulosic biomass Biores. Technol. 96: 673-686.

Mojovic L, Nikolić S, Rakin M, Vukasinović M (2006). Production of bioethanol from corn meal hydrolyzates. Fuel 85: 1750-1755.

Najafpour GD (1990). Immobilization of microbial cells for the production of organic acids. J Sci Islam Repub Iran 1: 172–176.

Nigam JN (2001). Ethanol production from wheat straw hemicellulose hydrolysate by *Pichia stipitis*. J. Biotechnol. 87: 17-27.

Nigam JN (2002). Bioconversion of water-hyacinth (*Eichhornia crassipes*) hemicellulose acid hydrolysate to motor fuel ethanol by xylose–fermenting yeast. J. Biotechnol. 97: 107-116.

Nigam JN (2000) Continuous ethanol production from pineapple cannery waste using immobilized yeast cells. J. Biotechnol. 80: 189-93.

Nilvebrant N, Reimann A, Larsson S, Jonsson LJ (2001). Detoxification of lignocellulose hydrolysates with ion exchange resins. Appl. Biochem. Biotechnol. 91-93: 35-49.

Nyguyen QA, Saddler JN (1991). An integrated model for technical and economic evaluation of an enzymatic biomass conversion process. Biores. Technol. 35: 275-282.

Oh KK, Kim SW, Jeong YS, Hong SI (2000). Bioconversion of cellulose into ethanol by nonisothermal simultaneous saccharification and fermentation. Appl. Biochem. Biotechnol. 89: 15–3.

Olsson L, Hahn-Hägerdal B (1996). Fermentation of lignocellulosic hydrolysates for ethanol production. Enzyme Microb. Technol. 18: 312-331.

Palmarola-Adrados B, Galbe M, Zacchi G (2005). Pretreatment of barley husk for bioethanol production. J. Chem. Technol. Biotechnol. 80: 85-91.

Palmqvist E, Hahn-Hägerdal B, Galbe M, Zacchi G (1996). The effect of water-soluble inhibitors from steam-pretreated willow on enzymatic hydrolysis and ethanol fermentation. Enzyme Microb. Technol. 19: 470-476.

Palmqvist E, Greg H, Meinander NQ, Hanhaggerdal B (1999). Main and interaction effects of acetic acid, furfural, parahydroxybenzoic acid on growth and ethanol productivity of yeasts . Biotechnol. Bioeng. 63: 46-55

Palmqvist E, Hahn-Hagerdal B (2000). Fermentation of lignocellulosic hydrolysates. I: inhibition and detoxification and II:inhibitors and mechanisms of inhibition. Bioresour. Technol. 74: 17–33.

Ravindranath NH, Somashekar HI, Nagaraja MS, Sudha P, Sangeetha G, Bhattacharya SC and Abdul Salam P (2005). Assessment of sustainable non-plantation biomass resources potential for energy in India. 29: 178-190.

Rodrigues RCLB, Felipe MGA, Almeida e silva JB, Vitola M, Gomez PV (2001). The influence of pH, temperature and hydrolysate concentration on the removal of volatile and nonvolatile compounds from sugarcane bagasse hemicellulosic hydrolysate treated with activated charcoal before or after vacuum evaporation. Brazilian J. Chem. Eng. 18: 299-311.

Saha BC, Iten LB, Cotta MA, Wu YV (2005). Dilute acid pretreatment, enzymatic saccharification and fermentation of wheat straw to ethanol. Proc. Biochem. 40: 3693-3700.

Sanchez G, Pilcher L , Roslander C, Modig T, Galbe M, Liden G (2004). Dilute-acid hydrolysis for fermentation of the Bolivian straw material *Paja brava.* Biores. Technol. 93: 249-256.

Sawada T, Nakamura Y, Kobayashi F, Kuwahara M, Watanabe T (1995). Effect of fungal pretreatment and steam explosion pretreatment on enzymatic saccharification of plant biomass. Biotechnol. Bioeng. 48: 719-724.

Schneider HI (1989). Conversion of pentoses to ethanol by yeast and fungi .Crit. Rev. Biotechnol. 9: 1-40.

Schugerl K (1987). Bioreaction Engineering (volume 1), John Wiley

Sharma SK (2000). Saccharifixcation and bioethanol production from sunflower stalks and hull. Ph.D thesis. Department of Microbiology, Punjab Agriculture University, Ludhiana, India.

Sharma SK, Kalra KL, Kocher S (2004). Fermentation of enzymatic hydrolysate of sunflower hulls for ethanol production and its scale-up. Biomass Bioenergy 27: 399-402.

Shama G (1988). Developments in bioreactors for fuel ethanol production. Proc. Biochem. 23: 138-145.

Sheehan J, Aden A, Paustian K, Killian K, Brenner J, Walsh M, Nelson R (2003). Energy and environmental aspects of using corn stover for fuel ethanol. J. Ind. Ecol. 7: 117-146.

Singh A, Kumar PK (1991). Fusarium oxysporum: status in bioethanol production. Crit. Rev. Biotechnol. 11(2): 129-47.

Sree NK, Sridhar M, Suresh K, Rao LV, Pandey A (1999). Ethanol production in solid substrate fermentation using thermotolerant yeast. Proc Biochem 34:115-119.

Sree NK, Sridhar M, Suresh K, Banat IM, Rao LV (2000). Isolation of thermotolerant, smotolerant, flocculating *Saccharomyces cerevisiae* for ethanol production. Biores. Technol. 72: 43-46.

Sridhar M, Sree NK, Rao LV (1999). Utilization of damaged sorghum and rice grains for ethanol production by simultaneous saccharification and fermentation Biores Technol 68:301-304.

Sridhar M, Sree NK, Rao LV (2002). Effect of UV radiation on thermotolerance, ethanol tolerance and osmotolerance of *Saccharomyces cerevisiae* VS_1 and VS_3 strains. Biores. Technol. 83: 199-202.

Sun Y, Cheng J (2002). Hydrolysis of lignocellulosic materials for ethanol production: a review Biores Technol 83: 1-11.

Sun Y and Cheng J (2004). Hydrolysis of lignocellulosic materials for ethanol production: a review. Biores. Technol. 83:1-11.

Subramanian KA, Singal SK, Saxena M, Singhal S (2005). Utilization of liquid biofuels in automotive diesel engines: An Indian perspective. Biomass Bioenergy 29: 65-72.

Taherjadeh M (1999). Ethanol from Lignocellulose: Physiological Effects of Inhibitors and Fermentation Strategies. Ph.D.Thesis. Lund University, Lund, Sweden.

Toivola A, Yarrow D, Van-den-bosch E, Van-dijken JP, Sheffers WA (1984). Alcoholic fermentation of D-xylose by yeasts. Appl. Microb. Biotechnol. 47: 1221-1223.

Toman M, Cazorla M (1998). The Clean Development Mechanism: A Primer," *Weathervane*, Resources for the Future, Washington, D.C. Source:

Tonon SMT, Brown F, Luchi A, Mirandola A, Stoppato, Ulgiati S (2006). An integrated assessment of energy conversion processes by means of thermodynamic, economic and environmental parameters Energy. 31: 149-163.

Tucker MP, Kim KH, Newman MM and Nguyen QA (2003). Effects of temperature and moisture on dilute-acid steam explosion pretreatment of corn stover and cellulase enzyme digestibility. Appl Biochem Biotechnol 10: 105–108.

Vidal PF, Molinier J (1988). Ozonolysis of lignin – improvement of *in vitro* digestibility of poplar sawdust. Biomass 16: 1–17.

von Sivers M, Zacchi G, Olsson L and Hahn-Hägerdal B: Cost analysis of ethanol production from willow using recombinant *Escherichia coli*. Biotechnol Prog 1994. 10: 555-560.

Wahlbom CF, Hahn-Hägerdal B (2002) Furfural, 5-hydroxymethyl furfural, and acetoin act as external electron acceptors during anaerobic fermentation of xylose in recombinant *Saccharomyces cerevisiae*. Biotechnol. Bioeng. 78: 172-178.

Wayman M, Parekh SR (1990) Biotechnology of biomass conversion; Fuels and chemicals from renewable resources. Open University Press Milton Keynes.

Wilke CR, Yang RD, Scamanna AF and Freitas RP (1981) Raw material evaluation and process development studies for conversion of biomass to sugars and ethanol. Biotechnol. Bioeng. 23: 163–183

Wingren A, Galbe M, Zacchi G (2003). Techno-Economic evaluation of producing ethanol from softwood: comparison of SSF and SHF and identification of bottlenecks. Biotechnol. Prog. 19: 1109-1117.

Wright JD (1988). Ethanol from lignocellulosics: An overview. Energy Prog. 84: 71-80

Wyman CE, Hinman ND (1990). Ethanol. Fundamentals of production from renewable feedstocks and use as transportation fuel. Appl Biochem. Biotechnol. 24/25: 735-75.

Wyman CE, Dale BE, Elander RT, Holtzapple M, Ladisch MR and Lee YY (2005) Comparative sugar recovery data from laboratory scale application of leading pretreatment technologies to corn stover. Biores Technol. 96 (18): 2026-2032.

Wooley R, Ruth M, Sheehan J, Ibsen K, Majdeski H, Galvez A (1999). Lignocellulosic biomass to ethanol process design and economics utilizing co-current dilute acid prehydrolysis and enzymatic hydrolysis: current and futuristic scenarios. NREL/TP-580-26157, 1999, National Renewable Energy Laboratory, Golden, CO.

Yamada T, Fatigati MA, Zhang M (2002). Performance of immobilized Zymomonas mobilis 31821 (pZB5) on actual hydrolysates produced by Arkenol technology. Appl. Biochem. Biotechnol. 98: 899–907.

Yu ZS, Zang HX (2004). Ethanol fermentation of acid hydrolysed cellulosic pyrolysate with *Saccharomyces cerevisiae*. Biores. Technol. 93: 199-204.

Zheng YZ, Lin HM, Tsao GT (1998). Pretreatment for cellulose hydrolysis by carbon dioxide explosion. Biotechnol. Prog. 14: 890–896.

2

Phytochemical antioxidants for health and medicine – A move towards nature

Duduku Krishnaiah*, Rosalam Sarbatly and Awang Bono

Department of Chemical Engineering, School of Engineering and Information Technology
University Malaysia Sabah, 88999, Kota Kinabalu, Malaysia.

Oxidants or reactive oxygen species are produced in our body during aerobic metabolism leading to many diseases such as cancer, cardiovascular diseases etc. Antioxidants are the chemicals that neutralize these oxidants. Natural antioxidants are the secondary metabolites of phytochemicals and are preferred over synthetic antioxidants, which are found to impose side effects. A review has been done on the effect of oxidants on human health and their neutralization by natural antioxidants. Major natural antioxidants and their plant sources, different recovery processes for antioxidants from plant matrices and the advantage of indirect supercritical fluid extraction over other processes are presented.

Keywords: Free radical, secondary metabolites, oxidative stress, recovery processes, process parameters.

Table of content

1. Introduction
2. Formation of Oxidants
3. Exogenous Sources
4. Endogenous Sources and Characteristics of oxygen radicals
5. Antioxidant and its mechanism
6. Antioxidant and Cancer
7. Antioxidant extraction processes
8. Future prospects
9. Conclusions
10. References

INTRODUCTION

Phytochemicals are the chemicals extracted from plants. These chemicals are classified as primary or secondary constituents, depending on their role in plant metabolism. Primary constituents include the common sugars, amino acids, proteins, purines and pyrimidines of nucleic acids, chlorophyll's etc. Secondary constituents are the remaining plant chemicals such as alkaloids (derived from amino acids), terpenes (a group of lipids) and phenolics (derived from carbohydrates) (Walton et al., 1999).

Antioxidants are secondary constituents or metabolites found naturally in the body and in plants such as fruits and vegetables. An antioxidant can be defined in simple terms as anything that inhibits or prevents oxidation of a susceptible substrate. Plants produce a very impressive array of antioxidant compounds that includes carotenoids, flavonoids, cinnamic acids, benzoic acids, folic acid, ascorbic acid, tocopherols and tocotrienols to prevent oxidation of the susceptible substrate (Hollman, 2001). Common antioxidants include vitamin A, vitamin C, vitamin E, and certain compounds called carotenoids (like lutein and beta-carotene) (Hayek, 2000). These plant-based dietary antioxidants are believed to have an

*Corresponding author. E-mail: Krishna@ums.edu.my

important role in the maintenance of human health because our endogenous antioxidants provide insufficient protection against the constant and unavoidable challenge of reactive oxygen species (ROS; oxidants) (Fridovich, 1998).

Generation of free radicals or reactive oxygen species (ROS) during metabolism and other activities beyond the antioxidant capacity of a biological system gives rise to oxidative stress (Mikulikova and Popes, 2001). Oxidative stress plays a role in heart diseases, malaria, neurodegenerative diseases, AIDS, cancer and in the aging process (Sian et al., 2003). This concept is supported by increasing evidence that oxidative damage plays a role in the development of chronic, age-related degenerative diseases, and that dietary antioxidants oppose this and lower risk of disease (Atoui et al., 2005; Alasalvar et al., 2005) and thus there arises a necessity to extract these antioxidants from the plant matrices. In a recent study (Grigonisa, 2005) different extraction techniques, such as dispersed-solids, percolation, Soxhlet, microwave assisted extraction and supercritical fluid extraction have been used to isolate antioxidants from the plants. Supercritical fluid extraction (SFE) is found to be the feasible and sophisticated technology for the extraction of the antioxidants (Nguyen et. al. 1994). The methodology of solid phase extraction (SPE) with super critical fluids such as CO_2 is found to yield higher purity antioxidants.

This review discusses the effect of oxidants on human health and their neutralization by antioxidants. Different types of antioxidants, their properties and their extraction processes from plant matrices have been dealt in detail. Recent advances in supercritical extraction of antioxidants are presented.

Formation of oxidants

Oxygen, an essential element for life, can also be a reason for the destruction of tissue and/or impair its ability to function normally (Kehrer et al., 1993). Oxidants or free radicals or oxygen-free radicals (OFR) or more generally called as reactive oxygen species (ROS) are formed due to various exogenous and endogenous factors. A free radical contains one or more unpaired electrons and is capable of independent existence. The formation of oxygen radicals could be the reason for the damaging effects of O_2. A class of enzymes called superoxide dismutases (SODS) is responsible for the catalytic removal of superoxide free radical, O_2^- (Lee et al., 2001). An average person has around 10,000–20,000 free radicals attacking each body cell every day. In some cases, ROS are produced specifically to serve essential biological functions, whereas in other cases, they are the byproducts of metabolic processes (Shigenaga et al., 1994).

Exogenous sources

Exposure to radiation from the environment and manmade sources is the exogenous source for formation of oxidants. Low-wavelength electromagnetic radiation such as gamma rays splits water in the body to generate hydroxyl radical, OH^-. The highly reactive OH^- thus formed begins to react vigorously with the nearby cells (Halliwell, 1994). Even though OH^- scavengers usually have rate constants more than 10^{10} M^{-1} sec^{-1} for reaction with OH^-, the most endogenous molecules react equally fast. The antioxidant systems that defend against damage by OH^- do so by preventing its formation and by repairing the damage it causes (Timothy et al., 2003).

It has been estimated that 1-3% of the oxygen we breathe in is used to make O_2^-. Since humans consume large quantities of O_2, a simple calculation shows that over 2 kg of O_2^- is made in the human body every year-people with chronic inflammations may make much more (Halliwell et al., 1994). These oxidants damage cellular macromolecules, including DNA, protein and lipid (Fraga et al., 1990) and accumulation of such damage may con-tribute to ageing and age related diseases.

Endogenous sources and characteristics of oxygen radicals

Other than the exogenous sources such as exposure to radiation, enzymatically or non-enzymatically mediated electron transfer reactions are the source of free radicals produced in the cells. Electron leakage that occurs from electron transport chains, such as those in the mitochondria and endoplasmic reticulum, to molecular oxygen are the major source of free radicals (Fridovich, 1986).

Oxidants are formed in the cells of our body mainly from the following four endogenous sources.

1. Consumption of O_2 by mitochondria during normal aerobic respiration to produce H_2O. Oxidants such as oxygen free radical, H_2O_2 and hydroxyl radical are the by products of this process
2. Destroying of bacteria and virus infected cells by phagogytic cells releases nitric oxide, hydrogen peroxide and oxygen free radical.
3. Degradation of fatty acids and other molecules by peroxisomes, the organelles produce hydrogen peroxide as byproduct, which is then degraded by catalase. The nondegraded peroxide gets into other compartments of nearby cell thereby leading to oxidative DNA damage (Ames et al., 1993). When two free radicals a nonradical is produced due to the formation of covalent bond between their unpaired electrons. But a radical is formed when a free radical reacts with a nonradical and thus can initiate a chain reaction in the body.
4. Oxidants produced during the course of p. 450 degra -dation of natural toxins.

Organisms have developed many defense mechanisms to limit the level of reactive oxidants and the damage inflicted by them (Sang et al., 2002). Despite the cell's antioxidant defense system to counteract oxidative damage from free radicals, radical-related damage of DNA and proteins have been proposed to play a key role in the development of age-dependent diseases such as cancer, arteriosclerosis, arthritis, eurodegenerative disorders and others (Ames, 1989). Reactive oxygen species interacts with cellular components including DNA bases and forms damaged bases or strand breaks (Atoui et al., 2005). Oxygen radicals oxidize lipids or proteins generating intemediates that react with DNA by forming adducts. Therefore, it is an utmost necessity to take antioxidants exogenously due to the changes in the environment for which man made activities such as deforestation, rise in carbon dioxide level in atmosphere etc., are also responsible factors.

Antioxidant and its mechanism

Antioxidants are defined as the substance that when prsent in low concentrations compared to those of an oxidisable substrate significantly delays or prevents oxidation of that substance (Halliwell and Guteridge, 1989). For the *in vivo* situation the concept of antioxidants includes antioxidant enzymes, iron binding and transport proteins and other compounds affecting signal transduction and gene expression (Gutteridge, 1989). In case of foods and beverages, antioxidants are related to the protection of specific oxidation substrates or the formation of specific oxidation.

Synergism, antagonism, co-antioxidants and oxidation retarders are the other useful concepts related to antioxidants. Synergism can be defined as the phenomenon in which a number of compounds, when present together in the same system, have a more pronounced effect than if they were alone (Uri, 1961). Antagonism can be defined likewise by substituting "more" with "less", whereas co-antioxidants may be defined by substituting "more" with "same". The compounds that reduce the rate of oxidation without showing a distinct lag phase of oxidation are retarders of oxidation. Antioxidant action is measured as a decrease in over-all rate of oxidation and as the length of the lag phase.

Antioxidants are divided into two classes: preventive antioxidants and chain breaking antioxidants. Preventive antioxidants inhibit oxidation by reducing the rate of chain initiation. In most cases hydroperoxide product, ROOH of the oxidation is the cause for the initiation process. Preventive antioxidants convert the hydroperoxides to molecular products that are not potential sources of free radicals (Burton et al., 1985). Most biological preventive antioxidants are also peroxide decomposers. Certain enzymes such as glutathione peroxidase can reduce H_2O_2 to H_2O and also lipid hyroperoxides to the corresponding alcohol as shown in the following equation (1).

$$ROOH \xrightarrow[\text{Glutathione peroxides}]{[2H]} ROH + H_2O \qquad (1)$$

Commercial chain breaking antioxidants are generally phenols or aromatic amines. They owe their antioxidant activity to their ability to trap peroxyl radicals are as shown in equation (2).

$$O_2^{\cdot} + \text{Antioxidant} \longrightarrow O_2 + \text{Antioxidant} + \text{heat} \qquad (2)$$

Antioxidants can also be manufactured synthetically. These belong to the class of synthetic antioxidants. The main disadvantage with these antioxidants is their side effect when taken in vivo (Chen et al., 1992). Most of the natural antioxidants are found to have higher antioxidant activity when compared with that of the synthetic ones. Several arguments suggest that the antioxidant components of fruits and vegetables contribute in the defense effect. Epidemiological studies and intervention trials on prevention of diseases such as cancer and cardiovascular disease in people have shown the positive effects of taking antioxidant supplements (Ames et al., 1993; Enstrom et al., 1992; Rimm et al., 1993).

Carotenoids, flavonoids, cinnamic acids, benzoic acids, folic acid, ascorbic acid, tocopherols and tocotrienols are some of the antioxidants produced by the plant for their sustenance. Some of the widely known antioxidants are beta-carotene, ascorbic acid and alpha tocopherol (McCall et al., 1999). Beta-carotene is known as a precursor to vitamin A; it is converted to vitamin A in the liver and the mucous membranes of the small intestine. Beta-carotene is found to be safer as it can be ingested in almost unlimited quantities without toxic effect to the body (Dagenais et al., 2000). Ascorbic acid has multi-functional properties. Based on conditions ascorbic acid can act as an antioxidant, pro-oxidant, a metal chelator, a reducing agent or an oxygen scavenger. Ascorbic acid can act as a pro-oxidant in aqueous systems containing metals, by reducing them, which become more active catalysts of oxidation in their lower valence state. In the absence of added metals, ascorbic acid is an effective antioxidant at high concentrations (Katsunari et al., 1999).

Vitamin E is a group of compounds with well known antioxidant functions. Among vitamin E compounds, tocopherol and especially alphtocopherol possesses the strongest biological activity. Tocopherol is prevalently found in mammalian tissue (Flohe et al., 2002)). Se is a naturally occurring antioxidant that preserves tissue elas-ticity by delaying oxidation of polyunsaturated fatty acids. Se is an essential component of glutathione peroxidase. Se deficiency has been implicated as contributing factor to the development of cardiovascular disease (congestive cardiomyopathy), accelerated atherosclerosis, skeletal muscle myopathy, increased cancer risk, aging, cataract and deranged immune function (Zima et al., 2004). Small molecule dietary antioxidants such as vitamin C (ascorbate), vitamin E (tocopherol), and cartoneoids have generated particular interest as anticarcinogens and as defenses against degenerative diseases (Leo, 1999). The

Table 1. Widely used antioxidants and their applications.

Antioxidant	Plant sources	Applications
Beta-Carotene $C_{40}H_{56}$	*Elaeis oleifera, Elaeis Guineensis Momordica Cochinchinnensis Spreng* *Eurycoma Longifolia* *Zanthoxylum Myriacanthum*	Reported to be anodyne, antidotal, aphrodisiac, diuretic, and vulnerary. Oil palm is a folk remedy for headaches, rheumatism and is used as a liniment for indolent tumors. Used as a coloring and flavoring agent in steamed glutinous rice, male aphrodisiac, stomach ache and antitumor agent.
Alpha-Tocopherol $C_{29}H_{50}O_2$	*Citrus Hystrix* *Calamus Scipronum* *Averrhoa Belimbi*	Fruit used as preservative, flavoring in both savory and sweet food. Leaves used as hair shampoo and as medicine. The buds of these canes are eaten as food and have medical and antiseptic properties. They are commonly used for treatment of fever and aches. The syrup of the fruit is useful in relieving thirst, febrile excitement, and also in some slight cases of hemorrhage from the bowels, stomach and internal hemorrhoids.
Ascorbic Acid $C_6H_8O_6$	*Apium Graveolens* *Sauropus Androgynous*	Arthritis, Back Pain (lower), Nervousness, Rheumatism. Insect and disease resistance.
Palmitic Acid $CH_3(CH_2)_{14}COOH$	*Elaeis Oleifera, Elaeis Guineensis*	Anodyne, antidotal, aphrodisiac, diuretic and vulnerary. Oil palm is source of palmitic acid and is a folk remedy for cancer, headaches, and rheumatism.
Beta Sitosterol $C_{29}H_{50}O$	*Morinda Citrifolia* *Alpinia Officinarum* *Sida Acuta*	Diabetes, high blood pressure, arthritis, skin afflictions, and conditions of aging Flatulence, dyspepsia, vomiting and sickness at stomach, and recommended as a remedy for stomach cancer. Entire plant for stomach ache.
Selenium	*Astragalus Membranaceus* *Valeriana Officinalis* *Achillea Millefolium*	Prevents severe side effects of chemotherapy in patients with cancer. Inhibits the growth of murine renal cell carcinoma. Activation of immune system. Sedative activity. General tonic for the cardio-vascular system, lowers blood pressure, and slows heartbeat.
Anthraquinone $C_{14}H_8O_2$	*Cassia Acutifolia*	Antihelminthic, antibacterial, laxative, diuretic, for treatment of snakebites and uterine disorders.
Tannic acid $C_{76}H_{52}O_{46}$	*Costus Spinosa*	Tanning of leather.
Quercetin$C_{15}H_{10}O_7$	*Blumea Balsamifera*	Treatment for the swelling of pancreas.

details of these antioxidants are shown in the Table 1 (Heinerman, 1996; Hashimoto et al., 2002; Cao, 1993; Nessa et al., 2004; Somchit, et al., 2003; Burton, et al., 1985; Halkes, et al., 2002; Burkill, 1993; Oomen and Grubben, 1998 and Zipser, et al., 1998).

Antioxidant and cancer

It has been estimated that one human cell is exposed to approximately 105 oxidative hits a day from hydroxyl radical and other such species of oxidants. ROS are normal oxidant by-products of aerobic metabolism, and under normal metabolic conditions about 2–5% of O_2 consumed by mitochondria is converted to ROS (Lopaczynsk and Zeisel, 2001 and Dreher et al., 1996). Oxidative stress thus created permanently modifies the genetic material leading to numerous degenerative or chronic diseases, such as atherosclerosis and cancer (Ames et al., 1993). Misrepair of DNA damage could result in mutations such as base substitution and deletion which could lead to carcinogenesis (Barcellos, 2005). Two different mechanisms are thought to play a role in oxidative damage and in the development of carcinogenesis.

The first mechanism is through the modulation of gene expression. Epigenetic effects on gene expression can lead to the stimulation of growth signals and proliferation (Crawford et al., 1995). Chromosomal rearrangements are thought to result from strand breakage misrepair, contributing to genetic amplifications, iterations in gene expression and loss of heterozygosity, which in turn may promote neoplastic progression (Bohr et al., 1995). Active oxygen species have been demonstrated to stimulate protein kinase and poly (ADP ribosylation) pathways, thus affecting signal transduction pathways. This further can lead to modulation of the expression of essential genes for proliferation and tumour promotion (Cerutti and Trump, 1991). There is a suggestion that free radical signal may be mediated through *ras* signal transduction pathways (Lander et al., 1995).

In the second mechanism, radicals induce genetic alterations, such as mutations and chromosomal rearrangements, which can play a role in the initiation of carcinogenesis (Guyton and Kensler, 1993; Cerda and Weitzman, 1997). Oxidative DNA damage results in a wide range of chromosomal abnormalities, causing a blockage of DNA replication and wide cytotoxicity (Bohr et al., 1995). Mutations can occur through misrepair or due to incorrect replication, while chromosomal rear-rangements can result from strand breakage misrepair.

It is known that repair mechanisms decay with age and thus DNA lesions accumulate with age (Jaruga and Dizdaroglu, 1996). The sequence specificity of DNA damage sites affects the mutation frequency (Dizdaroglu et al., 2002). Therefore, investigation of the sequence specificity of DNA damage would be beneficial for cancer prevention. Mutagenic potential is directly proportional to the number of oxidative DNA lesions that escape repair.

Antioxidant extraction processes

Plants contain a wide spectrum of metabolites, as many as 200,000 different compounds (Fiehn, 2002), although not every metabolite occurs in every species. These metabolites represent many different classes of compounds and their derivatives such as amino acids, fatty acids, carbohydrates, and organic acids. The physical-chemical properties of the metabolites are highly variable. Therefore appropriate extraction protocols have to be chosen, as the optimum extraction conditions differ widely for different types of compounds. The plant tissue must be homogenized properly in order to extract plant metabolites efficiently. Various techniques such as grind-ing with a mortar and pestle together with liquid nitrogen, milling in vibration mills with chilled holders, homogeni-zation with a metal pestle connected to an electric drill (Edlund et al., 1995) and ultra-turrax devices (Orth et al., 1999) are available. The degree of homogenization deter-mines the efficiency at which the solvent can penetrate the tissue, and therefore strongly influences the length of time required for solvent extraction. The most common way to extract metabolites is to shake the homogenized plant tissue at low or high temperatures in organic sol-vents, or mixtures of solvents (Fiehn et al., 2000). Methanol, ethanol, and water are the solvents mostly used for extracting polar metabolites, whereas chloroform is the most common solvent for non-polar ones. Alternative extraction techniques include subcritical water extraction (SWE) (Ozel, 2003), pressurized liquid extraction (PLE) (Rostagno et al., 2004), microwave-assisted extraction (MAE) (Shu et al., 2003) and supercritical fluid extraction (SFE) (Roger, 1999)

Different extraction techniques, such as Soxhlet, microwave assisted extraction (MAE) and supercritical fluid extractions (SFE) (Lopez-sebastian et al., 1998) have been used to isolate antioxidants from the plants. Each extraction has its own advantages and disadvantages. The main disadvantages of Soxhlet extraction are long extraction time, degradation of thermolabile compounds and limited solvent choice (Lao et al., 1996). Other conventional liquid–solid extraction procedures also are time-consuming, require large amounts of solvents that sometimes are hazardous compounds, and consequently require further clean-up and concentration steps. Recently MAE (Eskilson and Bjorklund, 2000) has been used as an alternative laboratory scale extraction method, which proved to be considerably faster. MAE also requires less solvent and provides higher recoveries compared to Soxhlet extraction. The major disadvantage with MAE is that it is usually performed at higher temperatures (110–150 ◦C). This temperature range may lead to the denaturation of the thermo labile compounds (Kaufmann and Christen, 2002).

Supercritical fluid extraction (SFE) with carbondioxide as supercritical fluid is a very attractive method for extraction (Guarise et al., 1994). This is because CO_2 is inert, non-flammable, non-explosive, inexpensive, odorless, colorless, clean solvent that leaves no solvent residue in the product. Moreover, the critical temperature of carbon dioxide being 304º K makes it attractive for the extraction of thermo labile compounds (Tonthubthimthong et al., 2001). However, carbon dioxide is restricted by its inadequate solvating power for highly polar analytes, which can, to some extent, be boosted by using an appropriate modifier (Zima et al., 2004). SFE modifiers such as ethanol are introduced at the levels of 1–10%; large modifier concentrations (10 – 50%) are also of interest in some applications. Optimization of the operating conditions such as pressure, percentage of modifier, the fluid pressure and temperature and the extraction time are generally considered as the most important factors for good recoveries (Goli et al., 2005).

The results of comparison of different extraction procedures for isolation of the antioxidant 5,8 dihydroxycoumarin from sweet grass (*Hierochlo¨e odorata*) were done by Grigonisa, et al., 2005. High yield of 0.58% and concentration of 40.4% were obtained for Soxhlet extraction. But this method of extraction is not always acceptable for Industrial applications due to long extraction time, large consumption of hazardous solvents. SFE extraction could be an alternative, as giving second best compound yield of 0.46% and a concentration of 20.3%. MAE was less successful because of lower 5, 8 dihydroxycoumarin extraction yield of 0.30%. When comparing extraction time it can be noted that MAE requires 15 min where-as SFE requires 1 to 2 h. However the former requires some time for extract to cool down; consequently total MAE and extract cooling time becomes longer. Soxhlet and microwave assisted extraction are unsuitable for thermolabile antioxidants. The major disadvantage of high modifier consumption in SFE can be significantly reduced by recycling. Therefore, SFE can be recommended as a suitable method to isolate antioxidants.

For medicinal plants the use of sampling techniques such as Soxhlet extraction, microwave assisted extraction (MAE), or pressurized liquid extraction (PLE) often results in non-selective extraction of relatively large amounts of undesirable components (e.g. lipids, sterols, chlorophylls), which can severely affect the quality of the product (Huie, 2002). Direct supercritical fluid extraction process, in which the chemicals are extracted directly from the plant matrices by the action of supercritical carbondioxide (Catchpole et al., 2002) is also in most cases is subjected to the additional clean-up procedures before the extract is made into useful product. Solid phase extraction is a popular and effective tool not only for the clean up, but also for the extraction and/or concentration of analytes.

Solid-phase extraction (SPE) is a simple preparation technique based on the principles used in liquid chromatography, in which the solubility and functional group interactions of sample, solvent, and adsorbent are optimized to affect the retention and elution (Sargenti and McNair, 1998). Moderately polar to polar analytes are extracted from non-polar solutions onto polar sorbents. Sorbents for normal phase are modified with cyano-, diol- or amino groups. Non-polar to-moderately polar analytes are extracted from polar solutions onto non-polar sorbents (Tekel and Hatrik, 1996). Chemically modified adsorbent materials such as silica gel and synthetic resins enable precise group separation on the basis of different types of physicochemical interaction. Methods combining solid phase extraction with supercritical fluid technology namely indirect supercritical fluid extraction (Khundker et al., 1995) have been employed to extract phytochemicals from aqueous matrices and are found to result in higher yield, concentration and purity of the antioxidant.

Future prospects

Cancer, cardiovascular diseases, diabetes etc., are the major diseases faced by humanity. Antioxidants are found to be preventive against these diseases. Plants are the major source of antioxidants. However, they can be manufactured synthetically. Synthetic antioxidant gives side effects when taken *in vivo*. The latest trend of returning to the natural sources for health and medicine has created a lot of development in the recovery of antioxidants from plants. Even though there are many extraction methodologies and techniques for the separation of antioxidants from plants, each one has its own advantage and disadvantage. Therefore, the choosing of an appropriate technology which is feasible at an industrial scale, economical and at the same time yielding higher concentration and purity of antioxidants is required. Indirect supercritical fluid extraction satisfies these criteria. Appropriate process parameters such as tempe-rature, pressure and exact modifier have to be chosen to get the desired results.

Conclusions

Reactive oxygen species (ROS) or oxidants formed in our body due to exogenous and endogenous factors are found to be responsible for many diseases. Day to day research is revealing the potential of phytochemical antioxidants as health benefactors. This is due to their ability to neutralize the free radicals or reactive oxygen species or oxidants responsible for the onset of cell damage. Synthetic antioxidants are found to be harmful to the health. Most of the natural antioxidants from plant source are safer to health and have better antioxidant activity. Various processes of extraction such as Soxhlet extraction, subcritical water extraction (SWE), pressurized liquid extraction (PLE), microwave-assisted extraction (MAE) and supercritical fluid extraction (SFE) are used to obtain antioxidants from plant matrices. Solid phase extraction with supercritical fluid technology or indirect supercritical fluid extraction is found to be an industrially feasible process.

REFERENCES

Alasalvar CM, Al-Farsi PC, Quantick F, Shahidi R, Wiktorowic Z (2005). Effect of chill storage and modified atmosphere packaging (MAP) on antioxidant activity, anthocyanins, carotenoids, phenolics and sensory quality of ready-to-eat shredded orange and purple carrots. Food Chem. 89: 69–76.

Ames BN (1989). Endogenous oxidative DNA damage, aging, and cancer. Free Radial Research Communication. 7:121–128.

Ames BN, Shigenaga MK, Hagen TM (1993). Oxidants, antioxidants, and the degenerative diseases of aging. Proceedings of the National Academy of Sciences of the United States of America. 90: 7915-7922.

Atoui AK, Mansouri A, Boskou G, Panagiotis K (2005). Tea and herbal infusions Their antioxidant activity and phenolic profile. Food Chem. 89: 27-36.

Barcellos-Hoff MH (2005). Integrative radiation carcinogenesis interactions between cell and tissue responses to DNA damage. Seminars in Cancer Biology.15: 138-148.

Bohr VA, Taffe BG, Larminat F (1995). DNA repair, oxidative stress and aging. In R.G. Cutler, L. Packer, A. Bertram, A. Mori (Eds). Oxidative stress and aging. Switzerland, BirkhauserVerlag Basel. 1995: pp.101–110.

Burkill IH (1993). A Dictionary of the economic products of the Malay Peninsula. 3rd printing. Malaysia, Publication Unit, Ministry of Agriculture.

Burton GW, Foster DO, Perly B, Slater TF, Smith ICP, Ingold, KU (1985). Biological antioxidants. philosophical society of royal transactions of London. Series B, Biol. Sci. 311: 565-576.

Cao Jh QY (1993). Studies on the chemical constituents of the herb huanghuaren (Sida acuta Burm. f.) leaves. China Journal of Chinese Material Medicia. 18: 681-2.

Catchpole OJ, Perry NB, De Silva BMT, Grey JB, Smallfield BM (2002). Supercritical extraction of herbs I: Saw Palmetto, St John's Wort, Kava Root, and Echinacea. Journal of Supercritical Fluids. 22: 129-138.

Crawford DR, Edbauer-Nechamen CA, Schools GP, Salmon S, Davies JM, Davies KJA (1995). Oxidant-modulated gene expression. In Davies and Ursini (Eds). The Oxygen Paradox, Italy, Kleup University press.pp. 327–335.

Cerda S, Weitzman SA (1997). Influence of oxygen radical injury on DNA methylation. Mutation Research. 386: 141–152.

Cerutti PA, Trump BF (1991). Inflammation and oxidative stress in carcinogenesis. Cancer Cells. 3: 1–7.

Chen CH, Pearson AM, Gray JI (1992). Effects of synthetic antioxidants (BHA, BHT and PG) on the mutagenicity of IQ-like compounds. Food Chem. 3: 177-183.

Dagenais GR, Marchioli R, Tognoni G, Yusuf S (2000). Beta-Carotene. vitamin C, and vitamin E and cardiovascular Diseases, Current Cardiology Reports.2: 293-299.

Dizdaroglu M, Jaruga P, Birincioglu M, Rodriguez H (2002). Free radical induced damage to DNA Mechanisms and measurement. Free Radical Biol. Med. 32: 1102–15.

Dreher D, Junod AF (1996). Role of oxygen free radicals in cancer development. European J. of Cancer. 32A: 30–8.

Edlund AS, Sundberg B, Moritz T, Sandberg GA (1995). Microscale technique for gas- chromatography mass-spectrometry measurements of pictogram amounts of indole-3-acetic-acid in plant-tissues. Plant Physiology.108: 1043–1047.

Enstrom JE, Kanim LE, Klein MA (1992). Vitamin C intake and mortality among a sample of the United States population. Epidemiology. 3: 194-202.

Eskilsson CS, Björklund E (2000). Analytical-scale microwave-assisted extraction. J. of Chromatography. 902: 227-250.

Fiehn O (2002). Metabolomics.the link between genotypes and phenotypes, Plant Molecular Biology. 48: 155–171.

Fiehn OJ, Trethewey RN, Willmitzer L (2000). Identification of uncommon plant metabolites based on calculation of elemental compositions using gas chromatography and quadrupole mass spectrometry. Analytical Chemistry. 72: 3573–3580.

Flohé RB, Frank J, Salonearn JT, Neuzil J, Zingg J, Azzi A (2002). The European perspective on vitamin E current knowledge and future research. American Journal of Clinical Nutrition. 76: 703-716.

Fraga CG, Shigenag AMK, Park JW, Degan P, Ames BN (1990). Oxidative damage to DNA during aging - 8-hydroxy-2'- deoxyguano-sine in rat organ DNA and urine Proceedings of National Academy of Science. 87: 4533-4537.

Fridovich I (1998). Oxygen toxicity, a radical explanation.The J. Experimental Biol. 201: 1203–1209.

Fridovich I (1986). Biological effects of the superoxide radical. Archives Biochemistry and Biophysics. 247: 1–11.

Goli AH, Barzegar MS, Mohammad A (2005). Antioxidant activity and total phenolic compounds of pistachio (Pistachia Vera) hullextracts. Food Chemistry. 92: 521-525.

Grigonisa D, Venskutonisa PR, Sivikb B, Sandahlb M, Eskilssonc CS (2005). Comparison of different extraction techniques for isolation of antioxidants from sweet grass (Hierochlo¨e odorata). J. Supercritical Fluids. 33:223–233.

Guarise GB, Bertucco A, Pallado P (1994). Carbon dioxide as a supercritical solvent in fatty acid refining. theory and practice. In Rizvi S. S. H. Supercritical fluid processing of food and biomaterials, Glasgow. Lackie Academic and Professional. pp. 27-43.

Gutteridge JM (1989). Iron and oxygen. a biologically damaging mixture. Acta Paediatrica Scandinavia. 36: 78-85.

Guyton KZ, Kensler TW (1993). Oxidative mechanisms in carcinogenesis. British Medical Bulletin. 49: 523–544.

Halkes BA, Vrasidas I, Rooijer GR, Van Den B, Albert JJ, Liskamp RMJ, Pieters RJ (2002). Synthesis and biological activity of polygaloyl-dendrimers as stable tannic acid mimics. Bioorganic and Medicinal Chemistry Letters. 12: 1567–1570.

Halliwell B (1994). Free radicals and antioxidants free radicals, antioxidants, and human disease: curiosity, cause, or consequence?. Lancet. 344: 721-4.

Halliwell B, Gutteridge JMC (1989). Free Radicals in Biology and Medicine, 2 Ed., Clarendon, UK, Oxford science publications. pp. 22–85.

Hashimoto H, Yoda T, Kobayashi T, Young AJ. (2002). Molecular structure of carotenoids as predicted by MNDO-AMI molecular orbital calculations. J. mole. struct. 604: 125-146.

Hayek MG (2000).Dietary vitamin E improves immune function in cats. In Reinhart G. A and Carey D. P. Eds. Recent Advances in Canine and Feline Nutrition, Iams Nutrition Symposium Proceedings.

Wilmington, OH, Orange Frazer Press. 3: 555-564.

Heinerman J (1996). Heinerman's Encyclopadia of healing herbs and spices. Englewood cliffs, New Jersey, Parker publishing company. p. 425.

Hollman PCH (2001). Evidence for health effects of plant phenols: local or systemic effects?. J. Sci. Food Agric. 81: 842–852.

Huie CW (2002). A review of modern sample-preparation techniques for the extraction and analysis of medicinal plants. Analytical and Bioanalytical Chemistry. 373: 23–30.

Jaruga P, Dizdaroglu M (1996). Repair of products of oxidative DNA base damage in human cells. Nucleic Acids Research. 24:1389–1394.

Katsunari AK, Ito I, Higashio JT (1999). Evaluation of antioxidative activity of vegetable extracts in linoleic acid emulsion and phospholipid bilayers. J. Sci. Food Agric. 979: 142010 – 2016.

Kaufmann B, Christen P (2002). Recent extraction techniques for natural products. microwave-assisted extraction and pressurized solvent extraction, Phytochemical Analysis. 13:105–113.

Kehrer JP (1993). Free-radicals as mediators of tissue-injury and disease. Critical Reviews in Toxicology. 23: 21–48.

Khundker S, Dean JR, Jones PA (1995). Comparison between solid phase extraction and supercritical fluid extraction for the determination of fluconazole from animal feed. Journal of pharmaceutical and biomedical analysis.12: 1441-1447.

Lander HM, Ogiste JS, Teng KK, Novogrodsk YA (1995). p21ras as a common signaling target of reactive free radicals and cellular redox stress. The Journal of Biological Chemistry, 270: 21195–21198.

Lao RC, Shu YY, Holmes J, Chiu C (1996). Environmental sample cleaning and extraction procedures by microwave-assisted process (MAP) technol. Microchem J. 53: 99-108.

Lee JH, Choi IY, Kim IS, Kim SY, Yang ES, Park JW (2001). Protective role of superoxide dismutases against ionizing radiation in yeast. Biochemical Biophysica Acta. 1526: 91-198.

Leo MA, Lieber CS (1999). Alcohol, vitamin A, and ß-carotene: adverse interactions, including hepatotoxicity and carcinogenicity. Am. J. Clin. Nut.. 69:1071-1085.

Lopaczynski W, Zeisel SH (2001). Antioxidants, programmed cell death, and cancer. Nutrition Research. 21:295-307.

Lopez-Sebastian S, Ramos E, Ibanez E, Bueno JM, Ballester L, Tabera J, Reglero G (1998). Dearomatization of antioxidant rosemary extracts by treatment with supercritical carbon dioxide. J. Agric. Food Chem.. 46:13-19.

Mc Call, MR Frei B (1999).Can antioxidant vitamins materially reduce oxidative damage in humans?. Free radical boil. med. 26: 1034–1053.

Mikulikova L, Popov P (2001). Oxidative stress, metabolism of ethanol and alcohol-related diseases. J. Biomed. Sci. 8: 59–70.

Nessa F, Ismail Z, Mohamed N, Hakim MR, Haris M (2004). Free radical-scavenging activity of organic extracts and of pure .flavonoids of Blumea balsamifera. Food Chem. 88: 243–252.

Nguyen N, Evans DA, Frakman G (1994). Natural antioxidants produced by supercritical fluid extraction. In Rizvi, S. S. H. Supercritical fluid processing of food and biomaterials; Glasgow. Lackie, Academic and Professional.

Oomen HAPC, Grubben GJH (1998). Tropical leaf vegetables in human nutrition. Amsterdam, Royal Tropical Institute & Orphan publishing Co.

Orth HC, Rentel C, Schmidt PC (1999). Isolation, purity analysis and stability of hyperforin as a standard material from Hypericum perforatum L, J. The J. Pharmacy Pharmacol. 5: 193–200.

Ozel, MZ, Gogus F, Lewis AC (2003). Subcritical water extraction of essential oils from Thymbra spicata. Food Chem. 82: 381-386.

Rimm EB, Stampfer MJ, Ascherio A, Glovannucci E, Colditz GA, Rosner B, Willett W C (1993). Vitamin E consumption and the risk of coronary heart disease in men. New England J. Med. 328: 1450-1456.

Roger MS (1999). Supercritical fluids in separation science – the dreams, the reality and the future J. Chromat. 856: 83-115.

Rostagno MA, Palma M, Barroso CG (2004) Pressurized liquid extraction of isoflavones from soybeans. Analytica Chimica Acta. 522: 169-177.

Sang SC, Stark RE, Rosen RT, Yang CS, Ho CT (2002). Chemical studies on antioxidant mechanism of tea catechins: analysis of radical reaction products of catechin and epicatechin with 2,2-diphenyl-1-picrylhydrazyl. Bioorgan. Med. Chem. 10: 2233-7.

Sargenti SR, Mcnair HM (1998). Comparison of solid-phase extraction and supercritical fluid extraction for extraction of polycyclic aromatic hydrocarbons from drinking water. Journal of Microcolumn Separations. 10: 1125 – 131.

Shigenaga KK, Tory MH, Bruce NA (1994). Oxidative damage and mitochondrial decay in ageing. Proceedings of National Science Academy. 91: 10771-10778.

Shu YY, Ko MY, Chang YS (2003). Microwave-assisted extraction of ginsenosides from ginseng root Microchem. J. 74:131-139

Sian BA (2003). Dietary antioxidants—past, present and future? Trends in Food Sci. Technol. 14: 93-98.

Somchit BN, Reezal I, Nur V, Mutalib AR (2003). In vitro antimicrobial activity of ethanol and water extracts of Cassia alata. J. Ethnopharmacol. 84: 1-4.

Tekel J, Hatrik S (1996). Review Pesticide residue analyses in plant material by chromatographic methods: clean-up procedures and selective detectors. J. Chromat. A. 754: 397-410.

Timothy WM, Charles DJ, David MB (2003). Reaction of OH_ radicals with H2 in sub-critical water Chemical Physics Letters. 7:144–149.

3

Telomerase: Roles in aging, cancer and hereditary disease

K Sreenivasulu[1]* and M Vijaya Lakshmi[2]

[1]Department of Biotechnology, KL University, Vaddeswaram 522502, India.
[2]Department of Botany, Acharya Nagarjuna University (ANU), Guntur, India.

Telomerase is an enzyme that adds DNA sequence repeats ("TTAGGG" in all vertebrates) to the 3' end of DNA strands in the telomere regions, which are found at the ends of eukaryotic chromosomes. This region of repeated nucleotide repeats called telomeres contain condensed DNA material and prevents constant loss of important DNA from chromosome ends. As a result, every time the chromosome is copied only a couple of telomeres are lost, which causes no damage to the organism. Telomerase is a reverse transcriptase that carries its own RNA molecule, which is used as a template when it elongates telomeres, which are shortened after each replication cycle. The existence of a compensatory shortening of telomere (telomerase) mechanism was first predicted by Soviet biologist Alexey Olovnikov in 1973, who also suggested the telomere hypothesis of aging and the telomere's connections to cancer. Telomerase was discovered by Carol W. Greider and Elizabeth Blackburn in 1984 in the ciliate Tetrahymena. Together with Jack W. Szostak, Greider and Blackburn were awarded the 2009 Nobel Prize in Physiology or Medicine for their discovery. This review focus on role of telomerase in various diseases.

Key words: Telomere, chromatin, aging, cancer, hereditary diseases.

INTRODUCTION

Telomeres, specialized structures at eukaryotic chromosome ends, were originally defined based on the observation that naturally occurring chromosome ends behave differently from induced double stranded DNA breaks (Muller, 1938). The protective function of telomeres is due to the formation of a nucleoprotein complex, which in mammals is comprised of 6 protein subunits (TRF1, TRF2, hRAP1,POT1, TIN2 and TPP1) called shelter in (de Lange, 2005), complex with the telomeric hexa nucleotide TTAGGG repeats. The shelter in complex sequesters, the end of the DNA molecule and prevents it from activating DNA damage pathways (de Lange, 2005). This is achieved by organizing telomeres into lariat structures, called telomere loops or t-loops, formed by invasion of the single-stranded overhang into the duplex telomeric repeats, thereby sequestering the chromosome end and providing a chromosome “cap” (Griffith et al., 1999).

The integrated view of telomere structure that is currently accepted came about piecemeal, as the protein components of the shelter in complex were identified and their interactions and activities were characterized. TRF1 was the first human telomeric binding protein to be biochemically isolated, cloned and characterized (Chong et al., 1995). The related telomeric binding protein TRF2, was identified shortly thereafter by virtue of sequence homology with TRF1 (Broccoli et al., 1997; Bilaud et al., 1997). Both proteins bind to telomeric repeats as homodimers via myb type DNA binding domains located at the car boxy-terminus (Broccoli et al., 1997). TIN2 and hRAP1 were identified in yeast two-hybrid screens asTRF1 (Kim et al., 1999) and TRF2 (Li et al., 2000) interacting proteins, respectively.

Pot1 was identified and cloned based on sequence homology with the end binding proteins of the ciliates *Oxytrichia nova* and *Euplotes crassus* (Baumann and Cech, 2001). POT1 is a single-stranded DNA binding protein that specifically binds to the G-rich strand of telomeric DNA (Baumann and Cech, 2001) through its

*Corresponding author. E-mail: sathwik.kamma@gmail.com.

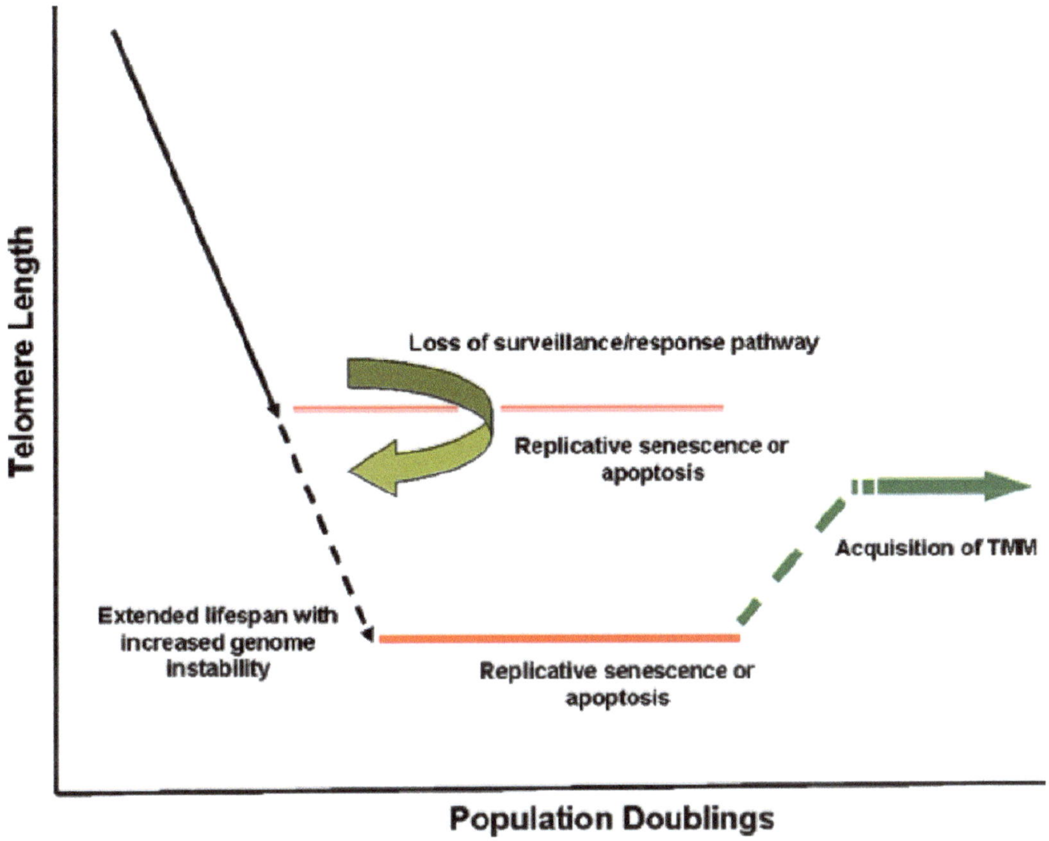

Figure 1. Telomere attrition limits the proliferative potential of cells. Loss of end- capping function leads to cell cycle arrest or apoptosis which can be bypassed by mutation of surveillance and response pathways. Continued cell division with dysfunctional telomeres lead to increased genome instability. Acquisition of a telomere maintenance mechanism stabilizes telomeric DNA arrays and supports unlimited cell division.

oligonucleotide/oligosaccharide binding motif, a motif common to telomeric end binding proteins. TPP1 was identified simultaneously by three laboratories as a TIN2 interacting protein that also interacts with TRF2 and POT1, linking the internal double-stranded telomeric binding proteins to the factor binding at the physical end of the chromosome (Houghtaling et al., 2004; Ye et al., 2004; Liu et al., 2004). Telomere dynamics are central to the processes of human aging and tumor genesis (Figure 1). Due in part to sequence loss as a result of replication, telomeres become shorter with each cell division (Harley et al., 1990). Eventually, in sufficient telomeric DNA remains to recruit enough protein to form a functional telomeric complex. The uncapped chromosome end resembles a double-stranded DNA break (d'Adda di Fagagna et al., 2003; Takai et al., 2003), a highly unstable entity that can give rise to chromosome rearrangements and which activates the cellular DNA damage response, resulting in replicative senescence or apoptosis. Human aging is a complex phenotype arising from a variety of factors. One aspect of aging is the accumulation of tissue damage, due at least in part to an inability to undergo indefinite cellular renewal.

This limited self-renewal capacity is in part a result of the finite division potential of most human cell types. The restricted proliferative capacity of human cells is recapitulated in vitro by primary fibroblasts that irreversibly exit from the cell cycle and undergo replicative senescence after a fixed number of divisions (Hayflick, 1965), driven largely by dysfunctional telomeres. Accumulating evidence supports a model linking telomere attrition with aging phenotypes *in vivo*. For example, premature aging progeroid syndromes such as Werner Syndrome and disorders that involve renewal cell depletion such as Dyskeratosis Congenita have been linked to aberrations in components required for telomere homeostasis (Crabbe et al., 2007; Wong and Collins, 2006; Savage et al., 2006; Savage et al., 2008). Proliferation-associated telomere attrition acts as a tumor suppressor mechanism by limiting the total number of divisions any given cell can undergo.

However, if a mutation in a surveillance or response pathway prevents the cell from responding appropriately to critically shortened telomeres, then continued cellular proliferation will lead to elevated chromosome instability (Chin et al., 1999). Infact, passage through telomere crisis, a period characterized by a heightened rate of genome rearrangements arising from loss of telomere function, may be a common feature of human cancer cells (DePinho and Polyak, 2004). Ultimately, for a tumor to arise, the initiating cell(s) must circumvent the telomere length-dependent limitation on subsequent cellular divisions. In most carcinomas, this is achieved through the inappropriate activity of a specialized reverse transcriptase called telomerase (Shay and Bacchetti, 1997), which uses an RNA template to copy telomeric sequences *de novo* onto the 3′ ends of existing DNA molecules. A second pathway, called Alternative Lengthening of Telomeres (ALT), relies upon a recombination-based mechanism and is observed in certain classes of malignancies, most notably sarcomas (Costa et al., 2006; Johnson et al., 2005; Ulaner et al., 2003) and glioblastoma multiforme (Hakin-Smith et al., 2003).

In addition to its role in maintaining telomere length, telomerase may also impact on telomere stability by contributing to end-capping. The first hint that telomerase might provide a protective function came from careful observation of telomere length in cells engineered to over-express hTERT (Zhu et al., 1999). As expected, hTERT conferred unlimited replicative potential upon these cells. Surprisingly however, telomere length continued to decrease during the first-40 population doublings of these cells and stabilized at a length shorter than that observed in control cultures without telomerase. Thus, telomerase supported the presence of shorter telomeres by contributing in some way to telomere capping. In agreement, mutant alleles of telomerase that do not affect proliferation-associated sequence loss are capable of significantly extending the lifespan of human primary fibroblasts (Kim et al., 2003) suggesting the presence of attenuated hTERT, which protects shortened telomeres and delays the activation of the cellular senescence response.

Furthermore, in yeast deficient for telomerase, the telomeres are much more likely to fuse to a double strand break regardless of starting telomere length (Chan and Blackburn, 2003). Characterization of the structure of the fusion products indicated that the telomeres involved, had undergone dramatic sequence loss. Recent work suggests that similar short telomeres are present in human cancer cells and may be protected by telomerase (Xu and Blackburn, 2007). Telomerase has also been proposed to have telomere maintenance independent roles, enhancing proliferation (Sarin et al., 2005) and attenuating the DNA damage response (Masutomi et al., 2005). Early studies in the budding yeast *Saccharomyces cerevisiae* demonstrated that telomeres are organized into higher order chromatin structures that repress the expression of genes placed in proximity (Gottschling et al., 1990). Similar studies in mammalian cells initially failed to find convincing evidence of a heterochromatin-like structure at telomeres which affected gene expression (Bayne et al., 1994; Sprung et al., 1996).

However, as molecular markers diagnostic of chromatin states were identified,it became clear that mammalian telomeres also contained epigenetic marks consistent with heterochromatin. Recent studies have demonstrated that telomeric chromatin organization plays a critical role in determining telomere structure which impacts telomere length regulation (Benetti et al., 2007; Garcia-Cao et al., 2004; Gonzalo et al., 2005). These studies provide a molecular link between telomeric chromatin structure, telomere length regulation and processes implicated in human aging and tumorigenesis. Here, we discuss progress in understanding the role of telomeric chromatin in aging and human disease.

TELOMERIC CHROMATIN

Much of our understanding of the organization of telomeric chromatin and its effects on telomere metabolism are derived from studies in model organisms. In particular, studies in *S. cerevisiae* have been very informative and have provided the foundation for many of the subsequent studies in mammalian systems. Any discussion of telomeric chromatin must distinguish between the telomeric DNA sequences, e.g., those sequences that interact with specific proteins to form the chromosome end-capping structure and sub-telomeric sequences which are located internally adjacent to the telomeric repeats. Sub-telomeric regions are often repetitive, and may contain shared sequence elements among more than chromosome end and are highly variable (Riethman, 2008; Louis, 1995).

Accumulating evidence, discussed subsequently, suggest the sub-telomeric regions may regulate telomere length, structure, and affect expression of adjacent genes. A great deal has been learned about how perturbation of telomeric chromatin affects telomere structure, length regulation and function by studying mouse models and has been recently reviewed (Blasco, 2007). Alterations in the level of epigenetic modifications, such as tri-methylation of H3K9 and H4K20 via knockout of the relevant modifying enzymes led to increased telomere length without effect on telomere end-capping function (Garcia-Cao et al., 2004). The major caveat to these experiments is that the effect of reducing chromatin-modifying enzymes on chromatin structure is not limited to the telomere. Thus, while it is straightforward to document alterations in telomeric chromatin and associate this with changes in telomere length, it is much more difficult to assess the indirect contributions to the phenotype as a result of non-telomeric changes in chromatin structure throughout the genome. This is exemplified by the increased genome instability caused by deletion of DNA methyl transferases,

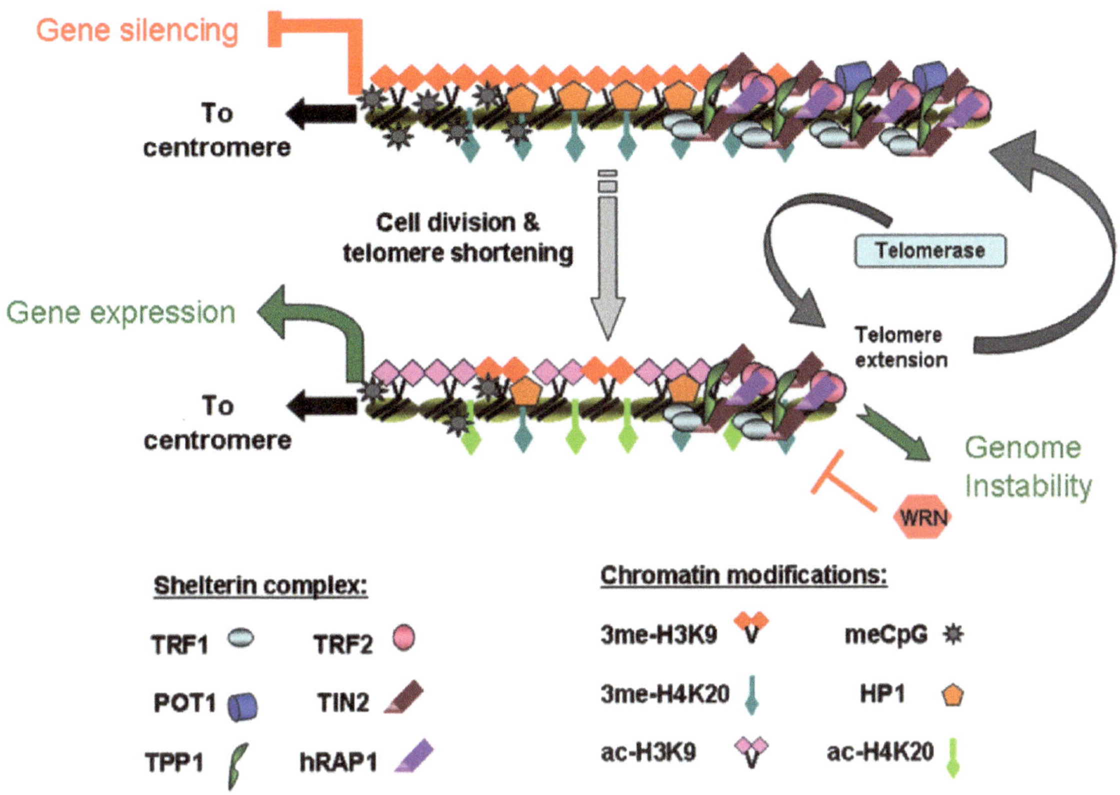

Figure 2. Telomeric DNA is packaged into nucelosomes, depicted as gold bars with black double lines.

which is present at, but not limited to the telomere (Gonzalo et al., 2006).

In an exciting new study, the first report of a histone modification that appears to directly target the telomere was identified. Mammals have seven Sir2 homologs termed SIRT1-7. Mice rendered nullizygous for the SIRT6 gene exhibit premature aging and genomic instability (Mostoslavsky et al., 2006); however, the molecular mechanism for these observed phenotypes was unknown. In this recent report, SIRT6 was found to be a histone H3K9 deacetylase that specifically targets the telomere (Michishita et al., 2008). Furthermore, SIRT6 modification of telomeric chromatin is required for stable association of the WRN helicase, which is mutated in the adult onset progeria Werner Syndrome. WRN has previously been shown to be essential for telomeric replication, preventing loss of telomeric DNA replicated by lagging strand synthesis (Crabbe et al., 2007).

Studies in the telomerase knock-out mouse demonstrated that as telomeres become shorter, the heterochromatic marks are lost and replaced by marks characteristic of open chromatin, such as increased acetylation of histone tails (Benetti et al., 2007), suggesting that a minimum telomere length is necessary to maintain the appropriate chromatin structure at chromosome ends. This change in telomeric chromatin may also alter the expression of adjacent genes. Historically viewed as gene-poor regions of the genome, recent work suggests that in fact, sub-telomeric regions may contain a higher density of genes than was previously appreciated (Linardopoulou et al., 2005; Riethman et al., 2001), raising the possibility that alterations in gene expression as a function of telomere length contribute to aging phenotypes.

ARE CHANGES IN TELOMERIC CHROMATIN STRUCTURE ASSOCIATED WITH HUMAN DISEASE?

As is apparent from the discussion stated earlier, alterations in telomeric chromatin have the potential to affect human health (Figure 2). Changes in chromatin structure at chromosome ends may affect telomere length regulation and end protection, contributing to genome instability and associated pathologies of premature aging and cancer. These perturbations may also affect expression of neighboring genes with concomitant phenotypic consequences. We discuss the evidence that these mechanisms are at work in human aging, cancer and in hereditary disease, as in Figure 2. Telomeric chromatin contains marks characteristic of heterochromatin such as HP1 (orange pentagram) 3me-H3K9 (red diamonds) and 3me-H4K20 (aqua diamonds). Sub-telomeric DNA is also methylated at CpG di nucleotides (grey suns). This repressive chromatin structure leads to silencing of adjacent genes.

Telomeres are assembled into a higher order structure via interaction with the shelter in complex, composed of TRF1 (light blue circle), TRF2 (pink circle), POT1 (dark blue barrel), TIN2 (brown triangle), TPP1 (green arc), hRAP1 (purple triangle). Upon telomere shortening, these marks are lost and replaced by marks consistent with open chromatin such as ac-H3K9 (pink diamonds) and ac-H4K20 (green diamonds). Changes in chromatin structure may lead to activation of gene expression and perturbation of telomere length regulation, either through increased access of telomerase or, in telomerase deficient cells, by inhibition of replication (e.g, increased H3K9 acetylation inhibiting WRN association with the telomere thereby hindering lagging strand replication resulting in increased genome instability).

Aging

The link between telomere length and aging is now well established. Given the observation that heterochromatic marks are lost at the short telomeres of late generation telomerase null mice (Benetti et al., 2007), one might predict a similar scenario in human cells as they approach replicative senescence, although this is yet to be tested. Changes in expression of telomere-proximal genes in relation to individual telomere length have been studied in the context of young versus senescent human fibroblasts (Ning et al., 2003). In this study, there was no significant correlation between telomere length and gene expression of the 34 telomeric genes analyzed. Indeed roughly equal numbers of genes had increased expression in young and senescent fibroblasts (9 and 7 respectively), with the majority of genes (18 of the 34 tested) exhibiting no change in expression level, regardless of final telomere length or amount of sequence lost.

However, alterations in chromatin structure at specific human telomeres as they become shorter with cellular age have not been assessed. While these experiments would be both labor intensive and quite expensive, they are theoretically feasible. For example, the ChIP-Chip approach used to demonstrate an increase in DNA damage factors at individual telomeres in senescent cells (d'Adda di Fagagna et al., 2003) might also detect differences in the level of heterochromatin marks that could then be correlated to levels of telomere-specific gene expression.

In summary, there is currently very little experimental evidence documenting alterations in telomeric chromatin in human cells with age and the biological consequences of these changes. The exception is the study discussed in detail previously, wherein SIRT6, a telomere-specific H3K9 deacetylase, is required for WRN association at telomeres. Mutation of WRN leads to premature aging, in part as a result of defective telomeric replication (Crabbe et al., 2007). Intriguingly, 22 out of 129 patients (17%) with atypical Werner's syndrome surveyed did not contain mutations in the WRN coding region or in LMNA encoding nuclear scaffold proteins lamin A (Chen et al., 2003), which is mutated in the human premature aging syndrome Hutchinson–Gilford Progeria (Eriksson et al., 2003). These data open the door to the possibility that mutations in pathways regulating the assembly and maintenance of telomeric chromatin can be linked with premature aging.

Cancer

Telomeres are tightly linked to tumor genesis. In particular, acquisition of a telomere maintenance mechanism is critical for the unlimited proliferative potential characteristic of most tumors. Studies using knockout mice imply altered telomeric chromatin might impact the efficiency of telomere maintenance (Benetti et al., 2007; Garcia-Cao et al., 2004; Gonzalo et al., 2005; Gonzalo et al., 2006). Two mechanisms for telomere maintenance have been described in human tumors: (1) telomerase activity and (2) ALT. Telomeric chromatin has been implicated in affecting both mechanisms. It has been suggested that the preferential elongation of short telomeres by telomerase may be a consequence of the more open chromatin, that is, loss of heterochromatic marks, at these telomeres (Blasco, 2007). Consistent with this is the observation that over-expression of twoHP1 isoforms, HP1 and HP1, impairs the association of hTERT with telomeres. Similarly, decreased DNA methylation of the sub telomeric region in DNA methyl transferase deficient mice is associated with characteristics of ALT, including long and hetero-geneously sized telomeres, increased telomeric recombination and the presence of ALT-associated PML bodies (APBs) (Gonzalo et al., 2006). APBs, nuclear structures in which the telomere binding proteins, such as TRF1 and TRF2, and telomeric DNA co-localize with the promyelocytic leukemia (PML) nuclear body (Yeager et al., 1999), are tightly correlated with activation of ALT (Yeager et al., 1999). These data suggest decreased methylation of sub-telomeric regions, might accompany ALT activation although this has not yet been directly tested in human tumors. Alterations in telomeric chromatin may also contribute to tumorigenesis by altering the level of cellular DNA damage.

In the study in which HP1 isoforms were over expressed in human cells, perturbed telomere structures lead to an increase in end-to-end fusions, an associated increased sensitivity to ionizing radiation and suppression of tumorigenicity in a xenograft model. In the simplest model, altering HP1 levels at telomeres inhibits hTERT access, thereby leading to telomere attrition and subsequent loss of end capping function. Again, it is important to note that while telomere phenotypes are observed, global levels of HP1 were altered in these experiments and may be responsible for the increased radio-sensitivity and suppression of tumorigenecity. In contrast to the HP1 study, perturbation of chromatin structure in mouse cells by reduction of chromatin modifying enzymes is not always associated with changes in telomere stability (Garcia-Cao et al., 2004). This could be a result of longer telomeres characteristic of inbred strains of mice or alternatively reflect the continued association of the shelter in complex with telomeres even in the absence of heterochromatic modification of histones.

Hereditary disease

Telomere associated chromosomal rearrangements are responsible for 3 to 5% of unexplained mental retardation syndromes (Ravnan et al., 2006). The chromosomal rearrangements are likely a consequence of the increased recombination that occurs in these regions (Rudd et al., 2007). The plasticity of the subtelomeric region has been implicated in evolution, specifically environmental adaptation, with the undesired corollary being increased mutation rate within this 0.1% of the genome. The relationship between telomeric chromatin structure and telomeric rearrangements in the context of idiopathic mental retardation has yet to be investigated. However, the observation that decreased subtelomeric or H4K20 methylation is associated with increased recombination (Benetti et al., 2007; Gonzalo et al., 2006) in this region implies that the two processes may be mechanistically linked. The role of TPE in human disease has been postulated for many years.

However, to date, this lingering hypothesis has not been supported by the data. The most widely touted disease possibility driven by TPE is the autosomal dominant Facio-Scapulo-HumoralDystrophy (FSHD). FSHD is associated with a decreased number of D4Z4 units, a 3.3 kb tandemly repetitive sequence located in the subtelomeric region of chromosome 4q (Fisher and Upadhyaya, 1997). Given the association between repetitive DNA and variegated gene expression, as well as the telomeric location of the locus, it is not surprising that TPE was invoked as a potential mechanism underlying the disease (Hewitt et al., 1994) in the early days following identification of this phenomenon in yeast and long before evidence supporting TPE in mammalian cells existed. Demonstration that the 4q telomere localizes to the nuclear periphery with constitutive heterochromatin led credence to the TPE hypothesis (Winokur et al., 1994). The double homeo box 4 gene (DUX4) is located within each D4Z4 repeat unit and has been recently demonstrated to be up regulated, along with its cognate target genes, in myoblasts of patients with FSHD (Tam et al., 2004). Additionally, to date there is no evidence supporting altered chromatin structure at the repetitive locus associated with the disease. In contrast, H4 acetylation levels were found to be similar to that observed in euchromatin, rather than in heterochromatin (Dixit et al., 2007). Despite the absence of direct demonstration of TPE contributing to human pathology, the possibility remains that altered gene expression as a result of telomere proximity may be found to play a role in human disease. To date, Dyskeratosis Congenita is the disease best documented to be linked to alteration of telomere maintenance (Jiang et al., 2003). Dyskeratosis Congenita is a progressive bone marrow failure syndrome. Additional features include mucocutaneous abnormalities, chromosome instability and cancer predisposition. The autosomal dominant form of Dyskeratosis Congenita has been demonstrated to be a result of mutations in telomerase, either the RNA (Kirwan and Dokal, 2008) or TERT (Vulliamy et al., 2001) components, or in the shelterin protein TIN2 (Savage et al., 2008).

Furthermore, anticipation, where onset of the disease becomes progressively earlier and symptoms more severe with each generation, has been documented for Dyskeratosis Congenita (Armanios et al., 2005). The anticipation was correlated with a shorter telomere length in each successive generation. These data support the model wherein the clinical pathological features of Dyskeratosis Congenita arise, as a result of impaired telomere maintenance leading to chromosome instability. Less severe presentation of Dyskeratosis Congenita asaplasticanemia has also been documented (Vulliamy et al., 2004).

Again, affected individuals had mutations in the telomerase RNA gene and disease was associated with severely shortened telomeres, suggesting that pathologies arising as a result of impaired telomere maintenance may be more widespread than previously anticipated. Indeed, mutations in telomerase components have also been documented in a percentage of patients with familial idiopathic pulmonary fibrosis (Fogarty et al., 2003). The patients did not exhibit additional clinical features characteristic of Dyskeratosis Congenita, suggesting that the pathology associated with impaired telomere maintenance may be varied. The factors impacting on clinical features associated with impaired telomere maintenance have yet to be elucidated. It is likely that with increased scrutiny, additional pathologies associated with alterations in telomere structure and maintenance will be identified.

REFERENCES

Armanios M, Chen JL, Chang YP, Brodsky RA, Hawkins A, Griffin CA, Eshleman JR, Cohen AR, Chakravarti A, Hamosh A, Greider CW (2005). Haploinsufficiency of telomerase reverse transcriptase leads to anticipation in autosomal dominant dyskeratosis congenital. Proc. Natl. Acad. Sci. U.S.A., 102: 15960-15964.

Baumann P, Cech TR (2001). Pot1, the putative telomere end-binding protein in fission yeast and humans. Science, 292: 1171-1175.

Bayne RA, Broccoli D, Taggart MH, Thomson EJ, Farr CJ, Cooke HJ, (1994). Sand- wiching of a gene within 12 kb of a functional telomere and alpha satellite does not result in silencing. Hum. Mol. Genet., 3: 539–546.

Benetti R, Gonzalo S, Jaco I, Schotta G, Klatt P, Jenuwein T, Blasco MA (2007). Suv4- 20h deficiency results in telomere elongation and derepression of telomere recombination, J. Cell Biol., 178: 925–936.

Bilaud T, Brun C, Ancelin K, Koering CE, Laroche T, Gilson E (1997). Telomeric localization of TRF2, a novel human telobox protein. Nat. Genet., 17: 236–239.

Blasco MA (2007). The epigenetic regulation of mammalian telomeres. Nat. Rev. Genet., 8: 299–309

Broccoli D, Smogorzewska A, Chong L, De Lange T (1997). Human telomeres con- tain two distinct Myb-related proteins TRF1 and TRF2. Nat. Genet., 17: 231–235.

Chan SW, Blackburn EH (2003). Telomerase and ATM/Tel1p protect telomeres from no homologous end joining. Mol. Cell, 11: 1379–1387.

Chen L, Lee L, Kudlow BA, Dos Santos HG, Sletvold O, Shafeghati Y, Botha EG, Garg A, Hanson NB, Martin GM, Mian IS, Kennedy BK, Oshima J (2003). LMNA mutations in atypical Werner's syndrome. Lancet, 362: 440–445.

Chin L, Artandi SE, Shen Q, Tam A, Lee SL, Gottlieb GJ, Greider CW, DePinho RA (1999). p53 deficiency rescues the adverse effects of telomere loss and cooperates with telomere dysfunction to accelerate carcinogenesis. Cell, 97: 527–538.

Chong L, van Steensel B, Broccoli D, Erdjument-Bromage H, Hanish J, Tempst P, De Lange T (1995). A human telomeric protein. Science, 270: 1663-1667.

Costa A, Daidone MG, Daprai L, Villa R, Cantu S, Pilotti S, Mariani L, Gronchi A, Henson JD, Reddel RR, Zaffaroni N (2006). Telomere maintenance mechanisms in liposarcomas: association with histologic subtypes and disease progression. Cancer Res., 66: 8918–8924.

Crabbe L, Jauch A, Naeger CM, Holtgreve-Grez H, Karlseder J (2007). Telomere dys- function as a cause of genomic instability in Werner syndrome, Proc. Natl. Acad. Sci. U.S.A, 104: 2205–2210.

d'Adda di Fagagna F, Reaper PM, Clay-Farrace L, Fiegler H, Carr P, Von Zglinicki T, Saretzki G, Carter NP, Jackson SP (2003). A DNA damage checkpoint response in telomere-initiated senescence. Nature, 426: 194–198.

De Lange T (2005). Shelterin: the protein complex that shapes and safeguards human telomeres. Genes Dev., 19: 2100–2110.

DePinho RA, Polyak K (2004). Cancer chromosomes in crisis. Nat. Genet., 36: 932–934.

Dixit M, Ansseau E, Tassin A, Winokur S, Shi R, Qian H, Sauvage S, Matteotti C, Van Acker AM, Leo O, Figlewicz D, Barro M, Laoudj-Chenivesse D, Belayew A, Coppee F, Chen YW (2007). DUX4, a candidate gene of facioscapulo- humeral muscular dystrophy, encodes a transcriptional activator of PITX1. Proc. Natl. Acad. Sci. U.S.A., 104: 18157–18162.

Eriksson M, Brown WT, Gordon LB, Glynn MW, Singer J, Scott L, Erdos MR, Robbins CM, Moses TY, Berglund P, Dutra A, Pak E, Durkin S, Csoka AB, Boehnke M, Glover TW, Collins FS (2003). Recurrent de novo point mutations in lamin A cause Hutchinson Gilford progeria syndrome. Nature, 423: 293–298.

Fabre E, Muller H, Therizols P, Lafontaine I, Dujon B, Fairhead C (2005). Comparative genomics in hemiascomycete yeasts: evolution of sex, silencing, andsubtelomeres, Mol. Biol. Evol., 22: 856–873.

Fisher J, Upadhyaya M (1997). Molecular genetics of facioscapulohumeral musculardystrophy (FSHD). Neuromuscul. Disord., 7: 55–62.

Fogarty PF, Yamaguchi H, Wiestner A, Baerlocher GM, Sloand E, Zeng WS, Read EJ, Lansdorp PM, Young NS (2003). Late presentation of dyskerato- sis congenita as apparently acquired aplastic anaemia due to mutations in telomerase RNA. Lancet, 362: 1628–1630.

Garcia-Cao M, O'Sullivan R, Peters AH, Jenuwein T, Blasco MA (2004). Epigenetic regulation of telomere length in mammalian cells by the Suv39h1 and Suv39h2 histone methyltransferases. Nat. Genet., 36: 94–99.

Gonzalo S, Jaco I, Fraga MF, Chen T, Li E, Esteller M, Blasco MA (2006). DNA methyltransferases control telomere length and telomere recombination in mammalian cells. Nat. Cell. Biol., 8: 416–424.

Gonzalo S, Garcia-Cao M, Fraga MF, Schotta G, Peters AH, Cotter SE, Eguia R, Dean DC, Esteller M, Jenuwein T, Blasco MA (2005). Role of the RB1 family in stabilizing histone methylation at constitutive heterochromatin. Nat. Cell Biol., 7: 420–428.

Gottschling DE, Aparicio OM, Billington BL, Zakian VA (1990). Position effect at S. cerevisiae telomeres: reversible repression of Pol II transcription, Cell, 63: 751–762.

Griffith JD, Comeau L, Rosenfield S, Stansel RM, Bianchi A, Moss H, De Lange T (1999). Mammalian telomeres end in a large duplex loop, Cell, 97: 503-514.

Hakin-Smith V, Jellinek DA, Levy D, Carroll T, Teo M, Timperley WR, McKay MJ, Reddel RR, Royds JA (2003). Alternative lengthening of telomeres and survival in patients with glioblastoma multiforme, Lancet, 361: 836–838.

Harley CB, Futcher AB, Greider CW (1990). Telomeres shorten during ageing of human fibroblasts. Nature, 345: 458–460.

Hayflick L (1965). The limited in vitro lifetime of human diploid cell strains, Exp. Cell. Res., 37: 614–636.

Hewitt JE, Lyle R, Clark LN, Valleley EM, Wright TJ, Wijmenga C, van Deutekom JC, Francis F, Sharpe PT, Hofker M (1994). Analysis of the tandem repeat locus D4Z4 associated with facioscapulohumeral muscular dystrophy. Hum. Mol. Genet., 3: 1287-1295.

Hottiger S, Hursting JC, Barrett L, Guarente R, Mulligan B, Demple GD, Yancopoulos FW (2006). Genomic insta- bility and aging-like phenotype in the absence of mammalian SIRT6. Cell, 124: 315–329

Houghtaling BR, Cuttonaro L, Chang W, Smith S (2004). A dynamic molecular link between the telomere length regulator TRF1 and the chromosome end protector TRF2, Curr. Biol., 14: 1621–1631.

Jiang G, Yang F, van Overveld PG, Vedanarayanan V, van der Maarel S, Ehrlich M (2003). Testing the position-effect variegation hypothesis for facioscapulo- humeral muscular dystrophy by analysis of histone modification and gene expression in subtelomeric 4q. Hum. Mol. Genet., 12: 2909–2921.

Johnson JE, Varkonyi RJ, Schwalm J, Cragle R, Klein-Szanto A, Patchefsky A, Cukierman E, Von Mehren M, Broccoli D (2005). Multiple mechanisms of telomere maintenance exist in liposarcomas, Clin. Cancer Res., 11: 5347–5355.

Kim M, Xu L, Blackburn EH (2003). Catalytically active human telomerase mutants with allele-specific biological properties, Exp. Cell. Res., 288: 277–287.

Kim SH, Kaminker P, Campisi J (1999). TIN2, a new regulator of telomere length in human cells. Nat. Genet., 23: 405–412.

Kirwan M, I Dokal (2008). Dyskeratosis congenita: A genetic disorder of many faces. Clin. Genet., 73: 103–112.

Li B, Oestreich S, de Lange T (2000). Identification of human Rap1: implications for telomere evolution. Cell, 101: 471–483.

Linardopoulou EV, Williams EM, Fan Y, Friedman C, Young JM, Trask BJ (2005) Human subtelomeres are hot spots of interchromosomal recombination andsegmental duplication. Nature, 437: 94–100.

Liu D, Safari A, O'Connor MS, Chan DW, Laegeler A, Qin J, Songyang Z (2004). PTOP interacts with POT1 and regulates its localization to telomeres. Nat. Cell. Biol., 6: 673–680.

Louis EJ (1995). The chromosome ends of *Saccharomyces cerevisiae*, Yeast, 11: 1553–1573.

Masutomi K, Possemato R, Wong JM, Currier JL, Tothova Z, Manola JB, Ganesan S, Lansdorp PM, Collins K, Hahn WC (2005). The telomerase reverse tran- scriptase regulates chromatin state and DNA damage responses. Proc. Natl. Acad. Sci. USA, 102: 8222–8227.

Michishita E, McCord RA, Berber E, Kioi M, Padilla-Nash H, Damian M, Cheung P, Kusumoto R, Kawahara TL, Barrett JC, Chang HY, Bohr VA, Ried T, Gozani O, Chua KF (2008). SIRT6 is a histone H3 lysine 9 deacetylase that modulatestelomeric chromatin. Nature, 452: 492–496.

Mostoslavsky R, Chua KF, Lombard DB, Pang WW, Fischer MR, Gellon L, Liu P, Mostoslavsky G, Franco S, Murphy MM, Mills KD, Patel P, Hsu JT, Hong AL, Ford E, Cheng HL, Kennedy C, Nunez N, Bronson R, Frendewey D, Auerbach W, Valenzuela D, Karow M, Muller HJ (1938). The remaking of chromosomes, The Collecting Net - Woods Hole, 13(1938): 181–195.

Ning Y, Xu JF, Li Y, Chavez L, Riethman HC, Lansdorp PM, Weng NP (2003). Telomere length and the expression of natural telomeric

genes in human fibroblasts. Hum. Mol. Genet., 12: 1329–1336.
Ravnan JB, Tepperberg JH, Papenhausen P, Lamb AN, Hedrick J, Eash D, Ledbetter DH, Martin CL (2006). Subtelomere FISH analysis of 11 688 cases: An evaluation of the frequency and pattern of subtelomere rearrangements in individuals with developmental disabilities. J. Med. Genet., 43: 478–489.
Riethman H (2008). Human telomere structure and biology. Annu. Rev. Genomics Hum. Genet., page number
Riethman HC, Xiang Z, Paul S, Morse E, Hu XL, Flint J, Chi HC, Grady DL, Moyzis RK (2001). Integration of telomere sequences with the draft human genomesequence. Nature, 409: 948–951.
Rudd MK, Friedman C, Parghi SS, Linardopoulou EV, Hsu L, Trask BJ(2007). Elevated rates of sister chromatid exchange at chromosome ends. PLoS Genet., 3: e32.
Sarin KY, Cheung P, Gilison D, Lee E, Tennen RI, Wang E, Artandi MK, Oro AE, Artandi SE (2005). Conditional telomerase induction causes proliferation of hair follicle stem cells. Nature, 436: 1048–1052.
Savage SA, Giri N, Baerlocher GM, Orr N, Lansdorp PM, Alter BP (2008). TINF2, a component of the shelterin telomere protection complex, is mutated in dyskeratosis congenita, Am. J. Hum. Genet., 82: 501–509.
Savage SA, Stewart BJ, Weksler BB, Baerlocher GM, Lansdorp PM, Chanock SJ, Alter BP(2006). Mutations in the reverse transcriptase component of telomerase (TERT) in patients with bone marrow failure. Blood Cells Mol. Dis., 37: 134–136.
Shay JW, Bacchetti S (1997). A survey of telomerase activity in human cancer, Eur. J. Cancer, 33: 787–791.
Sprung CN, Sabatier L, Murnane JP (1996). Effect of telomere length on telomeric gene expression. Nucleic Acids Res., 24: 4336–4340.
Takai H, Smogorzewska A, de Lange T (2003). DNA damage foci at dysfunctional telomeres, Curr. Biol., 13: 1549–1556.
Tam R, Smith KP, Lawrence JB (2004). The 4q subtelomere harboring the FSHD locus is specifically anchored with peripheral heterochromatin unlike most human telomeres. J. Cell. Biol., 167: 269–279.
Ulaner GA, Huang HY, Otero J, Zhao Z, Ben-Porat L, Satagopan JM, Gor- lick R, Meyers P, Healey JH, Huvos AG, Hoffman AR, Ladanyi M (2003). Absence of a telomere maintenance mechanism as a favorable prognostic factor in patients with osteosarcoma, Cancer Res., 63: 1759–1763.
Vulliamy T, Marrone A, Goldman F, Dearlove A, Bessler M, Mason PJ, Dokal I (2001). The RNA component of telomerase is mutated in autosomal dominant dyskeratosis congenital. Nature, 413: 432–435.
Vulliamy T, Marrone A, Szydlo R, Walne A, Mason PJ, Dokal I (2004). Disease anticipation is associated with progressive telomere shortening in families with dyskeratosis congenita due to mutations in TERC. Nat. Genet., 36: 447–449.
Winokur ST, Bengtsson U, Feddersen J, Mathews KD, Weiffenbach B, H Bailey, RP Markovich, JC Murray, JJ Wasmuth, MR Altherr (1994). The DNA rearrangement associated with facioscapulohumeral muscular dystrophy involves a heterochromatin-associated repetitive element: implications for a role of chromatin structure in the pathogenesis of the disease. Chromosome Res., 2: 225–234.
Wong JM, Collins K (2006). Telomerase RNA level limits telomere maintenance in X-linked dyskeratosis congenital. Genes Dev., 20: 2848–2858.
Xu L, Blackburn EH (2007). Human cancer cells harbor T-stumps, a distinct class of extremely short telomeres. Mol. Cell, 28: 315–327.
Ye JZ, Hockemeyer D, Krutchinsky AN, Loayza D, Hooper SM, Chait BT, de Lange T (2004). POT1-interacting protein PIP1: a telomere length regulator that recruits POT1 to the TIN2/TRF1 complex, Genes Dev., 18: 1649–1654.
Yeager T, Neumann A, Englezou A, Huschtscha L, Noble J, Reddel R (1999). Telomerase-negative immortalized human cells contain a novel type ofpromyelocytic leukemia (PML) body, Cancer Res., 59: 4175–4179.
Yeager TR, Neumann AA, Englezou A, Huschtscha LI, Noble JR, Reddel RR (1999). Telomerase-negative immortalized human cells contain a novel type of promyelocytic leukemia (PML) body, Cancer Res., 59: 4175–4179.
Zhu J, Wang H, Bishop JM, Blackburn EH (1999). Telomerase extends the lifespan of virus-transformed human cells without net telomere lengthening. Proc. Natl. Acad. Sci. USA., 96: 3723–3728.

4

Plant protease inhibitors: a defense strategy in plants

Huma Habib and Khalid Majid Fazili*

Department of Biotechnology, The University of Kashmir, P/O Naseembagh, Hazratbal, Srinagar -190006, Jammu and Kashmir, India.

Proteases, though essentially indispensable to the maintenance and survival of their host organisms, can be potentially damaging when overexpressed or present in higher concentrations, and their activities need to be correctly regulated. An important means of regulation involves modulation of their activities through interaction with substances, mostly proteins, called protease inhibitors. Some insects and many of the phytopathogenic microorganisms secrete extracellular enzymes and, in particular, enzymes causing proteolytic digestion of proteins, which play important roles in pathogenesis. Plants, however, have also developed mechanisms to fight these pathogenic organisms. One important line of defense that plants have to fight these pathogens is through various inhibitors that act against these proteolytic enzymes. These inhibitors are thus active in endogenous as well as exogenous defense systems. Protease inhibitors active against different mechanistic classes of proteases have been classified into different families on the basis of significant sequence similarities and structural relationships. Specific protease inhibitors are currently being overexpressed in certain transgenic plants to protect them against invaders. The current knowledge about plant protease inhibitors, their structure and their role in plant defense is briefly reviewed.

Key words: Proteases, enzymes, protease inhibitors, serpins, cystatins, pathogens, defense.

Table of content

1. Introduction
2. Families of PPIs
2.1 Serpin (Serine PI) family
2.2 Bowman Birk inhibitors (BBIs) family
2.3 Kunitz family
2.4 Squash inhibitors
2.5 Cereal trypsin/α-amylase inhibitors
2.6 Mustard (Sinapis) trypsin inhibitor (MSI)
2.7 Potato type I PIs (PI 1)
2.8 Potato type II PIs (PI 2)
2.9 Cysteine PIs (CYS), the cystatin superfamily
2.10 Aspartyl and metallocarboxypeptidase inhibitors
3. Plant protection
4. Transgenic plants
5. Resistance to inhibitors
6. Concluding remarks
7. Acknowledgement
8. References

INTRODUCTION

Proteolytic enzymes, also called proteases, are the enzymes that catalyse the hydrolytic cleavage of specific peptide bonds in their target proteins. These enzymes are widely distributed in nearly all plants, animals and microorganisms (Joanitti et al., 2006; Neurath, 1989; Valueva and Mosolov, 2004; Christeller, 2005; Haq et al., 2004; Supuran et al., 2002; Mosolov and Valueva, 2005; Mosolov et al., 2001; Lawrence and Koundal, 2002; Ryan, 1990). In higher organisms, nearly 2% of the genes code for these enzymes (Barrett et al., 2001). Being essentially indispensable to the maintenance and

*Corresponding author. E-mail: fazili@kashmiruniversity.net or khalid_fazili@yahoo.com.

Abreviations: BBI, Bowman-Birk inhibitor; BTI, Barley trypsin inhibitor; CPI, Cysteine protease inhibitor; CpTi, Cowpea trypsin inhibitor; Cmps, Cucurbita maxima phloem serpin; MSI, Mustard trypsin inhibitor; PI, Protease inhibitor; PPI, Plant PI; PI1, Potato type1PI; PI2, Potato type2 PI; PVy, Potato virus Y; SFTI, Sunflower trypsin inhibitor; TEV, Tobaco etch virus.

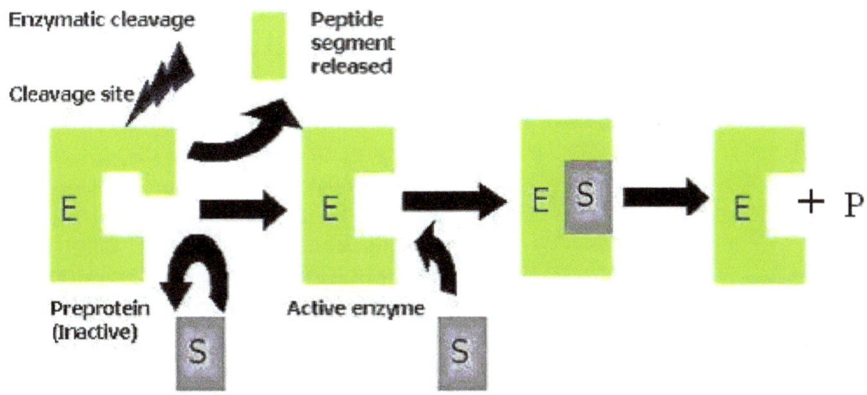

Figure 1A. A diagrammatic model for control of enzyme activity at the level of synthesis. Enzymes may be synthesized as inactive preproteins and are processed to be activated. E represents enzyme, S represents substrate and P represents the products.

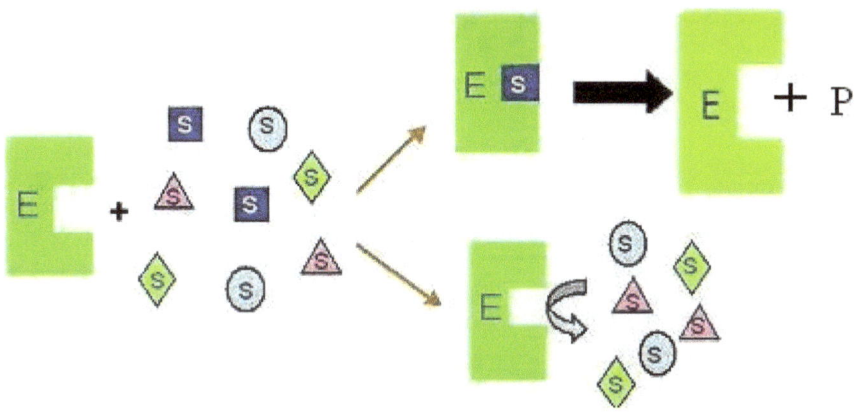

Figure 1B. Model showing enzymes activity regulation by virtue of substrate specificity. E,S and P have the same meaning as in Figure 1A.

survival of their host organism, proteases play key roles in many biological processes. The proteolytic events catalysed by these enzymes serve as mediators of signal initiation, transmission and termination in many of the cellular events such as inflammation, apoptosis, blood clotting and hormone processing pathways (Ivanov et al., 2006). Despite the fact that these enzymes are indispensable to the cells and organisms that host them, they may be potentially damaging when overexpressed or present in higher concentrations. For this reason the activities of these enzymes need to be strictly regulated and controlled (Rawlings et al. 2004a). The synthesis of these enzymes as inactive pre-proteins (Figure 1A), and their substrate specificity (Figure 1B) keeps a control on their activities, but it does not fulfill the desired level of regulation, and the fact remains, that cells and organisms require additional means of control. One important control mechanism involves interaction of the active enzymes with proteins that inhibit their activities (Figure 2). These inhibitors form less active or fully inactive complexes with their cognate enzymes, and are called protease inhibitors (PIs). Trypsin can be considered as a prototype of the class of enzymes synthesized as inactive precursors. Synthesised as trypsinogen, it requires proteolytic processing to be activated. Once activated trysin acts specifically only on peptide bonds whose carboxyl functions are contributed by lysine or arginine residues. Further check on the activity of trypsin is due to its interaction with antitrypsin, the protein inhibitor of the activated form (Laskowski and Qasim, 1999).

PIs are of very common occurrence. They have been isolated and characterized from a large number of organisms, including plants, animals and microorganisms (Valueva and Mosolov, 2004; Christeller, 2005; Haq et al., 2004; Supuran et al., 2002; Mosolov and Valueva, 2005; Mosolov et al., 2001). Naturally occurring PIs are essential for regulating the activity of their corresponding proteases and play key regulatory roles in many biological processes. For a few PIs, functions other than blocking protease action have also been found, such as

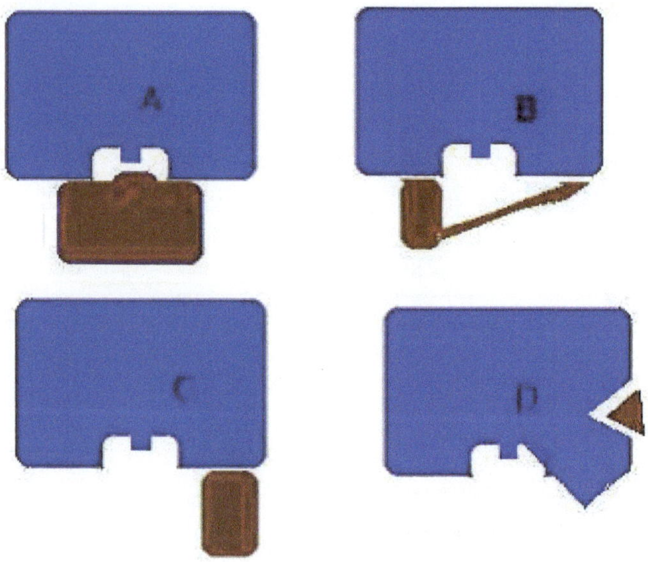

Figure 2. Diagrammatic representation of the different modes of protease inhibition (A) Direct blockage of the active center canonical inhibition of trysin like proteases (B) Indirect blockage of the active center (C) Adjacent or and exocite binding (D) Allosteric interaction.

growth factor activities, receptor clearance signaling or involvement in carcinogenesis (Qi et al., 2005). A number of inherited diseases such as emphysema, and epilepsy result from abnormalities in PIs (Lomas et al., 2002; Ritchie, 2003; Lehesjoki, 2003; Bitoun et al., 2002)

PIs are of common occurrence in the plant kingdom. Plant PIs (PPIs) are generally small proteins that have mainly been described as occurring in storage tissues, such as tubers and seeds, but they have also been found in the aerial parts of plants (De Leo et al., 2002). They are also induced in plants in response to injury or attack by insects or pathogens (Ryan, 1990). In plants, these PIs act as anti-metabolic proteins, which interfere with the digestive process of insects. One of the important defense strategies that are found in plants to combat predators involves PIs which are in particular effective against phytophagous insects and microorganisms. The defensive capabilities of PPIs rely on inhibition of proteases present in insect guts or secreted by micro-organisms, causing a reduction in the availability of amino acids necessary for their growth and development (Lawrence and Koundal, 2002)

With this background and the attached medical and therapeutic significance of PIs, the current review is an attempt to give a comprehensive presentation of the different families of PPIs. The structure of each class of PIs is briefly given with a description of their role in plant defence. The progress in the development of transgenic plants carrying PI genes I also reviewed briefly. Having enormous potential to intervene in a large number of human disorders, the scope for exploration of the natural PIs remains wide open.

Families of PIs

PIs have been grouped into families and subfamilies and into different clans on the basis of sequence relationship and the relationship of protein folds of the inhibitory domains or units. An inhibitor domain is defined as the segment of the amino acid sequence containing a single reactive site after removal of any parts that are not directly involved in the inhibitor activity. On the basis of sequence homologies of their inhibitor domains, PIs have been classified into 48 families (Rawlings et al., 2004b). Proteins containing a single inhibitor unit are termed simple inhibitors, and those that contain multiple inhibitor units are termed complex inhibitors. A total of 11 families belong to the latter category and contain between 2 and 15 inhibitory domains. Most of these are homotypic, containing inhibitor units from a single family, some are however heterotypic, and contain inhibitor units from different families (Richardson et al., 2001; Trexler et al., 2001, 2002). On the basis of tertiary structure, 31 of the 48 families have been assigned to 26 clans, indicating that a large proportion of families show no relationships in their three dimensional structures.

Peptidases have been classified on the basis of structural and evolutionary relationships into different families and clans (Rawlings and Barrett, 1993). These families have been divided into five groups based on chemistry of the catalytic site of enzymes (Barrett, 1986, 1994; Grudkowska and Zagdanska, 2004). The families of PIs could not, however, be grouped on the basis of the catalytic type of enzymes inhibited, since a number of families contain cross-class inhibitors (Rawlings et al., 2004b). The proteins in family 13 Kunitz-type PPIs generally inhibit serine peptidases, but they also include inhibitors of cysteine and aspartate proteases (Heibges et al., 2003). Family 14, the serpin family, mostly contain inhibitors of serine proteases, but it also contains inhibitors of some cysteine proteases. In the past, however, PIs have been classified into serine, cysteine, aspartate and metallocarboxy PIs. (De Leo et al., 2002; Laskowski et al., 2003; Koiwa et al., 1997). PIs that are active against all the mechanistic classes of proteases have been described in plants. A database for PPIs, coordinated by Luigi R Ceci at Centro di studio sui Mitocondri and Metabolismo Energetico-C.N.K, (via Amendola 165/A, Bari, Italy), contains information about 495 inhibitors with several isoinhibitors identified in 129 different plants (De Leo et al., 2002). The database describes nine families of PPIs based on sequence similarities. The families along with some representative examples are given in Table 1.

All these families except those of the cysteine-PI family and the metallocarboxypeptidase inhibitor family contain PIs of serine proteases. These may, however, also contain inhibitors active against other mechanistic classes of proteases. For example the plant Kunitz family (family 13) contains inhibitors of serine proteases (family S1), but also includes inhibitors of cysteine proteases (C1) and

Table 1. Families of plant protease inhibitors with some type examples.

Common name	MEROPS Family/subfamily	Type example	Source	Target Protease	Referencces
Kunitz (plant)	13A	soybean Kunitz trypsin inhibitor	Glycine max	Trypsin, Chymotrypsin	Laskowski and Kato (1980)
		barley subtilisin inhibitor	*Hordeum vulgare*	Subtilisin,Alpha-amylase	Vallee et al. (1998)
		winged-bean chymotrypsin inhibitor	*Psophocarpus tetragonolobus*	Alpha-chymotrypsin	Habu et al. (1992)
		Kunitz cysteine peptidase inhibitor 1	*Solanum tuberosum*	Cysteine proteases	Gruden et al. (1997)
Kunitz (plant)	I3B	proteinase inhibitor A inhibitor unit	Sagittaria sagittifolia	Trypsin, Chymotrypsin, Kallikerin	Laskowski and Kato (1980)
		Kunitz subtilisin inhibitor	*Canavalia lineata*	Subtilisin-type microbial serine proteases	Terada et al. (1994)
		cathepsin D inhibitor	*Solanum tuberosum*	Cathepsin D, Trypsin	Strukelj et al. (1992)
		trypsin inhibitor	*Acacia confusa*	Trypsin and alpha-chymotrypsin	Lin et al. (1991)
cereal	I6	ragi seed trypsin/α-amylase inhibitor	Eleusine coracana	Alpha amylase	Hojima et al. (1980)
		barley trypsin/factor XIIa inhibitor	*Hordeum vulgare*	Alpha-amylase, Trypsin	Lazaro et al. (1988)
		wheat trypsin/alpha-amylase inhibitor	*Triticum aestivum*	Alpha-amylase, Trypsin	Shewry et al. (1984)
		maize trypsin/factor XIIa inhibitor	*Zea mays*	Mammalian trypsin, activated hageman factor	Mahoney et al. (1984)
squash	I7	trypsin inhibitor MCTI-1	Momordica charantia	Pancreatic elastase	Wiezorek et al. (1985)
		trypsin inhibitor MCTI-II	*Momordica charantia*	Trypsin	Huang et al. (1992)
		macrocyclic squash trypsin inhibitor	*Momordica cochinchinensis*	Trypsin	Hernandez et al. (2000)
		trypsin inhibitor CSTI-IV	*Cucumis sativus*	Trypsin	Wieczorek et al. (1985)
Potato type I	I13	chymotrypsin inhibitor I	Solanum tuberosum	Chymotrypsin, Trypsin	Richardson (1974)
		glutamyl peptidase II inhibitor	*Momordica charantia*	Glu S.griseus protease , Subtilisin	Ogata et al. (1991)
		subtilisin-chymotrypsin inhibitor CI-1A	*Hordeum vulgare*	Subtilisin , Chymotrypsin	Greagg et al. (1994)
		wheat subtilisin/chymotrypsin inhibitor	*Triticum aestivum*	B.lichenoformis subtilisin, Alpha-chymotrypsin	Poerio et al. (2003)

Table 1. Contd.

mustard	I18	mustard trypsin inhibitor	Sinapis alba	Beta-trypsin	Menegatti et al. (1992)
		mustard trypsin inhibitor-2	*Brassica hirta*	Bovine beta-trypsin, Alpha-chymotrypsin	Ceci et al. (1995)
		rape trypsin inhibitor	*Brassica napus*	Trypsin, Chymotrypsin.	Ceciliani et al. (1994)
cystatin	I25B	onchocystatin	Onchocerca volvulus	Cysteine proteinase	Lustigman et al. (1992)
		ovocystatin	Gallus gallus	Thiol proteases	Laber et al. (1989)
		oryzacystatin II	Oryza sativa	Cysteine proteinases	Ohtsubo et al. (2005)
Kininogen	I25C	metalloprotease inhibitor	Bothrops jararaca	Atrolysin C, Jararhagin.	Cornwall et al. (2003)
		sarcocystatin	*Sarcophaga peregrina*	Cysteine proteinase	Saito et al. (1989)
Bowman-Birk	I12	Bowman–Birk plant trypsin inhibitor unit 1	Glycine max	Trypsin, Chymotrypsin	Odani and Ikenaka (1976)
		Bowman-Birk trypsin/chymotrypsin inhibitor	*Arachis hypogaea*	Trypsin, Chymotrypsin	Suzuki et al. (1987)
		sunflower cyclic trypsin inhibitor	*Helianthus annuus*	Trypsin, Cathepsin G, Elastase, Chymotrypsin and thrombin	Mulvenna et al. (2005)
Potato type II	I20	proteinase inhibitor II	Solanum tuberosum	Trypsin, Chymotrypsin.	Greenblatt et al. (1989)
		potato peptidase inhibitor II inhibitor unit 1	*Solanum tuberosum*	Trypsin, Chymotrypsin.	Keil et al. (1986)
		tomato peptidase inhibitor II inhibitor unit 1	*Solanum lycopersicum*	Trypsin, Chymotrypsin	Graham et al. (1985)
		tomato peptidase inhibitor II inhibitor unit 2	Solanum lycopersicum	Trypsin, Chymotrypsin	Barrette-Ng et al. (2003)

the aspartic protease cathepsin D. Similarly, the serpin family of PIs active against serine proteases also contains inhibitors of cysteine proteases (Heibges et al., 2003; Laskowski et al., 2003). Serine PIs belonging to arious families have been reported either in storage organs or in the vegetative cells of a wide variety of plants (Garcia et al., 1987). The inhibitors from at least four families belonging to serine PIs are induced sequentially in various plants. These families include potato (*Solanum tuberosum*) and tomato (*Lycopersicon esculentum*) inhibitors I and II in solanaceous plants (Melville and Ryan, 1972; Bryant et al., 1976; Plunkett et al., 1982; Valueva et al., 2001; Farran et al., 2002), Bowman-Birk inhibitors in alfa alfa (Brown and Ryan, 1984) and a Kunitz inhibitor in poplar trees (Bradshaw et al., 1989; Ledoigt et al., 2006).

Serpin (Serine PI) family

The serpin family is the largest and the most widespread superfamily of PIs. Serpin-like genes have been identified in nearly all types of organisms, including viruses, bacteria, plants and animals (Irving et al., 2000; Rawlings et al., 2004b; Christeller and Liang, 2005; Law et al., 2006; Gettins, 2002). Whereas prokaryotes generally have a single serpin gene (Irving et al., 2002a), large multicellular eukaryotic organisms, on the other hand, have several to many genes. Analysis of the thale cress (*Arabidopsis thaliana*) genome shows 29 serpin genes (Silverman at al., 2001).

Plant serpins have been purified and characterized from cereal seeds (Laskowski and Kato, 1980; Yoo et al., 2000; Tsybina et al., 2004), pollens and from the phloem exudates of *Cucurbita maxima* (Wieczorek et al., 1985). Plant serpins have been shown to inhibit model trypsin like proteins (Roberts et al., 2003), but there are no obvious targets for these inhibitors in plants, which may, apparently be involved in inhibiting proteases of plant pathogens (Hejgaard, 2005). Wieczorek and his coworkers have shown an inverse correlation between the up-regulation of squash phloem serpin-1 (cmps) and aphid survival. On the other hand Yoo and his coworkers have reported that feeding of purified serpin to aphids had no impact on insect survival. These data suggest a more complex role for plant serpins in defense (Wieczorek et al., 1985; Yoo et al., 2000).

The serpin 1 of Arabidopsis has been shown to act on metacaspase-like proteins *in vivo* and play a role in the plant immune response (Vercammen et al., 2006). It has been suggested that rather than directly interacting with pathogens, plant serpins may have a role in the complex pathways involved in up-regulating the host immune response (Law et al., 2006).

Plant serpins have molecular mass of 39 - 43 kDa, with amino acid and nucleotide homology with other well-characterized serpins. The majority of serpins inhibit serine proteases, but serpins that inhibit caspases (Ray et al., 1992) and papain like cysteine proteases (Schick et al., 1998; Irving et al., 2002b, c) have also been reported. Plant serpins exhibit differing and mixed specificities towards proteases (Al-Khunaizi et al., 2002). Barley (*Hordeum vulgae*) serpin is a potent inhibitor of trypsin and chymotrypsin at overlapping reactive sites (Dahl et al., 1996a). This inhibitor also inhibits thrombin, plasma kallikrein, Factor VIIa and Factor Xa (Dahl et al., 1996b). Wheat (*Triticum aestivum*) serpins inhibit chymotrypsin and cathepsin G and have glutamic acid, lysine or arginine at P1 site (Roberts et al., 2003). Two oat (*Avena sativa*) serpins show specificity for chymotrypsin and / or elastase, and another one has specificity for trypsin and chymotrypsin at overlapping loop sites (Irving et al., 2002b, c). Squash serpin Cmps-1 also inhibits elastase at two overlapping sites (Ligoxygakis et al., 2003).

Serpins are irreversible 'suicide' inhibitors. The cleavage of an appropriate peptide bond in the reactive centre loop of the inhibitor triggers a rapid conformational change so that catalysis does not proceed beyond the formation of an acyl-enzyme complex (Huntington et al., 2000).

Bowman Birk inhibitors (BBIs) family

On the basis of sequence homology, this forms another family of serine PIs. The family is named after D.E. Bowman and Y. Birk, who were the first to identify and characterise a member of this family from soybean (*Glycine max*) (Bowman 1946; Birk et al., 1963). The soybean inhibitor is now the most-well-studied member of this family and is often referred as the classic BBI. The inhibitors have been found in legumes and cereals (Laing and McManus, 2002; Tanaka et al., 1997; Norioka and Ikenaka, 1983) and in the grass family Poaceae (Odani et al. 1986). The inhibitors of this family are generally found in seeds, but are also wound-inducible in leaves (Eckelkamp, 1993). A small cyclic inhibitor has been identified in sunflower (*Helianthus annuus*) called sunflower trypsin inhibitor 1 (SFTI-1) (Korsinezky, 2001; Luckett et al., 1999).

BBIs have been classified on the basis of their structural features and inhibitor characteristics. The inhibitors from dicotyledonous plants consist of a single polypeptide chain with the molecular mass of 8 kDa. These are double-headed, with two homologous domains each bearing a separate reactive site for the cognate proteases. These inhibitors interact independently, but simultaneously, with two proteases, which may be same or different (Raj et al., 2002; Birk, 1985). The first reactive site in these inhibitors is usually specific for trypsin, chymotrypsin and elastase (Qi et al., 2005). The active-site configuration in these inhibitors is stabilized by the presence of seven conserved disulfide bonds (Chen et al., 1992; Lin et al., 1993).

BBIs from monocotyledonous plants are of two types. One group consists of a single polypeptide chain with a molecular mass of about 8 kDa. They have a single reactive site. Another group has a molecular mass of 16 kDa with two reactive sites (Tashiro et al., 1987, 1990; Prakash et al., 1996). It has been suggested that larger inhibitors have arisen from smaller ones by gene duplication (Odani et al., 1986).

In the case of double-headed BBIs, it has been found that the relative affinity of binding of proteases is altered when one site is already occupied. Peanut (*Arachis hypogoea*) inhibitor has been found to exhibit no activity against chymotrypsin when preoccupied with trypsin and vice versa (Tur et al., 1972). In the same way, the activity of soybean BBIs decreases 100-fold when trypsin is bound at the other site (Gladysheva et al., 1999).

The BBI family of protease inhibitors contains a unique disulfide-linked nine-residue loop that adopts a characterristic canonical conformation (Bode and Hubr, 1992). The loop is called protease-binding loop and binds the protease in a substrate-like manner (Lee and Lin, 1995).

SFTI-1 is a very strong naturally occurring BBI. It consists of a cyclic 14-amino-acid residue-long peptide structure and contains a disulfide-linked loop of nine-amino-acid residues that shares the sequence homology with the first residue loop of BBIs (Korsinezky et al., 2001; Luckett et al., 1999). Analysis of the three-dimensional structure of SFTI-1 in solution, and that of its crystal structure in complex with trypsin, has shown the structures are quite similar (Korsinezky et al., 2001). This has lead to the suggestion that the structure of SFTI-1 is rigid and pre-organized for potent binding, making it a stronger and potent inhibitor than other naturally occurring BBIs.

BBIs are cysteine-rich proteins with inhibitory activity against proteases that are widely distributed in monocot and dicot species (Lin et al., 2006). They have been shown to act as anticarcinogenic compounds. The soybean derived BBI with a well-characterized ability to inhibit trypsin and chymotrypsin is particularly effective in suppressing carcinogenesis in a variety of *in vivo* and *in vitro* systems (Kennedy 1998). BBI has been shown to reduce the proliferation of MCF7 breast cancer cells through accumulation of MAP kinase phosphatase-1 (Wen et al., 2005).

Kunitz family

On the basis of sequence homologies, Kunitz-type inhibitors form a separate family. The members of this family are mostly active against serine proteases, but may also inhibit other proteases (Laing and McManus, 2002; Ritonja et al., 1990). The inhibitors in this family are widespread in plants and have been described in legumes, cereals and in solanaceous species (Ishikawa et al., 1994; Laskowski and Kato, 1980). A 20.5 kDa Kunitz-type trypsin inhibitor with antifungal activity has been reported from the roots of punce ginseng (*Pseudostellaria heterophylla*) (Wang and Ng 2006).

Kunitz-type PIs are also produced under stress, as has been found in potato tubers (*S. tuberosum*) (Park et al., 2005; Ledoigt et al., 2006; Plunkett et al., 1982). The inhibitors usually have molecular mass of 18 - 22 kDa; have two disulfide bridges and one reactive site.

The members of this family are mostly active against serine proteases and have been shown to inhibit trypsin, chymotrypsin and subtilisin (Laing and McManus, 2002; Park et al., 2005), but they also inhibit other proteases, including the aspartic protease cathepsin D and the cysteine proteinase papain. These inhibitors are canonical and form a tight complex with the target protease, which dissociates very slowly (Ritonja et al., 1990).

Squash inhibitors

Squash-family inhibitors have been described only in plants and form yet another family active against serine proteases. The members of this family have been described from many cucurbit families (Lee and Lin, 1995; Hamato et al., 1995; Felizmenio et al., 2001). Seven serine PIs belonging to this family have been isolated and characterized from the seeds of wild cucumber (*Cyclanthera pedata*) (Kuroda et al., 2001). Recently two different but inter-convertible (cis-trans isomers) inhibitors have been isolated and characterized from seeds of wax gourd [*Benincasa hispida* (Thumb) cogn] (Atiwetin et al., 2006). The members of this family consist of a small single peptide chain containing between 28 and 30 amino acids with molecular mass of 3.0 - 3.5 kDa (Heitz et al., 2001; Le Nguyen et al., 1990). These inhibitors have three disulfide bridges and fold in a novel knottin structure (Hara et al., 1989). The small size of these inhibitors, combined with potential activity against important biological molecules such as Hageman factor, human leucocyte elastase and cathepsin G (Hojima et al., 1982; McWherter et al., 1989), has made them particularly attractive for studying proteinase and inhibitor interactions. Chemical synthesis of these inhibitors has created powerful tools for investigating their structure and function relationships (Kupryszewski et al., 1985, 1986; Rolka et al., 1992). The structures of squash inhibitors, and inhibitor and proteinase complexes have been determined by X-ray crystallography and NMR spectroscopy (Holak et al., 1989 a, b; Nilges et al., 1991; Thaimattam et al., 2002). These inhibitors have been shown to follow the standard mechanism for inhibition.

Cereal trypsin/α-amylase inhibitors

The members of this family have serine proteinase inhibitory activity and/or α-amylase- inhibitory activity (Gourinath et al., 2000). A large number of inhibitors in this family have only α-amylase-inhibitory activity; however inhibitors from barley (*Hordeum vulgare*), rye (*Secale cereale*) and tall fescue (*Festuca arundinacea*) are active against trypsin (Odani et al., 1983). Maize (*Zea mays*) and ragi (*Elusine coracana*) inhibitors show dual activities and can inhibit serine proteinases as well as α-amylase (Mahoney et al., 1984; Shivraj and Pattabiraman, 1981). The cereal trypsin/α-amylase inhibitors consist of a single polypeptide chain containing five disulfide bonds with a molecular mass of about 13 kDa (Christeller and Liang, 2005). The structure of the ragi inhibitor solved by NMR spectroscopy and that of its complex with yellow-mealworm (*Tenebrio molitor*) α-amylase by x-ray crystallography has shown that the proteinase-binding loop adopts a canonical conformation (Strobl et al., 1998).

Mustard (Sinapis) trypsin inhibitor (MSI)

These are small single polypeptide chain inhibitors with the molecular mass of about 7 kDa, found in the family Cruciferae and form yet another family of serine PIs (Laing and McManus, 2002; Menengatti et al., 1992). These inhibitors have been isolated and characterized from a number of species including white mustard (*Sinapis alba*) and tape (*Brassica napus*) (Ascenzi et al., 1999; Volpicella et al., 2000). These inhibitors are expressed in seeds during their development and are also wound-inducible (Ceci et al., 1995; De Leo et al., 2001). The inhibitors form a tight binding complex with trypsin and apparently follow the standard mechanism (Ceciliani et al., 1994).

Potato type I PIs (PI 1)

The inhibitors of this family are widespread in plants and have been described in many species, including potato tubers (Ryan and Balls, 1962), tomato fruit (Margossian et al., 1988, Wingate et al., 1989), squash phloem exudates (Murray and Christeller, 1995) and in tomato leaves in response to wounding (Lee et al., 1986). These inhibitors have the molecular mass of 8 kDa and are generally monomeric. While the inhibitors from cucurbit and potato tubers contain a single disulphide bond, the inhibitors in this family in general lack any disulphide bridges (Cai et al., 1995). The inhibitory mechanism in this family is considered to fit the standard model.

Potato type II PIs (PI 2)

The members of this group have been reportedonly from the members of Solanaceae family. Initially characterized from potato tubers (Christeller and Liang, 2005), these inhibitors have been found in leaves, flowers, fruit and phloem of other solanaceaous species (Iwasaki et al., 1971; Pearce et al., 1993). A low molecular-mass inhibitor of this family has been found to be constitutively present in Jasmme tobacco (*Nicotiana alata*) flowers (Atkinson et al., 1993). Six small wound-inducible proteinase inhibitors of this family have been reported from tobacco leaves (Pearce et al., 1993). An analysis of these inhibitors and genes has shown that they are composed of multiple repeat units varying between one and eight (Antcheva et al., 2001; Miller et al., 2000; Choi et al., 2000). Inhibitors in this family have been reported to inhibit chymotrypsin, trypsin, elastase, oryzin, Pronase E and subtilisin (Antcheva et al., 1996).

Cysteine PIs (CYS), the cystatin superfamily

The cystatin superfamily is composed of several families and includes proteins that are related in structure and function to an inhibitor of cysteine proteinase, first described in egg white and referred to as 'chicken egg-white cystatin' (Colella et al., 1989). The members of these families inhibit the activity of cysteine proteases and are called cysteine PIs or cystatins. They are widely distributed in plants, animals and microorganisms (Oliveira et al., 2003). These inhibitors are grouped into four families based on sequence relationships, molecular mass and disulfide-bond numbers and arrangements (Turk and Bode, 1991; Barrett, 1987).

Family-1 cystatins (stefin family)

The members of this group have a molecular mass of about 11 kDa. They are generally present in the cytosol and are devoid of any carbohydrate groups and disulfide bonds (Stato et al., 1990; Machleidt et al., 1983)

Family-2 cystatins (cystatin family)

These inhibitors consist of proteins having 120 – 126 amino acids and the molecular mass of 13.4 - 14.4 kDa. These inhibitors contain two disulphide bonds but are devoid of any carbohydrate groups (Grzonka et al., 2001). They also contain a signal sequence and are known to be secreted (Abrahanson et al., 1987). All the family-2 cystatin inhibitors contain a conserved tripeptide of sequence Phe-Ala-Val near the C-terminus and a conserved dipeptide, Phe-Tyr, near the N- terminus. These conserved sequences are important in binding to the target proteases (Machleidt et al., 1983; Turk et al., 1997)

Family-3 cystatins (kininogen family)

These inhibitors are glycoproteins and are of three different types. High Molecular Weight kininogens (HMW) with a molecular mass of 120 kDa and Low Molecular Weight kininogens (LMW) with molecular mass ranging between

60 and 80 kDa are known. A third type T kininogens with molecular mass of 68 kDa has also been reported. These proteins contain tandem domains that result from gene duplication of the family–2 cystatins. These proteins are also secreted and play key roles in blood coagulation (Otto and Schirmeister, 1997; Salvesen et al., 1986).

Family 1 and 3 cystatins contain a conserved pentapeptide sequence Gln-Val-Val-Ala-Gly and the family-2 members have the homologous peptide Gln-X-Val-Y-Gly, in which X and Y represent any amino acid.

Family–4 cystatins (phytocystatins)

This family includes nearly all the cysteine PIs described in plants. They have been identified in rice (Abe et al., 1987 a, b), maize (Abe et al., 1992), soybean (Hines et al., 1991; Botella et al., 1996), apple (*Malus*) fruit (Ryan et al., 1998), carnation (*Dianthus caryophyllus*) leaves (Kim et al., 1999) and several other monocotyledonous and dicotyledonous plants (Brown and Dziegielewska, 1997; Pernas et al., 1998; Sakuta et al., 2001). Celostatin, a cysteine PI from crested cock's comb (*Celosia cristata*) has recently been cloned and characterized (Gholizadeh et al., 2005).

Phytocystatins have sequence similarity to stefins and cystatins, but do not contain free cysteine residues (Fernandes et al., 1993; Zhao et al., 1996). The unique feature of this superfamily, however, is a highly conserved region of the G58 residue, the glu-val-val-ala-gly (QVVAG) motif and a pro-trp (PW) motif. The studies on the papain inhibitory activity of oryzacystatin and its various truncated forms have identified the conserved QVVAG motif as a primary region of interaction between the inhibitor and its cognate enzyme The PW motif is believed to act like a cofactor (Arai et al., 1991; Abe et al., 1988).

Phytocystatins, on the basis of protein structure, can be divided into two groups. One group consists of single-domain proteins and includes a majority of these inhibitors (Abe et al., 1987 a, b; Pernas et al., 1998). Another group contains multiple-domains and includes the cysteine PIs isolated from potato tubers and tomato leaves (Walsh TA and, Strictland JA 1993; Bolter, 1993).

Plant cysteine PIs are encoded by gene families (Fernandes et al., 1993) and show different expression patterns during development and defense response to biotic and abiotic environmental stress (Felton and Korth, 2000). Expression is usually limited to specific organs or to specific phases during development, such as germination (Botella et al., 1996), early leaf senescence (Huang et al., 2001) cold and salt stress (Van der- Vyver et al., 2003; Pernas et al., 2000).

Aspartyl and metallocarboxypeptidase inhibitors

Plants contain two families of metalloproteinase inhibitors, the metallocarboxypeptidase inhibitor family in potato and tomato plants (Graham and Ryan, 1997; Rancour and Ryan, 1968) and a cathepsin D inhibitor family in potatoes (Keilova and Tomasek, 1976).

The inhibitors that bind to metallocarboxypeptidases have been identified in solanaceaous plants, in the medicinal leech (*Hirudo medicinalis*), in the intestinal parasite roundworm *Ascaris suum*, in the blood tick *Rhipicephalus bursa* and in rat and human tissues (Arolas et al., 2005; Homandberg et al., 1989; Reverter et al., 1998; Normant et al., 1995; Liu et al., 2000). The plant inhibitors have been described in tomato and potato. These inhibitors are small peptide inhibitors consisting of 38 - 39 amino acid residues and have the molecular mass of about 4.2 kDa (Hass et al., 1975; Hass and Hermodson, 1981). These inhibitors inhibit strongly, but competitively, a broad spectrum of carboxypeptidases from both animals and microorganisms, but do not inhibit serine carboxypeptidases from yeast and plants (Havkioja and Neuvonen, 1985). A metallocarboxypeptidase inhibitor is found to accumulate in potato tuber tissues during development, along with the potato inhibitor I and II families of serine PIs. The inhibitor also accumulates in potato leaf tissues, along with the inhibitors of other families, as a response to wounding (Ryan, 1990).

Aspartyl PIs have been described in sunflower, barley and cardoon (*Cynara cardunculus*) flowers and in potato tubers (Park et al., 2000; Kervinen et al., 1999; Lawrence and Koundal, 2002; Marres et al., 1989; Wolfson and Murdock, 1987). The cathepsin D inhibitor, an aspartyl PI described in potato tubers shares considerable amino acid sequence homology with soybean trypsin inhibitor. It is a 27 kDa protein and inhibits serine proteases trypsin and chymotrypsin in addition to the aspartyl protease cathepsin D, but does not inhibit pepsin, cathepsin E and rennin, which are all aspartyl proteases (Lawrence and Koundal, 2002). Pepstatin, a powerful and strong inhibitor of aspartyl proteases has been shown to inhibit proteolysis of the midgut enzymes of Colorado potato beetle (*Lptinotarsa decemlineata*) (Wolfson and Murdock 1987).

Plant protection

Plant PIs are small proteins generally present in high concentration in storage tissues, contributing upto 10% of the total protein content, they are also detectable in leaves in response to the attack of insects and pathogenic microorganisms (Ryan, 1990). PPIs have been shown to play a potent defensive role against predators and pathogens. Many PIs have been shown to act as defensive compounds against pests by direct assay or by expression in transgenic crop plants, and a body of evidence for their role in plant defense has continued to accumulate (Krattiger, 1997).

The possible role of PIs in plant protection was initially noticed when it was observed that the larvae of certain insects were unable to develop normally on soybean products. Subsequently the trypsin inhibitors present in soybean were shown to be toxic to the larvae of the flour beetle (*Tribolium confusum*) (Lawrence and Koundal, 2002). Following these early studies, there have been many examples of PIs that are active against certain insect species. Studies involved both *in vitro* assays against insect gut proteases (Pannetier et al., 1997; Koiwa et al., 1998) and *in vivo* artificial diet bioassays (Urwin et al., 1997; Vain et al., 1998). PIs also exhibit a very broad spectrum of activity against pathogenic nematodes. Cowpea trypsin inhibitor (CpTi) inhibits the growth of nematodes, *Globodera tabacum*, *Globodera pallida* and *Meloidogyne incognita* (Williamson and Hussey, 1996). The expression of rice BBI from *Oryza sativa* is upregulated and induced by pathogens or insects during germination of rice seeds (Lin et al., 2006).

The buckwheat (*Fagopyrum sculentum*) trypsin/chymotrypsin inhibitor interferes with spore germination and mycelium growth of the tobacco brown-spot fungus *Alternaria alternata* (Dunaevskii et al., 1997). Cysteine PIs from pearl millet (*Pennisetum glaucum*) inhibit growth of many pathogenic fungi, including *Trichoderma reesei* (Joshi et al., 1998). These advantages make protease inhibitors an ideal choice to be used in developing transgenic crops resistant to insect pests.

Expression of the PI genes is usually limited to specific organs or to particular phases during plant growth, such as germination (Botella et al., 1996), early leaf senescence (Huang et al., 2001) and drought (Van der- Vyver et al., 2003; Pernas et al., 2000). Wounding or treatment with methyl jasmonate evokes a similar pattern of gene expression. Further, the cytosolic localization of the inhibitors also suggests that they are involved in plant defense against insects (Zhao et al., 1996).

Kazal-type inhibitors play important roles in maintenance of normal cellular processes in animals (Magert et al., 2002; Kreutzmann et al., 2004) and the pathogenesis of mammalian parasitic apicomplexan protozoa and plant pathogenic oomycete fungi (Pszenny et al., 2002; Tian et al., 2004). Typical Kazal domains contain six cysteine residues forming a 1-5, 2-4, 3-6-disulfide bond pattern (Lee and Lin, 1995). However, a novel class of Kazal domains in which the third and sixth cysteine residues are missing has been described. These typical Kazal domains are ubiquitous in serine PIs of plant pathogenic oomycetes. Two of the Kazal like inhibitors namely EP11 and EP110 of the potato late-blight fungus *Phytophora infestans* target the defense-related protease P69B of the host plant tomato (Magert et al., 1999; Tian and Kamoun, 2005).

It has been found that potato tubers treated with elicittors, jasmonic, salicylic or arachidonic acids are able to excrete potatin and three chymotrypsin inhibitors.

Wounding and water stress prompts the secretion of two kinds of Kunitz-type PIs by potato tubers. These inhibitors are closely associated with other secreted polypeptides and would protect them against degradation by extracellular chymotrypsin like protease. The secreted inhibitors could therefore interact with plant defense system (Valueva et al., 2001; Ledoigt et al., 2006).

Oryzacystatin is found to prevent the growth of rice weevil (*Sitophilus oryzae*) by inhibiting the cysteine proteases in the gut of this organism (Hosoyama et al., 1994).

Two extracellular cysteine protease inhibitors (ECIP-1 and ECIP-2) isolated from species of the unicellular green alga *Chlorella* seem to have a role in protecting the cells from attacks by viruses and insects (Ishihara et al., 1999, 2000).

Phytostatins are involved in the control of endogenous cysteine proteinases during maturation and germination of seeds (Abe and Arai, 1991) and play a role in the apoptosis required in plant development and senescence (Solomon et al., 1999). Oryzacystatins have been shown to inhibit the α,β and γ cysteine proteinases that are produced during seed germination (Watanabe et al., 1991) Zeins and maize proteinases are inhibited by maize cystatins, suggesting a role for these inhibitors in the endogenous defense mechanism (Steller, 1995; Hoorn and Jones, 2004). Phytostatins from various plants inhibit the activity of gut cysteine proteinases involed in protein digestion in the gut of various members of the Coleoptera (beetles) attacking these plants, and thus play a role in the exogenous defense system of these plants (Oliveira et al., 2003)

The overexpression of cystatin in soybean cell suspensions blocked programmed cell death (PCD) (Solomon et al., 1999). The overexpression of a cystatin that inhibits papain activity in *Arabidopsis* cell cultures blocked cell death in response to avirulent bacteria and nitric oxide. The overexpression of this inhibitor in tobacco plants blocked the hypersensitive response induced by avirulent bacteria (Hoorn and Jones, 2004; Belenghi et al., 2003).

The use of cystatins has served as a specific approach to control insect predation and diseases in plants (Sakuta et al., 2001). Transgenic rice expressing maize cystatin has been shown to exhibit enhanced resistance against insect predation (Irie, 1996). The rice cystatins have been reported to confer resistance against polyviruses in transgenic tobacco and sweet-potato (*Ipomoea batatas*) plants (Campos et al., 1999).

Transgenic plants

Several transgenic plants expressing PIs have been produced in the last 15 years and tested for enhanced defensive capabilities, with particular efforts directed against insect pests (Valueva et al., 2001). Since the economically important orders of insect pests namely Lepidoptera, Diptera and Coleoptera, use serine and cysteine proteinases in their digestive system to degrade proteins in the ingested food, efforts have generally been directed at genes encoding PIs active against these mechanistic classes of proteases for developing transgenic plants. The PI genes have been particularly useful in developing transgenic plants resistant to insect pests, as they require the transfer of a single defensive gene, and can be expressed from the wound-inducible or constitutive promoters of the host (Boulter, 1993).

The first PI gene to be successfully transferred was that coding for CpTi and produced transgenic tobacco with significant resistance against tobacco hornworm (*Manduca sexta*) (Hilder et al., 1987). The efficiency of transgenic tobacco plants expressing CpTi was tested against armyworm (*Spodoptera litura*) in feeding trials under laboratory conditions. Reduction to the extent of 50% was observed in the biomass of army worm larvae fed on transgenic leaves expressing 3 - 5 μg of CpTi/g of fresh leaves (Sane et al., 1997).

Potato PI-II gene from potato was introduced into several japonica rice varieties to produce transgenic rice plants shown to be insect resistant in greenhouse trials. Wound-inducible PI-II promoter with the first intron of rice actin I gene was able to give high-level expression of PI-II gene in transgenic rice plants. These transgenic plants were resistant to pink stem borer (*Sesamia inferens*) (Duan et al., 1996).

The transformation of plant genomes with PI-encoding cDNA clones appears attractive not only for the control of plant pests and pathogens, but also as a means to produce PIs useful in alternative systems, and the use of plants as factories for the production of heterologous proteins (Sardana et al., 1998). The plant derived PI genes have been used for developing insect-resistant transgenic crops. Several transgenic plants expressing PIs have been produced, and these plants are found to be more resistant against insect pests. Recently, protease inhibitors have also been used to engineer resistance against viruses in transgenic plants (Ussuf et al., 2001). Bean α-amylase inhibitor 1 in transgenic peas (*Pisum sativum*) provides complete protection from pea weevil (*Bruchus pisorum*) under field conditions (Roger et al., 2000).

When both soybean BBI and Kunitz inhibitors were introduced and expressed in sugar-cane (*Saccharum officinarum*), the growth of neonate larvae of sugar-cane borer (*Diatraea saccharalis*) feeding the leaf tissues was significantly retarded as compared to larvae feeding on leaf tissues from untransformed plants (Falco and Silva, 2003).

Transgenic potato expressing two cystatin genes developed resistance to a nematode, coleopteran insects (Cowgill et al., 2002) and transgenic rape plants expressing rice cystatin 1 were resistant to aphid (Rahbe et al., 2003).

Pearl millet cysteine protease inhibitor (CPI) has been found to possess anti-fungal activity in addition to its anti-feedent (protease inhibitory) activity (Joshi et al., 1998). The presence of anti-fungal and anti-feedent activity on a single protein opens up a new possibility of raising a transgenic plant resistant to pathogens, as well as pests, by transfer of a single CPI gene.

The transgenic wheat (*Triticum aestivum*) carrying barley trypsin inhibitor gene (BTI) showed a significant reduction of infection with Angoumois grain moth (*Sitotroga cerealella*). However, only early-instar larvae were inhibited in transgenic seeds, and expression of BTI protein in transgenic leaves did not have a significant protective effect against leaf-feeding insects (Altpeter et al., 1999). Expression of oryzacystatin, the rice cysteine-proteinase inhibitor, into the tobacco plant induced significant resistance against two important polyviruses, tobacco etch virus (TEV) and potato virus Y (PVY). A good correlation existed between the level of oryzacystatin protein and resistance to TEV and PVY at all levels tested. These results suggest that plant cystatins can be used against different potyviruses and potentially also against other viruses whose replication involves cysteine proteinase activity (Campos et al., 1999).

Resistance to inhibitors

PIs are highly specific for a particular class of digestive enzymes. However, insects have shown enough flexibility to switch the proteinase composition of their guts to overcome the particular PI expressed in the transgenic plants (Jongsma et al., 1995). It has been observed that insects can adapt to the ingestion of PIs. Insects belonging to both the Lepidoptera and Coleoptera can overexpress existing gut proteinases or induce the production of new types that are insensitive to the introduced PIs to overcome the deleterious effect of PI ingestion. This might be a contributing factor to the decreased effectiveness of the PIs expressed in transgenic plants. In a recent study it was shown that high level expression of soybean-trypsin-inhibitor gene in transgenic tobacco plants failed to confer resistance against *Helicoverpa armigera*. It is known that gut digestive enzymes are not the only targets affected by PI they can also affect water balance, molting and enzyme regulation of insects (Boulter, 1993).

A number of phytophagous insects including *Helicoverpa zea* a common pest of many solanaceous plants such as potato have adapted to the protease inhibitors of their host plants. Their survival and larval development is not affected by the presence of such molecules in their diet. They seem to have the ability to express specific proteases that are insensitive to the inhibitors, depending upon the specific inhibitor repertoire of the host plant (Jongsma et al., 1995). It has been found that trypsin that is sensitive to the inhibitor differs only marginally in amino acid sequence and substrate specificity from the insensitive form. They do, however, differ in their response to the trypsin inhibitor. It has been shown in the case of the tomato fruitworm that a B-type carboxypeptidase developes resistance to the potato carboxypeptidase inhibitor as a result of rearrangement of two small regions that otherwise stabilizes the enzyme-inhibitor complex. This leads to a displacement of the active-site entrance, which impairs a proper interaction between the protease and its inhibitor (Bayes et al., 2005).

Conclusion

Proteases play key roles in pathogenesis. A large number of human disorders result from an imbalance in proteolytic activity. In this context, PIs are key players in the endogenous defense system, as they help regulate and balance protease activities. In plants these inhibitors are also important participants in the exogenous defense. The importance of PIs has been realized for some time now, and many transgenic plants overexpressing different PIs have been produced with resistance against different pathogenic organisms. This is, however, yet to be fully appreciated, and it can have important consequences beyond their recognized scope. PIs have been generally thought to counter tumour progression and metastasis. They also have the potential to counteract many of the inherited disorders such as emphysema and epilepsy. These inhibitors can also interfere with the life cycle of many viruses and may help prevent many viral disorders. Synthetic PIs currently form a part of the combinational therapy against AIDS, and have potential to be used as drugs against many other diseases. Although plant PIs have been isolated and characterized from a large number of sources, and that the natural inhibitors have been made available by gene therapy and through transgenic plants overexpressing specific inhibitors with therapeutic significance, the potential for the natural inhibitors in medicine and agriculture is enormous, awaiting full-scale exploration.

ACKNOWLEDGEMENTS

We thank Professor K. I. Andrabi and Dr E. Haq of the Department of Biotechnology and Dr M. A. Zargar of the Clinical Biochemistry Department of University of Kashmir, Srinagar, India for the fruitful discussions and suggestions during the preparation of this review. We gratefully acknowledge the suggestions from Dr Stuart M. Hobday Executive Editor, Biotechnolgy and Applied Biochemistry that considerably improved the quality of this manuscript. The financial support to HH from University of Kashmir, Srinagar, India is also acknowledged.

REFERENCES

Abe K, Emori Y, Kondo H, Arai S, Susuki K (1988). The NH2-terminal 21 amino acid residues are not essential for the papain-inhibitory activity of oryzacystatin, a member of the cystatin suprfamily. J. Biol. Chem., 263(16): 7655-7659.

Abe K, Emori Y, Kondo H, Susuki K, Aria S (1987a). Molecular cloning of a cysteine proteinase inhibitor of rice (oryzacystatin)-Homology with animal cystatins and transient expression in the ripening process of rice seeds, J. Biol. Chem., 262(35): 16793-16797.

Abe M, Kondo, H, Arai S (1987b). Purification and characterization of a rice cysteine proteinase inhibitor, Agric. Biol. Chem., 51: 2763-2768.

Abe M, Abe K, Kudora M, Arai S (1992). Corn Kernel cysteine proteinase inhibitor as novel cystatin superfamily member of plant origin. Eur. J. Biochem., 209, 933-937.

Abe M, Arai S (1991). Some properties of a cysteine proteinase inhibitor from corn endosperm, Agric. Biol. Chem., 55(9): 2417-2418.

Abrahanson M, Ritonja A, Brown MA, Grubb A, Machleidt W, Barrett AJ (1987). Identification of the probable inhibitory reactive sites of the cysteine proteinase inhibitors, human cystatin C and chicken cystatin, J. Biol. Chem., 262: 9688-9494.

Al-Khunaizi M, Luke CJ, Cataltepe S, Miller D, Mills DR, Tsu C, Brömme D, Irving JA, Whisstock JC, Silverman GA (2002). The serpin SON-5 is a dual mechanistic class inhibitor of serine and cysteine proteinases. Biochemistry, 41(9): 3189-3199.

Altpeter F, Diaz I, McAuslane H, Gaddour K, Carbonero P, Vasil IK (1999). Increased insect resistance in transgenic wheat stably expressing trypsin inhibitor Cme. Mol. Breed., 5(1): 53-63.

Antcheva N, Patthy A, Athanasiadis A, Tchorbanov B, Zakhariev S, Pongor S (1996). Primary structure and specificity of a serine proteinase inhibitor from paprika (*Capsicum annuum*) seeds. Biochim. Biophys. Acta., 1298(1): 95-101.

Antcheva N, Pintar A, Patthy A, Simoncsits A, Barta E, Tchorbanov B, Pongor S (2001). Proteins of circularly permuted sequence present within the same organism: the major serine proteinase inhibitor from *Capsicum annuum* seeds. Protein Sci., 10: 2280-2290.

Arai S, Watanabe H, Kondo H, Emori Y, Abe K (1991). Papain-inhibitory activity of oryzacystatin, a rice seed cysteine proteinase inhibitor, depends on the central Gln-Val-Val-Ala-Gly region conserved among cystatin superfamily members. Biochem., 109: 294-298.

Arolas JL, Lorenzo J, Rovira A, Castella J, Aviles FX, Sommerhoff CP (2005). A carboxypeptidase inhibitor from the Tick *Rhipicephalus bursa*, Isolation, cDNA cloning, recombinant expression and characterization, J. Biol. Chem., 280(5): 5441-5448.

Ascenzi P, Ruoppolo M, Amoresano A, Pucci P, Consonni R, Zetta L, Pascarella S, Bortolotti F, Menegatti E (1999). Characterization of low-molecular-mass trypsin isoinhibitors from oil-rape (*Brassica napus* var. oleifera) seed. Eur. J. Biochem., 261(1): 275-84.

Atiwetin P, Harada S, Kamei K (2006). Serine protease inhibitor from Wax Gourd (*Benincasa hispida*[Thumb] seeds). Biosci. Biotech. Biochem., 70(3): 743-745.

Atkinson AH, Heath RL, Simpson RJ, Clarke AE, Anderson MA (1993). Proteinase inhibitors in *Nicotiana alata* stigmas are derived from a precursor protein which is processed into five homologous inhibitors. Plant Cell, 5: 203-213.

Barrett AJ, Rawlings ND, Brien EA (2001). The MEROPS database as a protease information system, J. Struct. Biol., 14: 95-102.

Barrett AJ (1986). The classes of proteolytic enzymes. In Plant Proteolytic Enzymes. Dalling MJ, eds. CRC Press, Boca Raton, Fl. 1: 1-16.

Barrett AJ (1987). The Cystatins, a new class of peptidase inhibitors, Trend Biochem. Sci., 12: 193-196.

Barrett AJ (1994). Classification of peptidases. Methods Enzymol., 244: 1-15.

Barrette-Ng IH, Ng KKS, Cherney MM, Pearce G (2003). Structural basis of inhibition revealed by a 1:2 complex of the two-headed tomato inhibitor-II and subtilisin Carlsberg. J. Biol. Chem., 278: 24062-24071.

Bayes A, Comellas-Bigler M, Vega MR, Maskos K, Bode W, Aviles FX, Jongsma MA, Beekwilder J, Vendrell J (2005). Structural basis of the resistance of an insect carboxypeptidase to plant protease inhibitors: PNAS, 102(46): 16602-16607.

Belenghi B, Acconcia F, Trovato M, Perazzoli M, Bocedi A, Poticelli F, Ascenzi P, Delledonne M (2003). Atcyssi, a cystatin from *Arabidopsis thaliana*, suppresses hypersensitive cell death. Eur. J. Biochem., 270: 2593-2604.

Birk Y, Gertler A, Khalef S (1963). A pure trypsin inhibitor from soybean. J. Biochem. 87: 281-284.

Birk Y (1985). The Bowman-Birk inhibitor. Trypsin and chymotrypsin inhibitor from soybeans, Int. J. Pept. Protein Res., 25: 113-131.

Bitoun E, Chavanas S, Irvine AD, Lonie L, Bodemer C, Paradisi M,

Hamel-Teillac D, Ansai S, Mitsuhashi Y, Taïeb A, de Prost Y, Zambruno G, Harper JI, Hovnanian A (2002). Netherton syndrome: disease expression and spectrum of SPINK5 mutations in 21 families. J. Invest. Dermatol., 118(2): 352-361.

Bode W, Hubr R (1992). Natural protein proteinase inhibitors and their interaction with proteinases. Eur. J. Biochem., 204: 433-451.

Bolter CJ (1993). Methyl jasmonate induces papain inhibitors in tomato leaves. Plant Physiol., 103: 1347-135.

Botella MA, Xu Y, Prabha TN, Zhao Y, Narasimhan ML, Wilson KA, Nielsen SS, Bressan RA, Hasegawa PA (1996). Differential expression of soybean cysteine proteinase inhibitor genes during development and in response to wounding and methyl jasmonate. Plant Physiol., 112: 1201-1210.

Boulter D (1993). Insect pest control by copying nature using genetically engineered crops. Biochem., 34: 1453-1466.

Bowman DE (1946). Differentiation of soybean antitryptic factor. Proc. Soc. Exp. Biol. Med., 63: 547-550.

Bradshaw HD, Hollick JB, Parsons TJ, Clarke HRG, Gorson MP (1989). Systematically wound-responsive genes in poplar trees encode proteins similar to sweet potato sporramins and legume Kunitz trypsin inhibitors. Plant Mol. Biol., 14: 51-59.

Brown WE, Ryan CA (1984). Isolation, characterization of a wound induced trypsin inhibitor from alfalfa leaves. Biochemistry, 23: 3428-3422.

Brown WM, Dziegielewska KM (1997). Friends and relations of the cystatin super family - new members and their evolution. Protein Sci., 6: 5-12.

Bryant J, Green TR, Gurusaddaiah S, Ryan CA (1976). Proteinase inhibitor II from potatoes: isolation and characterization of its promoter components. Biochemistry, 15: 345-378.

Cai M, Gong Y, Kao JL, Krishnamoorthi R (1995). Three-dimensional solution structure of Cucurbita maxima trypsin inhibitor-V determined by NMR spectroscopy, Biochemistry 34(15): 5201-11.

Campos RG, Acosta JAT, Aria LJS, Lim MAG (1999). The use of cysteine proteinase inhibitors to engineer resistance against potyviruses in transgenic tobacco plants, Nat. Biotechnol., 17: 1223-1226.

Ceci LR, Spoto N, de Virgilio M, Gallerani R (1995). The gene coding for the mustard trypsin inhibitor-2 is discontinuous and wound-inducible. FEBS. Lett., 364: 179-181.

Ceciliani F, Bortolotti F, Menengatti E, Ronchi S, Ascenzi P, Pakmieri SA (1994). Purification, inhibitory properties, amino acid sequence and identification of the reactive site of a new serine proteinase inhibitor from oil-rape (*Brassica napus*) seed, FEBS Lett. 342: 221-224.

Chen P, Rose J, Love R, Wei CH, Wang BC (1992). Reactive sites of an anticarcinogenic Bowman-Birk proteinase inhibitor are similar to other trypsin inhibitors. J. Biol. Chem., 267: 1990-1994.

Choi D, Park JA, Seo YS, Chun YJ, Kim WT (2000). Structure and stress-related expression of two cDNAs encoding proteinase inhibitor II of *Nicotiana glutinosa,* Biochim. Biophys. Acta, 1492(1): 211-215.

Christeller J Liang W (2005). Plant serine protease inhibitors. Protein and Peptide Letters, 12: 439-447.

Christeller JT (2005). Evolutionary mechanisms acting on proteinase inhibitor variability. FEBS Journal, 272: 5710–5722.

Colella R, Sakaguchi Y, Nagase H, Bird JWC (1989). Chicken egg white cystatin. J. Biol. Chem. 264: 17164-17117.

Cornwall GA, Cameron A, Lindberg I, Hardy DM, Cormier N, Hsia N (2003). The cystatin-related epididymal spermatogenic protein inhibits the serine protease prohormone convertase 2. Endocrinol. inhibits 144: 901-908.

Cowgill SE, Wright C, Atkinson HJ (2002). Transgenic potatoes with enhanced levels of nematode resistance do not have altered susceptibility to non target aphids. Mol. Ecol. 11: 821-827.

Dahl SW, Rasmussen SK, Hejgaard J (1996a). Heterologous expression of three plant serpins with distinct inhibitory specificities. J. Biol. Chem. 271: 25083-25088.

Dahl SW, Rasmussen SK, Petersen LC, Hejgaard J (1996b). Inhibition of coagulation factors by recombinant barley serpin BSZX. FEBS Lett., 394: 165-168.

De Leo F, Ceci LR, Jouanin L, Gallerani R (2001). Analysis of mustard trypsin inhibitor-2 gene expression in response to developmental or environmental induction. Planta 212: 710-717.

De Leo F, Volpicella M, Licciulli F, Liuni S, Gallerani R, Ceci LR (2002). PLANT-PIs: a database for plant protease inhibitors and their genes. Nucleic Acids Res. 30 (1): 347-348.

Duan X, Li X, Xue Q, Abo-El-Saad M, Xu D, Wu R (1996). Transgenic rice plants harboring an introduced potato proteinase inhibitor II gene are insect resistant. Nat. Biotechnol. 14 (4): 494-498.

Dunaevskii YE, Gladysheva IP, Pavlukova EB, Beliakova GA, Gladysheva DP, Papisova AI, Larionova NI, Belozersky MA (1997). The anionic protease inhibitor BBWI-I from buckwheat seeds. Kinetic properties and possible biological role. Physiol. Plant. 100: 483-488.

Eckelkamp C, Ehmann B, Schopfer P (1993). Wound-induced systemic accumulation of a transcript coding for a Bowman-Birk trypsin inhibitor-related protein in maize (*Zea mays* L.) seedlings, FEBS Lett. 323: 73-76.

Falco MC, Silva FMC (2003). Expression of soybean proteinase inhibitors in transgenic sugarcane plants. Effects on natural defense against *Diatraea saccharalis.* Plant Phys. Biochem., 41: 761-766.

Farran I, Sanchez SJI, Medina JF,Prieto J, Mingo-Castel AM (2002). Targeted expression of human serum albumin to potato tubers. Transgenic Res. 1: 337-346.

Felizmenio-QME, Daly NL, Craik DJ (2001). Circular proteins in plants: solution structure of a novel macrocyclic trypsin inhibitor from Momordica cochinchinensis. J. Biol. Chem. 276: 22875-22882.

Felton GW, Korth KL (2000). Trade-offs between pathogen and herbivore resistance. Curr. Opin. Plant Biol. 3: 309-314.

Fernandes KVS, Sabelli PA, Barrett DHP, Richardson M, Xavier FJ, Shevery PR (1993). The resistance of cow pea seeds to bruchid beetles is not related to level of cysteine proteinase inhibitors. Plant Mol. Biol. 23: 215-219.

Garcia OF, Salcedo G, Sanchez MR, Gomez L, Carbonero RJ (1987). Plant proteinaceous inhibitors of proteinases and amylase. Oxf. Surv. Plant Mol. Cell. Biol. 4: 275-334.

Gettins PG (2002). Serpin structure, mechanism, and function. Chem Rev. 102: 4751-4804.

Gholizadeh A, Santha IM, Kohnehrouz BB, Lodha ML, Kapoor HC (2005). Cystatins may confer viral resistance in plants by inhibition of a virus-induced cell death phenomenon in which cysteine proteinases are active: cloning and molecular characterization of a cDNA encoding cysteine proteinase inhibitor (celostatin) from celosia cristata (crested cock's comb), Biotechnol. Appl. Biochem. 42: 197-204.

Gladysheva IP, Zamolodchikova TS, Sokolova EA, Larionova NI (1999). Interaction between duodenase, a proteinase with dual specificity, and soybean inhibitors of Bowman-Birk and Kunitz type. Biochemistry 64: 1244-1249.

Gourinath S, Alam N, Srinivasan A, Betzel C, Singh TP (2000). Structure of the bifunctional inhibitor of trypsin and alpha-amylase from ragi seeds at 2.2 A resolution. Acta Crystallographica Section D. 56(3): 287-293.

Graham RAA, Holland-Czytko H, Andersen JL, Ryan CA (1985). Vacuolar localization of wound induced carboxypeptidase inhibitor in potato leaves. Plant Physiol. 78: 76-79.

Graham JS, Ryan CA (1997). Accumulation of metallocarboxy-peptidase inhibitor in leaves of wounded potato plants. Biochem. Biophys. Res. Comm. 101: 1164-1170.

Greagg MA, Brauer AB, Leatherbarrow RJ (1994). Expression and kinetic characterization of barley chymotrypsin inhibitors 1a and 1b. Biochim. Biophys. Acta 1222: 179-186.

Greenblatt HM, Ryan CA, James MNG (1989). Structure of the complex of Streptomyces griseus proteinase B and polypeptide chymotrypsin inhibitor-1 from Russet Burbank potato tubers at 2.1-A resolution. J. Mol. Biol. 205: 201-228.

Gruden K, Strukelj B, Ravnikar M, Poljsak-Prijatelj M (1997). Potato cysteine proteinase inhibitor gene family: molecular cloning, characterisation and immunocytochemical localisation studies. Plant Mol. Biol. 34: 317-323.

Grudkowska M and Zagdanska B (2004). Multifunctional role of plant cystiene proteinases. Acta Biochem. Pol., 51(3): 609-624.

Grzonka Z, Jankowska E, Kasprzykowska R, Lankiewicz L, Wiczk W, Wieczerzak E, Ciarkowski J, Drabik P, Janowski R, Kozak E, Jaskólski M, Grubb A (2001). Structural studies of cysteine proteases

and their inhibitiors. Acta. Biochem. Pol. 48: 1-20.
Habu Y, Peyachoknagul S, Umemoto K, Sakata Y, Ohno T (1992). Structure and regulated expression of Kunitz chymotrypsin inhibitor genes in winged bean [*Psophocarpus tetragonolobus* (L.) DC.] J. Biochem. 111: 249-258.
Hamato N, Koshiba T, Pham TN, Tatsumi Y, Nakamura D, Takano R, Hayashi K, Hong YM, Hara S (1995). Trypsin and elastase inhibitors from bitter gourd (*Momordica charantia* LINN) seeds: purification, amino acid sequences, and inhibitory activities of four new inhibitors. J. Biochem. 117: 432-437.
Haq SK, Atif SM, Khan RH (2004). Protein proteinase inhibitor genes in combat against insects, pests, and pathogens: natural and engineered phytoprotection. Arch. Biochem. Biophys. 43(1): 145-159.
Hara S, Makino J, Ikenaka T (1989). Amino acid sequences and disulfide bridges of serine proteinase inhibitors from bitter gourd (*Momordica charantia* Linn) seeds. J. Biochem. 105: 88-92.
Hass GM, Nau H, Biemann K, Grahn DT, Ericsson LH, Neurath H (1975). The amino acid sequence of a carboxypeptidase inhibitor from potatoes. Biochem. 14: 14-142.
Hass GM, Hermodson MA (1981). Amino acid sequence of a carboxypeptidase inhibitor from tomato fruit. Biochemistry, 20: 2256-2260.
Havkioja E, Neuvonen L (1985). Induced long- term resistance to birch foliage against defoliators: defense or incidental. Ecology 66: 1303-1308.
Heibges A, Salamini F, and Gebhardt C (2003). Functional comparison of homologous membranes of three groups of Kunitz-type enzyme inhibitors from potato tubers (*Solanum tuberosum* L). Mol. Genet. Genomics 269: 535-541.
Heitz A, Hernandez JF, Gagnon J, Hong TT, Pham TT, Nguyen TM, Le Nguyen, D, Chiche L (2001). Solution structure of the squash trypsin inhibitor MCOTI-II. A new family for cyclic knottins. Biochemistry 40: 7973-7983.
Hejgaard J (2005). Inhibitory plant serpins with a sequence of three glutamine residues in the reactive centre. Biol. Chem. 386: 1319-1323.
Hernandez JF, Gagnon J, Chiche L, Nguyen TM (2000). Squash trypsin inhibitors from *Momordica cochinchinensis* exhibit an atypical macrocyclic structure. Biochemistry 39: 5722-5730.
Hilder VA, Gatehouse AMR, Sherman SE, Barker RF, Boulter D (1987). A novel mechanism of insect resistance engineered into tobacco. Nature 300: 160-163.
Hines ME, Osuala CI, Nielsen SS (1991). Isolation and partial characterization of a soybean cystatin cysteine proteinase inhibitor of Coleopteran digestive proteolytic activity. J. Agric. Food. Chem. 39: 1515-1520.
Hojima Y, Pierce JV, Pisano JJ (1982). Pumpkin seed inhibitor of human factor XIIa (activated Hageman factor) and bovine trypsin. Biochemistry. 21: 3741-3746.
Hojima Y, Pirce JV, Pisano JJ (1980). Hageman factor fragment inhibitor in corn seeds: purification and characterization. Thromb. Res., 20: 149-162.
Holak TA, Bode W, Huber R, Otlewski J, Wilusz T (1989a). Nuclear magnetic resonance solution and X-ray structures of squash trypsin inhibitor exhibit the same conformation of the proteinase binding loop. J. Mol. Biol. 210: 649-654.
Holak TA, Gondol D, Otlewski J, Wilusz T (1989b). Determination of the complete three-dimensional structure of the trypsin inhibitor from squash seeds in aqueous solution by nuclear magnetic resonance and a combination of distance geometry and dynamical simulated annealing. J. Mol. Biol., 210: 635-648.
Homandberg GA, Litwiller RD, Peanasky RJ (1989). Carboxypeptidase inhibitors from *Ascaris suum:* the primary structure. Arch. Biochem. Biophys. 270: 153-161.
Hoorn RAL, Jones DG (2004). The plant proteolytic machiney and its role in defense. Current Opinion in Plant Biology, 7: 400-407.
Hosoyama H, Irie K, Abe K, Arai S (1994). Oryzacystatin exogenously introduced into proplasts and regeneration of transgenic rice. Biosci. Biotechnol. Biochem., 58: 1500-1505.
Huang Q, Liu S, Tang Y, Zeng F, Qian R (1992). Amino acid sequencing of a trypsin inhibitor by refined 1.6 A X-ray crystal structure of its complex with porcine beta-trypsin. FEBS. Lett., 297: 143-146.
Huang YJ, To KY, Yap MN, Chiang WJ, Suen DF, Chen SCG (2001). Cloning and characterization of leaf senescence up-regulated genes in sweet potato, Physiol. Plant, 113: 384-391.
Huntington JA, Read RJ, Carrell RW (2000). Structure of serpin protease complex shows inhibition by deformation. Nature, 407(6806): 923-926.
Irie K (1996). Transgenic rice established to express corn cystatin exhibits strong inhibitory activity against insect gut proteinases. Plant. Mol. Biol., 30: 149-157.
Irving JA, Steenbakkers PJ, Lesk AM, Camp HJ, Pike RN, Whisstock JC (2002a). Serpins in prokaryotes, Mol. Biol. Evol., 19(11): 1881-1890.
Irving JA, Pike RN, Dai W, Bromme D, Worrall DM, Silverman GA, Coetzer TH, Dennison C, Bottomley SP, Whisstock JC (2002b). Evidence that serpin architecture intrinsically supports papain-like cysteine protease inhibition: engineering alpha (I) antitrypsin to inhibit cathepsin proteases. Biochemistry, 41: 4998-5004.
Irving JA, Shushanov SS, Pike RN, Popova EY, Bromme D, Coetzer TH, Bottomley SP, Boulynko IA, Grigoryev SA, Whistock JC (2002c). Inhibitory activity of a heterochromatin-associated serpin (MENT) against papain like cysteine proteinases affects chromatin structure and blocks cell proliferation. J. Biol. Chem., 277: 13192-13201.
Irving JA, Pike RN, Lesk AM, Whisstock JC (2000). Phylogeny of the serpin superfamily: implications of patterns of amino acid conservation for structure and function. Genome Res., 10(12): 1845-1864.
Ishihara M, Atta K, Tawata S, Toyama S (1999). Purification and characterization of intracellular cysteine protease inhibitor from *Chlorella* sp. Food Sci. Technol. Res, 5: 210-213.
Ishihara M, Morine M, Tiara T, Toyama S (2000). Purification and characterization of intracellular cysteine protease inhibitor from *Chlorella* sp. Food Sci. Technol. Res, 6: 161-165.
Ishikawa A, Ohta S, Matsuoka K, Hattori T, Nakamura K (1994). A family of potato genes that encode Kunitz type proteinase inhibitors: structural comparisons and differential expression. Plant Cell Physiol., 35: 303-312.
Ivanov D, Emonet C, Foata F, Affolter M, Delly M, Fisseha M, Blum-Sperisen S, Kochhar S, Arigoni F (2006). A serpin from the gut bacterium Bifidobacterium longum inhibits eukaryotic elastase like serine proteases. J. Biol. Chem. 281(25): 17246-52.
Iwasaki T, Kiyohara T, Yoshikawa M (1971). Purification and partial characterization of two different types of proteinase inhibitors (inhibitors II-a and II-b) from potatoes. J. Biochem., 70: 817-826.
Joanitti GA, Freitas SM, Silva LP (2006). Proteinaceous Protease Inhibitors: Structural Features and Multiple Functional Faces. Current Enzyme Inhibition, 2(3): 199-217.
Jongsma MA, Bakker PL, Peters J, Bosch D, Stiekema WJ (1995). Adaptation of Spodoptera exigua larvae to plant proteinase inhibitors by induction of gut proteinase activity insensitive to inhibition. Proc. Natl. Acad. Sci., 92(17): 8041-8045.
Joshi B, Sainani M, Bastawade K, Gupta VS Ranjekar PK (1998). Cysteine protease inhibitor from pearl millet: a new class of antifungal protein. Biochem. Biophys. Res. Comm. 246: 382-387.
Keil M, Sanchez-Serrano J, Schell J, Willmitzer L (1986). Primary structure of a proteinase inhibitor II gene from potato (*Solanum tuberosum*). Nucleic Acids Res., 14: 5641-5650.
Keilova H, Tomasek V (1976). Isolation and properties of cathepsin D inhibitor from potatoes. Collect Czech Chem. Commun., 41: 489-497.
Kennedy AR (1998). The Bowman-Birk inhibitor from soybeans as an anticarcinogenic agent American Journal of Clinical Nutrition, 68(6 Suupl), 1406S-1412S.
Kervinen J, Tobin GJ, Costa J, Waugh D, S, Wlodawer A, Zdanov A (1999). Crystal structure of plant aspartic proteinase prophytepsin: inactivation and vacuolar targeting. J. Eur. Mol. Biol. Organ. 18(14): 3947-3955.
Kim JY, Chung YS, Paek KH, Park YI, Kim JK, Yu SN, Oh BJ, Shin JS (1999). Isolation and characterization of a cDNA encoding the cysteine proteinase inhibitor, induced upon flower maturation in carnation using suppression subtractive hybridization. Mol. Cells., 9(4): 392-397.
Koiwa H, Bressan RA, Hasegawa PM (1997). Regulation of protease

inhibitors and plant defense. Trends in Plant Sci., 2: 379-384.

Koiwa K, Shade RE, Zhu-Salzman K, Subramanian L, Murdock LL, Nielsen SS, Bressan RA, Hasegawa PM (1998). Phage display selection can differentiate insecticidal activity of soybean cystatins. Plant J., 14: 371-379.

Korsinezky ML, Schirra HJ, Rosengern KJ, West J, Condie BA, Otvos L, Anderson MA, Craik DJ (2001). Solution structures by 1H NMR of the novel cyclic trypsin inhibitor SFTI-from sunflower seeds and an acyclic permutant. J. Mol. Biol., 311: 579-591.

Krattiger AF (1997). Insect resistance in crops: a case study of Bacillus thuringiensis (Bt) and its transfer to developing countries. ISAAA Briefs, 2: 42.

Kreutzmann P, Schulz A, Standker L, Forssmann WG, Magert HJ (2004). Recombinant production, purification and biochemical characterization of domain 6 of LEKTI: a temporary Kazal-type-related serine proteinase inhibitor. J Chromatogr B Analyt Technol Biomed Life Sci, 803(1): 75-81.

Kupryszewski G, Ragnarsson U, Rolka K, Wilusz T (1985). Synthesis of trypsin inhibitor CMTI-III from squash seeds (*Cucurbita maxima*). Experientia, 41: 1422-1423.

Kupryszewski G, Ragnarsson U, Rolka K, Wilusz T (1986). Solid phase synthesis of trypsin inhibitor CMTI-III from squash seeds (*Cucurbita maxima*). Int J Pept Protein Res., 27: 245-250.

Kuroda M, Kiyosaki T, Matsumoto I, Misaka T, Arai S, Abe K (2001). Molecular cloning, characterization and expression of wheat cystatins, Biosci. Biotechnol. Biochem., 65: 22-28.

Laber B, Krieglstein K, Henschen A, Kos J, Turk V (1989). The cysteine proteinase inhibitor chicken cystatin is a phosphoprotein. FEBS. Lett., 248: 162-168.

Laing WA, McManus MT (2002). In Protein Protein interactions in plants, (McManus MT, Laing WA and Allan AC eds.) Sheffield Academic Press. 7: 77-119.

Laskowski Jr M, Qasim MA (1999). What can the structures of enzyme-inhibitor complexes tell us about the structures of enzyme substrate complexes: Biochem. Biophys. Acta. 1477: 324 -337.

Laskowski Jr M, Qasim MA, Lu SM (2003). Interaction of standard mechanism, canonical protein inhibitors with serine protinases. In Protein-Protein Recognition (Kleanthous, C, ed). Oxford University Press, Oxford. pp. 228-279.

Laskowski M and Kato I (1980). Protein inhibitors of proteinases. Annu. Rev. Biochem., 49: 593-626.

Law R, Zhang Q, McGowan S, Buckle AM, Silverman GA, Wong W, Rosado CJ, Langendorf CG, Bird PI, Whisstock JC (2006). An overview of the serpin superfamily. Genome Biol., 7(5): 216.

Lawrence PK, Koundal KR (2002). Plant protease inhibitors in control of phytophagous insects. Electron. J. Biotechnol., 5(1): 93-109.

Lazaro A, Rodriguez-Palenzuela P, Marana C, Carbonero P, Garcia-Olmedo F (1988). Signal peptide homology between the sweet protein thaumatin II and unrelated cereal alpha-amylase/trypsin inhibitors. FEBS. Lett. 239: 147-150.

Coletti-Previero MA (1990). Molecular recognition between serine proteases and new bioactive microproteins with a knotted structure. Biochimie, 72: 431-435.

Ledoigt G, Griffaut B, Debiton E, Vian C, Mustel A, Evray G, Maurizis JC, Madelmont JC (2006). Analysis of secreted protease inhibitors after water stress in potato tubers. Int. J. Biol. Macromols. 38: 268-271.

Lee CF, Lin JY (1995). Amino acid sequences of trypsin inhibitors from the melon Cucumis melo. J. Biochem. 118(1): 18-22.

Lee JS, Brown WE, Graham JS, Pearce G, Fox EA, Dreher TW, Ahern KG, Pearson GD, Ryan CA (1986). Molecular characterization and phylogenetic studies of a wound-inducible proteinase inhibitor I gene in *Lycopersicon* species). PNAS 83(19): 7277-7281.

Lehesjoki AE (2003). Molecular background of progressive myoclonus epilepsy. EMBO J. 22: 3473-3478.

Ligoxygakis P, Roth S, Reichhart JM (2003). A serpin regulates dorsal-ventral axis formation of the Drosophila embryo. Curr. Biol. pp. 2097-2102.

Lin GD, Bode W, Huber R, Chi CW, Engh RA (1993). The 0.25nm X-ray structure of the Bowman-Birk type inhibitor from mung bean in ternary complex with porcine trypsin. Eur. J. Biochem., 212: 549-555.

Lin JY, Chu SC, Wu HC, Hsieh YS (1991). Trypsin inhibitor from the seeds of *Acacia confusa*. J. Biochem., 110: 879-883.

Lin YH, Li HT, Huang YC, Hsieh YC, Guan HH, Liu MY, Chang T, Wang AHJ, Chen CJ (2006). Purification, crystallization and preliminary X-ray crystallographic analysis of rice Bowman-Birk inhibitor from *Oryza sativa*. Acta Cryst. F62: 522-524.

Liu Q, Yu L, Gao J, Fu Q, Zhang J, Zhang P, Chen J, Zhao S (2000). Cloning, tissue expression pattern and genomic organization of latexin, a human homologue of rat carboxypeptidase A inhibitor. Mol. Biol. Rep., 27: 241-246.

Lomas DA, Lourbakos A, Cumming SA, Belorgey D (2002). Hypersensitive mousetraps, alpha 1- antitrypsin deficiency and dermentia. Biochem. Soc. Trans., 30: 89-92.

Luckett S, Garcia RS, Barker JJ, Konarev AV, Shewry PR, Clarke AR, Brady RL (1999). High resolution structure of a potent, cyclic proteinase inhibitor from sunflowers. J. Mol. Biol., 290: 525-533.

Lustigman S, Brotman B, Huima T, Prince AM, McKerrow JH (1992). Molecular cloning and characterization of onchocystatin, a cysteine proteinase inhibitor of *Onchocerca volvulus*. J. Biol. Chem., 267: 17339-17346.

Machleidt W, Borchart U, Fritz H, Brzin J, Ritonja A, Truck V (1983). Protein inhibitors of cysteine proteinase II. Primary structure of stefin, a cytosolic inhibitor of cysteine proteinases from human polymorphonuclear granulocytes. Hoppe-Seyler's Physiol. Chem., 364: 1481-1486.

Magert HJ, Kreutzmann P, Standker L, Walden M, Drogemuller K, and Forssmann WG (2002). LEKTI: a multidomain serine proteinase inhibitor with pathophysiological relevance. Int. J. Biochem. Cell Biol, 43: 573-576.

Magert HJ, Standker L, Kreutzmann P, Zucth HD, Reinecke M, Sommerhoff CP, Fritz H, Forssmann WG (1999). LEKTI, a novel 15-domain type of human serine proteinase inhibitor. J. Biol. Chem., 274: 21499-21502.

Mahoney WC, Hermodson MA, Jones B, Powers DD (1984). Amino acid sequence and secondary structural analysis of the corn inhibitor of trypsin and activated Hageman factor. J. Biol. Chem. 259: 8412-8416.

Margossian LJ, Federman AD, Giovannoni JJ, Fischer RL (1988). Ethylene-regulated expression of a tomato fruit ripening gene encoding a proteinase inhibitor I with a glutamic residue at the reactive site. Proc. Natl. Acad. Sci., 85(21): 8012-8016.

Marres M, Meloun B, Pavlik M, Kostka V, Baudys M (1989). Primary structure of cathepsin D inhibitor from potatoes and its structural relationship to trypsin inhibitor family. FEBS. Lett., 251: 94-98.

McWherter CA, Malkenhorst WF, Campbell J, Glover GI (1989). Novel inhibitors of human leukocyte elastase and cathepsin G: sequence variants of squash seed protease inhibitor with altered protease selectivity. Biochem. 28: 5708-5714.

Melville CJ, Ryan CA (1972). Chymotrypsin inhibitor I from potatoes: large scale preparation and characterization of its subunit components. J. Biol. Chem., 247(11): 3445-3453.

Menengatti E, Tedeschi G, Ronchi S, Bortolotti F, Ascenzi P, Thomas RM, Bolognesi M, Pakmieri SA (1992). Purification, inhibitory properties and amino acid sequence of a new serine proteinase inhibitor from white mustard (*Sinapis alba* L.) seed. FEBS. Lett. 301: 10-14.

Miller EA, Lee MC, Atkinson AH, Anderson MA (2000). Identification of a novel four-domain member of the proteinase inhibitor II family from the stigmas of *Nicotiana alata*. Plant. Mol. Biol., 42(2): 329-333.

Mosolov VV, Grigor'eva LI, Valueva TA (2001). Plant proteinase inhibitors as polyfunctional proteins (a review) Prikl. Biokhem. Mikrobiol. 37(6): 643-50.

Mosolov VV, Valueva TA (2005) Proteinase inhibitors and their function in plants: a review. Prikl Biokhim Mikrobiol. 41(3): 261-82.

Mulvenna JP, Foley FM, Craik DJ (2005). Discovery, structural determination, and putative processing of the precursor protein that produces the cyclic trypsin inhibitor sunflower trypsin inhibitor 1. J. Biol. Chem. 280: 32245-32253.

Murray C, Christeller JT (1995). Purification of a trypsin inhibitor (PFTI) from pumpkin fruit phloem exudate and isolation of putative trypsin and chymotrypsin inhibitor cDNA clones. Biol. Chem. Hoppe-Seyler, 376(5): 281-287.

Neurath H (1989). Proteolytic processing and physiological regulation.

Trends Biochem. Sci. 14: 268-271.
Nilges M, Habazettl J, Brunger AT, Holak TA (1991). Relaxation matrix refinement of the solution structure of squash trypsin inhibitor. J. Mol. Biol. 219: 499-510.
Norioka S, Ikenaka T (1983). Aminoacid sequence of trypsin chymotrypsin inhibitors (A-I, A-II, B-I and B-II) from peanut (*Arachis hypogaea*). A discussion on the molecular evolution of legume Bowman-Birk type inhibitor. J. Biochem., 94: 589-599.
Normant E, Martres MP, Schwartz JC, Gros C (1995). Purification, cDNA cloning, functional expression, and characterization of a 26-kDa endogenous mammalian carboxypeptidase inhibitor. PNAS, 92: 12225-12229.
Odani S, Koide T, Ono T (1983). The complete amino acid sequence of barley trypsin inhibitor. J. Biol. Chem., 258: 7998-8003.
Odani S, Koide T, Ono T (1986). Wheat gram trypsin inhibitor. Isolation and structural characterization of single-headed and double-headed inhibitors of the Bowman-Birk type. J. Biochem. 100(4): 975-983.
Odani S, Ikenaka T (1976). The amino acid sequences of two soybean double headed proteinase inhibitors and evolutionary consideration on the legume proteinase inhibitors. J. Biochem. 80: 641-643.
Ogata F, Miyata T, Fujii N, Yoshida N, Noda K (1991). Purification and amino acid sequence of a bitter gourd inhibitor against an acidic amino acid-specific endopeptidase of *Streptomyces griseus*. J. Biol. Chem., 266: 16715-16721.
Ohtsubo S, Kobayashi H, Noro W, Taniguchi M, Saitoh E (2005). Molecular cloning and characterization of oryzacystatin-III, a novel member of phytocystatin in rice (*Oryza sativa* L. japonica). J. Agric. Food Chem., 53: 5218-5224.
Oliveira AS, Filho JX, Sales MP (2003). Cysteine proteinases cystatins. Braz. Arch. Biol. Technol., 46(1): 91-104.
Otto HH, Schirmeister T (1997) Cysteine Proteases and Their Inhibitors. Chem. Rev. 97: 133-171.
Pannetier C, Giband M, Couzi P, Letan V, Hazier M, Tourneur J, Hau B (1997). Introduction of new traits into cotton through genetic engineering: insect resistance as example. Euphytica 96: 163-166.
Park H, Yamanaka N, Mikkonem A, Kusakabe I, Kobayashi H (2000). Purification and characterization of aspartic proteinase from sunflower seeds. Biosci. Biotechnol. Biochem. 64: 931-999.
Park Y, Choi BH, Kwak JS, Kang CW, Lim HT, Cheong HS, Hahm KS (2005). Kunitz-type serine protease inhibitor from potato (*Solanum tuberosum* L. cv. Jopung). J. Agric. Food. Chem., 53: 6491-6496.
Pearce G, Johnson S, Ryan CA (1993). Purification and characterization from Tobacco (*Nicotiana tabacum*) Leaves of Six small, Wound-Inducible, Proteinase Isoinhibitors of the Potato inhibitor II Family. Plant Physiol., 102: 639-644.
Pernas M, Sanchez MR, Gomez L, Salcedo G (1998). A chestnut seed cystatin differentially effective against cysteine proteinases from closely related pests. Plant Mol. Biol. 38: 1235-1242.
Pernas M, Sanchez-Mong R, Salcedo G (2000). Biotic and abiotic stress can induce cystatin expression in chestnut. FEBS. Lett. 467(2-3): 206-210.
Plunkett G, Senear DF, Zuroske G, Ryan CA (1982). Proteinase inhibitors I and II from leaves of wounded tomato plants: purification and properties. Arch. Biochem. Biophys. 213: 463-472.
Poerio E, Di Gennaro S, Di Maro A, Farisei F (2003). Primary structure and reactive site of a novel wheat proteinase inhibitor of subtilisin and chymotrypsin. Biol. Chem. 384: 295-304.
Prakash B, Selvaraj S, Murthy MRN, Sreerama YN, Rao DR, Gowda LR (1996). Analysis of the aminoacid sequences of plant Bowman-Birk inhibitors. J. Mol. Evol., 42: 560-569.
Pszenny V, Ledesma BE, Matrajt M, Duschak VG, Bontempi EJ, Dubremetz JF, Angel SO (2002). Subcellular localization and post secretory targeting of TgPI, a serine proteinase inhibitor from *Toxoplasma gondii*. Mol. Biochem. Parasitol. 121: 283-286.
Qi RF, Song Z, Chi C (2005). Structural features and molecular evolution of Bowman-Birk protease inhibitors and their potential application: Acta Biochemica et Biophysica. 37(5): 283-292.
Rahbe Y, Deraison C, Bonade-Bottino M, Girard C, Nardon C, Jouanin L (2003). Effects of the cysteine protease inhibitor oryzacystatin on different expression (OC-I)n different aphids and reduced performance of Myzus persicae on OC-I expressing transgenic oilseed rape. Plant Sci., 164: 441-450.
Raj SS, Kibushi E, Kurasawa T, Suzuki A, Yamane T, Odani S, Iwasaki Y, Ashida T (2002). Crystal structure of bovine trypsin and wheat germ trypsin inhibitor (I-2b) complex (2:1) at 2.3 a resolution. J. Biochem., 132: 927-933.
Rancour JM, Ryan CA (1968). Isolation of a carboxypeptidase B inhibitor from potatoes. Arch. Biochem. Biophys., 125: 380-382.
Rawlings ND, Barrett AJ (1993). Evolutionary families of peptidases, Biochem. J. 299: 205-218.
Rawlings ND, Tolle DP, Barrett AJ (2004a). MEROPS: the peptidase database. Nucleic Acid Res, 32 Database issue: D. pp. 160-164.
Rawlings ND, Tolle DP, Barrett AJ (2004b). Evolutionary families of peptidase inhibitors, J. Biochem., 378: 705-716.
Ray CA, Black RA, Kronheim SR, Greenstreet TA, Sleath PR, Salvesen GS, Pickup DJ (1992). Viral Inhibition of inflammation, cow pox virus encodes an inhibitor of interleukin-1beta converting enzyme. Cell, 69: 597-604.
Reverter D, Vendrell J, Canals F, Horstmann J, Aviles FX, Fritz H, Sommerhoff CP (1998). A carboxypeptidase inhibitor from the medical leech *Hirudo medicinalis*. Isolation, sequence analysis, cDNA cloning, recombinant expression, and characterization. J. Biol. Chem., 273: 32927-32933.
Richardson M (1974). Chymotryptic inhibitor I from potatoes. The amino acid sequence of subunit A. Biochem. J., 137: 101-112.
Richardson RT, Sivashanmugam P, Hall SH, Hamil KG, Moore PA, Ruben SM, French FS, O'Rand M (2001). Cloning and sequencing of human Eppin: a novel family of protease inhibitors expressed in the epididymis and testis. Gene, 270: 93-102.
Ritchie BC (2003). Protease inhibitors in the treatment of hereditary angioedema Transfus, Apheresis. Sci., 29: 259-267.
Ritonja A, Krizaj I, Mesko P, Kopitar M, Lucovnik P, Strukelj B, Pungercar J, Buttle DJ, Barrett AJ, Turk V (1990). The amino acid sequence of a novel inhibitor of cathepsin D from potato: FEBS. Lett., 267: 1-15.
Roberts TH, Marttila S, Rasmussen SK, Hejgaard J (2003). Differential gene expression for suicide-substrate serine proteinase inhibitors (serpins) in vegetative and grain tissues of barley. J. Exp. Bot., 54: 2251-2263.
Roger LM, Hart ES, Kaye SB, Maarten JC, Eric A, Thomas JVH (2000). Bean α-amylase inhibitor 1 in transgenic peas (*Pisumsativum*) provides complete protection from pea weevil (*Bruchus pisorum)* under field conditions. PNAS, 97(8): 3820-3825.
Rolka K, Kupryszewski G, Rozycki J, Ragnarsson U, Zbyryt T, Otlewski J (1992). New analogues of *Cucurbita maxima* trypsin inhibitor III (CMTI III) with simplified structure. Biol. Chem. Hoppe-Seyler, 373: 1055-1060.
Ryan CA, Balls AK (1962). An inhibitor of chymotrypsin from *Solanum tuberosm* and its behavior toward trypsin. Proc. Natl. Acad. Sci., 48: 1839-44.
Ryan CA (1990). Proteinase inhibitors in plants: genes for improving defenses against insects and pathogens. Annu. Rev. Phytopathol, 28: 425-449.
Ryan SN, Liang WA, McManus MT (1998). A cysteine proteinase inhibitor purified from apple fruit. Phytochemistry 49: 957-63.
Saito H, Suzuki T, Ueno K, Kubo T, and Natori S (1989). Molecular cloning of cDNA for sarcocystatin A and analysis of the expression of the sarcocystatin A gene during development of *Sarcophaga peregrina*. Biochemistry 28: 1749-1755.
Sakuta C, Konishi A, Yamakawa S, Kamada H, Satoh S (2001). Cysteine proteinase gene expression in the endosperm of germinating carrot seeds. Biosci. Biotechnol. Biochem., 65: 2243-2248.
Salvesen G, Parkes C, Abrahanson M, Grubb A, and Barrett AJ (1986). Human low-Mr kininogen contain three copies of a cystatin sequence that are divergent in structure and in inhibitory activity for cystine proteinases. Biochem. J., 234: 429-434.
Sane VA, Nath P, Aminuddin and Sane PV (1997). Development of insect resistant.transgenic plants using plant genes expression of cowpea trypsin inhibitor in transgenic tobacco plants. Curr. Sci. 72: 741-747.
Sardana RK, Ganz PR, Dudani AK, Tackaberry ES, Cheng X, Altosaar I (1998). Synthesis of recombinant human cytokine GMCSF in the seeds of transgenic tobacco plants. In. CUNNINGHAM, C and PORTER A J R, eds. Recombinant proteins from plants. Production

and isolation of clinically useful compounds. Totowa N J, Humama Press. pp. 77-87.

Schick C, Pemberton PA, Shi GP, Kamachi Y, Cataltepe S, Bartuski AJ, Gornstein ER, Bromme D, Chapman HA, and Silverman GA (1998). Cross class inhibition of cysteine proteinases cathepsins K, L and S by serpin squamous cell carcinoma antigen I: a kinetic analysis. Biochemistry, 37: 5258-5266.

Shewry PR, Lafiandra D, Salcedo G, Aragoncillo C (1984). N-terminal amino acid sequence of chloroform/methanol-soluble proteins and albumins from endosperm of wheat, barley and related species. FEBS. Lett., 175: 359-363.

Shivraj B, Pattabiraman TN (1981). Natural plant enzyme inhibitors. Characterization of an unusual alpha-amylase/trypsin inhibitor from ragi (*Eleusine coracana* Geartn.). Biochem. J. 193: 29-36.

Silverman GA, Bird PI, Carrell RW, Church FC, Coughlin PB, Gettins PG, Irving JA, Lomas DA, Luke CJ, Moyer RW, Pemberton PA, O'Donnell ER, Selvesen GS, James Travis and Whisstock JC (2001). The sepins are an expanding superfamily of structurally similar but functionally diverse proteins, evolution, mechanism of evolution, novel functions, and revised nomenclature. J. Biol. Chem. 276: 33293-33296.

Solomon M, Belenghi B, Delledonne M, Menachem E, Levine A (1999). The involvement of cysteine protease inhibitor genes in the regulation of programmed cell death in plants. Plant Cell, 11: 431-443.

Stato N, Ishidoh K, Uchiyama E, Kominami E (1990). Molecular cloning and sequencing of cDNA from rat ystatin B: Nucleic Acids Res., 18(22): 6698.

Steller H (1995). Mechanisms and genes of cellular suicide. Science, 267: 1445-1449.

Strobl S, Maskos K, Wiegand G, Huber R, Gomis-Ruth FX, Glockshuber R (1998). A novel strategy for inhibition of alpha-amylases: yellow meal worm alpha-amylase in complex with the Ragi bifunctional inhibitor at 2.5 A resolution. Structure, 6: 911-921.

Strukelj B, Pungercar J, Mesko P, Barlic-Maganja D (1992). Characterization of aspartic proteinase inhibitors from potato at the gene, cDNA and protein levels. Biol. Chem. Hoppe-Seyler 373: 477-482.

Supuran CT, Scozzafava A, Clare BW (2002). Bacterial protease inhibitors. Med Res Rev. 22(4): 329-72.

Suzuki A, Tsunogae Y, Tanaka I, Yamane T, Ashida T (1987). The structure of Bowman-Birk type protease inhibitor A-II from peanut (*Arachis hypogaea*) at 3.3-A resolution. J. Biochem. 101: 267-274.

Tanaka AS, Sampaio MU, Marangoni S, de Oliveira B, Novello JC, Oliva ML, Fink E, Sampaio CA (1997). Purification and primary structure determination of a Bowman-Birk trypsin inhibitor from Torresea cearensis seeds. Biol. Chem. 378: 273-281.

Tashiro M, Asao T, Hitara C, Takahashi K, Kanamori M (1990). The complete aminoacid sequence of a major trypsin inhibitor from seeds of foxtail millet (*Setaria italica*). J. Biochem. 108: 669-672.

Tashiro M, Hashino, K, Shiozaki, M, Ibuki, F, Maki, Z (1987). The complete aminoacid sequence of rice bran trypsin inhibitor, J. Biochem. 102: 297-306.

Terada S, Fujimura S, Katayama H, Nagasawa M, Kimoto E (1994). Purification and characterization of two Kunitz family subtilisin inhibitors from seeds of *Canavalia lineata.* J. Biochem. 115: 392-396.

Thaimattam R, Tykarska E, Bierzynski A, Sheldrick GM, Jaskolski M (2002). Atomic resolution structure of squash trypsin inhibitor: unexpected metal coordination. Acta. Crystallogr. D. Biol. Crystallogr. 58: 1448-1461.

Tian M, Kamoun S (2005). A two disulfide bridge Kazal domain from Phytophthora exhibits stable inhibitory activity against serine proteases of the subtilisin family. BMC Biochem. 6: 15.

Tian M, Huitema E, Da Cunha L, Torto-Alalibo T, Kamoun S (2004). A Kazal like extracellular serine protease inhibitor from Phytopthora infestans targets the tomato pathogenesis related protease P69B. J. Biol. Chem. 279: 26370-26377.

Trexler M, Banyai L, Patthy L (2001). A human protein containing multiple types of protease-inhibitory modules. Proc. Natl. Acad. Sci., 98: 3705-3709.

Trexler M, Banyai L, Patthy L (2002). Distinct expression pattern of two related human proteins containing multiple types of protease-inhibitory modules. Biol. Chem. 383: 223-228.

Tsybina T, Dunaevsky Y, Musolyamov A, Egorov T, Larionova N , Popykina N , Belozersky M (2004). New protease inhibitors from buckwheat seeds: properties, partial amino acid sequences and possible biological role: Biological Chemistry, 385(5): 429-434.

Tur Sinal A, Birk Y, Gertler A, Rigbi M (1972). A basic trypsin and chymotrypsin inhibitor from groundnuts (*Arachis hypogaea*). Biochim. Biophys. Acta., 263: 666-672.

Turk B, Turk V, Turk D (1997). Structural and functional aspects of papain like cysteine proteinase and their protein inhibitors. Biol. Chem. Hoppe-Seyler, 378: 141-150.

Turk V, Bode W (1991). The Cystatins: protein inhibitors of cysteine proteinases. FEBS. Lett., 285: 213-219.

Urwin PE, Lilley CJ, Mcpherson MJ, Atkinson HJ (1997). Resistance to both cyst and rootnot nematodes conferred by transgenic Arabidopsis expressing a modified plant cystatin. Plant journal, 12(2): 455-461.

Ussuf KK, Laxmi NH Mitra R. (2001). Proteinase inhibitors. Plant derived genes of insecticidal protein for developing insect-resistant transgenic plants. Current Sci. 80, 7(10): 47-853.

Vain P, Worland B, Clarke MC, Richard G, Beavis M, Liu H, Kohli A, Leech M, Snake J, Christou P (1998). Expression of an engineered cysteine proteinase inhibitor (Oryzacystatin-I delta D86) for nematode resistance in transgenic rice plants. Theor. Appl. Genet. 96: 266-271.

Vallee F, Kadziola A, Bourne Y, Juy M, Rodenburg KW (1998). Barley alpha-amylase bound to its endogenous protein inhibitor BASI: crystal structure of the complex at 1.9-A resolution. Structure, 6: 649-659.

Valueva TA, Revina TA, Gvozdeva EL, Gerasimova NG, Ilhinskaia LI, Ozretskovakaia OL (2001). Effects of Elicitors on the Accumulation of Proteinase Inhibitors in Injured Potato Tubers. Appl. Biochem. Microbiol. 37: 512-516.

Valueva TA, Mosolov VV (2004). Role of Inhibitors of Proteolytic Enzymes in Plant Defense against Phytopathogenic Microorganisms. Biochemistry, 69(11), 1305-1309.

Van derVyver C, Schneidereit J, Driscoll S, Turner J, Kunert K, Foyer C (2003). Oryzacystatin I expression in transformed tobacco produces a conditional growth phenotype and enhances chilling tolerance. Plant Biotech. J., 1(2): 101- 112.

Vercammen D, Belenghi B, van de Cotte B, Beunens T, Gavigan JA, De Rycke R, Brackenier A, Inze D, Harris JL, Van Breusegem F (2006)

Serpin1 of Arabidopsis thaliana is a suicide inhibitor for metacaspase 9. J. Mol. Biol. 364(4): 625-636.

Volpicella M, Schipper A, Jongsma MA, Spoto N, Gallerani R, Ceci LR (2000). Characterization of recombinant mustard trypsin inhibitor 2 (MTI2) expressed in Pichia pastoris. FEBS. Lett., 468(2-3): 137-41.

Walsh TA and Strictland JA (1993). Proteolysis of the 85 kDa crystalline cysteine proteinase inhibitor from tomato release functional cystatin domains. Plant Physiol. 103(4): 1227-1234.

Wang HX, Ng TB (2006). Concurrent isolation of a Kunitz-type trypsin inhibitor with antifungal activity and a novel lectin from *Pseudostellaria heterophylla* roots. BBRC 342(1): 349-353.

Watanabe H, Abe K, Emori Y, Hosoyama H, Arai S (1991). Molecular cloning and gibbrllin-induced expression of multiple cystine protease of rice seds (*orizains*). J. Biol. Chem. 266: 16897-16902.

Wen CY, Huang SC, Lin-Shiau1 SY, Lin JK (2005). Bowman--Birk inhibitor abates proteasome function and suppresses the proliferation of MCF7 breast cancer cells through accumulation of MAP kinase phosphatase-1. Carcinogenesis, 26(7): 1296-1306.

Wieczorek M, Otlewski J, Cook J, Parks K, Leluk J (1985). The squash family of serine proteinase inhibitors. Amino acid sequences and association equilibrium constants of inhibitors from squash, summer squash, zucchini, and cucumber seeds. Biochem. Biophys. Res. Commun., 126: 646-652.

Williamson VM, Hussey RS (1996). Nematode pathogenesis and resistance in plants. Plant cell, 8(10): 35-1745.

Wingate VP, Broadway RM, Ryan CA (1989). Isolation and characterrization of a novel, developmentally regulated proteinase inhibitor I protein and cDNA from the fruit of a wild species of tomato. J. Biol. Chem., 264(30): 17734-17738.

Wolfson JL, Murdock LL (1987). Suppression of larval Colorado beetle growth and development by digestive proteinase inhibitors. Entomologia Experimentalis et. Applicata, 44: 235-240.

Yoo BC, Aoki K, Xiang Y, Campbell LR, Hull RJ, Xoconostle-Cazares B,

Monzer J, Lee JY, Ullman DE, Lucas WJ (2000). Characterization of Cucurbita maxima phloem serpin-1 (CmPS-1). A developmentally regulated elastase inhibitor. J. Biol. Chem., 275: 35122-35128.
Zhao Y, Botella MA, Subramanian L, Niu X, Nielsen SS, Bressan RA, Hasegawa PM (1996). Two wound-inducible soybean cysteine proteinase inhibitors have greater insect digestive proteinase inhibitory activities than a constitutive homolog. Plant Physiol., 111(4): 1299-1306.

Biodegradation alternative in the cleanup of petroleum hydrocarbon pollutants

Anthony I Okoh,

Department of Biochemistry and Microbiology, University of Fort Hare, Private Bag X1314, Alice 5700, South Africa.
Email: aokoh98@yahoo.com

Extensive petroleum hydrocarbon exploration activities often result in the pollution of the environment, which could lead to disastrous consequences for the biotic and abiotic components of the ecosystem if not restored. Remediation of petroleum-contaminated system could be achieved by either physicochemical or biological methods. However, the attendant negative consequences of the physicochemical approach are currently directing greater attention to the exploitation of the biological alternatives. This paper provides a review of the menace of petroleum hydrocarbon pollution and its biodegradation in the environment with the view of understanding the biodegradation processes for better exploitation in bioremediation challenges.

Key words: Petroleum hydrocarbon, pollution, environment and biodegradation.

Table of Contents

1.0 Introduction
2.0 Chemistry and Biodegradability of petroleum hydrocarbon
3.0 Factors influencing petroleum hydrocarbon biodegradation
4.0 Distribution of petroleum hydrocarbon utilising microorganisms
5.0 Genetic aspects of petroleum hydrocarbon degradation
6.0 Conclusion
7.0 References

1.0 Introduction

Claude U. Sable had as far back as 1946, recognised that many microorganisms have the ability to utilise hydrocarbons as the sole source of carbon and energy, and that such microorganisms are widely distributed in nature. He further recognised that the microbial utilisation of hydrocarbons was highly dependent on the chemical nature of the components within the petroleum mixture, and environmental determinants (Atlas 1981).

However, it was only after the sinking of the super tanker *Torney Canyon* in the English Channel that the attention of the scientific community was drawn towards the problems of oil pollution. Thereafter, several studies have examined the fate of petroleum in various ecosystems (Boehm et al., 1995; Whittaker et al., 1999). The development of petroleum industry into new frontiers, the apparent inevitable spillages that occur during routine operations, and records of acute accidents during transportation has called for more studies into oil pollution problems (Timmis et al., 1998), which has been recognized as the most significant contamination problem on the continent (Snape et al., 2001). Also, the extensive use of petroleum products leads to the contamination of almost all compartments of the environment, and biodegradation of the hydrocarbons by natural populations of microorganisms has been reported to be the main process acting in the depuration of hydrocarbon-polluted environments (Challain et al., 2004), the mechanism of which has been extensively studied and reviewed (van Hamme et al., 2003).

Crude oil can be accidentally or deliberately released into the environment leading to serious pollution problems (Thouand et al., 1999). Even small releases of petroleum hydrocarbons into aquifers can lead to concentrations of dissolved hydrocarbons far in excess of regulatory limits (Spence et al., 2005). These pollution problems often result in huge disturbances of both the biotic and abiotic components of the ecosystems (Mueller et al., 1992), more so that some hydrocarbon components have been known to belong to a family of carcinogenic and neurotoxic organopollutants (Hallier-Soulier et al., 1999). The processes leading to the eventual removal of hydrocarbon pollutants from the environment has been extensively documented and involves the trio of physical, chemical and biological alternatives.

The currently accepted disposal methods of incineration or burial in secure landfills (USEPA 2001; ITOPF 2006) can become prohibitively expensive when the amounts of contaminants are large. This often results in cleanup delays while the contaminated soil continues to pollute groundwater resources if on land, and death of aquatic life if on waterways (Pye and Patrick 1983), thus necessitating speedy removal of the contaminants.

The biodegradation of oil pollutants is not a new concept as it has been intensively studied in controlled conditions (Sugiura et al., 1997; Chaillan et al., 2004) and in open field experiments (Chaıneau et al., 2003; Gogoi et al., 2003), but it has acquired a new significance as an increasingly effective and potentially inexpensive clean-up technology. Its potential contribution as a countermeasure biotechnology for decontamination of oil-polluted systems could be enormous. The initial interest was in the fate and persistence of pesticides in soils (Prince 1993). However, the field has expanded in recent years to encompass a wide variety of chemicals and a broad array of issues. Some technologies are being developed that markedly enhance microbial destruction or degradation of organic pollutants that otherwise would have persisted at the cleanup of many polluted groundwater and soils using the orthodox physical and chemical methods (Alexander 1994).

Bioremediation, which employs the biodegradative potentials of organisms or their attributes, is an effective technology that can be used to accomplish both effective detoxification and volume reduction. It is useful in the recovery of sites contaminated with oil and hazardous wastes (Caplan 1993). Besides, bioremediation technology is believed to be non-invasive and relatively cost effective (April et al., 2000). In some cases it may not require more than the addition of some degradation enhancers to the polluted system. It could end up being the most reliable and probably least expensive option for exploitation in solving some chemical pollution problems (Mesarch et al., 2000).

Westlake (1982) noted that no single microbial species has the enzymatic ability to metabolise more than two or three classes of compounds typically found in crude oil. A consortium composed of many different bacterial species is thus required to degrade crude oil significantly. The use of a bacterial consortium provides certain advantages over biostimulation in cases where pollutant toxicity or a lack of appropriate microorganisms (both quantity and quality) is important. Determination of the potential success of application of bacterial consortium requires an understanding of the survival and activity of the added microorganism(s) or their genetic materials, and the general environmental conditions that control the degradation rates such as the peculiarity of the contaminated site, for example, water or soil systems (Vogel 1996). These factors may very well vary from place to place and from organism to organism.

It is a common stance that many farmers in the oil exploration areas in developing countries are experiencing tremendous difficulties in restoring the fertility of pollution devastated farmlands due to lack of knowledge on appropriate remediation procedures. This problem could be attended to if adequate attention is given to the need for baseline data for the evaluation of the application of bioremediation technology in the peculiar localities, using indigenous isolates of microorganisms such as reported by Ebuehi et al. (2005). The non-chalant attitude to the problem of oil pollution is particularly of serious concern for food safety in such neglected areas as the Niger delta regions of Nigeria as persistence of the pollution could result in the release of toxic pollutants into the food chain and water products (Bradley et al., 1997).

The success of the bioremediation efforts in the cleanup of the oil tanker *Exxon Valdez* oil spill of 1989 (Atlas and Bartha, 1998) in Prince William Sound and the Gulf of Alaska has created tremendous interest in the potential of biodegradation and bioremediation technology. Many countries are currently convinced of the full potential of this technology and are, therefore, encouraging research into improving and optimising the procedures.

2.0 Chemistry and Biodegradability of petroleum hydrocarbon

Petroleum has been known for several years to occur in the surface seepage and was first obtained in pre-Christian times by the Chinese. The modern petroleum industry had its beginning in Romania and in a well-sunk in Pennsylvania by Colonel E. A. Drake in 1859 (Alloway and Ayres 1993). The principal early use of the product of the petroleum industry was for the replacement of expensive whale oil for lighting. Today, its consumption as a fuel and its dominance in the world market as a source of chemicals has diversified tremendously.

Petroleum is defined as any mixture of natural gas, condensate, and crude oil. Crude oil which is a heterogeneous liquid consisting of hydrocarbons comprised almost entirely of the elements hydrogen and carbon in the ratio of about 2 hydrogen atoms to 1 carbon atom. It also contains elements such as nitrogen, sulphur and oxygen, all of which constitute less than 3% (v/v). There are also trace constituents, comprising less than

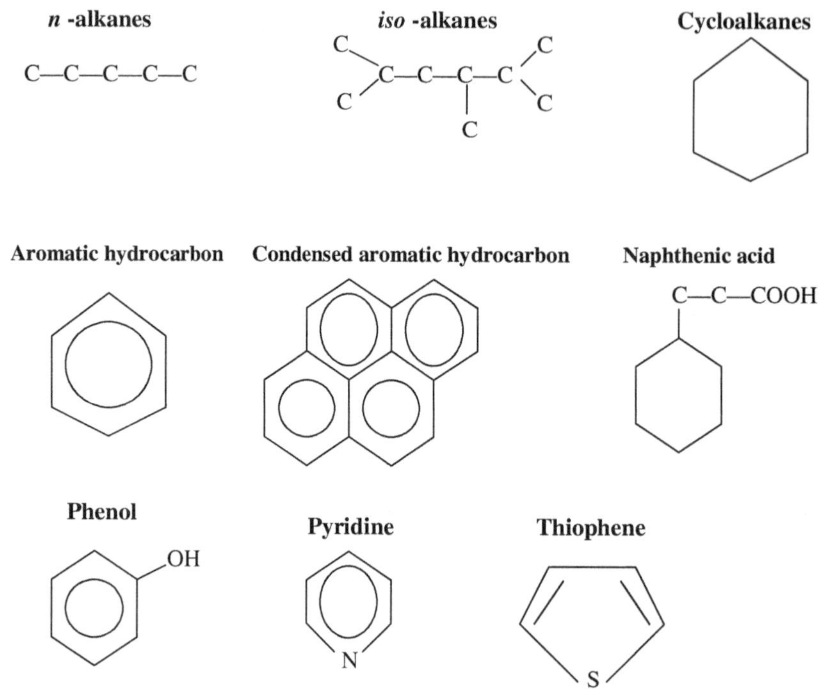

Figure 1. Structural classification of some crude oil components (Alloway and Ayres 1993).

1% (v/v), including phosphorus and heavy metals such as vanadium and nickel. Crude oils could be classified according to their respective distillation residues as paraffins, naphthenes or aromatics and based on the relative proportions of the heavy molecular weight constituents as light, medium or heavy. Also, the composition of crudes may vary with the location and age of an oil field, and may even be depth dependent within an individual well. About 85% of the components of all types of crude oil can be classified as either asphalt base, paraffin base or mixed base. Asphalt base contain little paraffin wax and an asphaltic residue (Atlas 1981). The sulphur, oxygen and nitrogen contents are often relatively higher in comparison with paraffin base crudes, which contain little or no asphaltic materials. Mixed crude oil contains considerable amount of oxides of nitrogen and asphalt.

Petroleum oil biodegradation by bacteria can occur under both oxic and anoxic conditions (Zengler et al., 1999), albeit by the action of different consortia of organisms. In the subsurface, oil biodegradation occurs primarily under anoxic conditions, mediated by sulfate reducing bacteria (e.g., Holba et al., 1996) or other anaerobes using a variety of other electron acceptors as the oxidant. On a structural basis, the hydrocarbons in crude oil are classified as alkanes (*normal* or *iso*), cycloalkanes, and aromatics (Figure 1). Alkenes, which are the unsaturated analogs of alkanes, are rare in crude oil but occur in many refined petroleum products as a consequence of the cracking process. Increasing carbon numbers of alkanes (homology), variations in carbon chain branching (*iso*-alkanes), ring condensations, and interclass combinations e.g., phenylalkanes, account for the high numbers of hydrocarbons that occur in crude oil. In addition, smaller amounts of oxygen – (phenols, naphthenic acids), nitrogen- (pyridine, pyrrole, indole), and sulfur- (alkylthiol, thiophene) containing compounds, collectively designated as "resins" and partially oxygenated, highly condensed asphaltic fraction occur also in crude but not in refined petroleum (Atlas and Bartha 1973).

The inherent biodegradability of these individual components is a reflection of their chemical structure, but is also strongly influenced by the physical state and toxicity of the compounds. As an example, while *n*-alkanes as a structural group are the most biodegradable petroleum hydrocarbons, the $C_5 - C_{10}$ homologues have been shown to be inhibitory to the majority of hydrocarbon degraders. As solvents, these homologues tend to disrupt lipid membrane structures of microorganisms. Similarly, alkanes in the $C_{20} - C_{40}$ range, often referred to as "waxes", are hydrophobic solids at physiological temperatures. Apparently, it is this physical state that strongly influences their biodegradation (Bartha and Atlas 1977).

Primary attack on intact hydrocarbons always requires the action of oxygenases and, therefore, requires the presence of free oxygen. In the case of alkanes, monooxygenase attack results in the production of alcohol. Most microorganisms attack alkanes terminally whereas some perform sub-terminal oxidation (Figure 2). The alcohol product is oxidised finally into an aldehyde

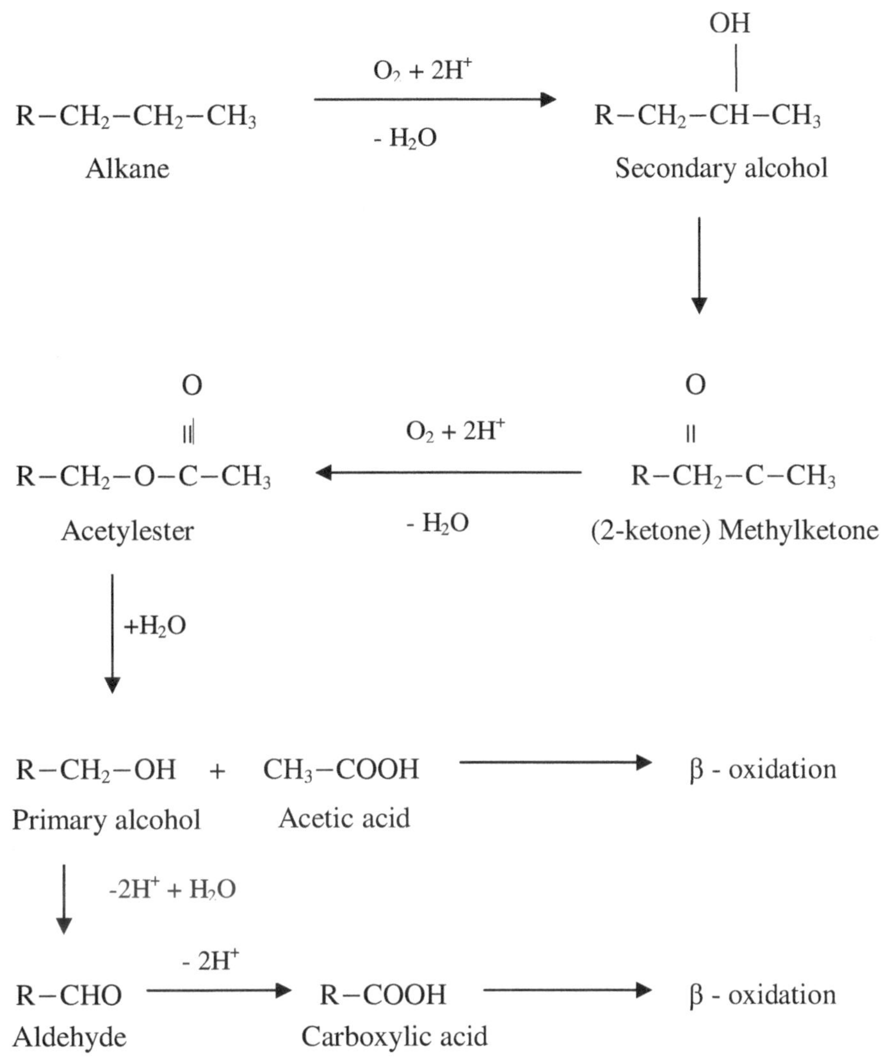

Figure 2. Pathways, through which subterminal oxidation of alkanes yield two fatty acid moieties, which are metabolized further by beta-oxidation (Atlas and Bartha 1998).

and finally, to a fatty acid. The latter is degraded further by *beta*-oxidation.

Extensive methyl branching interferes with the beta-oxidation process and necessitates diterminal attack (Figure 3) or other bypass mechanisms. Therefore, *n*-alkanes are degraded more readily than iso alkanes. Cycloalkanes are transformed by a not fully characterized oxidase system to a corresponding cyclic alcohol, which is dehydrated to ketone. Then, a monooxygenase system lactonises the ring, which is subsequently opened by a lactone hydrolase. These two oxygenase systems usually never occur in the same organisms and hence, the frustrated attempts to isolate pure cultures that grow on cycloalkanes (Bartha 1986b). However, synergistic actions of microbial communities are capable of dealing with degradation of various cycloalkanes quite effectively.

As in the case of alkanes, the monocyclic compounds, cyclopentane, cyclohexane, and cycloheptane have a strong solvent effect on lipid membranes, and are toxic to the majority of hydrocarbon degrading microorganisms. Highly condensed cycloalkane compounds resist biodegradation due to their structure and physical state (Bartha 1986a).

Prokaryotes convert aromatic hydrocarbons by an initial dioxygenase attack, to *trans*-dihydrodiols that are further oxidised to dihydroxy products, e.g., catechol in the case of benzene (Atlas and Bartha, 1998). Eucaryotic microorganisms use monooxygenases, producing benzene 1, 2-oxide from benzene, followed by the addition of water, yielding dihydroxydihydrobenzene (*cis*-dihydrodiol). This is oxidised in turn to catechol, a key intermediate in biodegradation of aromatics, which is then opened by *ortho*- or *meta*-cleavage, yielding muconic acid or 2-hydroxymuconic semialdehyde, respectively.

Condensed polycyclic aromatics are degraded, one ring at a time, by a similar mechanism, but biodegrada-

CH_3-$(CH_2)_n$-CH_3
n-alkane
↓$(1/2O_2)$
CH_3-$(CH_2)_n$-CH_2OH
Primary fatty alcohol
↓$(-2H^+)$
CH_3-$(CH_2)_n$-CHO
Fatty aldehyde
↓$(1/2O_2)$
CH_3-$(CH_2)_n$-COOH
Monocarboxylic fatty acid
↓$(1/2O_2)$
$HOCH_2$-$(CH_2)_n$-COOH
ω-hydroxy fatty acid
↓ $(- 2H^+)$
HOC-$(CH_2)_n$-COOH
ω -aldehyde fatty acid
↓$(1/2O_2)$
HOOC-$(CH_2)_n$-COOH
Dicarboxylic acid fatty acid

Figure 3. Pathway of diterminal alkane oxidation (Atlas 1984).

bility tend to decline with the increasing number of rings and degree of condensation (Atlas and Bartha 1998). Aromatics with more than four condensed rings are generally not suitable as substrates for microbial growth, though, they may undergo metabolic transformations. Biodegradation process also declines with the increasing number of alkyl substituents on the aromatic nucleus. Asphaltics tend to increase during biodegradation in relative and sometimes absolute amounts. This would suggest that they not only tend to resist biodegradation but may also be formed *de novo* by condensation reactions of biodegradation and photodegradation intermediates.

In crude petroleum as well as in refined products, petroleum hydrocarbons occur in complex mixtures and influence each other's biodegradation. The effects may go in negative as well as positive directions. Some *iso*-alkanes are apparently spared as long as *n*-alkanes are available as substrates, while some condensed aromatics are metabolised only in the presence of more easily utilisable petroleum hydrocarbons, a process referred to as co-metabolism (Wackett 1996).

3.0 Factors influencing petroleum hydrocarbon biodegradation

Successful application of bioremediation technology to contaminated systems requires knowledge of the characteristics of the site and the parameters that affect the microbial biodegradation of pollutants (Sabate et al., 2004). However, a number of limiting factors have been recognised to affect the biodegradation of petroleum hydrocarbons. Typical examples of these factors are as listed in Table 1. Biodegradability is inherently influenced by the composition of the oil pollutant. For example, kerosene, which consists almost exclusively of medium chain alkanes is, under suitable conditions, totally biodegradable. Similarly, crude oil is biodegradable quantitatively, but for heavy asphaltic-naphthenic crude oils, only about 11% may be biodegradable within a reasonable time period, even if the conditions are favourable (Bartha, 1986a). Okoh et al. (2002) reported that between 8.8 and 29% of the heavy crude oil *Maya* was biodegraded in soil microcosm by mixed bacterial consortium in 15 days, although major peak components of the oil was reduced by between 6.5 and 70% (Okoh et al., 2003). Also, about 89% of the same crude oil was biodegraded by axenic culture of *Burkholderia cepacia* RQ1 in shake flask (Okoh et al., 2001) within similar time frame, although petroleum biodegradation has been reported to be mostly enhanced in presence of a consortium of bacteria species compared to monospecies activities (Ghazali et al., 2004).

The composition and inherent biodegradability of the petroleum hydrocarbon pollutant, therefore, is the first and most important consideration when the suitability of a cleanup approach is to be evaluated. We have reported elsewhere (Okoh, 2002) that heavier crude oils are generally much more difficult to biodegrade than lighter ones, just as heavier crude oils could be suitable for inducing increased selection pressure for the isolation of petroleum hydrocarbon degraders with enhanced efficiency. Also, Okoh et al. (2002) noted that the amount of heavy crude oil metabolised by some bacterial species increased with increasing concentration of starter oil up to 0.6% (w/v), while degradation rates appeared to be more pronounced between the concentrations of 0.4 and 0.6% (w/v) oil. In another report (Rahman et al., 2002), the percentage of degradation by the mixed bacterial consortium decreased from 78% to 52% as the concentration of crude oil was increased from 1 to 10%.

Although, there has been a reported case of lack of correlation between degradation rates, specific growth rates and concentration of the starter oil (Thouand et al., 1999), in such a case, it would appear that biomass was required only to a particular threshold enough to produce the appropriate enzyme system that carry through the degradation process even when biomass production had ceased (Okoh 2002), a phenomenon completely at systems (Pitter and Chudoba 1990), where production of variance with the theory of microbial growth in batch cells is totally dependent on the consumed carbon source. However, such scenario is relevant especially in the light of public clamour for the use of cell free systems in bioremediation works against the scepticism that has befallen the application of live organisms, especially

Table 1. Some common factors affecting petroleum hydrocarbon biodegradation.

Limiting factor	Explanation or examples
Petroleum hydrocarbon- composition (PHC).	Structure, amount, toxicity.
Physical state	Aggregation, spreading, dispersion, adsorption.
Weathering	Evaporation, photooxidation.
Water potential	Osmotic and matrix forces, exclusion of water from hydrophobic aggregates
Temperature	Influence on evaporation and degradation rates.
Oxidant	O_2 required to initiate oxidation, NO_3^- or SO_4^{2-} to sustain, PHC biodegradation.
Mineral nutrients	N, P, Fe may be limiting
Reaction	Low pH may be limiting
Microorganisms	PHC degraders may be absent or low in numbers

Source: Bartha 1986a.

GMOs in bioremediation of polluted environment. Also, although many reports on the effect of sunlight irradiation so far published have focused on the physicochemical changes on intact crude oil other than to biodegraded crude oil (Jacquot et al., 1996; Nicodem et al., 1998), recent studies have reported that photooxidation increases the biodegradability of petroleum hydrocarbon by increasing its bioavailability and thus enhancing microbial activities (Maki et al., 2005). In a related study assessing the bioremediation efficiency of a weathered and recently contaminated soil in Brazil (Trindale et al., 2005), the authors reported low biodegradation efficiencies in the weathered soil contaminated with a high crude oil concentration compared to recently contaminated soil. Also, both soils (weathered and recently contaminated) submitted to bioaugmentation and biostimulation techniques presented biodegradation efficiencies approximately twice higher than the ones without natural attenuation.

Temperature plays very important roles in biodegradation of petroleum hydrocarbons, firstly by its direct effect on the chemistry of the pollutants, and secondly on its effect on the physiology and diversity of the microbial milieu. Ambient temperature of an environment affects both the properties of spilled oil and the activity or population of microorganisms (Venosa and Zhu, 2003). At low temperatures, the viscosity of the oil increases, while the volatility of toxic low-molecular-weight hydrocarbons is reduced, delaying the onset of biodegradation (Atlas 1981). Temperature also variously affects the solubility of hydrocarbons (Foght et al., 1996). Although hydrocarbon biodegradation can occur over a wide range of temperatures, the rate of biodegradation generally decreases with decreasing temperature. Highest degradation rates generally occur in the range of 30–40 °C in soil environments, 20–30 °C in some freshwater environments, and 15–20 °C in marine environments (Bossert and Bartha 1984; Cooney 1984). However, the bulk of information on hydrocarbon degradation bothers on activities of mesophiles, although significant biodegradations of hydrocarbons have been reported in psychrophilic environments in temperate regions (Yumoto et al., 2002; Pelletier et al., 2004; Delille et al., 2004). Nevertheless, documented research on the environmental consequences of terrestrial spills in cold regions is still scarce, even though petroleum contamination is recognized as a significant threat to polar environments. Full-scale in situ remediation of petroleum contaminated soils has not yet been used in Antarctica for example, partly because it has long been assumed that air and soil temperatures are too low for an effective biodegradation (Delille et al., 2004).

Microbial utilisation of hydrocarbons (being fully reduced substrates) requires an exogenous electron sink. In the initial attack, this electron sink has to be molecular oxygen. In the subsequent steps too, oxygen is the most common electron sink. In the absence of molecular oxygen, further biodegradation of partially oxygenated intermediates may be supported by nitrate or sulphate reduction. Little or no hydrocarbon metabolism occurs in strictly anoxic sediments (Dibble and Bartha 1976). Inadequate mineral nutrient, especially nitrogen, and phosphorus, often limits the growth of hydrocarbon utilisers in water and soils. Iron was reported to be limiting only in clean, offshore seawater (Swannell et al., 1996). Sulphur, in form of sulphate ions, is plentiful in seawater, but could be limiting in some freshwater environments. The slight alkaline pH of seawater seems to be quite favourable for petroleum hydrocarbon degradation, but in acidic soils liming to pH 7.8 – 8.0 had a definite stimulatory effect. The diversity of petroleum hydrocarbon degraders in most natural environments is usually high, but in the absence of a previous pollution history, their numbers may be low (Swanell et al., 1996).

Nutrients are very important ingredients for successful biodegradation of hydrocarbon pollutants, especially nitrogen, phosphorus and in some cases iron (Cooney 1984). Depending on the nature of the impacted environment, some of these nutrients could become limiting thus affecting the biodegradation processes. When a major oil spill occurs in marine and freshwater environments, the supply of carbon is dramatically increased and the availability of nitrogen and phosphorus generally becomes the limiting factor for oil degradation (Atlas 1984). This is more pronounced in marine environments, due to the low background levels of nitrogen and phosphorus in seawater (Floodgate 1984), unlike in freshwater systems that regularly fluctuate in nutrient status as result of perturbations and receipt of industrial and domestic effluents and agricultural runoff (Cooney 1984). Freshwater wetlands are typically considered to be nutrient limited, due to heavy demand for nutrients by the plants, and they could also be nutrient traps, as a substantial amount of nutrients may be bound in biomass (Mitsch and Gosselink 1993). Hence the additions of nutrients are necessary to enhance the biodegradation of oil pollutants (Choi et al., 2002; Kim et al., 2005). Pelletier et al. (2004) assessed the effectiveness of fertilizers for crude oil bioremediation in sub-Antarctic intertidal sediments over a one-year and observed that chemical, microbial and toxicological parameters demonstrated the effectiveness of various fertilizers in a pristine environment. In another study using poultry manure as organic fertilizer in contaminated soil, biodegradation was reported to be enhanced in the presence of poultry manure alone, but the extent of biodegradation was influenced by the incorporation of alternate carbon substrates or surfactants (Okolo et al., 2005). However, excessive nutrient concentrations can inhibit the biodegradation activity (Challain et al., 2006), and several authors have reported the negative effect of a high NPK levels on the biodegradation of hydrocarbons (Oudot et al., 1998; Chaıneau et al., 2005) and more especially on the aromatics (Carmichael and Pfaender, 1997).

Uptake and utilisation of water insoluble substrates, such as petroleum alkanes in crude oil, require specific physiological adaptations of the microorganisms. Synthesis of specific amphiphilic molecules, i.e., biosurfactants, has often been taken as a prerequisite for either specific adhesion mechanisms to large oil drops or emulsification of oil, followed by uptake of submicron oil droplets. Various species of bacteria have been observed to develop different strategies to deal with water insoluble substrates, such as hydrocarbons (Rosenberg 1991). Hence, to facilitate hydrocarbon uptake through the hydrophilic outer membrane, many oil-utilising microorganisms produce cell wall-associated or extracellular surface-active agents (Haferburg et al., 1986). This includes such low molecular weight compounds such as fatty acids, triacylglycerols and phospholipids, as well as the heavier glycolipids, examples of which include *emulsan* and *liposan* (Cirigliano and Carman 1984).

It is clear that the introduction of external nonionic surfactants, e.g., the main components of oil spill dispersants, will influence the alkane degradation rate (Bruheim and Eimhjelle, 1998; Rahman et al., 2003). Experience so far indicates that the use of surfactants in situations of oil contamination may have a stimulatory, inhibitory, or neutral effect on the bacterial degradation of the oil components (Liu et al., 1995). The need to accurately characterise the roles of chemical and biological surfactants has been proposed in order that performance in biological systems may be predicted (Rocha and Infante 1997; Lindstrom and Braddock 2002). However, in contrast to chemical dispersants, which caused ecological damage after application for abatement of spilled oil in marine ecosystems (Smith 1968), biosurfactants from soil or freshwater microorganisms are less toxic and partially biodegradable (Poremba et al., 1991).

4.0 Distribution of petroleum hydrocarbon utilising microorganisms

Hydrocarbon degrading bacteria and fungi are widely distributed in marine, freshwater, and soil habitats. Similarly, hydrocarbon degrading cyanobacteria have been reported (Challana et al., 2004; Lliros et al., 2003), although contrasting reports indicated that growth of mats built by cyanobacteria in the Saudi coast led to preservation of oil residues (Barth 2003). Typical bacterial groups already known for their capacity to degrade hydrocarbons include *Pseudomonas, Marinobacter, Alcanivorax, Microbulbifer, Sphingomonas, Micrococcus, Cellulomonas*, *Dietzia*, and *Gordonia* groups (Brito et al., 2006). Molds belonging to the genera *Aspergillus, Penicillium, Fusarium, Amorphoteca, Neosartorya, Paecilomyces, Talaromyces, Graphium* and the yeasts *Candida, Yarrowia* and *Pichia* have been implicated in hydrocarbon degradation (Chaillan et al., 2004). However, reports in literature on the actual numbers of hydrocarbon utilisers are at variance with one another because of the methodological differences used to enumerate petroleum-degrading microorganisms. The initial method involved the use of hydrocarbons incorporated into agar-based medium (Horowitz et al., 1978). This approach has its problems. In some cases, a high correlation has been found between growth on agar and media containing hydrocarbons as the sole carbon source, and the ability to rigorously demonstrate hydrocarbon utilisation by isolates from these media in liquid culture. In other studies, only a low percentage of isolates from agar-based media could be demonstrated to be capable of hydrocarbon utilisation.

The use of silica gel-oil medium for the enumeration of petroleum degrading microorganisms has been recommendded (Walker and Colwell 1976), suggesting that counts of petroleum degraders be expressed as a percentage of the total population rather than as total numbers of petroleum degraders *per se*. Also, the ability to utilise hydrocarbons is widespread, even in environments not subjected to high levels of hydrocarbon pollution. Atlas (1978) reported that quantitative differentces in the distribution of hydrocarbon utilizers were relatively unimportant over large geographic distances.

The Most Probable Number (MPN) procedure has also been tried as a substitute for the plate count technique for the estimation of hydrocarbon utilising microorganisms, since it eliminates the need for a solidifying agent and permits direct assessment of the ability to actually utilise hydrocarbons. The use of liquid media for MPN also permits removal of trace organic contaminants and allows for the chemical definition of a medium with a hydrocarbon as a sole source of carbon. This technique thus incorporates the specificity for counting only hydrocarbon utilizers and eliminates the problem of counting organisms growing on other trace organic contaminants (Braddock and Catterall 1999).

The problems of culture techniques arise from the fact that most (90–99%) of the species making up competent degrading communities do not form colonies when current laboratory-based culture techniques are used (MacNaughton et al., 1999). However, the appli-cation of molecular techniques for the analysis of the microbial communities that take part in *in situ* hydro-carbon biodegradation activities is helping to address these problems. The measurement of lipid biomarkers, specifically, phospholipid fatty acids (PLFA), together with nucleic acid–based molecular techniques for fingerprint-ting the 16S ribosomal DNA (rDNA) component of microbial cells is a powerful combination of techniques for elucidating the microbial ecology of actively bioremediating communities (Stephen et al., 1999). Lipid biomarker-based techniques measure the lipid profiles of microbes in the environment irrespective of culturability, thereby avoiding culture bias (White et al., 1998). These methods provide insight into several important characteristics of microbial communities, especially the viable biomass, community structure, and nutritional status or physiological stress responses of the gram-negative bacteria (White at al. 1998).

Molecular genetic techniques allow researchers to examine microbial communities without cultivation using universal primers 16S rRNA gene primers (Borneman et al., 1996). PCR has been particularly useful for detecting genes involved in the degradation of xenobiotic compounds (Joshi and Walia 1996). Despite the shortcomings of cultivation-based techniques, standard culture methods are still adequate for site evaluation to determine whether indigenous bacteria are capable of degrading the contaminants.

5.0 Genetic aspects of petroleum hydrocarbon degradation

Several new methodologies have enabled recent studies on the microbial biodegradation mechanisms of organic pollutants. Culture-independent techniques for analysis of the genetic and metabolic potential of natural and model microbial communities that degrade organic pollutants have identified new metabolic pathways and enzymes for aerobic and anaerobic degradation (Pieper et al., 2004).

The study of the genetics of hydrocarbon utilising micro-organisms has essentially been stepped up within the last decades. Before now, our insight into the genetics of simple aromatic hydrocarbon utilising microorganisms surpassed our knowledge and understanding of the genetics of alkane utilising microorganism. This is due to the availability of precise and definite database for aromatics metabolism compared to their *n*-alkane coun-terpart (Singer and Finnerty 1984).

Genetic studies have focused mainly on aerobic pathways, and many details of these metabolic routes have been documented (van der Meer et al., 1992), although Widdel and Rabus (2001), Heider et al. (1998) and Sporman and Widdel (2000) presented some detailed and comprehensive reviews on anaerobic biodegradation of hydrocarbons and the mechanisms involved. A general comparison of the major pathways for catabolism of aromatic compounds in bacteria has revealed that the initial conversion steps are carried out by different enzymes but that the compounds are transformed to a limited number of central intermediates, such as protocatechuate and (substituted) catechols (Chaudry and Chapalamadugu 1991). These dihydroxylated intermediates are channelled into one of two possible pathways, either a *meta*-cleavage- type pathway or an *ortho*-cleavage type pathway (Figure 4). Both types of pathways lead to intermediates of central metabolic routes, such as the tricarboxylic acid cycle. This generalised scheme of catabolic pathways for aromatic compounds suggests that microorganisms have extended their substrate range by developing peripheral enzymes, which are able to transform initial substrates into one of the central intermediates (van der Meer et al., 1992).

Genetic factors play important roles in conferring biodegradation potentials on microorganisms. Plasmids probably play leading role in this aspect. The ability to degrade more recalcitrant components of petroleum such, as the aromatic fractions are generally plasmid mediated (Cerniglia 1984).

Exposures of a microbial community to hydrocarbons have been shown to result in an increase in the number of bacterial plasmids types (Leahy et al., 1996). Catabolic plasmids are non-essential genetic elements in so far as viability and reproduction of an organism is concerned, but they do provide a metabolic versatility not normally present in the cell. Such genetic potential allows for the evolution of integrated and regulated pathways for the degradation of hydrocarbons. The observed increase in the study of the genetics of such systems has closely paralleled the development of advances in molecular biology, particularly the application of recombinant DNA technology (Singer and Finnerty 1984), gene probes (Barriault and Sylvestre 1993), and polymerase chain reaction (PCR) technology. Many bacterial catabolic pathways are specified by conjugative plasmids (Frantz

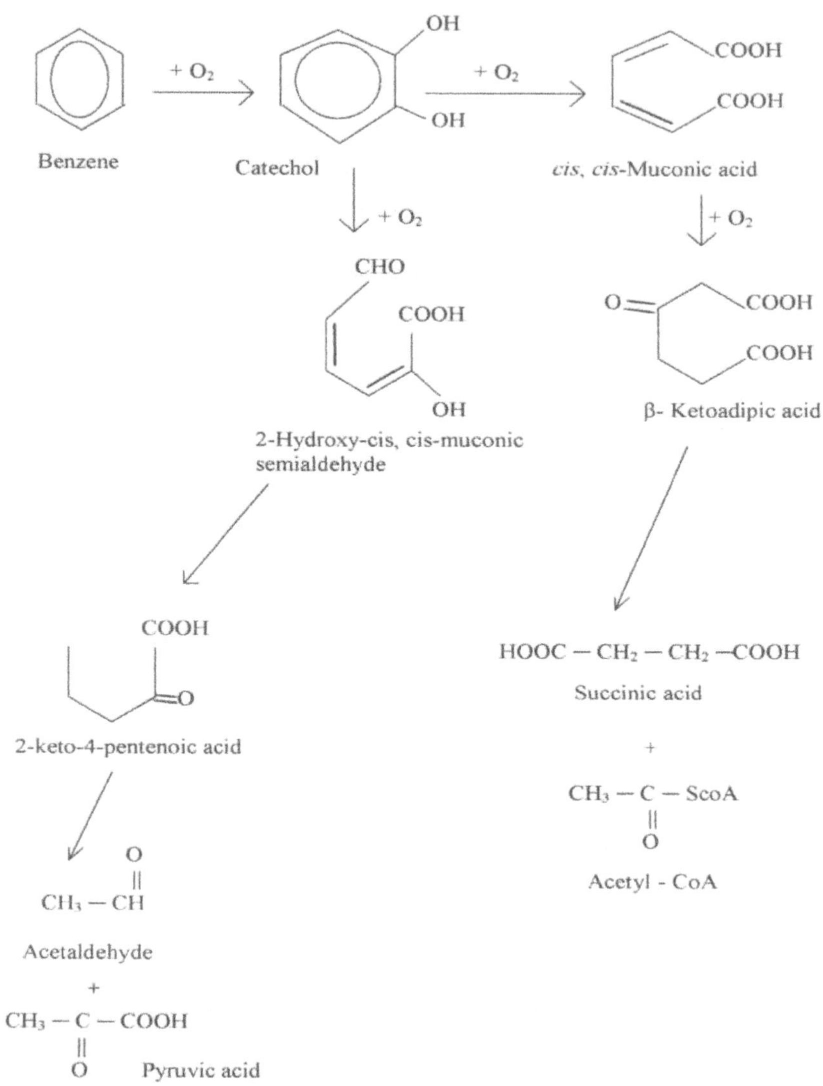

Figure 4. Microbial metabolism of the aromatic ring by *meta* or *ortho* cleavage as shown for benzene (Atlas and Bartha 1998).

and Chakrabarty 1986; Table 2). These plasmids are readily transferred laterally into new host bacteria, thereby enhancing the metabolic potential of other members of an ecosystem. Conjugative plasmids are thus important agents of genetic changes and evolution in bacteria, and could be picked up from or brought together in different organisms as groups of genes, which through mutations and recombination can specify new metabolic functions (Lessie and Gaffney 1986). For example, five aerobic toluene degradative pathways are characterized in pseudomonads. The best characterized of these pathways is encoded by the TOL plasmid (pWW0) of *P. putida* PaW1 (Worsey and Williams 1975), which converts toluene to benzyl alcohol, benzaldehyde, benzoate, and catechol, which further undergoes *meta* cleavage by an extradiol dioxygenase, or catechol 2,3-dioxygenase (C230). *Pseudomonas putida* F1 metabolizes toluene to 3-methylcatechol, which undergoes *meta*-cleavage by a C230 (Gibson et al., 1970).

Three other pathways for the catabolism of toluene were observed in *Burkholderia cepacia* G4 (Shields et al., 1989), *P. pickettii* PK01 (Kukor and Olsen 1990), and *P. mendocina* KR1 (Whited and Gibson 1991), in which toluene is respectively converted to *ortho*-, *meta*-, and *para*-cresol. Peripheral oxygenases of these five toluene degradative pathways possessed well-distinguished substrate specificity (Duetz et al., 1994). Particularly, xylenes can only be degraded by the *xyl* pathway encoded by TOL plasmid.

In a study of environmental *P. aeruginosa* strains isolated in Morroco, Belhaj et al. (2002) reported the presence of genetic information for several alkane monooxygenases. Most strains do carry genes very simi-

Table 2. Plasmids encoding catabolic functions.

Plasmid	Host	Component(s) catabolised
TOL	*Pseudomonas putida*	Toluene, *p*- and *m*-xylene
NAH	*Pseudomonas putida*	Naphthalene
SAL	*Pseudomonas putida.*	Salicylate
pND50	*Pseudomonas putida*	*p*-cresol
pWW31	*Pseudomonas putida.*	Phenylacetate
pJP1	*Alcaligenes paradoxa*	2,3- dichlorophenoxyacetic acid
pJP4	*Alcaligenes eutrophus*	2,4- dichlorophenoxyacetate and 3- chlorobenzoate
pKF1	*Acinetobacter* sp.	4- chlorobiphenol
pAC21	*Pseudomonas* sp.	4- chlorobiphenol
pRE1	*Pseudomonas putida*	Isopropyl benzene
pCIT1	*Pseudomonas* sp.	Aniline
pAC25	*Pseudomonas putida*	3-chlorobenzoate
pWR1	*Pseudomonas* sp.	3-chlorobenzoate
pCS1	*Pseudomonas diminuta*	Parathion

Source: Frant and Chakrabarty 1986.

lar to *alkB1* and *alkB2* of strain PAO1, suggesting that two alkane hydroxylase complexes are commonly involved in the degradation of long chain alkanes in *P. aeruginosa*. A few strains carry genetic information for a third alkane hydroxylase complex, with a monooxygenase that is very similar to AlkB of strain GPo1. These latter strains have an extended spectrum of alkane utilization, from C_{12}-C_{22} to C_6-C_{22} alkanes, and can degrade toxic water-soluble alkanes. Whilst saturated alkanes with C_{12} to C_{18} chains or longer are readily attacked by a large variety of bacteria. The degradation of water-soluble short chain alkanes such as pentane, hexane, heptane and octane, which are toxic for the environment, is less frequent (Bouchez-Naïtali et al., 1999). The first step of alkane degradation is the oxidation of the methyl group to alcohol by the "alkane hydroxylase system" that is composed of a membrane-bound mono-oxygenase, the alkane hydroxylase per se (45.8 kDa) encoded by *alkB*, and a soluble electron transport system composed of two rubredoxins and an NADH-dependent rubredoxin reductase encoded by *alkG*, *alkF* and *alkT* respectively (Belhaj et al., 2002).

In an environment rich in a particular organic compound, a selective pressure may lead to acquisition and maintenance of a plasmid that specifies a corresponding catabolic pathway. Many degraders of exotic compounds have been isolated from soil or water contaminated with such compounds. In some cases, the same catabolic genes may be located on a plasmid in one organism and on the chromosome in another, and these catabolic genes may influence the expression of other set of catabolic genes present in the same cell (Haramaya and Timmis 1989). In at least one instance of the TOL plasmid, transposition of catabolic genes from the plasmid to the chromosome has been demonstrated in the laboratory (Tsuda and Iino 1987).

6.0 CONCLUSION

Petroleum hydrocarbon especially in the form of crude oil has been a veritable source of economic growth to society from the point of view of its energy and industrial importance. These realizations, which have become more pronounced in the last decade, have resulted in extensive exploration for more oil reserves. The resultant effects of these exploratory activities have been the extensive pollution of the environment. Bioremediation, which exploits the biodegradative abilities of live organisms and their attributes have proven to be the preferred alternative in the long-term restoration of petroleum hydrocarbon polluted systems, with the added advantage of cost efficiency and environmental friendliness. Although extensive investigations have been carried out regarding hydrocarbon biodegradation, these studies have been exhaustive, not exhausted. Nevertheless, the effectiveness of this technology has only rarely been convincingly demonstrated, and in the case of commercial bioremediation products, the literature is virtually completely lacking in supportive evidence of success (Zhu et al., 2004). Most existing studies have concentrated on evaluating the factors affecting oil bioremediation or testing favored products and methods through laboratory studies (Mearns, 1997). Only limited numbers of pilot-scale and field trials, which may provide the most convincing demonstrations of this technology, have been reported in the peer-reviewed literature (Venosa et al., 2002). The scope of current understanding of oil bioremediation is also limited because the emphasis of most of these field studies and reviews has been on the evaluation of bioremediation technology for dealing with large-scale oil spills on marine shorelines (Zhu et al., 2004). Some shortcomings are evident in petroleum hydrocarbons degradation studies. The identification of active strains is not always ascertained to a sufficient degree, and misidentifications or incomplete identifications are sometimes reported. Molecular techniques for the identification of hydrocarbon-degrading bacteria have been only rarely used in environmental studies (Röling et al., 2002), and the biodegradation activities are not always confirmed by chemical analyses of the degraded Hydrocarbon (Challain et al., 2004). Much need still exist for the optimization of the process conditions for more efficient application of biological degradation of oil pollutants under different climatic conditions and other diverse environmental milieu.

7.0 REFERENCES

Alexander M (1994). Biodegradation and Bioremediation. Publishers Academic Press, Inc. California, USA, pp. xi.

Alloway BJ, Ayres DC (1993). Organic Pollutants. In: Chemical Principles of Environmental Pollution. 1st edition. Chapman and Hall, India, Publishers, p. 201.

April TM, Foght JM, Currah RS (2000). Hydrocarbon-degrading filamentous fungi isolated from flare pit soils in northern and western Canada. Can. J. Microb. 46(1): 38-49.

Atlas RM (1978). An assessement of the biodegradation of petroleum in the Arctic. In: *Microbial Ecology.* Loutit MW, Miles JAR (editors.), Springer – Verlag, Berlin, pp. 86-90.

Atlas RM (1981). Microbial degradation of petroleum hydrocarbons: an environmental perspective. Microb. Rev. 45:180-209.

Atlas RM (1984). Pathways of hydrocarbon degradation. In: Petroleum Microbiology. Macmillan Publishing Company, New York, USA, pp. 1-15.

Atlas RM, Bartha R (1973). Simulated biodegradation of oil slicks using oleophilic fertilisers. Environ. Sci. Tech. 7:538-541.

Atlas RM, Bartha R (1998). Fundamentals and Applications. In: Microbial Ecology. 4th edition. Benjamin/Cummings Publishing Company, Inc. California, USA, pp. 523-530.

Barriault D, Sylvestre M (1993). Factors affecting PCB degradation by an implanted bacterial strain in soil microcosms. Can. J. Microb. 39:594-602.

Barth H-J (2003). The influence of cyanobacteria on oil polluted intertidal soils at the Saudi Arabian gulf shores. Mar. Pollut. Bull. 46:1245-52.

Bartha R (1986a). Microbial ecology: fundamentals and applications. Addisson_Wesley Publ., Reading, Mass.

Bartha R (1986b). Biotechnology of Petroleum pollutant biodegradation. Microb. Ecol. 12:155-172.

Bartha R, Atlas RM (1977). The microbiology of aquatic oil spills. Adv. Appl. Microb. 22: 225-266.

Belhaj A, Desnoues N, Elmerich C (2002). Alkane biodegradation in *Pseudomonas aeruginosa* strains isolated from a polluted zone: identification of *alkB* and *alkB*-related genes. Res. Microbiol. 153:339-344.

Borneman J, Skroch PW, O'Sullivan KM, Palus S, Rumjanek NG, Jansen JL (1996). Molecular microbial diversity of an agricultural soil in Wisconsin. Appl. Environ. Microb. 62:1935-1943.

Bossert I, Bartha R (1984). The Fate of Petroleum in Soil Ecosystems. In Petroleum Microbiology, RM Atlas (ed.), Macmillan, New York, pp. 453-473.

Bouchez-Naïtali M, Rakatozafy H, Marchal R, Leveau JY, Vandecasteele JP (1999). Diversity of bacterial strains degrading hexadecane in relation to the mode of substrate uptake, J. Appl. Microbiol. 86:421-428.

Boehm PD, Douglas GS, Brown JS (1995). Advanced chemical fingerprinting for oil spill identifications and natural resource damage assessments. Proceedings of the American petroleum institute oil spill conference, API, Long Beach, California.

Braddock JF, Catterall PH (1999). A Simple Method for Enumerating Gasoline- and Diesel-Degrading Microorganisms. Bioremed. J. 3(2):81-84.

Bradley SN, Hammill TB, Crawford RL (1997). Biodegradation of Agricultural chemicals. In: Manual of Environmental Microbiology, American Society of Microbiology, Washington, DC, USA, pp. 815-821.

Brito EMS, Guyoneaud R, Goñi-Urriza M, Ranchou-Peyruse A, Verbaere A, Crapez MAC, Wasserman JCA, Duran R (2006). Characterization of hydrocarbonoclastic bacterial communities from mangrove sediments in Guanabara Bay, Brazil. Research in Microbiology. In press.

Bruheim P, Eimhjelle K (1998). Chemically emulsified crude oil as substrate for bacterial oxidation: differences in species response. Can. J. Microb. 44(2):195-199.

Caplan JA (1993). The world-wide bioremediation industry: prospects for profit. Trds. Biotech. 11:320-323.

Carmichael LM, Pfaender F.K (1997). The effect of inorganic and organic supplements on the microbial degradation of phenanthrene and pyrene in soils. Biodegradation 8, 1-13.

Cerniglia CE (1984). Microbial degradation of polycyclic aromatic hydrocarbons. Adv. Appl. Microb. 30:31-71.

Chaillana F, Flècheb A, Burya E, Phantavonga Y-hui, Saliot A, Oudot J (2004). Identification and biodegradation potential of tropical aerobic hydrocarbon-degrading microorganisms. Res. Microb. 155(7):587-595.

Chaillan F, Chaıneau C.H, Point V, Saliot A, Oudot J (2006). Factors inhibiting bioremediation of soil contaminated with weathered oils and drill cuttings. Environmental Pollution. In Press.

Chaıneau CH, Yepremian C, Vidalie JF, Ducreux J, Ballerini D (2003). Bioremediation of a crude oil-polluted soil: biodegradation, leaching and toxicity assessments. Wat. Air. Soil. Poll. 144: 419-440.

Chaıneau CH, Rougeux G, Yepremian C, Oudot J (2005). Effects of nutrient concentration on the biodegradation of crude oil and associated microbial populations in the soil. Soil. Biol. Biochem. 37:1490-1497.

Chaudhry GR, Chapalamadugu S (1991). Biodegradation of halogenated organic compounds. Microb. Rev. 55:59-79.

Choi S-C, Kwon KK, Sohn JH, Kim S-J (2002). Evaluation of fertilizer additions to stimulate oil biodegradation in sand seashore mescocosms. J. Microbiol. Biotechnol. 12:431-436

Cirigliano MC, Carman GM (1984). Isolation of a bioemulsifier from *Candida lipolytica.* Appl. Environ. Microb. 48:747-750.

Cooney JJ (1984). The fate of petroleum pollutants in freshwater ecosystems. K 03099 Pollution; J 02905 Water; P 2000 Freshwater Pollut.

Delille D, Coulon F, Pelletier E (2004). Effects of temperature warming during a bioremediation study of natural and nutrient-amended hydrocarbon-contaminated sub-Antarctic soils. Cold Reg. Sci.Tech. 40:61-70.

Dibble JT, Bartha R (1976). The effect of iron on the biodegradation of petroleum in seawater. Appl. Environ. Microb. 31:544-550.

Duetz AW, de Jong C, Williams PA, van Andel JG (1994). Competition in chemostat culture between *Pseudomonas* strains that use different pathways for the degradation of toluene. Appl. Environ. Microb. 60:2858-2863.

Ebuehi OAT, Abibo IB, Shekwolo PD, Sigismund KI, Adoki A, Okoro IC (2005). Remediation of Crude Oil Contaminated Soil by Enhanced Natural Attenuation Technique. J. Appl. Sci. Environ. Mgt. 9 (1):103-106.

Floodgate GD (1984). The fate of petroleum in marine ecosystems. In RM Atlas, Petroleum microbiology. pp. 355-397. New York: MacMillan.

Foght JM, Westlake D, Johnson WM, Ridgway HF (1996). Environmental gasoline-utilising isolates and clinical isolates of *Pseudomonas aeruginosa* are taxonomically indistinguishable by chemotaxonomic and molecular techniques. Microbiol. 142: 1333-

2340.
Frantz B, Chakrabarty AM (1986). Degradative plasmids in Pseudomonas aeruginosa. In: The Bacteria. Sokatch JR and Ornston LN (editors), Academic Press, New York, USA, 10: 295-323.
Ghazali FM, Rahman RNZA, Salleh AB, Basri M (2004). Biodegradation of hydrocarbons in soil by microbial consortium. Intl. Biodeter. Biodegrad. 54: 61-67.
Gogoi BK, Dutta NN, Goswami P, Mohan, TR (2003). A case study of bioremediation of petroleum-hydrocarbon contaminated soil at a crude oil spill site. Adv. Environ. Res. 7:767-782.
Haferburg D, Hommel R, Claus R, Kleber HP (1986). Extracellular microbial lipids as biosurfactants. Adv. Biochem. Eng. Biotech. 33:53-93.
Hallier-Soulier S, Ducrocq V, Mazure N, Truffaut N (1999). Detection and quantification of degradative genes in soils contaminated by toluene. FEMS Microb. Ecol. 20:121-133.
Harayama S, Timmis KN (1989). Catabolism of Aromatic Hydrocarbons by Pseudomonas. In: Genetics of Bacterial Diversity, Hopwood, DA and Chater, KF (editors), Florida, USA, pp. 152-174.
Heider J, Spormann AM, Beller HR, Widdel F (1998). Anaerobic bacterial metabolism of hydrocarbons. FEMS Microb. Rev. 22 (5):459-.465.
Holba AG, Dzou IL, Hickey JJ, Franks SG, May SJ, Lenney T (1996). Reservoir Geochemistry of South Pass 61 Field, Gulf of Mexico: Compositional Heterogeneities Reflecting Filling History and Biodegradation: Org. Geochem. (24): 1179-1198.
Horowitz A, Sexstone A, Atlas RM (1978). Hydrocarbons and microbial activities in sediment of an Actic lake one year after contamination with leaded gasoline. Arctic. 31:180-191.
International Tanker Owners Pollution Federation Limited (ITOPF) (2006). Disposal of Oil and Debris. In: Response Strategies. http://www.itopf.com/index.html.
Jacquot F, Guiliano M, Doumenq P, Munoz D, Mille G (1996). *In vitro* photooxidation of crude oil maltenic fractions: Evolution of fossil biomarkers and polycyclic aromatic hydrocarbons. Chemosph. 33: 671-681.
Joshi B, Walia S (1996). PCR amplification of catechol 2,3-dioxygenase gene sequences from naturally occurring hydrocarbon degrading bacteria isolated from petroleum hydrocarbon contaminated groundwater. FEMS Microb. Ecol.19: 5-15.
Kim S, Choi DH, Sim DS, Oh Y (2005). Evaluation of bioremediation effectiveness on crude oil-contaminated sand. Chemosph. 59: 845-852.
Lessie TG, Gaffney T (1986). Catabolic potential of *Pseudomonas cepacia*. In: The Bacteria. Sokatch JR, Ornston LN (editors), Academic Press, New York, USA. 10: 439- 481.
Lindstrom JE, Braddock JF (2002). Biodegradation of petroleum hydrocarbons at low temperature in the presence of the dispersant Corexit 9500. Mar. Poll. Bull. 44: 739-747.
Liu Z, Jacobson AM, Luthy RG (1995). Biodegradation of naphthalene in aqueous nonionic surfactant systems. Appl. Environ. Microb. 61: 45-151.
Lliros M, Munill X, Sole A, Martinez-Alonso M, Diestra E, Esteve I (2003). Analysis of cyanobacteria biodiversity in pristine and polluted microbia mats in microcosmos by confocal laser scanning microscopy (CLSM). In: Mendez-Vilas A (Ed.), Science, Technology and Education of Microscopy: An Overview. Badjz. Formt. p. 483–9.
MacNaughton SJ, Stephen JR, Venosa AD, Davis GA, Chang C-J, White DC (1999). Microbial population changes during bioremediation of an experimental oil spill. Appl. Environ. Microb. 65(8):3566-3574.
Maki H, Sasaki T, Haramaya S (2005). Photooxidation of biodegradable crude oil and toxicity of the photooxidized products. Chemosph. 44:1145-1151.
Mesarch MB, Nakatsu CH, Nies L (2000). Development of catechol 2,3 –Deoxygenase-specific primers for monitoring bioremediation by competitive quantitative PCR. Appl. Environ. Microb. 66(2):678-690.
Mitsch WJ, Gosselink JG (1993). Wetlands, 2nd ed. John Wiley publ, New York.
Mueller JG, Resnick SM, Shelton ME, Pritchard PH (1992). Effect of inoculation on the biodegradation of weathered Prudhoe Bay crude oil. J. Indst. Microb.10: 95-102.
Nicodem DE, Guedes CLB, Correa RJ (1998). Photochemistry of petroleum I. Systematic study of a brazilian intermediate crude oil. Mar. Chem. 63: 93-104.
Okoh AI (2002). Assessement of the potentials of some bacterial isolates for application in the bioremediation of petroleum hydrocarbon polluted soil. Ph.D. thesis, Obafemi Awolowo University, Ile – Ife, Nigeria.
Okoh AI (2003). Biodegradation of Bonny light crude oil in soil microcosm by some bacterial strains isolated from crude oil flow stations saver pits in Nigeria. Afr. J. Biotech. 2(5): 104-108.
Okoh AI, Ajisebutu S, Babalola GO, Trejo-Hernandez MR (2001). Potentials of *Burkholderia cepacia* strain RQ1 in the biodegradation of heavy crude oil. Intl. Microb. 4: 83-87.
Okoh AI, Ajisebutu S, Babalola GO, Trejo - Hernandez MR (2002). "Biodegradation of Mexican heavy crude oil (*Maya*) by *Pseudomonas aeruginosa*". J. Trop. Biosci. 2(1): 12-24.
Okolo JC, Amadi EN, Odu CTI (2005). Effects of soil treatments containing poultry manure on crude oil degradation in a sandy loam soil. Appl. Ecol. Environ. Res. 3(1) : 47-53.
Oudot J, Merlin FX, Pinvidic P (1998). Weathering rates of oil components in a bioremediation experiment in estuarine sediments. Mar. Environ. Res. 45: 113-125.
Pelletier E., Delille D, Delille B (2004). Crude oil bioremediation in sub-Antarctic intertidal sediments: chemistry and toxicity of oiled residues. Mar. Environ. Res. 57: 311-327.
Pieper HD, Martins dos Santos VAP, Golyshin PN (2004). Genomic and mechanistic insights into the biodegradation of organic pollutants. Curr. Opin. Biotechn. 15 :215-224.
Pitter P, Chudoba J (1990). *Biodegradability of organic substances in the aquatic environment*. CRC Press, Boca Raton, Florida, USA, pp. 7-83.
Poremba K, Gunkel W, Lang S, Wagner F (1991). Marine biosurfactants. III. Toxicity testing with marine microorganisms and comparison with synthetic detergents. Zeitschrift. Natureforsch. 46c:210-216.
Prince RC (1993). Petroleum spill bioremediation in marine environments. Crit. Rev. Microb. 19: 217-242.
Pye VI, Patrick R (1983). Groundwater contamination in the United States. Science. 221: 713-718.
Rahman KSM, Thahira-Rahman J, Lakshmanaperumalsamy P, Banat IM (2002). Towards efficient crude oil degradation by a mixed bacterial consortium. Biores. Tech. 85: 257-261.
Rahman KSM, Rahman TJ, Kourkoutas Y, Petsas I, Marchant R, Banat IM (2003). Enhanced bioremediation of n-alkane in petroleum sludge using bacterial consortium amended with rhamnolipid and micronutrients. Biores Tech. 90(2): 159-168.
Rocha C, Infante C (1997). Enhanced oily sludge biodegradation by a tensio-active agent isolated from *Pseudomonas aeruginosa* USB-CS1. Appl. Microb. Biotech. 47: 615-619.
Röling WFM, Milner MG, Jones DM, Lee K, Daniel F, Swannell RJP, Head IM (2002). Robust hydrocarbon degradation and dynamics of bacterial communities during nutrient-enhanced oil spill bioremediation. Appl. Environ. Microbiol. 68: 5537-5548.
Rosenberg E (1991). The hydrocarbon-oxidizing bacteria. In: *The Prokaryotes.* Vol 1. 2nd edition. Bolows A, Truger HG, Dworkin M, Schleifer KH, Harder W (editors), Springer – Verlag, New York, USA, pp. 446-458.
Sabate J, Vinas M, Solanas A.M (2004). Laboratory-scale bioremediation experiments on hydrocarbon-contaminated soils. Internl. Biodeter. Biodegrad. 54:19-25.
Shields MS, Montgomerry SO, Chapman PJ, Cuskey MS, Pritchard PH (1989). Novel pathway of toluene catabolism in trichloroethylene-degrading bacterium G4. Appl. Environ. Microb. 55:1624-1629.
Singer JT, Finnerty WR (1984). Genetics of hydrocarbon-utilising microorganisms. In: Petroluem Microbiology. Atlas RM (editor), Macmillan Publishing. Company. Inc., New York, USA, pp. 299-354.
Snape I, Riddle MJ, Stark JS, Cole CM, King CK, Duquesne S, Gore DB (2001). Management and remediation of contaminated sites at Casey Station, Antarctica. Polar Rec. 37:199-214.
Smith J (1968). Torrey Canyon – pollution and marine life. Report by the Plymouth Laboratory of the Marine Biological Association of the United Kingdom, London. Cambridge University Press, London,

pp 1906.
Spence JM, Bottrell SH, Thornton SF, Richnow HH, Spence KH (2005). Hydrochemical and isotopic effects associated with petroleum fuel biodegradation pathways in a chalk aquifer. J. Contam. Hydrol. 79: 67-88.
Spormann AM, Widdel F (2000). Metabolism of alkylbenzenes, alkenes, and other hydrocarbons in anaerobic bacteria. Biodegrad. 11:85-105.
Stephen JR, Chang C-J, Gan D, Peacock A, Pfiffner SM, Barcelona MJ, White D C, MacNaughton SJ (1999). Microbial characterisation of JP-4 fuel contaminated site using a combined lipid biomarker/PCR-DGGE based approach. Environ. Microb. Biotech. 1: 231-243.
Sugiura K, Ishihara M, Shimauchi T, Harayama S (1997). Physicochemical properties and biodegradability of crude oil. Environ. Sci. Tech. 31:45-51.
Swanell RPJ, Lee K, McDonagh M (1996). Field evaluation of marine oil spill bioremediation. Microb. Rev. (June): 342-365.
Thouand G, Bauda P, Oudot J, Kirsch G, Sutton C, Vidalie JF (1999). Laboratory evaluation of crude oil biodegradation with commercial or natural microbial inocula. Can. J. Microb. 45(2): 106-115.
Timmis KN, Yakimov MM, Golyshin PN (1998). Hydrocarbonoclastic Bacteria: Novel Marine Bacteria that grow only on oil. Scientific Annual Report, German National Research Centre for Biotechnology (GBF), Germany, pp. 21-26.
Trindade PVO, Sobral LG, Rizzo ACL, Leite SGF, Soriano AU (2005). Bioremediation of a weathered and a recently oil-contaminated soils from Brazil: a comparison study. Chemosph. 58: 515-522.
Tsuda M, Iino T (1987). Genetic analysis of a transposon carrying toluene degrading genes on a TOL plasmid pWWO. Mol. Gen. 210: 270-276.
USEPA (2001). United States Environmental Protection Agency. Understanding Oil Spills and Oil Spill Response. http://www.fundyforum.com/profile_archives/profile3.html.
van der Meer JR, de Vos MW, Haramaya S, Zehnder AJB (1992). Molecular mechanism of genetic adaptation to xenobiotic compounds. Microb. Rev. 56(4):677-694.
van Hamme JD, Singh A, Ward OP (2003). Recent advances in petroleum microbiology. Microb. Mol. Biol. Rev. 67:503-549.
Venosa AD, Lee K, Suidan MT, Garcia-Blanco S, Cobanli S, Moteleb M, Haines JR, Tremblay G, Hazelwood M (2002). Bioremediation and biorestoration of a crude oil contaminated freshwater wetland on the St. Lawrence River. Bioremed. J. 6 (3). DOI: 10.1080/10889860290777602.
Venosa AD, Zhu X (2003). Biodegradation of Crude Oil Contaminating Marine Shorelines and Freshwater Wetlands. Spill Sci. Tech. Bull. 8(2):163-178.
Vogel TM (1996). Bioaugmentation as a soil bioremediation approach. Curr. Opin. Biotechn. 7:311-316.
Wackett LP (1996). Co-metabolism: is the emperor wearing any clothes? Curr. Opin. Biotechn. 7:321-325.
Walker JD, Colwell RR (1976). Petroleum degradation by estuarine microorganisms. In: Proceedings of the Third International biodegradation symposium. Sharpley JM, Kaplan AM (editors), Applied Science Publishers, Ltd., London, pp. 197-204.
Westlake DWS (1982). Microorganisms and the degradation of oil under northern marine conditions. In: Oils and dispersants in Canadian seas – research appraisal and recommendations. Publication EPS-3-EC-82-2. Environmental Protection Service Canada, Sprague JB, Vandermeulen JH, Wells PG (editors), Ottawa, Canada, pp. 47-50.
White DC, Flemming CA, Leung KT, MacNaughton SJ (1998). *In situ* microbial ecology for quantitative assessment, monitoring and risk assessment of pollution remediation in soils, the subsurface, the rhizosphere and in biofilms. J. Microbiol. Meth. 32:93-105.
Whited GM, Gibson DT (1991). Separation and partial characterisation of the enzymes of the toluene-4-monooxygenase catabolic pathway in *Pseudomonas mendocina* KR1. J. Bacteriol., 173:3017-3020.
Whittaker M, Pollard SJT, Risden G (1999). The fate of heavy oil wastes in soil microcosms II: a performance assessment of source correlation indices. Sci. Tot. Environ. 2:23-34.
Widdel F, Rabus R (2001). Anaerobic biodegradation of saturated and aromatic hydrocarbons. Curr. Opin.Biotech. 12:259-276.
Worsey MJ, Williams PA (1975). Metabolism of toluene and xylenes by *Pseudomonas putida* (arvilla) Paw1: evidence for a new function of the TOL plasmid. J. Bacteriol. 124:552-560.
Yumoto I, Nakamura A, Iwata H, Kojima K, Kusumuto K, Nodasaka Y, Matsuyama H (2002). *Dietzia psychralcaliphila* sp. nov., a novel facultatively psychrophilic alkaliphile that grows on hydrocarbons. Int. J. Syst. Evol. Microbiol. 52:85-90.
Zengler K, Richnow HH, Rossello-Mora R, Michaelis W, Widdel F (1999). Methane formation from long-chain alkanes by anaerobic microorganisms: Natr. (401):266-269.
Zhu X, Venosa AD, Suidan MT (2004). Literature review on the use of commercial bioremediation agents for cleanup of oil-contaminated estuarine environments. EPA/600/R-04/075.

6

Cationic lipids used in non-viral gene delivery systems

Sarder Nasir Uddin

Biotechnology and Genetic Engineering Disciplne, Khulna University, Khulna-9208, Bangladesh. E-mail: opunasir@gmail.com.

To perform gene transfer two types of vectors are available, (i) viral vectors (ii) nonviral vectors. Among the vectors, nonviral vectors have proved less toxic and safe compare to viral vectors through clinical trials. No single vector is proved suitable for every gene transfection experiment. Cationic lipids are experimentally established as non-viral vector with higher transfection efficiency. Identifying the barriers for transfection and the possible solutions, stability improvement research is needed for better clinical performance of cationic vectors. The newly described liposomal preparation, Liposomes-Protamine-DNA (LPD), has shown superiority over conventional Liposomes-DNA complexes (lipoplexes). In future it is also possible to find new conjugates as cationic lipids bearing many properties, suitable to bound with DNA and penetrate cell. Considering all these properties, these works reviews the most recent studies highlighting cationic lipids used in nonviral gene delivery systems.

Key words: Gene therapy, non-viral, lipids, cation, delivery systems.

Table of contents

1. Introduction
1.1. General aspects
1.2. Gene therapy vectors
1.3. Non-viral Vectors
2. Cationic lipids
2.1 Liposomes
2.1.1 Basic nature and transfection mechanism
2.1.2 Liposome mediated gene therapy in kidney
2.2 Liposome complexes
2.3 LID Vector complexes
3. Barriers of using cations in non viral gene therapy
4. Improvement of transfection efficiency
5. Conclusion
6. References

INTRODUCTION

General aspects

Gene therapy represents a new and promising therapeutic modality. The underlying principle is based on the introduction of genetic material into cells to generate a curative biological effect (Miller, 1992; Mulligan, 1993). Gene therapy is not limited to hereditary diseases but can be used for a broad variety of different acquired diseases, such as infections, degenerative disorders, and cancer. The most challenging issues for successful application ofgene therapy to human diseases concern are

(1) The choice of the relevant therapeutic gene.
(2) The choice of promoter and regulatory sequences driving expression of the transgene.
(3) The vector used for delivery of the transgene into target cells.

Promoter, regulatory elements and vector characteristics determine transduction efficacy (the number of target cells expressing the transgene and the intensity of gene expression per cell), specificity of the transduction, time of transgene expression, the host's immune response against the vector, and eventually undesired side effects. At present, efforts are being focused on the search for vectors with less toxicity and prolonged and controlled transgene expression, thereby widening the potential application of gene therapy to a high spectrum of medical fields. To perform gene transfer two types of vectors are available: viral vectors (retorviruses, adeneviruses, herpesviruses, lentiviruses and hybrid/ retoradeno viruses etc), and nonviral vectors (cations e.g. polymers and lipids, naked DNA, artificial chromosomes, plasmid, conjugates and complexes) (Uddin and Islam, 2006).

Despite the impressive progress in biomedical sciences during the last decades, the therapy of many liver diseases remains unsatisfactory. This applies not only to primary and metastatic liver tumours, hepatic cirrhosis, and hereditary metabolic diseases but also to a high proportion of cases with chronic viral hepatitis that do not respond to current antiviral therapy. There is evidently an urgent need for efficient alternative therapeutic approaches. In recent years gene therapy has emerged as a new and promising method to treat human diseases (Schmitz et al., 2005)

The genetic materials used for gene therapy are natural or chimeric genes (that direct the synthesis of therapeutic proteins inside the transduced cells) or subgenomic DNA and RNA molecules (that act directly to modify expression of endogenous genes). Natural genes are genomic sequences that encode natural proteins such as enzymes, costimulatory molecules, antigens etc. Chimeric genes are genomic constructs that do not exist in nature that encode for such molecules as transdominant negative proteins or single chain antibodies (Herskowitz, 1987).

Subgenomic DNA and RNA sequences include ribozymes, antisense molecules, and RNA decoys, among others. A ribozyme is an RNA molecule that hybridizes specifically with a messenger RNA and possesses a catalytic active domain which cleaves the target mRNA (Haseloff et al., 1988; Symons and Ribozymes, 1994). Antisense molecules are DNA or RNA constructs which bind complementary sequences by hybridization and interfere with pre-mRNA processing and messenger RNA translation and accelerate mRNA degradation (Crooke, 1998; Eguchi, 1991). RNA decoys are RNA sequences, which ligate transactivating proteins as a substitute for their original binding domain, thus, blocking the function of the regulatory proteins (Sullenger et al., 1991).

Gene therapy vectors

Gene therapy vehicles can be categorized into two groups: biological and non-biological systems. Each group has its own advantages and limitations. The main objective in gene therapy is the development of efficient, non-toxic gene carriers that can encapsulate and deliver foreign genetic materials into specific cell types. During the past two decades, enormous research in the area of gene delivery has been conducted worldwide, in particular for cancer gene therapy application (Uddin and Islam, 2006). Viral vectors are biological systems derived from naturally evolved viruses capable of transferring their genetic materials into the host cells. Many viruses including retrovirus, adenovirus, herpes simplex virus (HSV), adeno-associated virus (AAV) and pox virus have been modified to eliminate their toxicity and maintain their high gene transfer capability. The limitations associated with viral vectors, however, in terms of their safety, particularly immunogenicity, and in terms of their limited capacity of transgenic materials, have encouraged researchers to focus on non-viral vectors as an alternative to viral vectors. Non-viral vectors are generally cationic in nature. They include cationic polymers such as polyethylenimine (PEI) and poly L-lysine (PLL), cationic peptides and cationic liposomes. The newly described liposomal preparation LPD (liposomes/protamine/DNA), for example, has shown superiority over conventional liposomes/DNA complexes (lipoplexes) (Aneed, 2004). However, physical properties such as size and zeta potential play a critical role in their efficiency. In either delivery system, selected modifications that can produce safe efficient and targetable gene carriers are desirable. Although non-viral vectors are less efficient than viral ones, they have the advantages of safety, simplicity of preparation and high gene encapsulation capability. This article reviews the most recent studies highlighting the advantages and the limitations of nonviral cationic types of gene delivery systems used in gene therapy (Aneed, 2004).

Both viral and non-viral vectors are used to transfer genetic material to the inside of target cells. In general, non-viral vectors have lower transduction efficiency and allow for shorter duration of transgene expression than viral vectors. The ideal vector should have;

(i) Low antigenic potential, high capacity, high transduction efficiency.
(ii) Allow controlled and targeted transgene expression.
(iii) Reasonable expense.
(iv) Safe for both patient and the environment.

Gene delivery vehicles must be selected according to the specific therapeutic aim. But the perfect vector covering all therapeutic and safety requirements does not exist and much work is needed in this important research field (Prince, 1998).

Non-viral vectors

Non-viral systems are cationic in nature. They interact with negatively charged DNA through electrostatic interactions. The total charge however maintains a positive net value. This will enable the carrier of efficiently interacting with the negatively charged cell membranes and internalizes into the cell, which occurs mainly through the endocytosis pathway (Behr, 1994). The cationic nature of the non-viral vectors helps them to interact with

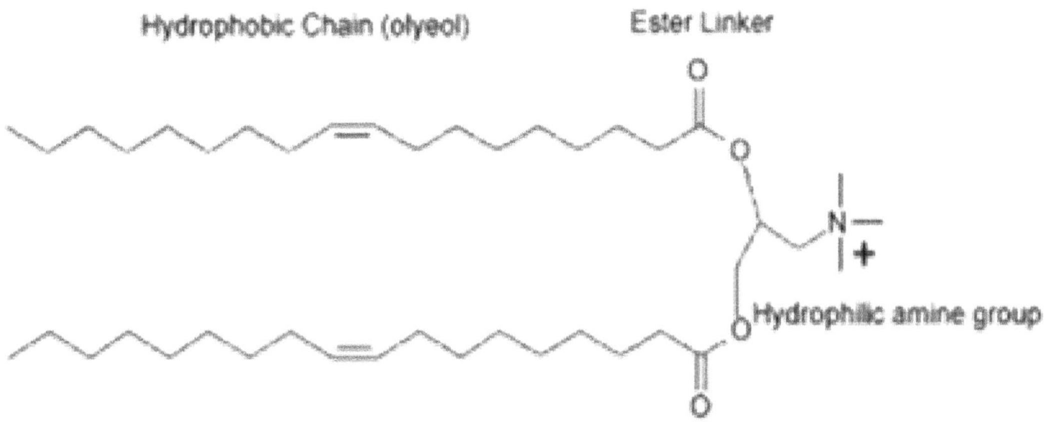

Figure 1. Schematic representation of the cationic lipid DOTAP. The hydrophobic part (olyel), the ester linker and the hydrophilic portion are presented. Most other cationic lipids share the same general structure (Aneed, 2004).

the negatively charged DNA through electrostatic interactions. The reaction complex also bears positive net values, which enable the carrier to react efficiently with negatively charged cell membranes. Nonviral gene transfer vectors have been actively studied in the past years in order to obtain structural entities with minimum size and defined shape. The final size of a gene transfer vector, which is compacted into unimolecular complexes, is directly proportional to the mass of the nucleic acid to be compacted. Expression of compacted ssDNA was observed in hepatoma cell lines. Firstly, galactosylated ssDNA complexes were successfully delivered into cells and then expression of the asialoglycoprotein receptor via receptor-mediated endocytosis. The reduced size and biophysical behaviour of ssDNA vectors may provide an advantage for transfection of eukaryotic cells (Behr, 1994).

Cationic lipids

Since their introduction as gene carriers in 1987 (Felgner et al., 1987). Liposomes have become one of the most studied non-viral vectors. They include a group of positively charged lipids at the physiological pH. As other nonviral vectors, they interact with the negatively charged DNA through electrostatic attractions. Cationic lipids were used mainly in the form of liposomes. More recently, however, cationic lipid emulsions have been described and evaluated as possible non-viral gene carriers (Yi et al., 2000; Choi et al., 2002). In a variety of cationic lipids synthesized, DC-Chol, *N*-[1-(2,3-dimyristyloxy)propyl]-*N, N*-dimethyl-*N*-(2-hydroxyethyl) ammonium bromide and N,N,N-trimethyl-2-bis[(1-oxo-9-octadecenyl)oxy]-(Z,Z)-1propanaminium methyl sulfate have been used in human clinical trails by local injection (Hara et al., 2003).

Liposomes

Basic nature and transfection mechanism

Cationic lipids are amphiphilic molecules composed of one or two fatty acid side chains (acyl) or alkyl, a linker and a hydrophilic amino group. The example of cationic lipid 1,2-dioleoyl-3-trimethylammonium-propane (DOTAP) Figure 1, which consists of two unsaturated diacyl side chains (oleoyl), ester linker and propyl ammonium group (Lasic, 1997). Other lipids may contain different linkers such as the more stable ether linkages, e.g. N [1-(2,3-dioleyloxy) propyl]- N,N,N-trimethylammonium chloride (DOTMA)(Mayer et al. 1986). The hydrophobic part can also be cholesterol derived moieties (Gao and Hui, 2001). Cationic liposomes spontaneously interact with the negatively charged DNA to form a stable complex that promotes the gene transfer to cells. The mode of formation and the size of cationic liposomes/DNA complexes were investigated using the atomic force microscopy (AFM). Also the most important physical-chemical factors involved in cationic liposome-mediated gene transfection, e.g. size and lipidic composition, were evaluated through the transfection of complexes with different liposomes/DNA molar ratio into three types of cultured cells. Cationic liposomes, composed of a neutral lipid (phosphatidilcoline), a cationic lipid dimethyldioctadecylammonium bromide (DDAB), a co-lipid-1,2-dioleoyl-sn-glycero-3-phosphatidylethanolamine (DOPE) and a phospholipid derivative of polyethylene glycol (DSPE-mPEG) at different molar ratio, were mixed with a plasmid pCMVbeta to form liposomes/DNA complexes.

In 2003, Ruozi et al. have demonstrated that the complexes were made by complicated structures in which the liposomes tend to aggregate and the DNA is surrounded by lipidic material. *In vitro* transfection efficiency by liposomes/plasmid pCMVbeta complexes was found to depend on the kind of lipid associated in the liposomes and the liposomes/DNA mixing ratio. The importance of associating DOPE in cationic liposomes was confirmed; this co-lipid is able to improve the ability of cationic liposomes to transfect cells but in addition, the AFM images and the EtBr fluorescence experiments have suggested that this lipid can also play an important role to facilitate the formation of stable liposomes, which efficaciously protect the DNA by nuclease digestion (Ruozi et al., 2003).

In aqueous media, cationic lipids are assembled into a bilayer vesicular-like structure (liposomes). Liposomes are first arranged into multilamellar vesicles. Unilamellar vesicles can then be obtained by sonication (Huang, 1969), detergent removal (Jiskoot et al., 1986) or extrusion through porous membranes (Hope et al., 1985; Mayer et al., 1986). Liposomes/DNA complex is usually termed a lipoplex. Based on freeze-fracture electron micrographs and x-ray diffraction studies, it was suggested that DNA is sandwiched between many liposomal particles (Radler et al., 1997; Sternberg et al., 1994). This structure is in agreement with the increase of particle size after the addition of DNA to cationic liposomes (Almofti et al., 2003). Negatively charged DNA will neutralize cationic liposomes resulting in aggregation and continuous fusion with time while DNA being entrapped during this process. Because of poor stability (that is, continuous aggregation), lipoplexes are usually administered directly after their formation. Many physical factors influence stability, complex formation and transfection efficiency of lipoplexes such as particle size, zeta potential, DNA/liposomes ratio and ionic strength of the medium (Almofti et al., 2003). Producing the favorable, stable, small lipoplex particles was obtained with the development of the novel liposomal formulation liposomes/protamine/DNA (LPD), are discussed as follows.

Liposome mediated gene therapy in kidney

Among non-viral vectors, the liposome system is a promising procedure for kidney-targeted gene therapy. Using cationic liposome, tubular cells were effectively transduced by retrograde injection of liposome/cDNA complex (U). Although transgene expression was reportedly modest using cationic liposomes, this method improved renal disease models such as carbonic anhydrase II deficiency and unilateral ureteral obstruction (U). In contrast, HVJ-liposome system is an effective transfection method to glomerular cells using intra-renal arterial infusion and improved glomerular disease models such as glomerulonephritis and glomerulosclerosis. In addition, intra-renal pelvic injection of DNA by HVJ-liposome system showed transgene expression in interstitial fibroblasts. In kidney-targeted gene therapy, liposome-mediated gene transfer is an attractive method because of its simplicity and reduced toxicity. In spite of modest transgene expression, several renal disease models were successfully modulated by liposome system (U). Although one limitation of liposome-mediated gene delivery is the duration of transgene expression, the liposome/cDNA complex can be repeatedly administered due to the absence of an immune response (Ito et al., 2003).

Transfer of single gene vs multiple genes:

Liposomal gene transfer is an effective therapeutic approach for the treatment of several pathophysiologic states. In 2004, Jeschke and Klein showed that exogenous gene transfer of multiple cDNA sequences have an additive effect on intracellular and biological responses when compared to the same gene administered as a single cDNA sequence. Our findings demonstrate that gene therapy with multiple genes is feasible, and that the gene transfer of multiple genes can enhance and accelerate physiologic and biological effects (Jeschke and Klein, 2004).

Transfer condition with extracellular factors

Liposome-mediated adenomatous polyposis coli (APC) to the duodenum is feasible even in the presence of bile and varying pH, raising the potential of future gene therapy for this extremely difficult to treat condition (Lee et al., 2004). Cationic polymers and liposomes are used to warp DNA into complexes that promote its cellular uptake. Polyanionic glycosaminoglycans (GAGs) on the cell surface interact with the cationic DNA complexes and influence transfection. The uptake of DNA complexes varies depending on carrier, cell type and amounts of cell surface heparan sulfate (HS), chondroitin sulfate (CS) and hyaluronan (HA) where as GAGs inhibit the transgene expression. This implies that cell-surface GAGs probably direct complexes into intracellular compartments that do not support transcription (Ruponen et al., 2004).

Liposome complexes

Reconstituted chylomicron remnants (RCR): Lipoproteins are naturally occurring biological emulsions and serve as carriers of cholesterol and other lipids in systemic circulation. Dietary lipids absorbed by the intestine are packed into triglyceride-rich lipoproteins termed chylomicrons. In the blood circulation, chylomicrons are remolded to chylomicron remnants by hydrolysis of the core triglycerides by lipoprotein lipase and by accepting apolipoproteins. Finally, the remnants are taken up by liver parenchymal cells via apolipoprotein-specific receptors (Hong et al., 2004).

Recently reconstituted chylomicron remnants (RCR) were established using commercially lipids and shown to be taken up by hepatocytes following intravenous injection. If the therapeutic DNA can be incorporated into the RCR, the resulting nonviral particles could be an efficient vector for hepatic gene therapy. The incorporation of DNA into RCR by means of the formation of a hydrophobic DNA complex with a quaternary ammonium derivative of cholesterol is already reported. Intraportal injection of the resulting particle leads to highly efficient expression of foreign genes into the liver of mice. The therapeutic potential of the novel vector is demonstrated using human 1-antitrypsin (hAAT) gene and its production in the liver (Hara et al., 2003).

Tat/liposome/DNA (TLD)

HIV-1 Tat protein transduction domain (PTD) peptide facilitates gene transfer in combination with cationic liposomes. Incorporation without covalent linkage of a 17-amino acid PTD peptide into gene delivery lipoplexes improves gene transfer. Tat/liposome/DNA (TLD) enable transfection of highly recalcitrant primary cells in the form of air/liquid interface cultures of sheep tracheal epithetlium. Treatment with chloroquine increased, and incubation at low temperature decreased, TLD transfection (Hyndman et al., 2004).

DOTAP, DOPE and cholesterol (DDC): A new ternary cationic liposome formulated with dioleoyl trimethylammonium propane (DOTAP), 1,2-dioleoyl-3-phosphatidylethanolamine (DOPE) and cholesterol (Chol) (DDC), It has been recently found to have a selective high gene transfer ability in ovarian cancer cells. The factors controlling cationic lipid/DNA (CL-DNA) complexes-mediated gene transfer depend not only on the formulations of the cationic liposomes and their thermodynamic phase, but also significantly on the cell properties (Caracciolo et al., 2003). In an experiment, wild-type p53 DNA was transfected into the ovarian cancer cells, using the DDC as a nonviral vector and the expression and activity of p53 gene were evaluated both *in vitro* and *in vivo*. DDC liposomes were prepared by mixing DOTAP: DOPE: chol in a 1:0.7:0.3 molar ratios using the extrusion method. Plasma DNA (pp53-EGFP) and DDC complexes were transfected into ovarian carcinoma cells (OVCAR-3 cells) and gene expression was determined by reverse transcription-polymerase chain reaction and western blot analysis. The transcription of liposome complexed p53 gene resulted in a high level of wild-type p53 mRNA and protein expression OVCAR-3 cells. The DDC-mediated p53 DNA delivery may have the potential for clinical application as nonviral vector-mediated ovarian cancer therapy due to its effective induction of apoptosis and tumor growth inhibition (Kim et al., 2003).

Anionic liposomal vectors

Anionic liposomal vectors based on the composition of retroviral envelops (artificial viral envelops, AVEs) can be used for transgene expression in many cell lines. AVE liposomes, dispersed in 10% serum containing growth medium, efficiently delivered plasmid DNA to HuH-7 (Human hepatoma cell line) cells. Small unilamellar AVE vesicles containing 15 mol % digalactosyl diglyceride (DGDG) are efficiently targeted to the liver via the hepatic asialoglycoprotein receptor (Mady et al., 2004).

Structure and activity relationship

Despite the numerous studies that focus on the possible relationship between cationic lipid structure/composition and lipoplex transfection activity, solid conclusions are rarely obtained. This is mainly due to the complex factors affecting gene transfer. The levels of transfection among different cell lines, for example, may significantly vary with the same lipoplex formulations (Zou et al., 2004; Balasubramaniam et al., 1996). More important, the correlation between *in vitro* and *in vivo* experiments is not always obtained (Gorman et al., 1997; Solodin et al., 1995). Therefore, it is expected that empirical findings will remain the main source for structure/activity relationships. It has, however, been proven that the addition of neutral lipids (colipids) will increase the transfection ability of lipoplexes both *in vitro* and *In vivo*. The most used colipids are cholesterol (Liu et al., 1997; Bennett et al., 1995) and dioleylphos-phatidylethanolamine (DOPE) (Hui et al., 1996; Farhood et al., 1995). While the former is more efficient *in vivo*, the latterenhances *in vitro* lipoplex transfection (Simberg et al., 2001; Hong et al., 2004). It was demonstrated that lipoplexes enter cells through endocytosis and fusion pathways (Farhood et al., 1995). Neutral lipids facilitate conformational changes from a bilayer structure into hexagonal arrangement at the endosomal level (Hafez et al., 2001). This change will trigger the release of the encapsulated DNA into the cytoplasm and not reaching the destructive lysosomal environment.

Liposomes-protamine-DNA (LPD)

Protamine is arginine-rich peptide, which can condense negatively charged DNA before being complexed with cationic lipids. The polycation PLL (which is less efficient) was also tested for LPD preparation in early studies (Gao and Huang, 1996).

Liposomes will interact with condensed DNA resulting in lipid rearrangement and the formation of compact liposomes/ DNA complex (LPD) (Li and Huang, 1999). In comparison of proposed complex structures between LPD and conventional lipoplexes (Figure 2). DOTAP and

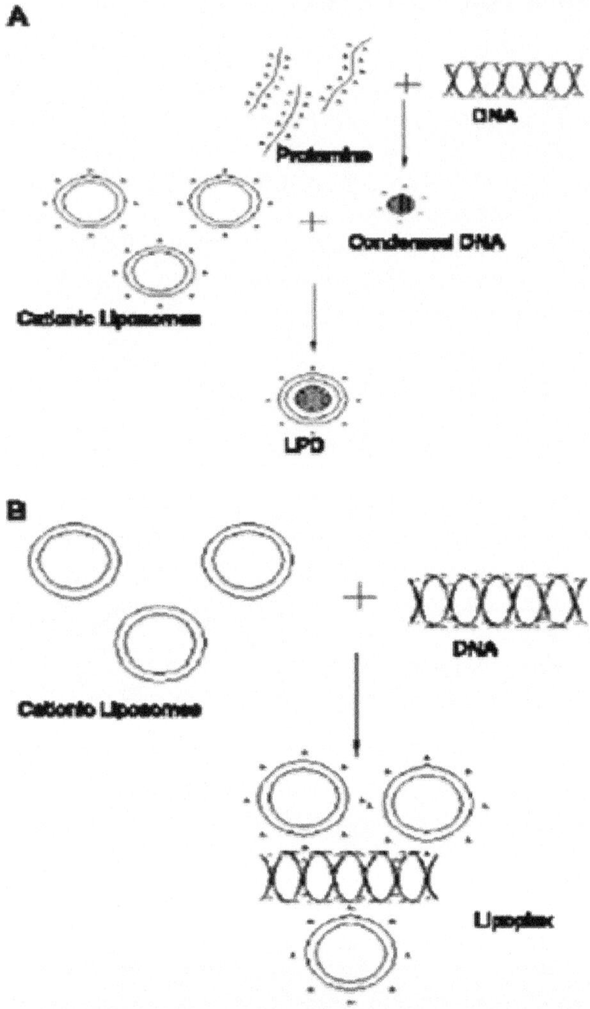

Figure 2. Schematic comparison between LPD and lipoplex particles. In (A): DNA is neutralized and condensed with the protamine before the formation of the complex LPD. In (B): DNA is sandwiched between liposomal particles forming the 'conventional' lipoplex. (Aneed, 2004).

3h-[N-(NV,NV-dimethylaminoethane)-carbamoyl]cholesterol (DC-Chol) were used for the preparation of LPD with the most common co-lipids: cholesterol and DOPE, respectively (Tan et al., 2002). Particle size distribution of LPD ranged from 100 to 250 nm, which is almost three to five times less than conventional lipoplexes (Gao and Huang, 1996; Ueno et al., 2002; Sorgi et al., 2001).

Stability with no compromise in the transfection ability was maintained for 4 months when LPD was stored at 4℃ or at room temperature after lyophylization (freeze-drying or spray-drying) (Seville et al., 2002; Li et al., 2000). Both *in vitro* and *in vivo* studies showed superiority of LPD-mediated gene transfer over conventional liposomes (Sorgi et al., 2001; Li et al., 1998). It is believed that the small size of LPD will facilitate endocytosis and increase the *in vivo* circulating half life. Like other non-modified liposomes, LPD tends to accumulate preferably in the lungs after injection from the tail vein of a mouse (Li and Huang, 1999; Li et al. 1998). In addition to the lungs, reporter gene expression, however, was also detected in other organs such as the kidney, the spleen and the liver (Li et al., 1998; Li and Huang, 1997). For therapeutic applications, that is, administration of DOTAP: Chol LPD carrying tumor suppressor genes Rb or E1A in cancer animal models resulted in apoptosis induction, tumor size reduction and life span increase in the treated animals (Ueno et al. 2002; Nikitin et al. 1999). As expected, antitumoral synergistic effects were obtained when E1A LPD treatment was combined with the chemotherapeutic agent paclitaxel (Lee et al. 1996).

For targeting purposes, LPD was recently coated with the liver targeting ligand asialofetuin through charge–charge interactions, which significantly increases HepG2, liver cancer cells, uptake of the encapsulated DNA (Arangoa et al., 2003). LPD composed of DC-Chol/DOPE was also tested in clinical settings when two children with Canavan disease (a fatal CNS disease characterized by spongy degradation of cerebral white matter) were treated with ASPA gene via intracerebral application; both subjects showed some clinical improvements (Leone et al., 2000).

LID vector complex

An integrin-targeted synthetic vector system was developed for the transfection of haematopoietic cell lines and dendritic cells. The vector consists of a cationic liposome, Lipofectin (L), a peptide that both targets integrins and binds to DNA (I) and plasmid DNA (D). These components interact electrostatically to form the LID vector complex. Targeted gene delivery was demonstrated by comparing transfected luciferase reporter gene levels using LID complexes containing integrin-targeting peptide sequences with a control peptide. Under optimized conditions transfection efficiencies of 19% for TF-1 cells, 28% for Jurkat cells and 10% for primary dendritic cells were achieved. The LID vector may thus find application for gene-transfer experiments in haematopoietic cell lines and for the development of genetic vaccines using transfected dendritic cells (Uduehi et al., 2003).

Solid lipid nanoparticles (SLN)

Cationic solid lipid nanoparticles (SLN) for gene transfer are formulated using the same cationic lipids as for liposomal transfection agents. SLN is smaller in diameter than the corresponding liposomes (88 vs.148 nm) and atomic force microscopy (AFM) supported some structural differences. Cationic lipid composition seems to be more dominant for *in vitro* transfection performance than the kind of colloidal structure it is arranged in. Hence, cationic SLN extend the range of highly potent non-viral transfection agents by one with favourable and distinct technological properties (Tabatt et al., 2004).

Peptide mediated gene transfer: Ligand-targeted plasmid based gene delivery systems was developed for gene transfer to tumor endothelium. The single and double disulfide lipopeptides were then tested for gene transfer to HUVECs (human umbilical derived endothelial cell) using DOTMA: Cholesterol cationic liposomes. The polyplexes were formed by rapidly mixing plasmid DNA with DOTMA:CHOL liposomes at a 3:1 charge ratio in 2% ethanol, 10% lactose. The ethanol was removed by lyophilization and upon rehydration; the lipoplexes had a mean diameter of approximately 100nm. HUVEC transfection studies showed that increasing the mol% of the single disulfide RGD lipopeptide to 20 mol% increased gene transfer by 10 fold (Anwer et al., 2004).

Barriers of using cations in non viral gene therapy

Extracellular and intracellular barriers: Complexes of DNA with cationic lipids and cationic polymers are frequently used for gene transfer. Extracellular interactions of the complexes with anionic glycosaminoglycans (GAGs) may interfere with gene transfer. Interactions of GAGs with carrier DNA complexes have been studied using tests for DNA relaxation (ethidium bromide intercalation), DNA release (electrophoresis), and transfection (pCMVbGal transfer into RAA smooth muscle cells). Several cationic lipid formulations (DOTAP, DOTAP/Chol, DOTAP/DOPE, DOTMA/DOPE, DOGS) and cationic polymers (fractured dendrimer, polyethylene imines 25 and 800 kDa, polylysines 20 and 200 kDa) were tested. Polycations condensed DNA more effectively than monovalent lipids. Hyaluronic acid did not release or relax DNA in any complex, but it inhibited transfection by some polyvalent systems (PEI, dendrimers, DOGS). Gene transfer by other carriers was not affected by hyaluronic acid. Sulfated GAGs (heparan sulfate, chondroitin sulfates B and C) completely blocked transfection, except in the case of liposomes with DOPE. Sulfated GAGs relaxed and released DNA from some complexes, but these events were not prerequisites for the inhibition of transfection. Furthermore, preliminary results suggest that cell surface GAGs, particularly heap-ran sulfate, inhibit gene transfer by cationic lipids and polymers (Ruponen, et al. 2003).

The additional hydrogen bonding or covalent interactions of the head group with the plasmid DNA, leading to higher binding affinity of the cationic lipids to pDNA, results in higher transfection. This hypothesis is supported by TEM observations where elongated complexes were observed and more lipids were seen associated with the DNA (Narang et al., 2005).

Effect of human plasma proteins

In situ gene expression assay indicated that both the ratio of DNA and liposome and the dose of DNA could affect the gene transfection efficiency. Naked endostatin plasmid intratumoral injection can get similar gene transfection efficiency to liposome-DNA complex when used in situ (Ma, et al. 2004).

Interaction of cationic lipid/DNA complex with the plasma is a limiting step for the cationic lipid-mediated intravenous gene transfer and expression process. Most of the plasma components that interact with the complex and inhibit its transfection efficiency are still unknown. In an experiment, human plasma proteins and lipoproteins that bind to a cationic lipid/DNA complex were isolated on a sucrose density gradient and identified by 2-D gel electrophoresis. Protein binding did not result in complex dissociation or DNA degradation. The effects of several complex-binding plasma components on the transfection efficiency were studied using lung endothelial cells cultured *in vitro*. Lipoprotein particles caused a drastic loss of the transfection efficiency of the complex. Surprisingly, fibrinogen was found to activate the transfection process. The roles of these complex-binding plasma components on the complex uptake efficiency were quantitatively assessed using radiolabeled plasmid DNA and qualitatively evaluated using fluorescence microscopy. A good correlation was found between the effects of the complex-binding plasma components on the transfection and on cell uptake efficiencies. In contrast to what was generally believed, our data suggest that disruption of the complex does not occur when it is in contact with the plasma and therefore could not be responsible for the loss of transfection activity. Instead, coating of complexes with plasma components seems to be responsible for reduced uptake by cells, which in turn results in reduced transfection (Tandia et al., 2003).

Improvement of transfection efficiency

Efficiency increasing by surface coating: Systemic gene delivery systems are needed for therapeutic applications. For systemic circulation, masking the surface charge of DNA complexes has to be accomp-lished to avoid interactions with plasma components, erythrocytes and the reticuloendothelial system. Polyplexes based on polyethylenimine (PEI), shielded with polyethylene glycol (PGE) and linked to the receptor binding ligands transferrin (Tf) or epidermal growth factors (EGF) have been developed. Complexes were found to mediate efficient gene transfer into tumor cell lines in a receptor dependent and cell-cycle dependent manner. Systemic administration of surface shielded Tf-PEI polyplexes into the tail vein of mice resulted in preferential gene into distantly growing subcutaneous tumors. In contrast, application of positively charged PEI polyplexes directed gene transfer primarily to the lung (Ogris and Wagner, 2002).

Delivery efficiency and gene expression efficiency increasing action

Tumor targeting is an important issue in cancer gene therapy. A gene transfection method, based on light inducible photochemical internalization (PCI) of a transgene, is already developed to improve gene delivery and expression selectively in illuminated areas, for example in tumors. PCI improve the non-viral vector polyethylenimine (PEI)-mediated transfection of therapeutic gene, the "suicide" gene encoding herpes simplex virus thymidine kinase (HSV+K). Photochemical transfection allows selective enhancement in gene expression and gene mediated biological effects (cell killing by the HSV+K/GCV approach) in response to illumination (Prasmickaite et al., 2004).

Systemic gene therapy vectors must be designed to safety and efficiently escort DNA from outside the cell to the nucleus and to overcome several physiological barriers that are obstacles to internalization, escape from endocytic vesicles, movement through the cytoplasm and transport into the nucleus. Chloroquine appears to facilitate PLL-mediated gene delivery by a mechanism other than buffering of endocytic compartments. Additionally, PEI does not appear to buffer endocytic compartments but requires exposure to an acidic environment for efficient gene delivery (Forrest and Pack, 2002).

Conclusion

Gene therapy is one of the youngest and most promising fields of medicine. In the course of one decade the basic research in recombinant DNA has been translated into a number of applied projects aiming at curing the diseases at its root, the gene. Though no clinically acceptable protocols have yet been developed, the researcher's ability to solve many technical problems of gene therapy has been greatly improved. Several viral and nonviral gene therapy vectors have been evaluated for a range of conditions in animal models and in clinic. Both physical and immunological barriers exist, and these must be tackled. Cationic lipids showed a great potential among the vector implied for gene therapy, and it also proved that they are safer than viral vectors. Many types of cationic lipids are tested but it is still a vital field of research to invent new one.

From this review, it has emerged that a single vector is highly unlikely to be optimal for all gene therapy applications. To improve gene transfer efficiency, the appropriate vector must be selected for transfection of the required cell type. Ultimately all disease targets will benefit from an improved understanding of the biology of vector delivery, uptake and expression. Improvements in vector design will also reap rewards in the clinic. Research into vector design and targeting needs to proceed in parallel with early clinical studies, where proof of concept and identification of barriers to clinical gene transfer can inform the next step to basic research.

REFERENCES

Almofti MR, Harashima H, Shinohara Y, Almofti A, Baba Y, Kiwada H (2003). Cationic liposome-mediated gene delivery: biophysical study and mechanism of internalization, Arch. Biochem. Biophys. 410: 246–253.

Aneed EA (2004). An overview of current delivery systems in cancer gene therapy. J. Controlled Release 94(1): 1-14.

Anwer K, Kao G, Rolland A, Driessen WH, Sullivan SM (2004). Peptide-mediated gene transfer of cationic lipid/plasmid DNA complexes to endothelial cells. J. Drug Target. 12(4): 215-221.

Arangoa MA, Duzgunes N, Tros de Ilarduya C, (2003). Increased receptor-mediated gene delivery to the liver by protamine-enhanced-asialofetuin lipoplexes, Gene Ther. 10: 5–14.

Balasubramaniam RP, Bennett MJ, Aberle AM, Malone JG, Nantz MH RW (1996) Malone, Structural and functional analysis of cationic transfection lipids: the hydrophobic domain. Gene Ther. 3: 163–172.

Behr JP (1994). Gene transfer with synthetic cationic amphiphiles: prospects for gene therapy, Bioconjug. Chem. 5: 382–389.

Bennett MJ, Nantz MH, Balasubramaniam RP, Gruenert DC, Malone RW (1995). Cholesterol enhances cationic liposome-mediated DNA transfection of human respiratory epithelial cells, Biosci. Rep. 15: 47–53.

Caracciolo G, Pozzi D, Caminiti R, Congiu Castellano A (2003). Structural characterization of a new lipid/DNA complex showing a selective transfection efficiency in ovarian cancer cells. Eur. Phys. J. E Soft Matter 10(4): 331-336.

Choi BY, Chung JW, Park JH, Kim KH, Kim YI, Koh YH, Kwon JW, Lee KH, Choi HJ, Kim TW, Kim YJ, Chung H, Kwon IC, Jeong SY, (2002). Gene delivery to the rat liver using cationic lipid emulsion/DNA complex: comparison between intra-arterial, intraportal and intravenous administration, Kor. J. Radiol. 3: 194–198.

Crooke ST (1998). Basic principles of antisense therapeutics. In: Crooke ST, eds. Antisense research and application. Handbook of experimental pharmacology, Berlin: Springer , 131: 1–50.

Eguchi Y (1991). Antisense RNA. Annu. Rev. Biochem. V. 60: 631–652.

Farhood H, Serbina N, Huang L (1995). The role of dioleoyl phosphatidylethanolamine in cationic liposome mediated gene transfer, Biochim. Biophys. Acta 1235: 289–295.

Felgner PL, Gadek TR, Holm M, Roman R, Chan HW, Wenz M, orthrop JP, Ringold GM, Danielsen M, (1987) Lipofection: a highly efficient, lipid-mediated DNA-transfection procedure, Proc. Natl. Acad. Sci. U.S.A. 84: 7413–7417.

Forrest ML, Pack DW (2002). On the kinetics of polyplex endocytic trafficking: implications for gene delivery vector design. Mol. Ther. 6(1): 57-66.

Gao, H, Hui KM (2001). hesis of a novel series of cationic lipids that can act as efficient gene delivery vehicles through systematic heterocyclic substitution of cholesterol derivatives, Gene Ther. 8: 855–863.

Gao X, Huang L (1996) Potentiation of cationic liposome-mediated gene delivery by polycations, Biochemistry 35: 1027–1036.

Gorman CM, Aikawa M, Fox B, Fox E, Lapuz C, Michaud B, Nguyen H, Roche E, Sawa T, Wiener-Kronish JP, (1997). Efficient in in vivo delivery of DNA to pulmonary cells using the novel lipid EDMPC, Gene Ther. 4: 983–992.

Hafez IM, Maurer N, Cullis PR, (2001) On the mechanism whereby cationic lipids promote intracellular delivery of polynucleic acids,

Gene Ther. 8: 1188–1196.
Haseloff J, Gerlach WL (1988). Simple RNA enzymes with new and highly specific endoribonuclease activities. Nature 334: 585–591.
Herskowitz I (1987). Functional inactivation of genes by dominant negative mutations. Nature. 329: 219–22.
Hong K, Zheng W, Baker A, Papahadjopoulos D, (1997). Stabilization of cationic liposome-plasmid DNA complexes by polyamines and poly(ethylene glycol)-phospholipid conjugates for efficient in vivo gene delivery, FEBS Lett. 400: 233–237.
Hope MJ, Bally MB, Webb G, Cullis PR, (1985) Production of large unilamellar vesicles by a rapid extrusion procedure. Characterization of size distribution, trapped violume and ability to maintain a membrane potential, Biochim. Biophys. Acta 812: 55–65.
Huang CH (1969). Studies on phosphatidylcholine vesicles. Formation and physical characterization, Biochemistry 8: 344–351.
Hui SW, Langner M, Zhao YL, Ross P, Hurley E, Chan K (1996). The role of helper lipids in cationic liposome-mediated gene transfer, Biophys. J. 71: 590–599.
Hyndman L, Lemoine JL, Huang L, Porteous DJ, Boyd AC, Nan X. (2004) HIV-1 Tat protein transduction domain peptide facilitates gene transfer in combination with cationic liposomes. J. Controlled Release 99(3): 435-444.
Ito K, Chen J, Asano T, Vaughan ED Jr, Poppas DP, Hayakawa M, Felsen D (2004). Liposome-mediated gene therapy in the kidney. Hum Cell. 17(1): 17-28.
Jeschke MG, Klein D (2004). Liposomal gene transfer of multiple genes is more effective than gene transfer of a single gene. Gene Ther. 11(10): 847-855.
Jiskoot W, Teerlink T, Beuvery EC, Crommelin DJ (1986). Preparation of liposomes via detergent removal from mixed micelles by dilution. The effect of bilayer composition and process parameters on liposome characteristics, Pharm. Week bull. Sci. Ed. 8: 259–265.
Kim CK, Choi EJ, Choi SH, Park JS, Haider KH, Ahn WS (2003). Enhanced p53 gene transfer to human ovarian cancer cells using the cationic nonviral vector, DDC. Gynecol. Oncol. 90(2): 265-272.
Kim JS, Kim BI, Maruyama A, Akaike T, Kim SW. (1998) A new non-viral DNA delivery vector: the terplex system. J Controlled. Release 53(1-3): 175-182.
Lasic DD, (1997) Liposomes in Gene Delivery, CRC Press, Boca Raton F pp. 231-238.
Lee J, Hargest R, Wasan H, Phillips RK (2004). In vitro model for liposome-mediated adenomatous polyposis coli gene transfer in a duodenal model. Dis. Colon Rectum. 47(2): 219-26.
Lee M, Nah JW, Kwon Y, Koh JJ, Ko KS, Kim SW (2001). Water-soluble and low molecular weight chitosan-based plasmid DNA delivery. Pharm. Res. 18(4): 427-31.
Lee ER, Marshall J, Siegel CS, Jiang C, Yew NS, Nichols MR, Nietupski, RJ Ziegler, MB Lane, KX Wang, NC Wan, RK Scheule, DJ Harris JB, Smith AE (1996). Cheng, Detailed analysis of structures and formulations of cationic lipids for efficient gene transfer to the lung, Hum. Gene Ther. 7: 1701–1717.
Lee H, Jeong JH, Park TG, (2002). PEG grafted polylysine with fusogenic peptide for gene delivery: high transfection efficiency with low cytotoxicity, J. Controlled Release 79: 283– 291.
Leone P, Janson CG, Bilaniuk L, Wang Z, Sorgi F, L Huang, R Matalon, R Kaul, Z Zeng, A Freese, S.W. McPhee, E Mee, M.J. During, L. Bilianuk, (2000) Aspartoacylase gene transfer to the mammalian central nervous system with therapeutic implications for Canavan disease, Ann. Neurol. 48: 27–38.
Li Z, Huang L (2004). Sustained delivery and expression of plasmid DNA based on biodegradable polyester, poly(D,L-lactide-co-4-hydroxy-L-proline). J Controlled Release. 98(3): 437-446.
Li, B, Li S, Tan Y, Stolz DB, Watkins SC, Block LH, Huang L, (2000) Lyophilization of cationic lipid – protamine –DNA (LPD) complexes, J. Pharm. Sci. 89: 355–364.
Li, S, Huang L, (1997). *In vivo* gene transfer via intravenous administration of cationic lipid – protamine –DNA (LPD) complexes, Gene Ther. 4: 891–900.
Li, S., Huang L (1999). Novel lipidic vectors for gene transfer, in: L. Huang MC, Hung, Wagner E (1999). (Eds.), Nonviral Vectors for Gene Therapy, Academic Press, San Diego pp. 289– 309.
Li, S, Rizzo MA, Bhattacharya S, Huang L (1998). Characterization of cationic lipid– protamin–DNA (LPD) complexes for intravenous gene delivery, Gene Ther. 5: 930– 937.
Liu, Y, Mounkes LC, Liggitt HD, Brown CS, Solodin I, Debs R, Heath TD (1997). Factors influencing the efficiency of cationic liposome-mediated intravenous gene delivery, Nat. Biotechnol. 15: 167–173.
Ma CH, Zhang Y, Wang XY, Gao LF, Liu H, Guo C, Liu SX, Cao YL, Zhang LN, Sun WS (2004). Human endostatin gene transfer, either naked or with liposome, has the same inhibitory effect on growth of mouse liver tumor cells *in vivo*. World J. Gastroenterol. 10(19): 2874-2877.
Mady MM, Ghannam MM, Khalil WA, Repp R, Markus M, Rascher W, Muller R, Fahr A (2004) Efficient gene delivery with serum into human cancer cells using targeted anionic liposomes. J Drug Target. 12(1): 11-18.
Mayer LD, Hope MJ, Cullis PR (1986). Vesicles of variable sizes produced by a rapid extrusion procedure, Biochim. Biophys. Acta 858: 161–168.
Miller AD (1992) Human gene therapy comes of age. Nature 357: 455–60.
Mulligan RC (1993). The basic science of gene therapy. Science 260: 926–32.
Narang AS, Thoma L, Miller DD, Mahato RI (2005). Cationic lipids with increased DNA binding affinity for nonviral gene transfer in dividing and nondividing cells. Bioconjug Chem. 16(1): 156-68.
Nikitin AY, Juarez-Perez MI, Li S, Huang L, Lee WH, (1999). RB-mediated suppression of spontaneous multiple neuroendocrine neoplasia and lung metastases in Rb+/_ mice, Proc. Natl. Acad. Sci. USA 96: 3916– 3921.
Ogris M, Wagner E (2002). Tumor-targeted gene transfer with DNA polyplexes. Somat Cell Mol Genet. Nov;27(1-6): pp 85-95.
Prasmickaite L, Hogset A, Olsen VM, Kaalhus O, Mikalsen SO, Berg K (2004). Photochemically enhanced gene transfection increases the cytotoxicity of the herpes simplex virus thymidine kinase gene combined with ganciclovir. Cancer Gene Ther. 11(7): 514-23.
Prince HM (1998) Gene transfer: A review of methods and applications. Pathology, 30: 335–347.
Radler JO, Koltover I, Salditt T, Safinya CR (1997). Structure of DNA-cationic liposome complexes: DNA intercalation in multilamellar membranes in distinct interhelical packing regimes, Sci. 275: 810–814.
Ruozi B, Forni F, Battini R, Vandelli MA (2003). Cationic liposomes for gene transfection. J Drug Target. 11(7): 407-414.
Ruponen M, Honkakoski P, Ronkko S, Pelkonen J, Tammi M, Urtti A (2003) Extracellular and intracellular barriers in non-viral gene delivery. J Control Release. 93(2): 213-217.
Ruponen M, Honkakoski P, Tammi M, Urtti A (2004). Cell-surface glycosaminoglycans inhibit cation-mediated gene transfer. J Gene Med. 6(4): 405-414.
Schmitz V, Qian C, Ruiz J, Sangro B, Melero I, Mazzolini G, Narvaiza I, Prieto J (2004). Gene therapy for liver diseases: recent strategies for treatment of viral hepatitis and liver malignancies. J Control Release. 99(2): 313.
Seville PC, Kellaway IW, Birchall JC (2002). Preparation of dry powder dispersions for non-viral gene delivery by freezedrying and spray-drying, J. Gene Med. 4: 428– 437.
Simberg D, Danino D, Talmon Y, Minsky A, Ferrari ME, Wheeler CJ, Barenholz Y (2001). Phase behavior, DNA ordering, and size instability of cationic lipoplexes. Relevance to optimal transfection activity, J. Biol. Chem. 276: 47453– 47459.
Solodin I, Brown CS, Bruno MS, Chow CY, Jang EH, Debs RJ, Heath TD (1995). A novel series of amphiphilic imidazolinium compounds for *in vitro* and *in vivo* gene delivery, Biochemistry 34: 13537– 13544.
Sorgi FL, Bhattacharya S, Huang L, (1997). Protamine sulfate enhances lipid-mediated gene transfer, Gene Ther. 4: 961– 968.
Sternberg B, Sorgi FL, Huang L, (1994). New structures in complex formation between DNA and cationic liposomes visualized by freeze-fracture electron microscopy, FEBS Lett. 356: 361– 366.
Sullenger BA, Gallardo FH, Ungers GE (1991). Analysis of trans-acting response decoy RNA-mediated inhibition of human immunodeficiency virus type 1 transactivation. J. Virol. 65: 6811–6816.

Symons RH (1994). Ribozymes. Curr Opin Struct Biol. 4: 322–30.
Tabatt K, Kneuer C, Sameti M, Olbrich C, Muller RH, Lehr CM, Bakowsky U (2004). Transfection with different colloidal systems: comparison of solid lipid nanoparticles and liposomes. J. Controlled Release. 97(2): 321-332.
Tan Y, Whitmore M, Li S, Frederik P, Huang L (2002). LPD nanoparticles– novel nonviral vector for efficient gene delivery, Methods Mol. Med. 69: 73– 81.
Tandia BM, Vandenbranden M, Wattiez R, Lakhdar Z, Ruysschaert JM, Elouahabi A. (2003). Identification of human plasma proteins that bind to cationic lipid/DNA complex and analysis of their effects on transfection efficiency: implications for intravenous gene transfer. Mol. Ther. 8(2): 264-273.
Hara T, Yadi Tan (2003). Leaf Huang. *In vivo* gene delivery to the liver using reconstituted chylomicron remnants as a novel nonviral vector. Proc. Natl. Acad. Sci. USA 94: 14547-14552.
Uddin SN, Islam KK (2006). Cationic Polymer and its Uses in Non-Viral Gene Delivery Systems: A Conceptual Research, Trends in Medical Research, 1(2): 86-99Uduehi A, Mailhos C, Truman H,Thrasher AJ, Kinnon C, Hart SL. .(2003) Enhancement of integrin-mediated transfection of haematopoietic cells with a synthetic vector system. Biotechnol. Appl. Biochem. 38(Pt 3): 201-209.
Ueno NTC, Bartholomeusz W, Xia P, Anklesaria EM, Bruckheimer E, Mebel R, Paul S, Li GH, Yo L, Huang MC (2002). Hung, Systemic gene therapy in human xenograft tumor models by liposomal delivery of the E1A gene, Cancer Res. 62: 6712– 6716.
Yi SW, Yune TY, Kim TW, Chung H, Choi YW, Kwon IC, Lee EB, Jeong SY (2000). A cationic lipid emulsion/DNA complex as a physically stable and serum-resistant gene delivery system, Pharm. Res. 17: 314–320.
Zou Y, Zong G, Ling YH (2000). Perez-Soler, Development of cationic liposome formulations for intratracheal gene therapy of early lung cancer, Cancer Gene Ther. 7: 683–696.

7

Status of biotechnology in Eastern and Central Africa

Godliving Y. S. Mtui

Department of Molecular Biology and Biotechnology, University of Dar es Salaam, P. O. Box 35179, Dar es Salaam, Tanzania. E-mail: gmtui@udsm.ac.tz or gmtui@hotmail.com

This work examines trends of both conventional and modern biotechnologies in selected Eastern and Central African countries namely Kenya, Uganda, Tanzania, Ethiopia, Rwanda, Burundi and Democratic Republic of Congo, with the aim of giving an up-to-date assessment of their national policies, institutional capacities, and the activities being carried out. Agricultural biotechnology seems to take the lead while biotechnologies related to health, industries and environment are lagging behind. Kenya leads the region with its biotechnology policy framework in place and more on-going biotechnology related activities, followed by Uganda. Tanzania has already developed its biotechnology policy but is slower to translate it into practice especially on matters related to modern biotechnology. The rest of the countries are yet to formulate their biotechnology policies but efforts are underway to achieve that goal. Plant tissue culture is done in all the countries and some projects have already been commercialised. Transgenic crops/animals projects are mainly at the field trial stage and none has been commercialised. The main constraints facing the biotechnology industry in the region are poorly skilled human resources, lack of modern facilities, poor public perception and weak political will by some governments. More vigorous practical actions are needed in order for biotechnology to benefit the people of this region in terms of food security, economic growth, improved health and environmental protection.

Key words: Biotechnology, genetically modified organisms, molecular assisted selection, tissue culture.

INTRODUCTION

Biotechnology is a branch of applied bioscience and technology which involves the practical application of biological organisms, or their sub-cellular components in agriculture, health, manufacturing and service industries, and in environmental management (Kasonta et al., 2002). It utilizes bacteria, yeasts, fungi, algae, plant cells or cultured mammalian cells as constituents of industrial processes (Persley, 1992). Successful application of biotechnology integrates multiplicity of scientific disciplines including microbiology, biochemistry, genetics, molecular biology, chemistry and chemical and process engineering. Some commercial products of biotechnology include beverages, vinegar, cheese, yogurt, biogas, compost and organic acids (citric acid, amino acids, acetic acid). Others are solvents (alcohols, acetone), gums, plastics, detergents, perfumes, hormones, peptide and steroids. Enzymes (hydrolases, proteases, amylases, peroxidases and oxilases) and vaccines, antibiotics, high fructose corn syrup, cellulose and single cell proteins are also products of biotechnology (Mtui, 2007).

In plant biotechnology, three applications of broad fields of study are plant tissue culture, plant molecular markers and genetic engineering, ranging from simple to sophisticated technologies (Brink et al., 1998). Tissue culture is the cultivation of plant cells or tissues on specifically formulated nutrient media. Under optimal conditions, a whole plant can be regenerated from a single cell. It is a rapid and essential tool in modern breeding for producing disease-free plants (Kumar and Naidu, 2006). Molecular markers are identifiable DNA sequences found at specific location of the genome. By determining location and likely actions of plant genes, scientists can quickly and accurately identify plants carrying desirable characteristics, hence breeding can be conducted with greater precision (Mneney et al., 2001). Molecular markers can be used in plant breeding in the following ways: increase the response and accuracy of selection by using marker assisted selection; increase the speed and efficiency of the introduction of new genes through marker assisted introgression; study of genetic diversity and taxonomic relationships between plant species; and studies of biological processes such as mating systems, pollen or disease dispersal (Johanson and Ives, 2001). Biotechnology enables the development of disease diagnostic kits for use both in the laboratory and in the field. Furthermore, diseases that are caused by microbes can be identified by a unique feature of the microbe, such as its DNA of a specific protein (Kumar and Naidu, 2006). Conventional biotechnology will do much better if blended with modern biotechnology.

On the other hand, modern biotechnology is a term which refers to biotechnological techniques for the manipulation of genetic material and the fusion of cells beyond normal breeding barriers. The most obvious example is genetic engineering to create genetically modified organisms through "transgenic technology" involving the insertion or deletion of genes. In genetic engineering or genetic transformation, the genetic material is modified by artificial means. It involves isolation of a gene and cutting it at a precise location by using restriction enzymes. Selected DNA fragments can then be transferred into the cells of the target organism, and that organism becomes 'genetically modified'. The commonly used method in genetic engineering is the use of a bacterium, *Agrobacterium tumafaciens*, as a vector (Johanson and Ives, 2001). Another method is ballistic impregnation method - attaching a DNA to be introduced onto a minute gold or tungsten particle, then 'firing' it into the plant tissue (Morris, 2011).

MODERN BIOTECHNOLOGY IN THE WORLD STAGE

Worldwide, modern biotechnology is centred in agriculture, with commercialised genetically modified maize, canola, soybean and cotton being the dominant crops in the developed countries (Johanson and Ives, 2001; Mugabe, 2003; Clive, 2010). In the developing countries, genetic and biotechnological improvements of 'neglected' food species are confined to specific crop centres in those countries and/or in specific collaborations with the agricultural research institutes in the industrialized countries. The goal of modern biotechnology is to produce genetically modified organisms with the following traits:

1) Increased resistance to pests and diseases – insects, viruses, fungi, bacteria and nematodes
2) Increased tolerance to environmental stresses – drought, flooding, soil acidity and alkalinity heavy metals and extreme temperatures
3) Increased yield
4) Reduced post-harvest losses
5) Improved nutritional contents of foods and feeds (Johanson and Ives, 2001).

Agricultural biotechnology presents many opportunities and challenges for the developing world. It promises to help meet food security needs, reverse declining per capita food production and improve the incomes and livelihoods of farmers. The international community acknowledges the potential of biotechnology to change lives. Overall, biotechnology promises to make a significant contribution in enabling the development of better health care, enhanced food security through sustainable agricultural practices, improved supplies of potable water, more efficient industrial development processes for transforming raw materials, and support for sustainable methods of reforestation and detoxification of hazardous wastes. Biotechnology also offers new opportunities for global partnerships. Biotechnology has grown into a global industry affecting many aspects of life. Globally, the biotechnology industry was estimated to have generated US$34.8 billion in revenues and employed about 190,000 persons in publicly traded firms worldwide in 2001, and an estimated 4,200 public and private biotechnology firms were in operation (Konde, 2006). In 2005, GM crops farm income alone was estimated at US$5.6 billion (Brookes and Barfoot, 2006), which jumped to US$10.1 billion in 2007 and US$ 44.1 billion for the period of 12 years of commercialised genetically modified organisms (GMO) crops (Brookes and Barfoot, 2009). In addition to high income generation, biotechnology offers comparative advantage compared to non-biotechnology processes: They are cheaper (that is less energy demanding); they offer longterm benefits as means of solving some major world problems such as those related to food security, novel medical and industrial products, pollution control and the development of new energy sources. They also hold a bright future in terms of market potential for new products (Johanson and Ives, 2001; Mtui, 2007).

TRENDS OF BIOTECHNOLOGY IN AFRICA

Substantial biotechnology related research is being carried out in most African countries, but few of these countries have reached the take-off stage on development of local and regional market products. Development of biotechnology, mostly plant biotechnology, in African countries relies on established biotechnological centres, either of a regional or international character, that specialize in *in vitro* cultivation of cash crops such as banana, coffee, cocoa, palm-oil, vanilla; and food crops such as maize, millet, cassava and cowpea (Massola, 1992). Plant-based biotechnology have so far achieved the following: Production of high quality biofertilizers; bioprospection of new nitrogen fixing species of bacteria and mycorrhizae; creation of novel genetic and hybrid variability; *in vitro* cloning of plants of ornamental and economic significance; and diversification of bio-industrial production of plant metabolites of medical significance like reserpine (*Rauwolfia serpentina*), glycyrrhetic acid (*Abrus precatorius*) and rotenone (*Tephrosia vogelii*). South Africa is the leader of biotechnology in Africa, as it has already developed and commercialised transgenic products such as maize and cotton. Other South African countries (Zimbabwe, Malawi and Madagascar) are doing micro-propagation of disease-free banana, rice, maize, groundnuts and tropical woody trees. North African countries (Morocco and Tunisia) are doing biological research and preliminary trials on palms, potatoes, tomatoes, maize and forest trees (Morris, 2011). West African countries (Burkina Faso, Cameroon, Cote d'Ivoire, Ghana, Nigeria, Gabon and Senegal) are doing various projects related to biological nitrogen fixation; production of legume inoculants; fermented foods; medicinal plants; plant tissue culture of cocoa trees, rubber trees, coffee trees, yams, oil-palm, pineapple, cotton, tea, banana, cassava, ginger, eucalyptus and acacia; and production of mycorrhizal-based bio-fertilizers for rural markets (Brink et al., 1998). Out of 53 countries of the African Union, only 16 have laws, regulations, guidelines or policies related to modern biotechnology. Of these, only South Africa, Egypt and Burkina Faso have experiences in commercialisation of GMO crops (Makinde et al., 2009). The main problems facing biotechnology experts in Africa is the scarcity of resources and slow passage of GM crops from experimental to commercial stages and difficulties in meeting regulatory requirements (Michaelis, 1993; Lynam, 1995; Nyira, 1995; Thomson, 2004). Africa's under-development in terms of insufficient good quality food, poor health care, unreliable sources of energy and degraded environments would be greatly reversed if the continent embraces biotechnology.

BIOTECHNOLOGY IN EASTERN AND CENTRAL AFRICAN COUNTRIES

This work attempted to carry out a facts-and-figures review on biotechnology activities in the Eastern and Central African countries namely Kenya, Uganda, Tanzania, Ethiopia, Rwanda, Burundi and Democratic Republic of Congo (DRC). While Kenya, Uganda and Tanzania have already developed and operationalized their biotechnology policies and guidelines, the rest are yet to do so (Table 1). Burundi is in a process of formulating its policy while the rest have only managed to put up biosafety frameworks, and efforts are underway to develop their biotechnology policies. Currently, biotechnology projects that have already been commercialised are mainly on conventional tissue culture and plant/animal breeding, while GMO related projects are still undergoing contained research in laboratories or screen houses, and confined experiments are being carried out in the trial fields (Olembo et al., 2010).

Table 1. Status of biotechnology policy frameworks in some Eastern and Central African countries (Republic of Kenya, RoK, 1998; Republic of Uganda, RoU, 2004); United Republic of Tanzania, URT, 2005; Republic of Rwanda, RoR, 2005, Republic of Burundi, RoB; 2006, Democratic Republic of Congo, DRC, 2007; Kassa 2011).

Country	Policy framework
Kenya	Biotechnology policy (2007)
Uganda	Biotechnology policy (2008)
Tanzania	Biotechnology policy (2009
Ethiopia	None
Rwanda	None
Burundi	In preparation
DRC	None

Status of biotechnology in Kenya

Kenya is the leading economy in East Africa. Apart from tourism, Kenya's economy is dependent on agriculture where tea, coffee, sisal and pyrethrum are the leading export crops, while the staple food crops are maize and wheat. Coconuts, pineapples, cashew nuts, cotton, sugarcane, sisal and corn are grown in the low-lying areas. Biotechnology is taking roots in Kenya. After the government had passed the national biotechnology policy in 2007 that was developed by the Kenya National Council for Science and Technology (NCST; www.ncst.go.ke, cited on 13th August 2011), many biotechnology projects got a big boost. Conventional biotechnology procedures such as tissue culture (TC) are widely used for production of planting materials for pyrethrum, banana, sugarcane, potato, strawberry, cassava, vanilla, oil palm and flowers. Marker assisted selection (MAS) is used for characterization and mapping of maize streak virus and grey leaf spot resistance genes in maize; development of drought tolerant wheat and aphid resistant maize and wheat; smut resistant sugarcane, breeding for desirable traits in cassava, rice and sorghum; and characterization of indigenous species of cattle, forages and tsetse flies. In case of livestock, biotechnology is directed at development of recombinant DNA vaccines for Newcastle disease, Rift valley fever and rind pest fever (Karembu, 2007; Olembo et al., 2010).

Research and development (R and D) institutions are engaged in the GMO research projects, with the Kenya Agricultural Research Institute (KARI) taking the leading role. The GMO projects on virus-resistant sweet potatoes and cassava; insect resistant maize, cotton, cowpea and sweet potatoes; drought resistant maize; and bio-fortified rice are among the projects which are at different stages of permit applications, laboratory and screen house containments, or fields trials (Thompson, 2004; Macharia, 2010; Olembo et al., 2010; Mugo et al., 2011). At the

Kenyatta University, research on striga-resistant sorghum is on-going and it has reached a confined field trial stage (Sithole-Niang, 2004; Kingiri and Ayele, 2009). Furthermore, the plant transformation laboratory at Kenyatta University is currently exploiting the availability of genes and technologies to improve drought tolerance, nutritional value and adaptability of major food crops (Muoma, 2010).

The International Livestock Research Institute (ILRI; http://www.ilri.org, cited on 13th August 2011) in Nairobi conducts research in systems analysis and impact assessment; people, livestock and environments; livestock policy analysis; livestock health; livestock feeds and nutrition; livestock genetics and genomics; strengthening partnerships for livestock research; smallholder dairy; and smallholder livestock systems. Being part of the consultative group on international agricultural research (CGIAR), ILRI works with partners worldwide to help poor people keeping their farm animals alive and productive. It also helps farmers to find profitable markets for their animal products. In the area of biotechnology, ILRI focuses on development of appropriate diagnostics to help identify disease threats and develop specific vaccines; identifying and using genetic adaptations such as disease resistance and developing appropriate marker technologies to facilitate delivery of genetic improvement into farmers' herds/flocks; and genetic adaptations to increase the quality of feeds (Olembo et al., 2010). Other R and D institutions involved in biotechnology research in Kenya include the University of Nairobi (Department of Biochemistry), Jomo Kenyatta University of Agriculture and Technology (JKUAT) and the National Potato Research Centre (NPRC). The private sector, including the genetic technologies international limited (GTIL) and MIMEA are involved in tissue culture and mass propagation activities. The on-going biotechnology-related research projects in Kenya are summarized in Table 2.

In Kenya, at least six genetically modified organisms (GMO) projects have been approved and they are under contained laboratory and screen house trials while others are in various stages of application (Table 3).

Kenya is moving fast up the biotechnology ladder. Concerted efforts by the government and other stakeholders in addressing insufficient financial, human resources and infrastructure challenges would sustain it as a leader of biotechnology in the Eastern and Central Africa region.

Status of biotechnology in Uganda

Uganda is endowed with significant natural resources, including ample fertile land, regular rainfall, and mineral deposits. Biotechnology in Uganda is set to contribute to the national goals of poverty eradication, improved healthcare, food security, industrialization and the protection of the environment. The Uganda National Biotechnology and Biosafety (BAB) policy was approved by the Ugandan Cabinet in 2008 after a thorough review and deliberation of various stakeholders. The policy, which was formulated by the Uganda National Council for Science and Technology (UNCST) in 2002, aims to build and strengthen national capacity in biotechnology through research and development, promote the utilization of biotechnology products and processes as tools for national development and provide a regulatory and institutional framework for safe and sustainable biotechnology development and application (www.absfafrica.org, cited on 25th August 2011). The objective of the BAB Policy is to provide regulatory and institutional framework for sustainable and safe application of biotechnology for national development (www.uncst.go.ug, cited on 25th August 2011).

Uganda has made an impressive start in biotechnology related activities which are mainly being carried out by R and D institutions. There are growing applications of biotechnologies related to agriculture, health, environment and industry such as tissue culture, production of transgenic crops, diagnostic tools, medicines, vaccines and hormones. Biotechnology is used in bioremediation, biofuels and production of enzymes (Olembo et al., 2010). The main public research and development institution in Uganda is the Makerere University with the following biotechnology-related departments: Department of Crop Science, Biochemistry, Animal Science, Parasitology and Microbiology, Food Science and Technology, Institute of Environment and Natural resources and the Medical School. Other R and D public institutions are the Natural Agricultural Research Organisation (NARO), which has the following research centres: Kawanda Agricultural Institute (KARI), Namulonge Agricultural and Animal Producing Research Institute (NAARI), Livestock Research institute (LIRI), Coffee Research Institute (CRI), Forest Research Institute (FRI) and Food Science and Research Institute (FSRI). Other public Institutes are the Uganda Virus Research Centre (UVRC) and the Joint Clinical Research Centre (JCRC). Private sector institutions are the Med-Biotech Laboratories (MBL) and Agro-Genetic Laboratories (AGL), while international institutions involved in biotechnology R&D in Uganda are the International Institute of Tropical Agriculture (IITA) and the International Network for Improvement of Banana and Plantain (INIBAP) (Olembo et al., 2010). The activities of these institutions are summarized in Table 4.

Several international and regional bodies have been supporting biotechnology projects in Uganda. They include the Rockefeller Foundation which has been supporting a project on disease-free banana and cassava and striga-resistant maize; the Mexico-based international maize and wheat improvement centre (CIMMYT) that supports insect and striga-resistant maize

Table 2. Institutions involved in biotechnology activities in Kenya (modified from Brink et al., 1998; Johanson and Ives, 2001; Glover, 2007, Onsongo, 2009; Olembo et al., 2010).

Institution and collaborators	Activity
KARI, Bill and Melinda Gates	Bio-fortified sorghum
KARI, CIMMYT	Insect resistant maize (leaves and seeds)
KARI, Monsanto	Insect resistant cotton
KARI, Monsanto	Virus resistant sweet potato
KARI, USAID	Virus- resistant cassava
KARI, and various collaborators	1) Production of disease free plants and micro-propagation of pyrethrum, bananas, potatoes, strawberries, sweet potato, citrus, sugar cane. 2) Micropropagation of ornamentals (Carnation, Alstromeria, Gerbera, Anthurium, leopard orchids) and forest trees. *3) In vitro* selection for salt tolerance in finger millet. 4) Transformation of tobacco, tomato and beans. 5) Transformation of sweet potato with proteinase inhibitor gene. 6) Tissue culture regeneration of papaya. *7) In vitro* long-term storage of potato and sweet potato. 8) Marker-assisted selection in maize for drought tolerance and insect resistance. 9) Well-established MIRCEN providing microbial biofertilizers in the East African region.
ILRI	1) Development of recombinant vaccines against East Cost fever. 2) Development of improved control interventions for *Peste des Petits* Ruminants virus. 3) Breeding for improved chicken production. 4) Development of molecular diagnostic and control tools and strategies for pork tapeworm cysticercosis. 5) Identifying appropriate germplasm for dairy development and delivery mechanisms. 6) Strain restricted immunity to parasitic protozoa that cause East Coast fever. 7) Developing second-generation vaccines for critical bovine diseases.
Nairobi University Biochem Department	Research on GM capripox virus, rinderpest recombinant vaccine production and production of transgenic sweet potato, MAS sorghum.
JKUAT	Tissue culture of various food crops and ornamentals.
Kenyatta University	Transgenic sorghum for resistance to Striga parasitic weed. Transgenic maize for drought tolerance.
NPRC	Tissue culture of Irish and sweet potatoes.

Table 2. Contd.

Moi University	Marker assisted selection in sorghum, maize
GTIL	Tissue culture of various plants including banana, potatoes, and sweet potatoes, coffee, pyrethrum, sugarcane, vanilla, fruits, trees and medicinal plants.
MIMEA	Mass propagation of cassava, sweet potatoes, banana and ornamentals.

project; and the Bill and Melinda gates which supports the WEMA Project (Olembo et al., 2010). Efforts towards commercialisation of GMOs in Uganda are underway. Already, 5 projects have been approved and are currently undergoing field trials while and 5 others are under review (Table 5).

The progress of biotechnology activities in Uganda is encouraging. With the Biotechnology and Biosafety Policy in place, the current research-based contained and confined trials will soon move a step further to realise commercial release of GMOs. Uganda is, however, faced with financial constraints, insufficient skilled manpower and inadequate facilities to fully engage in modern biotechnology operations. The government should, therefore, invest more in those lines in order to address the country's food security, improved people's health and overall socio-economic development challenges.

Status of biotechnology in Tanzania

Tanzania is faced with mass poverty and remains to be among the poorest countries of Sub-Saharan Africa and the 22nd poorest in the world with over 30% of its population living below the international poverty line, earning less than 1 USD per day (URT, 2005). The burden and incidence of poverty is more widespread in rural than urban areas, where over 60 percent of the people live and agriculture forms the main stay. Realizing that, it considers biotechnology to be one of the means to rid itself out poverty, ensure food security and public health (www.mst.go.tz, cited on 25th August 2011). The Tanzania Biotechnology Policy (TBP) was approved in 2009. Its mission and vision are "to achieve significant investment in harnessing biotechnology tools for generation of products, processes and technologies in food, agriculture and health for socioeconomic development". The general objective of the policy is to ensure that Tanzania has the capacity to capture the proven benefits arising from agriculture, health, industry and environmental applications of biotechnology while protecting and sustaining the safety of the community and the environment (Mneney, 2010). Biotechnology in Tanzania is mainly agricultural based. Activities involved include tissue culture and micro-propagation, marker-assisted breeding, disease diagnostics, livestock vaccines, and genetic engineering to produce GMOs (Mneney, 2001; Rutabanzibwa, 2004). In Tanzania, banana, cassava, sweet potato, pyrethrum, coconut and sisal have been tissue-cultured (Table 6). Regional projects are supporting marker assisted selection (MAS) research and contained field trials in the following crops: cassava, sweet potato, rice and maize, maize, sorghum, coconut and cashew nuts (Tairo et al., 2005, 2008). In health sciences, biotechnology is still at a research stage where molecular diagnostic tools for malaria, tuberculosis and HIV/AIDS are being investigated. The leading institutions in health related biotechnology are the national institute for medical research (NIMR; http://nimr.or.tz, cited on 5th October 2011) and the Ifakara Health Institute (IHI); www.ihi.org, cited on 5th October 2011). In veterinary sciences, projects on development of diagnostics and vaccines for east coast fever (ECF), New castle disease (NCD), Rift valley fever (RVF), contagious bovine pleuropneumonia (CBPP) and breeding for improved milk and meat production are underway at the animal diseases research institute (ADRI; www.mifugo.go.tz, cited on 5th October 2011). Various international projects have been supporting research in agricultural biotechnology and facilities at the University of Dar es Salaam (UDSM), Sokoine University of Agriculture (SUA), Mikocheni Agricultural Research Institute (MARI),other Agricultural Research Institutes (ARI) at Tengeru, Ukiriguru and Uyole; Tanzania Coffee Research Institute (TaCRI), Tanzania Pesticide Research Institute (TPRI), National Plant Genetic Resources Centre (NPGRC) and other R&D institutions. Contained research activities are carried out at MARI and UDSM. Confined field trial on

Table 3. Progress of GMO development in Kenya (modified from Thompson, 2004; Kingiri and Ayele, 2009; Macharia, 2010).

Genetically modified organisms (GMO) activity	Stage of development
Genetic engineering in sweet potato for disease resistance (feathery mottle virus).	Contained laboratory and confined field trials.
Genetic engineering in maize for resistance to African maize stem borer.	Contained laboratory and greenhouse; field trial.
Genetic engineering in cassava for cassava mosaic disease (CMD) resistance.	Contained greenhouse and confined field trials.
Genetic engineering in cotton for bollworm resistance (Bt cotton).	Contained greenhouse and confined field trials.
Genetic modification of sorghum for Striga (weed) resistance.	Contained laboratory and screen house.
Genetically modified water efficient maize (WEMA).	Contained and confined field trials.
Genetically modified sorghum for improved protein quality, digestibility and enhanced Iron and Zinc.	Application for contained greenhouse trials.
Genetically modified Maruca pod borer resistant cowpea.	Application for controlled trials.
Genetic engineering for bio-fortified (pro-vitamin A) cassava.	Application for confined trials.
Genetically modified vaccine for rinderpest marker vaccine.	Application for confined trials.
Genetically modified weevil resistant sweet potato.	Application for contained greenhouse trials.

Table 4. A summary of biotechnology activities in Uganda (modified from Ntawuruhunga et al., 2007; Mugoya, 1999; Olembo et al., 2010; Wamboga-Mugirya, 2010; Opuda-Asigo, (2010).

Institution	Biotechnology related activities
Makerere University	
Department of Animal Science	Marker assisted selection for local cattle breeds.
Department of Crop Science	MAS Research on disease/pest resistance; tolerance to abiotic stresses in maize; genetic engineering in banana and cassava; DNA fingerprinting in sweet potato; bean mosaic viruses and maize parasitic fungi.
Department of Veterinary Parasitology and Microbiology	Molecular diseases diagnosis; vaccine development and MAS for East Coast Fever virus.
Department of Food Science and Technology	Characterization of banana juice and starch, and characterization of lactic acid bacteria for yogurt production.
Medical School	Recombinant HIV vaccine, DNA sequencing.
Institute of Environment and Natural Resources	Genetic markers for wildlife; genetic characterization of microorganisms in wastewater treatment systems.
NARO	
KARI and NAARI	Protocols on somatic embryogenesis, molecular characterization of bean common mosaic; GM cassava varieties to resist the virulent cassava mosaic virus (CMV) and cassava brown streak virus (CBSV).
LIRI	Cloning and sequencing of trypanosome genes; diagnostic tools for bovine pleuropneumonia; East Coast fever vaccine improvement.

Table 4. Contd.

UVRI	Development of viral and bacterial diagnostic tools.
JCRC	Aids virus research/molecular diagnosis.
MBL	Molecular diagnostic tools and markers for malaria and cassava research; novel restriction enzymes.
AGT	Tissue culture production in coffee and banana.
IITA	Virus resistant sweet potatoes and cassava.
INIBAP	Banana transformation.
Coffee Research Centre (COREC)	Coffee wilt disease (CWD) resistant/tolerant coffee variety, with genes from wild relatives.
National Semi-Arid Resources Research Institute (NaSARRI)	BT and HT resistance to the African cotton bollworm and with genes from Bollgard II™ for herbicide-tolerance.
African Agricultural Technology Foundation (AATF)	CFT for drought -resistant maize (water efficient maize for Africa (WEMA).
National Agricultural Research Laboratories (NARL)	Bio-fortified banana with vitamin A, and Iron; the apple banana; Nakyitembe bananas modified with two genes; Sigatoga (fungal disease) resistant banana.

genetically modified water-efficient maize (WEMA) is planned at Makutopora, Dodoma. Research on industrial and environmental biotechnology (biogas, bioremediation, biofertilizers and biopesticides, are conducted at the Department of Molecular Biology and Biotechnology, UDSM (Parawira, 2009; Mneney, 2010).

Constrains facing biotechnology R and D in Tanzania include inadequate funding, limited capacity for biotechnology, lack of critical mass of highly trained scientists, technicians and entrepreneurs, lack of capacity to supply, service and repair scientific equipments and lack of inter-institutional networks (Mneney, 2010; Bull et al., 2011). Tanzania needs to invest more in biotechnology to take advantage of its richness in natural resources. It also needs to exploit new biotechnology innovations avenues such as novel DNA technologies, bioinformatics and genomics. The government should provide adequate funding by increasing budget for S and T; create enabling environments; and improve awareness and education in biotechnology.

Status of biotechnology in Ethiopia

Ethiopia is a country whose 41% of the economic activities depend on agriculture, including marketing, processing, and export of agricultural products. Principal crops include coffee, beans, oilseeds, cereals, potatoes, sugarcane, and vegetables (Kassa, 2011). Ethiopia does not have a national biosafety policy yet. It needs one in order to develop its capacity and apply the technology in Agriculture, environment, health and energy sectors. Ethiopia is one of the major genetic centers of origin with huge biodiversity of flora and fauna. A few examples of the legacy of Ethiopia to the rest of the world as a source of important genes include the yellow dwarf virus (BYDV) resistance gene found in barley (Niks et

Table 5. Progress on GM development in Uganda (Wamboga-Mugirya, 2010. cited on 27th August 2011).

Project	Status
Black sigatoga disease resistance in East African Highland Bananas (EAHBs	Confined field trial (CFT) data analysis
Herbicide tolerant cotton - RR Flex™	CFT
Bt cotton - Bollgard™	CFT
Biofortified banana (Iron, Pro-Vit A, Vit E)	CFT
Virus (CMV) resistant cassava	CFT
BXW resistant banana	CFT application under review
Drought tolerant maize	CFT application under review
Virus (CBSV) resistant cassava	CFT application under review
Virus resistant sweet potato	CFT application under review

Table 6. Institutions involved in biotechnology activities in Tanzania (modified from Mneney, 2010).

Name of institution	Type of biotechnology	Products
Kizimbani, Zanzibar	Tissue culture	Banana
SUA	Tissue culture	Banana and horticultural crops
	Disease diagnostics of ECF, NCD, RVF and CBPP	Disease resistant animals
	Training	Molecular biology, disease diagnostic tools.
NPGRC	Tissue culture	Banana
Central veterinary laboratory (CVL)	Disease (RVF, ECF) diagnostics	Disease resistant animals
TaCRI	Tissue culture	Coffee
Tengeru ARI	Tissue culture	Banana and horticulture crops.
Uyole ARI	Tissue culture	Pyrethrum and aeroponics.
MARI	Tissue culture	Cassava, banana, sweet potatoes cashew
	MAS	Cassava, sorghum, cashew.
	Disease diagnostics	Disease-free coconuts, cassava, rice, sweet potato and banana.
	GMO	Virus resistant cassava
NIMR, IHI	Disease diagnostics	Malarial, tuberculosis, HIV/AIDS diagnostic tools.
UDSM	Bioremediation	Biogas
	Biofertilizers	Nitrogen fixation, mycorrhizal inoculation
TPRI	Biotechnology regulation	Agricultural inspectors

al., 2004); mlo-11 and related genes for powdery mildew resistance (Piffanelli et al., 2004); the "stay-green" gene for drought tolerance (Haussmann et al., 2000; Mahalakshmi and Bidinger, 2002; Borrell et al., 2003, 2004); genes for cold tolerance, high lysine content, disease resistant sorghum (IBC, 2008) and drought tolerance in grass pea (Kassa, 2011). Currently, a total of 65 projects are underway in the country and most of them are in areas of agricultural, industrial, health and environmental biotechnology. These projects are on tissue culture, bio-fertilizers, molecular marker, embryo transfer, immunology, vaccine and diagnostic kit development and epidemiology, mainly for crops such as coffee, grass pea, teff and forest trees (Brink et al., 1998, Kassa, 2011). The institutes involved in biotechnology in Ethiopia include the Addis Ababa University (AAU), Ethiopia Institute of Agricultural Research (EIAR), Federal Research Centers (FRCs), Institute of Biodiversity Conservation and Research (IBCRI), National Veterinary Institute (NVI) and National Health and Nutrition Institute (EHNRI) (Table 7).

Biotechnology research and development in Ethiopia is at its infant stage compared to the neighboring countries like Kenya and Uganda where modern biotechnology projects are taking roots (Thomson, 2008). The lack of biosafety policy and guidelines; lack of proper assessment of the level of biotechnological capacity available in the country; poor coordination of the limited biotechnological capacity available in some institutions and research centers; lack of public's appreciation of opportunities provided by agro-biotechnology and financial constraints suggest that the government need to move steadfastly to address the situation (Abraham, 2009).

Table 7. Biotechnology activities in Ethiopia (modified from Otim-Nape, 2007; Kassa, 2011).

Institution	Activity
Addis Ababa University (AAU)	1) Research on Ethiopian flora including description of at least 60,000 Ethiopian plants. 2) Screening, isolation and characterization of key cereal crops using protoplasm fusion and somaclonal variations. 3) DNA sequencing for genetic characterization and micro-propagation of indigenous plants and tree species. 4) Localization on the chromosome quantitative trait loci (QTL) in Teff using RFLP techniques. 5) Genetic and biological approach towards producing clean and disease free Enset. 6) Micro-propagation of African redwood (Cusso) with reference to rooting. 7) Diagnosis and detection of viral pathogens in crops. 8) Biological nitrogen removal from tannery wastewater in Ethiopia. Development of various biological water treatment systems
EIAR	Tissue culture of banana, cardamom, grapevine, citrus fruits, garlic, potato, geranium, Enset, Teff, nigger, cabbage, coffee, sweet potato, ginger and cassava.
Jimma	Tissue culture – coffee, pineapple, spices and cassava
Holleta	Tissue culture – potato and enset
Melkassa	Tissue culture - Banana, sweet potato and garlic
Ebre Zeit	Tissue culture - grapevine, endod and Teff
IBCRI	Applying cytogenetic, biochemical and tissue culture techniques in the identification and characterization of collected germplasm.
EARO	Tissue culture and MAS to improve Teff, coffee, fruit cultivars and forest trees for commercial purposes.
NVI	Molecular and biochemical characterization of parasites
EHNRI	1) Screening of pathogenic bacteria such as Escherichia and Shigella, using DNA probes. 2) Development of animal vaccines against rinderpest, sheep pox, Newcastle, African horse sickness, foot and mouth disease and bovine pleuropneumonia.

Status of biotechnology in Rwanda

Rwanda, one of the least developed and densely populated countries in Sub-Saharan Africa, faces severe land shortage and land degradation due to the fragile ecosystem and poor land management. Yet, more than 90% of the population depends on agriculture (RoR, 2005; Rwanda ETOA, 2008). Therefore, adoption of biotechnology seems to be an inevitable strategy for achieving sustainable food security and poverty reduction. However, there is no clear biotechnology policy framework and guidelines to guide biotechnology R and D in Rwanda. In addition, there is no national strategy for biotechnology, reflecting limited appreciation of the importance of biotechnology in the socio-economic development of the country (RoR, 2005).

There is limited but steadily growing range of applications of conventional biotechnology in Rwanda covering various sectors including agriculture, health, environment and industry. Conventional biotechnology embraces tissue culture and tissue multiplication, production of various multipurpose plants, medicinal plant exploration and exploitation, brewing of beer, and production of juice and yogurt. Biotechnology is also used in diagnostics, production of bio-energy, and waste treatment. Rwanda has made modest progress in biotechnology applications mainly in the area of agriculture, animal husbandry (plant tissue culture, embryo transplants) and medicine (HIV/AIDS diagnostics, vaccine trials using recombinant DNA technology) as shown in Table 8. The R and D institutions involved in biotechnology in Rwanda include Institut des Scienses Agrinomiques du Rwanda (ISAR)-Rubona, ISAR-Musanze, the National University of Rwanda's (NUR; www.nur.ac.rw, cited on 25th August 2011), Kigali Institute of Science and Technology (KIST; www.kist.ac.rw,

Table 8. Summary of biotechnology activities in Rwanda (RoR, 2005).

Institution	Activity
National University of Rwanda (NUR) Research Commission	Biotechnology research related to plants, animals and human being in areas of tissue culture and disease diagnostics.
Kigali Institute of Science and Technology (KIST)	Research of food science and technology; applied microbiology; molecular biology and applied biotechnology; production of varnish and glue from locally available materials.
Ministry of Education, Science and Research Institutions	Production of Azolla and Rhizobia-based biofertilizers for rice cultivation; Tissue culture of medical plants and micropropagation of disease-free potato, banana and cassava.
Institut des Sciences Agrinomiques du Rwanda (ISAR)-Rubona, ISAR-Musanze.	Tissue culture of Irish potato, sweet potatoes, cassava, potatoes, banana and coffee. Disease diagnostics for cassava and sweet potato.

cited on 25th August 2011) and the Ministry of Education Science and Research Institutions (RoR, 2005).

The development and application of modern biotechnology in Rwanda is just emerging, and faces enormous challenges. There is very low human resource and infrastructure capacity coupled with lack of biotechnology facilities. To build the required capacities, the government of Rwanda needs to invest in biotechnology policy formulation, human resource training and infrastructure development. These efforts will require political will at the highest level to ensure that biotechnology is given the priority it deserves within sectoral and national development strategies and budgets. The development of the national biosafety framework for Rwanda (RoR, 2005) signifies a good start towards facilitation of safe development and application of modern biotechnology.

Status of biotechnology in Burundi

Burundi is a poverty stricken country, ranking 169th out 177 countries in terms of human development index (RoB, 2006). Around 90% of its population depends of agricultural sector. Having recognized the importance of using biotechnology as an important tool for development, it initiated a process of formulating biotechnology and biosafety policies for the purpose of improving agricultural productivity, health delivery and environmental conservation. In Burundi, there is only limited application of biotechnology, mainly at research stage in the R and D institutions (RoB, 2006). The University of Burundi's Faculty of Agromomic Sciences (FACAGRO), Faculty of Science (FoS) and the Higher Institute of Agriculture (ISA) are the main players. Other institutions include the Institute of Agronomic Science of Burundi (ISABU), the Agronomic and Zoootechnical Research Institute (IRAZ), and the National Centre for Food and Technology (CNTA). The types of research which is done there include *in vitro* tissue culture for mass propagation of cereals, roots and tuberous plants; ornamental plants, medicinal plants and microorganisms. Other plant biotechnology researches are onphytopathology, plant breeding, biofertilizers, single cell protein, mycoculture, conservation of germplasm and transformation of food products. Animal biotechnology at ISABU deals with bovine genetic improvement, embryo rescue and animal disease diagnostics. Biotechnology research related to human health is quite limited; only diagnostic techniques such as ELISA and Western blot techniques are being carried out. A summary showing biotechnology activities in Burundi and the institutions involved is shown in Table 9.

Biotechnology in Burundi is at its embryonic stage. The efforts that are being made by the government in formulating the biotechnology policy and guidelines are commendable. The joining of Burundi to the East African Community in 2007 puts it in a good position to advance technologically by forging close collaborations with other member countries such as Kenya and Uganda where biotechnology has already made more progress.

Status of biotechnology in Democratic Republic of Congo

The Democratic Republic of Congo (DRC) is a country endowed with vast natural resources. A country of 60 million people, it has 100 million h of arable land that supports a wide range of flora and fauna. The DRC does not have a stand-alone biotechnology policy but it has related policies such as agriculture policy, health policy and industrial development policy. The policies, however, do not address the issue of modern biotechnology. Even the scientific and technology policy fails to set any guidelines regarding biotechnology R and D. The DRC is under pressure from the southern African development community (SADC)'s agricultural research and development committee to set up harmonized legislation on biotechnology and biosafety (DRC, 2007). Biotechnology in the DRC is limited to conventional technologies such as production of beers and soft drinks,

Table 9. Summary of biotechnology activities in Burundi (RoB, 2006).

Name of institution	Type of biotechnology	Products
FoS, Burundi University	Tissue culture	Various crop and medicinal plants
FACAGRO	Phytopathology, Plant breeding, Bio-innoculation Mycoculture	Various *in vitro* crop plants Various crop and ornamental plants Biofertilizers Mushrooms
ISABU	Tissue culture Animal improvement Disease diagnostics	Cassava, sweet potatoes, Irish potatoes, yams, soya beans Cows – Ankole/Sahiwal crossing with exotic breed Cassava and sweet potato.
CNTA	Technology innovation	Transferred technologies
IRAZ	*In vitro* culture	Banana, potato, cassava, taro
Ministry of Agriculture and Livestock	Tissue culture and Animal diseases diagnostics	Various *in vitro* plants; Vaccines, medicines.
Ministry of Health Agro and Biotechnologies (AGROBIOTECH)	Disease diagnosis Tissue culture	Vaccines, molecular diagnostic tools Banana, Taro and potato.
Phytolab	Tissue culture	Sweet potatoes, cassava, Irish potato, taro, pineapple and tree species.

transformation and conditioning of dairy products, and traditional production of fermented foods. Its application is limited to a few experimental trials carried out by some research and teaching institutions such as universities and colleges. The agricultural sector dominates the DRC economy, with food crops such as cereals, roots, tubers,oilseeds, vegetables, and legumes; commercial crops such as coffee, cotton, tea, rubber, palm oil, cocoa, hevea, quinquina, onion, sugar cane, fruits and vegetables; forest products; and breeding of cattle, sheep, pigs, goats and poultry. In agricultural biotechnology, only a few trials have been carried out to date, such as transgenic banana and transformation of cassava to reduce the level of cyanide (Glover, 2007). In a total of 8 (all non-GMO) projects, 6 are related to crops, 1 on livestock and 1 on forestry (www.absfafrica.org, cited on 5th October 2011). These research activities are mainly carried out at the University of Kisangani and other R&D institutions as summarized in Table 10.

The constrains facing the biotechnology industry in DRC include the lack of a biotechnology policy, poor public awareness and participation in matters related to modern biotechnology and over dependence on donor-driven biotechnological projects. These shortfalls call for the DRC government to invest more on modern biotechnology.

CHALLENGES FACING MODERN BIOTECHNOLOGY IN THE EASTERN AND CENTRAL AFRICAN REGION

The development and application of biotechnology in the studied Eastern and Central African countries face the following challenges:

1) There is lack critical mass of skilled personnel, basic infrastructure and facilities, unreliable access to modern communication systems, unreliable power supply and poor availability of chemicals and consumables for research.
2) Indigenous 'neglected' crops attract funding only for exploitation outside Africa; and there is lack of basic research on 'neglected' or underutilized crops (Brink et al., 1998).
3) Linkages and networks (both internally and externally) are either weak or lacking.
4) Exotic crops are not well adapted to Africa, since imported lines or cultivars are not adapted to local conditions; these crops may be susceptible to local diseases and insects.
5) Lack of (national/regional) policies, strategies or legislation in biotechnology is evident in some countries in the Eastern and Central Africa Region. Some countries are yet to enact biosafety laws; such limitation constitutes

Table 10. Summary of biotechnology activities in the Democratic Republic of Congo (DRC, 2007).

Institution	Activity
University of Kisangani	Research on transgenic banana; Research on transformation of cassava with the application of bacteria and fungi to reduce the level of cyanide; *In vitro* propagation of potato, soybean, maize, rice and multipurpose trees, for example, *Acacia* and *Leucaena*. Production of rhizobial-based biofertilizers in experimental stage; Tissue culture of medical plants, for example, *Nuclea* sp *and Phyllanthus* sp.
National Institute for Agronomic Studies and Research (INERA)	Stability of geneotypes of cotton, cassava, maize, rice, plam coffee, oil seeds in various environments; Research on conservation of phlogenetic resources; Tissue culture of leguminous plants, root and tuber plants.
Agri-Food Research Centre (CRAA)	Industrial research in the agri-food science and technology (enzymology, bacteriology, quality control, spectroscopy)
National Seed Service (SENASEM)	Plant genetics and breeding centre including production multiplication and certification of seeds.
Maize Research Centre (CRM)	Maize breeding and general agronomic studies
National Livestock Development Authority (ONDE)	Zoo-sanitary protection, artificial insemination and animal disease diagnostics.
National institute of Biomedical Research Institute (INRB)	Human and veterinary biological and biochemical analyses. Research on malaria, trypanosomiasis, HIV/AIDS and medicinal plants.
National Natural Science Research Centre (CRSN-Lwiro)	Molecular biology and biotechnology research in agricultural, veterinary, entomology, limnology and marine biology.

a serious constraint that impairs the use, evaluation and release of genetically modified organisms.
6) Lack of protection of intellectual property rights hampers the development of new technologies, profitable inventions and investments, and initiatives by entrepreneurial biotechnologists.
7) Governments are not taking enough proactive political roles in promoting biotechnology (Sengooba et al., 2009).
8) Public acceptance is being frustrated by anti-modern biotechnology activism fuelled by internal and external pressure groups (Brink et al., 1998, Makinde et al., 2009).

Way forward

Investments in, and development of biotechnological research capacity in Africa would best be accomplished in phases. The first phase is conventional biotechnologies such as plant tissue culture and breeding, which is appropriate for Africa as many of the important food crops such as cassava, sweet potato, yam and banana are vegetatively propagated and could be crossbred to improve quality. The second phase would be the application of biotechnological tools, which can improve the efficiency of selection of varieties/cultivars, the techniques that include anther culture and embryo rescue, as well as molecular marker applications (diagnostics, fingerprinting and marker-assisted breeding). The third phase is the development of capacity to produce transgenic plants/animals, which would include gene isolation and cloning; gene insertion/transformation; regeneration of transgenic plants/animals followed by commercial release of GMO products (Lynam, 1995).

As a prerequisite, national biotechnology and biosafety policy frameworks are crucial so as to give directions on which way the technology should proceed (Keizire, 2000; Muraguri et al., 2003). In spite of the lack of support of national biosafety laws, all the Eastern and Central African countries covered in this study have formulated biosafety frameworks. Among the seven studied countries, only Kenya, Tanzania and Uganda have comprehensive biotechnology polices. It is suggested that the remaining countries should move fast in preparing their draft biotechnology policies so as to catch up with the rest. Burundi is moving faster in that direction, while Ethiopia, Rwanda and DRC are in the preparatory stages (RoK, 1998; RoU, 2004; RoR, 2005; URT, 2005, RoB; 2006; EPA, 2007; DRC 2007).

The successful application of biotechnology requires an adequate infrastructure, well trained technical expertise, and a critical mass of researchers with sufficient supporting sustainable funding to cover the high cost of modern biotechnology research. None of the studied countries have commercialised any GMO product yet, but Kenya and Uganda are far ahead of Tanzania, Ethiopia, Rwanda, Burundi and DRC towards that end. A South-South partnership with government agencies, universities, research institutions and private companies within the countries, and with other African countries like South Africa, Egypt and Burkina Faso (which have already commercialised genetically modified organisms) would be desirable. Whereas funding from the United States of America is generally supportive of GM technology, the opposite is true of funding from European sources. African countries are thus pulled in two different directions. The way out of this dilemma is for countries in the Sub-Saharan African region to develop strong South-South collaborations, whereby developing countries such as Argentina, Brazil and South Africa would partner with least developed countries to exploit biotechnological resources and develop new products (DaSilva et al., 2002, Morris, 2011). Where possible, in adopting modern technology, indigenous organisms should be utilized because they are adapted to local conditions. Moreover, countries' intellectual property rights (IPR) policies and regulations should be adhered to.

Strong leadership, effective priority-setting and adequate working opportunities for scientists are, however, required to provide incentives for the establishment of biotechnology-based economies. Participatory extension approach programs could be an ideal channel for the implementation of biotechnology products, as well as endowing resource-poor farmers with the confidence to develop and apply solutions to some of their problems. The Eastern and Central African region should take full advantage of its soil fertility, valuable biodiversity and favourable climatic conditions (Woodward et al., 1999; Makinde et al., 2009). It is emphasized that biotechnology alone will not solve the multitude of problems that Eastern and Central African countries face, but it has potential to contribute to transforming the livelihoods of its people from abject poverty to prosperity.

Conclusion

This study reviewed the current status of biotechnology-related policies and activities in Kenya, Uganda, Tanzania, Ethiopia, Rwanda, Burundi and Democratic Republic of Congo. Only Kenya, Uganda and Tanzania have biotechnology policy frameworks. Most of the biotechnology-related activities are being done by government agencies and public/private R and D institutions, with financial support from regional and international bodies. Conventional biotechnology activities such tissue culture and breeding are being practiced in all the countries, while modern biotechnology activities are mainly being field-tested in Kenya and Uganda. None of the countries studied has already commercialised any genetically modified product, but Kenya and Uganda will soon arrive to that goal. Biotechnology holds great potential in promoting socio-economic development in the Eastern and Central African region through increased productivity, improved resistance to biotic and abiotic stresses, reduction in pesticide usage, and enhancing nutritional quality, thereby contributing to food security and poverty alleviation. The Eastern and Central African region can benefit from previous experiences and results achieved in other developing regions in obtaining benefits from the applications of biotechnology. This can be done through proper planning, interactive cooperation among countries and network participants. The main challenges facing the region on matters pertaining to research and development in biotechnology are insufficient trained human resources, poor infrastructure and facilities and poor political will by some governments. The region stands to benefit from biotechnology if the governments invest more in terms of policies and budgets in order to speed up development and commercialisation of biotechnology products. It is envisaged that soon biotechnology will play an important role in food security, improved health and environmental management in the region. While it is not a panacea to all the socio-economic problems facing the Eastern and Central African countries, biotechnology will largely complement other approaches aimed at driving the region out of its poverty woes.

ACKNOWLEDGEMENTS

The author acknowledges the financial and moral support from the Swedish International Development Agency, through the International Science Program of Uppsala University, Sweden. The University of Dar es Salaam, Tanzania, is appreciated for logistical support.

REFERENCES

Abraham A (2009). Agricultural biotechnology research and development in Ethiopia. Afr. J. Biotechnol., 8(25): 7196-7204.

Borrell AK, Oosterom K, Erik van, Hammer G, Jordan D, Douglas A (2003). The physiology of "stay-green" in sorghum. In: "Solutions for a better environment". Proceedings of the 11th Australian Agronomy Conference, 2nd–6th February.

Borrell AK, Jordan D, Mullet J, Klein , Klein R, Nguyen H, Rosenow D, Hammer G, Douglas A, Henzell B (2004). Discovering Stay-Green Drought Tolerance Genes in Sorghum: A Multidisciplinary Approach. New Directions for a Diverse Planet: Proceedings of the 4th

International Crop Science Congress, Brisbane, Australia, 26th September–1st October 2004.
Brink JA, Woodward BR, Da Silva E (1998). Plant Biotechnology: A tool for development in Africa. Electronic J. Biotechnol., 1(3): Available online.
Brookes G, Barfoot P (2006). GM Crops: The first ten years – Global socio-economic and environmental impacts in the first ten years of commercial use. J. AgBioForum, 9(3): 139-151.
Brookes G, Barfoot P (2009). Global impact of biotech crops: Income and production effects, 1996-2007. J. AgBioForum, 12(2): 184-208.
Bull SE, Ndunguru J, Gruissem W (2011). Cassava: Constraints to production and the transfer of biotechnology to the African laboratories. Plant Res. Rep., 30: 779-787.
Clive J (2010). Global status of commercialized biotech/GM crops: ISAAA Brief No. 42. ISAAA: Ithaca, NY.
DaSilva EJ, Baydoun E, Badran A (2002). Biotechnology and the developing world. Electronic J. Biotechnol., 5(1). Available on line (ISSN 0717-3458).
DRC (2007). National Biosafety Framework in the Democratic Republic of Congo. Kinshasa, 107 pp.
Environmental Protection Authority of Ethiopia (EPA) (2007). Ethiopia Biosafety Framework. Addis Ababa, Ethiopia.
Glover D (2007). Agri-biotech in Sub-Saharan Africa: Facts and Figures. Science and Development Network. (www.scidev.net/en/agriculture-and-environment, cited on 25th July, 2011).
Haussmann BIG, Obilana AB, Ayiecho PO, Blum A, Schipprack W, Geiger HH (2000). Yield and yield stability of four population types of grain sorghum in a semi-arid area of Kenya, Crop Sci., 40: 319–329.
IBC (2008). Crop Genetic Resources Department. Institute of Biodiversity Conservation (IBC), Addis Ababa, Ethiopia. (http://www.ibc-et.org, cited on 25th July 2011).
Johanson A, Ives CL (2001). An inventory of the agricultural biotechnology for Eastern and Central Africa region. Michigan State University, p. 62.
Karembu, M (2007). Enhancing the diffusion of tissue culture banana to small-scale farmers in Kenya. ISAAA Afri Center Policy brief. 8pp. (www.isaaa.org/kc/cropbiotechupdate/documents/TC_Banana, cited on 25th August, 2011).
Kasonta JS, Lyaruu AE, Chuwa PC, Hosea KM, Kohi YM (Eds) (2002). Biotechnology research and development policy: The basis for socio-economic development in Tanzania. Proceedings of three National Biotechnology Policy Workshops in Tanzania during 2001/2 within the BIO-EARN. COSTECH, Dar es Salaam, Tanzania. (www.costech.or.tz, cited on 25th July, 2011).
Kassa DG (2011). The potential of biotechnology in Ethiopia: Food Chem. Toxicol., 49(3): 685-689.
Keizire BB (2000). Agricultural biotechnology and food security in Sub-Saharan Africa: Policy and institutional considerations. A paper presented at the conference on Global Dimensions of Food Security: Threats, challenges and responses. University College of Cork, Ireland, pp. 13-15.
Kingiri A, Ayele S (2009). Towards a smart biosafety regulation: The case of Kenya. Environ. Biosafety Res., 8: 133-139.
Konde V (2006). The biotechnology revolution and its implication for food security in Africa. ATPS Special Paper Series. No. 28. African Technology Policy Studies Network, Nairobi, Kenya, 34 pp.
Kumar V, Naidu MM (2006). Development in coffee biotechnology – *in vitro* plant propagation and crop improvement. Plant Cell Tiss. Organ. Cult., 87: 49-65.
Lynam JK (1995). Building biotechnology research capacity in African NARS. In: Turning priorities into feasible programs. Proceedings of a Regional Seminar on Planning Priorities and Policies for Agricultural Biotechnology, South Africa, pp. 33-40.
Macharia H (2010). Risk assessment for release of GMO plants: Kenya Country Report. A Paper presented at the Workshop on Theoretical Approaches and Their Practical Applications in Risk Assessment for Release of Genetically Modified Plants., Hermanus, South Africa, pp. 22-26.
Makinde D, Mumba L, Ambali A (2009). Status of Biotechnology in Africa: Challenges and opportunities. Asian Biotechnol. Rev., 11(3): 1-10.
Mahalakshmi V, Bidinger, FR (2002). Evaluation of Stay-Green Sorghum Germplasm Lines at ICRISAT. Crop Sci., 42: 965–974.
Massola R (1992). Plant biotechnology in Sub-Saharan African Today. In: Plant Biotechnology for Developing Countries. Sasson, A. and Costarini Eds., V. Publ CTA/FAO, Netherlands.
Michaelis AR (1993). The crisis in African agriculture. Interdisciplinary Sci. Rev., 18: 1-3.
Mneney EE, Mantel SH, Mark B (2001). Use of random amplified polymorphic DNA markers to reveal genetic diversity within and between populations of cashew (*Anacardium occidentale* L). J. Hort. Sci. Biotechnol., 77(4): 375-383.
Mneney EE (2010). Status of Biotechnology application in Tanzania: A Country Report. A Paper presented at the Workshop on Theoretical Approaches and Their Practical Applications in Risk Assessment for Release of Genetically Modified Plants., Hermanus, South Africa, pp. 22-26.
Morris EJ (2011). Modern biotechnology: Potential contribution and challenges for sustainable food production in Sub-Saharan Africa. Sustainability, 3: 809-822.
Mtui G (2007). Trends in industrial and environmental biotechnology research in Tanzania. Afr. J. Biotechnol., 6(25): 2860-2867.
Mugabe J (2003). International trends in modern biotechnology: Entry by and implications for African countries. ATPS Special Paper Series No. 15. Nairobi, Kenya. 28 pp (www.atpsnet.org), cited on 25th August, 2011).
Mugo SN, Mwimali M, Taracha CO, Songa JM, Gichuki ST, Tende R, Karaya H, Bergvinson DJ, Pellegrineschi A, Hoisington DA (2011). Testing public Bt maize events for control of stem borers in the first confined field trials in Kenya. Afr. J. Biotechnol., 10(23): 4713-4718.
Mugoya CF (1999). Status overview of biotechnology in Uganda. A Paper presented at a regional Workshop on Biotechnology Assessment. Nairobi, Kenya, pp. 27-29.
Muoma, J, Ombori O, Jesse M (2010). Signal transduction: One of the current molecular approaches in the management of drought stress in the Sub-Saharan region. Biotechnology, 9: 469-476.
Muraguri L, Jones RB, Freeman HAF (2003). Review of the policy, regulatory mechanisms and administration of biosafety in Eastern and Southern Africa: A study of Kenya, South Africa, Malawi and the ASARECA initiative. ICRISAT, Nairobi, Kenya. 32pp (www.oro.open.ac.uk/11420/1/icrisat_final, cited on 29th August 2011).
Olembo NK, M'mboyi F, Nyende B, Oyugi, K, Ambani L (2010). Status of biotechnology in Sub-Sahara Africa: A cross-country analysis. ABSF. Nairobi Kenya, p. 98.
Onsongo M (2009). Kenya: Agricultural biotechnology Annual Report. Global Agricultural Information Network. GAIN Report No. KE9023. (www.gain.fas.usda.gov, cited on 28th August, 2011).
Opuda-Asigo J (2010). Risk assessment for release of GMO plants: Uganda Country Report. A Paper presented at the Workshop on Theoretical Approaches and Their Practical Applications in Risk Assessment for Release of Genetically Modified Plants, Hermanus, South Africa, pp. 22-26.
Otim-Nape GW (2007). Harmonization of variety release procedures for potato and sweet potato in Ethiopia, Kenya, Rwanda and Uganda. National Potato Research Centre, Kenya. www.asareca.org, cited on 5th August, 2011).
Niks RE, Habekuss, A, Bekele B, Ordon F (2004). A novel major gene on chromosome 6H for resistance of barley against the barley yellow dwarf virus. Theor. Appl. Genet., 109: 1536–1543.
Ntawuruhunga P, Okao-Okuja G, Bembe A, Obambi M, Mvila JCA, Legg JP (2007). Incidence and severity of cassava mosaic disease in the Republic of Congo. Afr. Crop Sci. J., 15(1): 1-9.
Nyira ZM (1995). Need challenges and objectives for biotechnology and agriculture in Africa. In: Turning priorities into feasible programs. Proceedings of a Regional Seminar on Planning Priorities and Policies for Agricultural Biotechnology, South Africa, pp. 17-23.
Parawira W (2009). Biogas technology in Sub-Saharan Africa: Status, prospects and constraints. Rev. Environ. Sci. Biotechnol., 8: 187-200.
Persley GJ (1992). Beyond Mendel's Garden: Biotechnology in Agriculture. In: Biotechnology Enhancing Research on Tropical Crops in Africa. CTA/IITA co-publication, pp. 11-19.
Piffanelli P, Ramsay L, Waugh R, Benabdelmouna A, D'Hont A, Hollricher K, Jorgensen JH, Schulze-Lefert P, Panstruga R (2004). A

barley cultivation-associated polymorphism conveys resistance to powdery mildew. Nature, 430: 887–891.
RoB (2006). National Biosafety Framework in Burundi. Republic of Burundi, Bujumbura, Burundi, p. 59.
RoK (1998). Regulations and guidelines for biosafety in biotechnology for Kenya. Republic of Kenya. National Council for Science and Technology (NCST), No. 42.
RoR (2005). The National Biosafety framework for Rwanda. Republic of Rwanda. Kigali Rwanda.
RoU (2004). National Biotechnology and Biosafety Policy. Uganda National Council for Science and Technology. Republic of Uganda. Kampala, Uganda.
Rutabanzibwa AP (2004). Policies, regulation processes, monitoring and surveillance of biotechnology materials: In KM Hosea KM, Msaki ON, Swai F. (Eds). (2004). Genetically modified organisms in Tanzania. Envirocare, pp. 36-44. (http://www.envirocaretz.org, cited on 10th October, 2011).
Rwanda ETOA (2008). Rwanda Environmental Threats and Opportunities Assessment (ETOA) 2008 updates. US Agency for International Development, p. 80.
Sengooba T, Grumet R, Hancock, J; Zawedde, B, Kitandu, L, Karembu, M, Meredia K, Nampala P, James O. Ochanda JO, Quemada H, Rubindamayugi, M (2009). Biosafety education relevant to genetically engineered crops for academic and non-academic stakeholders in East Africa. Electronic J. Biotechnol. 12(1) (Online) http://www.ejbiotechnology.cl. Cited on 20th August, 2011.
Sithole-Niang I, Cohen J, Zambrano P (2004). Putting GM technologies to work: Public research pipelines in selected African countries. Afr. J. Biotechnol., 3(11): 564-571.
Tairo F, Mukasa SB, Jones RAC, Kullaya A, Rubaihayo, RP, Valkonen, JPT (2005). Unveiling the genetic diversity of the three main viruses involved in sweet potato virus disease (SPVD) and its implications. Mol. Plant Pathol., 6: 199-211.
Tairo F, Mneney, E, Kullaya A (2008). Morphological and agronomical characterization of sweet potato [*Ipomea batatus* (L). Lam.] germplasm collection from Tanzania. Afr. J. Plant Sci., 2(8): 77-85.
Thompson JA (2004). The status of plant biotechnology in Africa. AgForum, 7(1&2): 9-12.
Thomson JA (2008). The role of biotechnology for agricultural sustainability in Africa. Philos. Transcr. Royal Soc., 363: 905–913.
URT (2005). Biosafety frameworks for Tanzania. United Republic of Tanzania. Vice Presidents Office. Government Printer, Tanzania.
Wamboga-Mugirya P (2010). Status of agro-biotech research and development in Uganda. Scifode Foundation, pp. 21-28. (www.scifode-foundation.org, cited on 11th July 2011).
Woodward B, Brink J, Berger D (1999). Can agricultural biotechnology make a difference in Africa? AgBioForum, 2(3&4): 175-181. (http://www.agbioforum.org, cited on 25th August, 2011).

Lignocellulosic ethanol production: Current practices and recent developments

Bishnu Joshi[1]* Megh Raj Bhatt[1] Dinita Sharma[1] Jarina Joshi[1] Rajani Malla[1] and Lakshmaiah Sreerama[2]

[1]Central Department of Biotechnology, Tribhuvan University, Kirtipur, Nepal.
[2]Department of Chemistry, St Cloud State University, St Cloud, MN 56301, USA.

Production of renewable fuels, especially bio-ethanol from lignocellulosic biomass, holds remarkable potential to meet the current energy demand as well as to mitigate greenhouse gas emissions for a sustainable environment. Present technologies to produce bioethanol largely depend on sugarcane and/or starch based grains and tubers (mainly corn, potatoes). This is partly due to ease of substrate handling and processing. On the other hand, use of sugarcane and food grains to produce bio-ethanol has caused significant stress on food prices and food security. Accordingly, the recent focus has been on lignocellulosic materials as a source for bio-ethanol. In fact, many countries are moving towards developing or have already developed technologies to exploit the potential of lignocellulosic materials for the production of bioethanol. This process of ethanol production generally involves hydrolysis of lignocellulosic biomass to fermentable sugars followed by fermentation of such sugars to ethanol. Achieving fermentable levels of sugars from lignocellulosic biomass requires relatively harsh pretreatment processes. The pretreatment process has pervasive impact on the overall operation because the process depends on the choice of lignocellulosic source, the size reduction via grinding, chemical treatment, acid hydrolysis, neutralization and fermentation. Recent advances in the process technologies have made it possible to use simultaneous saccharification and fermentation. In this process cellulase enzyme is the critical reagent as well as the cost determining factor. The advances in biotechnology as related to bioethanol have focused on engineering organisms that are capable of producing ethanol from cellulose, hemicellulose and lignocellulose. Such organisms are expected to be capable of not only degrading cellulose, hemicellulose and lignocellulose to fermentable sugars, but also are able to utilize both pentose and hexose sugars to produce ethanol at a relatively high yield. More recent and emerging approaches in bioethanol production are focused on reducing production costs. This approach uses consolidated bioprocessing schemes in which cellulase production, substrate hydrolysis, and fermentation are all accomplished in a single step. Countries, such as Nepal, that totally depend on the import of fossil fuels cannot ignore the potential of bioethanol derived from lignocellulosic biomass. Nepal is rich in biodiversity and posses variety of energy crops. Accordingly, developing policies and mechanisms that promote bioethanol will go a long-way in reducing the fuel crises in the countries lacking oil resources.

Key words: Lignocellulosic biomass, bioethanol, saccharification and fermentation (SSF), consolidated bioprocessing (CBP).

INTRODUCTION

Ethanol is an oxygenated fuel with high octane value like that of petroleum fuels. Ethanol is known to run combustion engines at higher compression ratios and thus provides superior performance (Wheals et.al., 1999). The blending of ethanol into petroleum-based automobile fuels can significantly decrease petroleum use and release of greenhouse gas emissions. Further, ethanol can be a safer alternative to the common additive, methyl tertiary butyl ether (MTBE), in gasoline. MTBE is toxic and is a known contaminant in ground water (Wang et al.,

*Corresponding author. E-mail: joshi.bishnu@gmail.com.

1999). Thus, ethanol can be a substitute to mitigate the problems associated with the rising energy demands across the world as well as a way to reduce green house gas emissions to an extent of 85% (Perlack et al., 2005).

Ethanol may be produced either from petroleum products or from biomass. Today, most of the ethanol produced comes from renewable resources. Although currently most of the ethanol produced from renewable resources comes from sugarcane and starchy grains, significant efforts are being made to produce ethanol from lignocellulosic biomass (almost 50% of all biomass in the biosphere) such as agriculture residues (Bothast and Saha, 1997). The technological advances in recent years are promising to produce ethanol at low cost from lignocellulosic biomass.

The world population is estimated to increases from 6.7 billion to 8 billion by 2030 (USCB, 2008). On the other hand, global oil production is expected to decline from 25 billion barrels to 5 billion barrels by 2050 (Campbell and Laherree, 1998). Thus the energy demands of the future is likely to play a key role in geo-political economics. Given this reality, nations around the world are investing in alternative sources of energy, including bioethanol. The leading nations in bioethanol production are Brazil and the USA (Table 1), and USA is the world's largest producer of bioethanol (Carere et al., 2008). Asian countries altogether account for about 14% of world's bioethanol production.

Bioethanol production from sugarcane and starch rich feed stocks such as corn, potato, etc., is considered first generation process and it has already been developed. The long-term viability of this process is in question because it will require significantly increased amounts of cultivatable land and significant hike in food prices that will ultimately lead to food insecurity (Mitchell, 2008). Estimates clearly point to the fact that first generation ethanol production process can not sufficiently meet the global energy needs. Therefore, second generation processes to produce bioethanol are gaining momentum. The second generation processes will use lignocellulosic materials for this purpose and biosphere clearly has sufficient supplies of lignocellulosic materials. The production of ethanol from lignocellulosic biomass [corn stover, wheat straw, sugarcane bagasse, rice straw, rice hull, corn cob, oat hull, corn fiber, woodchips and cotton stalk; energy crops such as switch grass and Alfa Alfa, and various weeds such as *Saccharum spontaneum*, *Lantana camara*, *Eichhornia crassipes* (water hyacinth), etc.] has become one of the best alternatives, because these sources have widespread abundance and the cost of their procurement is relatively cheap.

Even though the lignocellulosic biomass is abundant, the commercialization of the process to produce bioethanol from it is limited due to insufficient research, especially the research related to minimization of production cost. Bioethanol production from lignocellulosic materials relies on technologies that will efficiently hydrolyze cellulosic biomass to fermentable sugars. The hydrolysis process produces byproducts that are toxic to yeast cells, thus interfere with fermentation process (Palmqvist, 2000a). Although several detoxification methods, such as activated charcoal adsorption and lime treatment process, have been devised, an appropriate strategy for efficient hydrolysis of cellulose to fermentable sugars is still lacking (Kaya et al., 2000; Aden et al., 2002).

CELLULOSIC BIOMASS AND THEIR SUGAR COMPOSITION

Lignocellulosic biomass consists of lignin, cellulose, hemicellulose, pectin and other components (Table 2).

Cellulose is the principle component of lignocellulosic biomass and its concentration ranges from 40 to 50% of dry weight. Cellulose is a homopolysaccharide composed of repeating β-D-glucopyranose units. The degree of polymerization and crystallinity of cellulose varies from species to species and this is shown to have a significant impact on hydrolytic process (acidic and enzymatic) (Zhang et al., 2004).

Hemicellulose is less complex, its concentration in lignocellulosic biomass is 25 to 35% and it is easily hydrolysable to fermentable sugars (Saha et al., 2007). Hemicellulose is a heteropoly saccharide composed of pentoses (D-xylose and D-arabinose), hexoses (D-mannose, D-glucose and D-galactose) and sugar acids. Softwood hemicellulose mainly contains mannose as a major constituent whereas hardwoods mainly contain Xylans (Balan et al., 2009).

Lignin is the third major component of lignocellulosic biomass and its concentration ranges for 20 to 35%. It is a complex polymer of phenyl propane (*p*-coumaryl, coniferyl and sinapyl alcohol). Lignin acts as cementing agent and an impermeable barrier for enzymatic attack (Howard et al., 2003). Lignin provides plants with the structural support and impermeability they need as well as resistance against microbial attack and oxidative stress. These properties of lignin may be attributed to its amorphous nature, water insolubility and optical inactivity. The later properties also make it tough to degrade it (Fengel and Wegener, 1984).

OVERVIEW OF LIGNOCELLULOSIC FERMENTATION

The lignin-hemicellulose-pectin complex forms one of the most stringent seals around cellulose. The first step in the overall process of lignocellulosic fermentation is breaking this barrier (pretreatment). This is the most important and rate limiting step in the overall process. Further steps involve isolation and hydrolysis of cellulose and hemicellulose to generate fermentable sugars (saccharification) followed by fermentation and distillation (Figure 1). The pretreatment processes involve the use of acids, alkalis and/or organic solvents. The aim of this

Table 1. Leading bioethanol producers in the world.

Country/group of countries	Ethanol produced in:	
	Million liters	MTOE
Brazil	19,000	10.44
Canada	1,000	0.55
China	1,840	1.01
India	400	0.22
USA	26,500	14.55
European Union	2,253	1.24
Others	1,017	0.56
World (Total)	52,000	28.57

*Source: Data from OECD-FAOA glink-Casimo database (2007). MTOE: Million tons of oil equivalents.

Table 2. Chemical composition of common lignocellulosic biomass.

Constituents	Hardwood (%)	Softwood (%)
Cellulose	40 to 50	40 to 50
Hemicellulose	25 to 35	25 to 30
Lignin	20 to 25	25 to 35
Pectin	1 to 2	1 to 2
Starch	Trace	Trace

*Source: Miller (1999).

process is to separate lignin, cellulose, hemicellulose and pectin from lignocellulosic biomass. Post pretreatment, the recalcitrant lignocellulosic biomass becomes susceptible to acid and/or enzymatic hydrolysis as the cellulosic microfibrils are exposed and/or accessible to hydrolyzing agents. In the pretreatment process, small amounts of cellulose and most of hemicellulose is hydrolyzed to sugar monomers; mainly D-xylose and D-arabinose. The pretreated biomass is then subjected to filtration to separate liquids (hemicellulose hydrolysate) and solid (lignin + cellulose). The liquid is sent to a xylose (pentose) fermentation column for ethanol production. Solids are subjected to hydrolysis (also called second stage hydrolysis). This process is mainly accomplished by enzymatic methods using cellulases. Mild acid hydrolysis using sulfuric and hydrochloric acids is an alternative procedure. The hydrolyzed sugars such as D-glucose, D-galactose, and D-mannose, can be readily fermented to ethanol using various strains of *Saccharomyces cerevisae.* The pentoses (D-xylose and D-arabinose) from hemicellulose hydrolysis are not easily utilized by saccharomyces strains; therefore, genetically modified strains of *Pichia stipitis, Zymomonas mobilis*, are used for their fermentation. *Candida shehatae* is capable of co-fermenting both pentoses and hexoses to ethanol and other value-added products at high yields.

PRETREATMENT OF LIGNOCELLULOSIC BIOMASS

Some of the important reasons for the pretreatment step are to (i) break the lignin-hemicellulose-pectin complex, (ii) disrupt/loosen-up the crystalline structure of cellulose and (iii) increase the porosity of the biomass. These changes in lignocellulosic materials make it easier for enzymatic saccharification (hydrolysis), results in higher fermentable sugars levels and will have a significant impact on the overall process (Mosier et al., 2005a; Sun and Cheng, 2007; Yang and Wyman, 2008). An ideal pretreatment process should yield high levels of pentoses, the hydrolysates will not have any inhibitory substances and the process is cost effective (Lynd, 1996). The current methods in practice include physical, chemical and biological processes. Physical pretreatment methods include combination of stem explosion and hydrothermolysis. Chemical pretreatment processes employ acids, alkali and organic solvents (Carillo et al., 2005). More recent processes have also employed pretreatment with hydrogen peroxide (Saha and Cotta, 2007), sulfite (Kuhad et al., 1999), ammonia (Teymouri et al., 2005) and sodium chlorite (Gupta et al., 2009). Biological pre treatments employ brown-, white- and soft-rot fungi (Basidiomycetes) to degrade lignin and hemicelluloses (Cardona et al., 2007).

Acid pretreatment

The common pretreatment practice in the last two decades has been to incubate lignocellulosic materials in the medium of dilute sulfuric acid (0.5 to 1.5% H_2SO_4) at 100 to 150°C (Wingren et al., 2003). In this process the

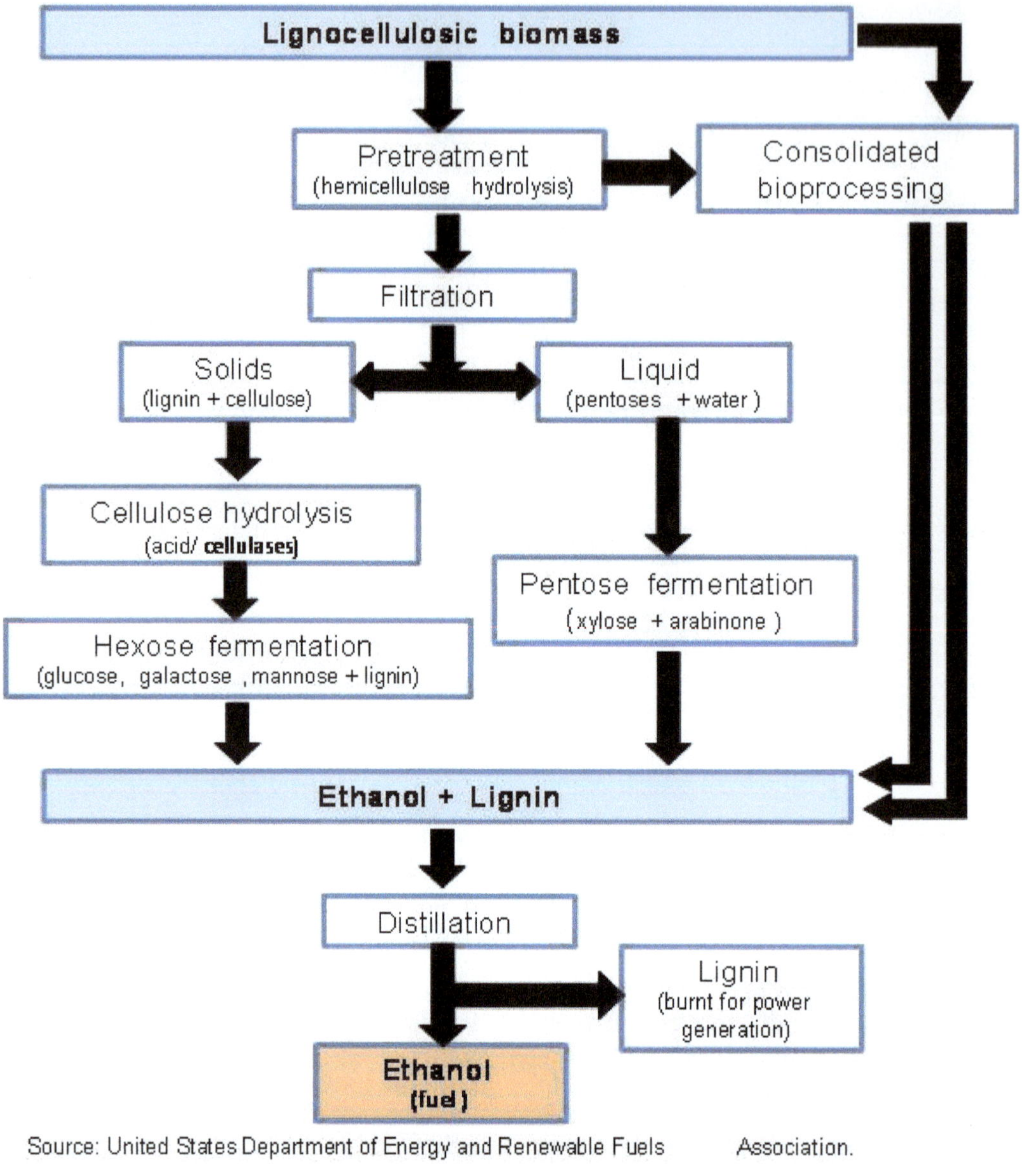

Figure 1. Overview of the process of ethanol production from lignocellulosic biomass.

hemicellulose is hydrolyzed into sugar monomers which are recovered in the liquid fraction after filtration. The residual solids will contain cellulose and lignin which will be further subjected to hydrolysis by cellulases (Taherzadeh and Karimi, 2008). This process has been shown to result in byproducts such as furfural (pentose sugars byproduct), and hydroxyl methyl furfural (HMF; hexose sugar byproduct), phenolic acid (lignin byproduct) and acetate (diacetylation product of hemicellulose). Each of these byproducts at concentration 5 mM or above have been shown to have significant inhibitory effect on the fermentation process (Martinez et al., 2000; Palmqvist et al., 2000b; Klinke et al., 2004). Use of higher concentrations of acid and higher temperatures yield higher reducing sugar concentrations, however, under such conditions reducing sugars are degraded at a much higher level (Zhu et al., 2009). Use of concentrated acid alone has been shown to (i) yield very high levels of sugar (90%), (ii) can handle diverse feedstock, (iii) is relatively rapid (10 to 12 hours), and (iv) causes less degradation. The latter requires more expensive corrosion resistant equipment (Hamelinck et al., 2005) and the hydrolysate have to be neutralized by the addition of lime and contaminants removed by treatment with activated charcoal (Gupta et al., 2009; Shleser,1994). In the later process, over-liming has been shown to catalyze condensation of lignin-derived compounds. The advantage of these reactions is that they detoxify the hydrolysate and these reactions occur at a much higher rate at higher pH. On the other hand, at higher pH, D-glucose is converted to HMF, for example, at pH 11 as much as 30% of glucose is converted to HMF. Additionally, over-liming the hydrolysate at an industrial scale and accurately controlling pH in fermenters pose a major challenge. Steam treatment of H_2SO_4 impregnated lignocellulose is saccharified at a much higher rate. The most favorable conditions for this process require that the samples be treated with 0.5% H_2SO_4 and hold the sample at 200°C for 4 to 8 min (Sassner et al., 2008). Yet another pretreatment method used a mixture of H_2SO_4 and acetic acid with an efficiency of 90% saccharification (DeMoraes-Racha et al., 2010).

Alkali pretreatment

Most commonly used alkali in the alkali pretreatment processes are NaOH and $Ca(OH)_2$. This process results in (i) the removal of all lignin and part of hemicellulose, and (ii) increased reactivity of cellulose in further hydrolysis steps (Hamelinck et al., 2005), especially, enzymatic hydrolysis. Effective removal of lignin minimizes adsorption of enzyme onto lignin and thus allows for effective interactions with cellulose (Aswathy et al., 2010). NaOH causes swelling of the fibers and with increase in internal surface area. This also increases the degree of polymerization and crystallinity of cellulose (Taherzadeh and Karimi, 2008). Pretreatment with NaOH increases the digestibility cellulose from 14 to 55% while decreasing the lignin content from 25 to 20 % (Kumar et al., 2009). Alkali pretreatment process show decreased sugar degradation and are more effective on agriculture residues as compared to wood materials (Kumar et al., 2009). Between NaOH and $Ca(OH)_2$, pretreatment with $Ca(OH)_2$ is preferable because it is less expensive, more safer as compared to NaOH and it can be easily recovered from the hydrolysate by reaction with CO_2 (Mosier et al., 2005b).

Pretreatment with oxidizing agents

In these methods, lignocellulosic materials were treated with powerful oxidizing agents such as per-acetic acid or hydrogen peroxide (Gould, 1984). For example, Teixeira and associates observed ethanol yields of 98% when the lignocellulosic biomass was pretreated with 21% per-acetic acid (Teixeira et al., 1999). Sugar cane bagasse was pretreated with 50% per-acetic acid solution in a liquid to solid ratio of 6:1 at 80°C for 2 h, 80% of the cellulose recovered was converted to glucose by cellulose (Teixeira et al., 1999). Additionally, pre-soaking lignocellulosic biomass with NaOH required significantly smaller concentrations of per-acetic acid to achieve the same results. The latter is likely due to partial removal of lignin prior to per-acetic acid treatment (Teixeira et al., 1999; Zhao et al., 2009).

Organic-solvent pretreatment

The organic solvents such as methanol, ethanol, acetone, ethyleneglycol, triethylene glycol, and tetrahydrofurfuryl alcohol have been used to pretreat lignocellulosic biomass. These solvents are especially effective because they solubilize lignin and some of hemicellulose. Typical pretreatment conditions require that the lignocellulosic biomass be treated with 40 to 60% organic solvent, at 160 to 190°C, for 30 to 60 min (Pan et al., 2005). Pretreatment of lignocellulosic materials with acidified organic solvents (mixture of 80% ethylene glycol 19.5% water and 0.5% HCl at 178°C for 90 min) has also been successfully used (Yamashita et al., 2010). The advantages of these methods include recovery and recycling of organic solvents as they can be easily distilled out. Further the recovery process has been shown to isolate lignin as a solid and carbohydrates as syrup (Lora and Aziz, 1985; Johansson et al., 1987; Aziz and Sarkanen, 1989). The disadvantages are that the process requires expensive high pressure equipment.

HYDROLYSIS

The goal of this process is to generate fermentable monomeric sugars from hemicellulose and cellulose content of lignocellulosic biomass. This can be accomplished by two different processes, namely, acid hydrolysis and enzymatic hydrolysis.

Acid hydrolysis

Mineral acids such as sulfuric acid, hydrochloric acid, hydrofluoric acid and nitric acid are widely employed for the hydrolysis of lignocellulosic biomass. Among these, the oldest and best understood process utilizes sulfuric acid. The sulfuric acid-based hydrolysis process is operated under two different conditions; (í) a process that uses high sulfuric acid concentration that operates at a lower temperature and, (íí) a process that uses low sulfuric acid concentration and operates at a higher temperature. Among the two, the latter is most commonly used. This process produces relatively large number of undesirable byproducts as compared to the concentrated acid process. On the other hand, the acid recovery and recycling process is more complicated and cost intensive. Use of hydrochloric acid, although not commonly used, has technical advantages over the sulfuric acid process, given it is relatively volatile and can be recovered by vacuum stripping methods. Following pretreatment, the acid hydrolysis process is applied to lignocellulosic biomass in a two-step (stage) process because the pentose sugars degrade/decompose more rapidly than hexose sugars. Accordingly, in the first stage hemicellulose is hydrolyzed (Figure 1) with dilute acid under ambient conditions and the more resistant cellulose is hydrolyzed at higher temperatures (213°C) and relatively higher concentration of acid (0.4%) in the second stage. Under the latter conditions, the recovery yields are 89% for mannose, 82% for galactose and 50% for glucose (Graf and Koehler, 2000; USDOE, 2003). In contrast, the concentrated acid methods result in high sugar recovery yields (90%), can handle diverse feed-stocks containing lignocellulose and the process is relatively rapid (10 to12 h) (Graf and Koehler, 2000; USDOE, 2003).

Enzymatic hydrolysis

Three major groups of enzymes are involved in the hydrolysis of cellulose, namely, 1,4-β-D-glucanglucanohydrolase (EC 3.2.1.3), 1,4-β-D-glucancellobiohydrolyase (EC 3.2.1.91) and β-D-glucosidase (EC 3.2.1.21) (Ladisch et al., 1983; Wright et al., 1988). These enzymes are commonly referred to as endoglucanase, exoglucanase and cellobiase, respectively. The endoglucanases attack randomly and cleave the cellulose chains to form glucose, cellobiose and cellotriose. The exoglucanases attack the non-reducing end of cellulose to form the cellobiose units. Finally, cellobiase converts cellobiose into D-glucose.

The factors affecting activity of cellulases include enzyme source and the concentration of enzyme. An effective concentration of enzyme for cellulose hydrolysis has been determined to be 10 to 60 FPU (filter paper units) per gram of dry cellulose or glucan- glucanase- β-D-glucosidase ratio of 1-75-2 IU (Kim et al., 2005). The yield of fermentable sugar levels obtained from pretreated biomass increases as the enzyme load increases and cellulose load decreases (Yang et al., 2002). The use of surfactants is known to block lignin and thus enhance enzymatic saccharification of cellulose (Tu et al., 2009). Addition of bovine serum albumin, polyethylene glycol (PEG) or Tween 20 have been shown to increase enzymatic saccharification and reduces the adsorption of cellulases on lignin and consequently improves the efficiency of cellulose hydrolysis (Pan et al., 2005; Yasmashita et al., 2010).

Both bacteria and fungi can produce glucanases (cellulases) that hydrolyze of lignocellulosic materials. These microorganisms can be aerobic or anaerobic and mesophilic or thermophilic. Bacteria belonging to genera of *Clostridium, Cellulomonas, Bacillus, Thermomonospora, Ruminococcus, Bacteriodes, Erwinia, Acetovibrio, Microbispora*, and *Streptomyces* are known to produce Cellulase (Bisaria, 1998). Anaerobic bacterial species such as *Clostridium phytofermentans, Clostridium thermocellum, Clostridium hungatei, and Clostridium papyrosolvens* produces cellulases with high specific activity (Duff and Murray, 1996; Bisaria, 1998). Most commercial glucanases (cellulases) are produced by *Trichoderma ressei* and β-D-glucosidase is produced from *Aspergillus niger* (Kaur et al., 2007). Fungi known to produce cellulases include *Sclerotium rolfsii, Phanerochaete chrysosporium* and various species of *Trichoderma, Aspergillus, Schizophyllum and Penicillium* (Sternberg, 1976; Fan et al., 1987; Duff and Murray, 1996). Among the fungi, *Trichoderma* species have been extensively studied for cellulase production (Sternberg, 1976).

FERMENTATION

During this process, both pentose and hexose sugars are fermented to ethanol under anaerobic/aerobic conditions. *Saccharomyces cerevisiae* is the most favored organism for ethanol production from hexoses. *P. stipitis* and *Candida shehatae* are capable of fermenting both hexose (glucose) and pentose (xylose) sugars to ethanol (Parekh et al., 1986).

An optimal process for fermentation uses a broth containing *saccharomyces cerevisiae* supplemented with 22% (w/v) sugar, 1% (w/v) of each of ammonium sulfate and potassium dihydrogen phosphate, and fermented at pH 5.0 and 30°C (Junior et al., 2009). Under such conditions a typical strain of *S. cerevisiae* is capable of producing 46.1 g ethanol/L broth (Maziar, et al., 2010).

Cane molasses conditioned with EDTA, ferrocyanide or zeolites, and fermented under similar conditions has been show to enhance ethanol production (Ergun et al., 1997). Further, addition of minimal concentrations of hops acids to the fermentation broth has been shown to prevent bacterial growth and thus enhances ethanol yields (Maye, 2006). Fermentation using immobilized yeast and broth supplemented with Mg, Zn, Cu or Ca-pantothenate has also been shown to increase fermentation efficiency by almost 20% (Nikolic, et al., 2009).

Among the pentose fermenting organisms, *P. stipitis* has been shown to have most promise for industrial applications (Agbogbo et al., 2006). For example, the hemicellulosic hydrolysates of *Prosopis juliflora* (18.24 g sugar/L broth) when fermented with *P. stipitis* produced 7.13 g/L ethanol (Gupta et al., 2009). Detoxified xylose rich hydrolysate of *L. camara* when fermented with *P. stipitis* 3498 at pH 5 and 30°C for 36 h resulted 0.33 g alcohol/g lignocellulose used (Kuhad et al., 2010). In yet another example, the detoxified water hyacinth hemicellulose acid hydrolysate (rich in pentose sugars) fermented with *P. stipitis* NCIM-3497 at pH 6.0 and 30°C resulted 0.425 g ethanol/g lignocellulose. *Candida tropicalis* is also capable of fermenting xylose (pentoses) under oxygen limited conditions in presence of increasing concentrations of polyethylene glycol (Hagerdal et al.,

1985). Genetically engineered strains of *Escherichia coli, S. cerevisiae,* and *Z. mobilis* have been developed to ferment xylose (Kim et al., 2005). Xylose fermentation has also been achieved by expressing xylose reductase and xylitol dehydrogenase together with overexpression of the endogenous xylulokinase and xylose isomerase in *P. stipitis* and other bacterial or fungal species (Eliasson et al., 2000; Kuyper et al., 2003; Karhumaa et al., 2005).

INTEGRATED FERMENTATION TECHNOLOGIES

Simultaneous saccharification and fermentation (SSF)

In simultaneous saccharification and fermentation (SSF) process both cellulose hydrolysis and fermentation of glucose are carried out in presence of fermentative microorganisms in a single step and the process optimally operates at 37 to 38°C. This technique reduces the number of steps in the process, and is a promising way for converting cellulose to ethanol (Lynd et al., 2005; Demain et al., 2005). In SSF process, the lignocellulosic biomass is first pretreated with a dilute acid (1.1% sulfuric acid at 160°C for 10 min) to breakdown lignin-hemicellulose-pectin complex. The resulting broth is filtered to drain the liquid from the system. The drained liquid containing pentose sugars is neutralized with lime and processed via xylose fermentation process (Figure 1). Remaining solids containing cellulose and lignin is then hydrolyzed and fermented simultaneously using by cellulase enzymes and yeast. The cellulase enzymes hydrolyze cellulose to D-glucose, which in turn is fermented to ethanol by yeast (Krishna et al., 2001). This combined process improves both kinetics of fermentation as well as economics of biomass processing via minimization of accumulation of hydrolysis product (glucose) that is inhibitory to cellulases. The disadvantages of using SSF process include (i) the operating temperature should be around 37 to 38°C and (ii) much of the sugar released by cellulose hydrolysis is used for the growth of yeast necessary to ensure good ethanol production.

Simultaneous saccharification and co-fermentation (SSCF)

In simultaneous saccharification and co-fermentation (SSCF), the pretreated lignocellulosic biomass is neutralized and directly exposed to different enzymes and microorganisms that are capable of hydrolyzing cellulose and hemicelluloses to fermentable sugars as well as ferment sugars to ethanol. The process is carried out using genetically engineered microorganisms. This technology is superior to SSF technology in terms of cost effectiveness, better yields, and shorter processing time (Wright, 1988; Lynd et al., 2005; Chandel et al., 2007).

Consolidated bioprocessing (CBP)

Consolidated bioprocessing (CBP) is an alternative processing strategy in which cellulase production, substrate hydrolysis, and fermentation are accomplished in a single step and in one reactor (Lynd et al., 2005; Demain et al., 2005; Vanzyl et al., 2007; Cardona and Sánchez, 2007). It is important to note that, in CBP, only one microbial consortium is employed for both the production of cellulase and fermentation. CBP offers the potential of lower production costs due to simpler lignocellulosic material processing, lower energy inputs and higher conversion efficiencies than SSF or SSCF based processes. CBP is an economically attractive near-term goal for processes involving “third generation” biofuel production (Lynd et al., 2005; Demain et al., 2005). There are two strategies for enabling consolidated bioprocessing; (i) engineering naturally occurring cellulolytic microorganisms to improve product-related properties such as yield and titer, and (ii) engineering non-cellulolytic organisms that exhibit high product yields and titers to express a heterologous cellulase system enabling cellulose utilization (Lynd et al., 2005). CBP is gaining increasing recognition as a potential breakthrough for low-cost biomass processing. The potential is limited by the fact that natural microorganisms exhibiting all the desired features for CBP are not readily available, although, a number of microorganisms, both bacteria and fungi, that possess some of the desirable properties have been identified (Vanzyl et al., 2007).

At present, there is no ideal CBP microorganism that can degrade lignocellulosic biomass proficiently and at the same time consume all the sugars released from biomass to yield ethanol. However, efforts are being made to engineer ethanol producing yeast (*S. cerevisiae*) capable of producing cellulases/hemicellulases and lignocellulose degrading bacteria (*C. thermocellum*) to be an efficient ethanol producer (Lynd et al., 2005). *S. cerevisiae* strains currently used in fermentation processes are incapabe of fermenting xylose to ethanol. The efforts refered to above via biotechnological engineering of yeast genome and protoplast fusion techniques have developed strains that show some promise (Demain et al., 2005). The genome of *C. phytofermentans* (ATCC 700394) encodes for the highest number of enzymes for degradation of lignocellulosic biomass among sequenced clostridial genomes (Weber et al., 2010). In one of our laboratories, this species has been used to ferment ligocellulosic biomass with and with-out pretreatment and observed production of ethanol under laboratory conditions (Mandal and Sreerama, 2011 unpublished observations). Above all, this species can ferment lignocellulosic biomass directly without any pretreatment procedures applied and the efficiency of ethanol production remains the same under either conditions. Thus, it has become a promising native anaerobic CBP microorganism. Recently published work shows that it secretes individual enzymes which not only degrade cellulose and hemicellulose to fermentable sugars but also can consume almost all the sugars present in lignocellulosic biomass and produce ethanol and acetate as the major products (Warnick et al., 2002; Weber et al., 2010).

GENETIC ENGINEERING PROSPECTS FOR BIOETHANOL PRODUCTION

Non-cellulolytic microorganisms with desired product formation properties, for example high yield and titer, can be starting points for CBP organism development. The primary objective of such developments will be to engineer a heterologous cellulase system that enables growth and fermentation on pretreated lignocellulosic biomass. Metabolic engineering via application of recombinant DNA techniques to direct the production of bioethanol is an emerging field. This can be achieved by improving cellular function through the modulation of enzymatic, transport, or other regulatory functions of the cell. Metabolic engineering often involves the introduction of heterologous genes or regulatory elements that are employed to confer novel metabolic configurations.

The cloning and expression of heterologous genes can serve several purposes, for example, (i) extending existing pathway to produce novel products, (ii) engineering arrays of enzymatic activities that synthesize novel structures, (iii) shifting metabolic flux towards synthesis of desired end-products, and (iv) accelerating a rate determining step (Bailey et al., 1991). The genes encoding alcohol dehydrogenase II and pyruvate decarboxylase from *Z. mobilis* have been inserted into *E. coli* under the control of a common promoter. This has resulted in high levels of expression of both enzymes in *E. coli* and resulted in increased cell growth and the production of ethanol (Ingram et al., 1987). To accomplish this, the coding sequences of pyruvate decarboxylase and alcohol dehydrogenase II from *Z. mobilis* were first cloned into the shuttle vector, pCB4, and then transformed into the *Cyanobacterium synechococcus* strain (PCC 7942) (Deng et al., 1999). It has been shown that the ability of *C. phytofermentans* to degrade cellulose relies upon a single Cphy3367 gene which encodes the sole family nine glycoside hydrolases (GH9). This was accomplished by interspecies conjugation with *E. coli* that allows for disruption in Cphy3367 gene (Tolonen et al., 2009). Heterologous gene transfer of pyruvate decarboxylase and alcohol dehydrogenase II from *Z. mobilis* and their expression in *C. cellulolyticum* shows an increase in growth by 2-fold, 1.5 fold increases in cellulose consumption, 93% increase in acetate formation, and a 53% increase in ethanol production as compared to wild type strains. This engineered bacterium also showed 75% increase in H_2 yields (Guedon et al., 1999; Guedon et al., 2002). Antisense RNA (as RNA) strategies allow for down-regulation of specific genes coding for native proteins and thus redirect metabolism (Desai et al., 1999). The feasibility of such technologies in a CBP organism could be extremely useful. This is further supported by models that combine fermentation bioenergetics and the kinetics of pretreated substrate hydrolysis with declining reactivity over the course of the reaction (Van Walsum et al., 1998). High cell density suspensions of the recombinant strains have also been shown to ferment amorphous cellulose, raw starch, and birch wood xylon to ethanol with relatively high yields (Hong et al., 2003). Given the above examples, genetic engineering clearly holds the future prospects of providing technologies that will not only lead to production of bioethanol efficiently, but also holds the potential to reshape bioethanol industry.

LIQUID BIOFUELS IN NEPAL: DEVELOPMENT, STATUS AND PROSPECTS

Nepal is a land-locked country with rich biodiversity and renewable resources. These resources have never been utilized to their full potential given the social and economic challenges the country faces. In Nepal, the traditional biomass such as firewood, cow dung and agricultural waste continues to supply virtually all energy demands. The rate at which the traditional biomass is consumed exceeds its replenishment. The increasing consumption pattern coupled with rising population and increasing per capita demand for energy has placed an unsustainable burden on the environment in Nepal. About 87% of the total energy consumption in Nepal is from non-monetized biomass such as firewood, cow dung and agricultural waste. The imported fossil fuels contribute to about 11% of all energy demands and Nepal invests 40% of its foreign exchange earnings (Pokharel, 2006) to import fossil fuels. Energy consumption in Nepal by fuel type includes 87% traditional fuel and 13% commercial fuel. The commercial fuels include petroleum products (9%), coal (2%), electricity (1.5%) and renewable fuels (0.54%) (WECS, 2006). For fossil fuel imports Nepal is entirely dependent on oil-rich countries and these are imported from middle-eastern countries via India. The price of petroleum products in Nepal are sky-rocketing, the number of motor vehicles are increasing and the demand for petroleum products is growing at a rate of 12% each year.

Although, Nepal is a resource rich country, the investments in renewable energy development is almost non-existent. On the other hand, the potential for the development of biofuels in Nepal is very high. It has been estimated that 30% of Nepal's land is climatically favorable for the cultivation of Jatropha (locally called

Sajiyon or Sajiba). Apart from Jatropha, there are many other non-edible oil seed-bearing plants such as soap nut (*Sapindus mukorossi*), Dhaka (*Aregmone Mexicana*), and Nageswhor (*Mesua ferrea*). In addition, the pine resin (turpentine) is another source of biofuels. The ethanol industry in Nepal is thriving however most of the alcohol is produced from grains and starch commodities. The government of Nepal has a policy on hand to blend 10% ethanol into gasoline (petrol). This has never been implemented due to unsettled disputes over ethanol prices and other vested interests. Studies have shown that bioethanol production from molasses and lignocellulosic materials in Nepal are financially feasible, commercially and economically viable, and environmentally friendly. The raw materials need to establish bioethanol industry using both first generation crops (sugarcane, maize) and second generation lignocellulosic biomass (corn stover, sugarcane bagasse, vegetable wastes, terrestrial weeds such *S. spontaneum, L. camara* (red sage), *Euphatorium adenophorum* (banmara), *Parthenium hysterophorus*, and *E. crassipes* (water hyacinth) are in large supply. However, currently there are no visible commercial applications of biodiesel or bioethanol in Nepal of any significance although credible adaptive research and development, and pilot/demonstration activities have been carried out since 1980's. Nepal has made headways in biotechnology education. Slowly but surely the biotechnology industry is taking a foot-hold.

CONCLUSIONS AND FUTURE PROSPECTS

Demand for transportation fuels across the globe is increasing. This demand is abnormally affecting developing counties in particular. The demand for transportation fuel may be mitigated by bioethanol in the scenario of shrinking energy resources. Lignocellulosic materials are in abundance and they are renewable. Hence, bioethanol production from lignocellulosic biomass holds tremendous potential in terms of meeting energy needs, and providing environmental benefits. Apart from bioethanol, a wide range of chemicals (olefins, plastics, and organic solvents) and value added products such as fermentable sugars, solvents and drink softeners can be produced from the lignocellulosic biomass. The latter needs technical knowledge and capabilities to deal with complexity of biomass. The improvements in pretreatment processes, improvement in efficacy of enzymatic hydrolysis via the development more efficient enzymes, development of efficient fermentation processes via the development of CBP microorganisms, efficient technologies to recover ethanol and removal of toxic byproducts will decrease the operating and capital costs. The reduction in cost of ethanol production can be achieved by reducing the cost of raw material and co-fermenting hexose and pentose sugars in the same tanks. Continued research to isolate enzymes from the natural resources such as termite gut, rumen and anaerobic soils in conjunction with developing more efficient enzymes using biotechnological tools is essential. CBP systems that will use lignocellulosic materials with little or no pretreatment to produce ethanol in a single step are currently evolving. In this regard, studies leading to metabolic engineering of specific strains of microorganisms that simultaneously utilize glucose and xylose have already yielded positive results. Commercialization of these technologies is forthcoming. Combination of strategies that include (meta) genomics, biodiversity studies, systems biology and metabolic engineering hold promising potential to improve biofuel yields and the establishment of renewable, non-polluting energy sources in the future.

REFERENCES

Aden A, Ruth M, Ibsen K, Jechura J,Neeves K, Sheehan J, Wallace B (2002). Lignocellulosic biomass to ethanol process design and economics utilizing co-current dilute acid prehydrolysis and enzymatic hydrolysis for corn stover. Technical Report: NREL/TP-510-32438. National Renewable Energy Laboratory. Golden, Colorado, USA., pp. 1-154.

Agbogbo FK, Coward-Kelly G, Torry-Smith M, Wenger KS (2006). Fermentation of glucose/xylose mixtures using *Pichia stipitis*. Process Biochem., 41: 2333-2336.

Aswathy US, Sukumaran RK, Lalitha D, Rajeshree KP, Singhania RR, Pandey A (2010). Bio-ethanol from water hyacinth biomass: An evaluation of enzymatic saccharification strategy. Bioresour. Technol., 101: 925-930.

Aziz S, Sarkanen K (1989).Organosolv pulping-A review. Tappi. J., 72: 169-175.

Bailey EJ (1991). Toward a science of metabolic engineering. Science, 252: 1668-1675.

Balan V, Sousa LDC, Chundawat SPS, Marshall D, Sharma LN, Chambliss CK, Dale BE (2009). Enzymatic digestibility and pretreatment degradation products of AFEX-treated hardwoods (*Populusnigra*). Biotechnol. Prog., 25: 365-375.

Bisaria VS (1998). Bioprocessing of agro-residues to value added products. In: Martin, A.M. (Ed.), Bioconversion of Waste Materials to Industrial Products, 2nd ed. Chapman and Hall, UK., pp. 197-246.

Bothast RJ, Saha BC (1997). Ethanol production from agricultural biomass substrates. Adv. Appl. Microbiol., 44: 261-286.

Campbell CJ, Laherree JH (1998). Preventing the next oil crunch-the end of cheap oil. Sci. Am., 278: 77-83.

Cardona CA, Sánchez OJ (2007). Fuel ethanol production: Process design trends and integration opportunities. Bioresour. Technol., 98: 2415-2457.

Carere CR, Sparling R, Cicek N, Levin DB (2008). Third generation biofuels via direct cellulose fermentation. Int. J. Mol., Sci. 9: 1342-1360.

Carillo F, Lis MJ, Colon X, Lopez-Mesas M, Valladeperas J (2005). Effect of alkali pretreatment on cellulose hydrolysis of wheat straw, kinetic study. Process Biochem., 40: 3360–3364.

Chandel AK, Chan EC, Rudravaram R, Narasu ML, Rao LV, Ravindra P (2007). Economics and environmental impact of bioethanol production technologies: An appraisal. Biotechnol. Mol. Biol. Rev., 2: 14-32.

Demain AL, Newcomb M, Wu JHD (2005). Cellulase, clostridia, and ethanol. Microbiol. Mol. Biol. Rev., 69: 124-154.

DeMoraes-Rocha GJ, Martin C, Soares IB, Maior AMS, Baudel HM, De Abreu CAM, (2010). Dilute mixed-acid pretreatment of sugarcane bagasse for ethanol production. Biomass Bioenergy, 35: 663-670.

Deng M, Coleman JR (1999). Ethanol synthesis by genetic engineering in Cyanobacteria. Appl. Environ. Microbiol., 2: 523-528.

Desai RP, Papoustakis ET (1999). Antisense RNA strategies for metabolic engineering of *Clostridium acetobutylicum*. Appl. Environ. Microbiol., 65(3): 936-945.

Duff SJB, Murray WD (1996). Bioconversion of forest products industry waste cellulosics to fuel ethanol: A review. Bioresour.Technol., 55: 1-33.

Eliasson A, Christensson C, Wahlbom CF, Hahn-Hägerdal B (2000). Anaerobic xylose fermentation by recombinant Saccharomyces cerevisiae carrying XYL1, XYL2, and XKS1 in mineral medium chemostat culture. Appl. Environ. Microbiol., 66(8): 3381-3386.

Ergun M, Mutlu SF, Gurel O (1997). Improved ethanol production by *Saccharomyces cerevisiae* with EDTA, ferrocyanide and zeolite X addition to sugar beet molasses. J. Chem. Technol. Biotechnol., 68: 147-150.

Fan LT, Gharpuray MM, Lee YH (1987). In: Cellulose Hydrolysis. Biotechnol. Monographs. Springer, Berlin, p. 57.

Fengel D, Wegener G (1984). Chemical composition and analysis of wood. In Wood: Chemistry, Ultrastructure, Reactions. Walter de Gruyter, Berlin, pp. 26-65.

Gould JM (1984). Alkaline peroxide delignification of agricultural residues to enhance enzymaticsaccharification. Biotechnol. Bioeng., 26: 46-52.

Graf A, Koehler T (2000). Oregon cellulose ethanol study, Oregon Office of Energy, Salem, OR, USA., pp. 1-36.

Guedon E, Desvaux M, Petitdemange H (2002). Improvement of cellulolytic properties of *Clostridium cellulolyticum* by metabolic engineering. Appl. Environ. Eng., 68: 53-58.

Guedon E, Payot S, Desvaux M, Petitdemange H (1999). Carbon and electron flow in *Clostridium cellulolyticum* grown in chemostat culture on synthetic medium. J. Bacteriol., 181: 3262-3269.

Gupta R, Sharma KK, Kuhad RC (2009). Separate hydrolysis and fermentation (SHF) of *Prosopis juliflora*, woody substrate for the production of cellulosic ethanol by *Saccharomyces cerevisiae* and *Pichia stipitis* NCIM 3498. Bioresour. Technol., 100: 1214-1220.

Hagerdal BH, Jonsson B, Vogel EL (1985). Shifting product formation from xylitol to ethanol in pentose fermentations using *Candida tropicalis* by adding polyethylene glycol (PEG). Appl. Microbiol. Biotechnol., 21: 73-175.

Hamelinck CN, Hooijdonk GV, Faaji APC (2005). Ethanol from lignocellulosic biomass: Techno-economic performance in short-, middle- and long-term. Biomass Bioenergy, 28: 384-410.

Hong J, Tamaki H, Yamamoto K, Kumagai H (2003). Cloning of a gene encoding thermo stable cellobiohydrolase from *Thermoascus-aurantiacus* and its expression in yeast. Appl. Microbiol. Biotechnol., 63: 42-50.

Howard RL, Abotsi E, Jansen van Rensburg EL, Howard S (2003). Lignocellulose biotechnology: Issues of bioconversion and enzyme production. Afr. J. Biotechnol., 2: 602-619.

Ingram LO, Conway T, Clark DP, Sewell GW, Preston JF (1987). Genetic engineering of ethanol production in *Escherichia coli*. Appl. Environ. Microbiol., 53: 2420-2425.

Johansson A, Aaltonen O, Ylinen P (1987). Organosolvent pulping method and pulp properties. Biomass, 13: 45-65.

Junior MM, Batistote M, Cilli EM, Ernandes JR (2009). Sucrose fermentation by Brazilian ethanol production yeast in media containing structurally complex nitrogen sources. J. Inst. Brewing, 115: 191-197.

Karhumaa K, Hahn-Hägerdal B, Gorwa-Grauslund MF (2005). Investigation of limiting metabolic steps in the utilization of xylose by recombinant *Saccharomyces cerevisiae* using metabolic engineering. Yeast, 22: 359-368.

Kaur J, Chandha BS, Kumar BA (2007). Purification and Characterization of β-Glucosidase from *Melanocarpus* Sp. MTCC3922. Elect. J. Biotechnol., 10: 260-270.

Kaya F, Heitmann JA, Thomas (2000). Influence of lignin and its degradation products on enzymatic hydrolysis of xylan. J. Biotechnol., 80: 241-247.

Kim S, Holtzapple MT (2005). Lime pretreatment and enzymatic hydrolysis of corn stover. Bioresour. Technol., 96: 1994-2006.

Klinke HB, Thomsen AB, Ahring BK (2004). Inhibition of ethanol-producing yeast and bacteria by degradation products produced during pre-treatment of biomass. Appl. Microbiol. Biotechnol., 66: 10-26.

Krishna SH, Reddy TJ, Chowdary GV (2001). Simultaneous saccharification and fermentation of lignocellulosic wastes to ethanol using a thermotolerant yeast. Bioresour. Technol., 77: 193-196.

Kuhad RC, Gupta R, Khasa YP, Singh A (2010). Bioethanol production from *Lantana camara* (red sage): Pretreatment, saccharification and fermentation. Bioresour. Technol., 101: 8348-8354.

Kuhad RC, Manchanda M, Singh A (1999). Hydrolytic potential of extracellular enzymes from a mutant strain of *Fusarium oxysporum*. Bioprocess. Eng., 20: 47-125.

Kumar R, Wyman CE (2009). Effects of cellulase and xylanase enzymes on the deconstruction of solids from pretreatment of poplar by leading technologies. Biotechnol. Prog., 25: 302-314.

Kuyper M, Harhangi HR, Stave AK, Winkler AA, Jetten MS, De Laat WT, Den Ridder JJ, Op Den Camp HJ, Van Dijken JP, Pronk JT (2003). High level functional expression of a fungal xylose isomerase: The key to efficient ethanolic fermentation of xylose by *Saccharomyces cerevisiae*. FEMS Yeast Res., 4: 69-78.

Ladisch MR, Lin KW, Voloch M, Tsao GT (1983). Process considerations in the enzymatic-hydrolysis of biomass. Enzyme Microbiol. Technol., 5: 82-102.

Lora JH, Aziz S (1985). Organosolv pulping: A versatile approach to wood refining. Tappi. J., 68: 94-97.

Lynd LR (1996). Overview and evaluation of fuel ethanol from cellulosic biomass: Technology, economics, the environment, and policy. Ann. Rev. Energy Environ., 21: 403-465.

Lynd LR, Van Zyl WH, McBride JE, Laser M (2005). Consolidated bioprocessing of cellulosic biomass: An update. Curr. Opin. Biotechnol., 16: 577-583.

Martinez A, Rodriguez ME, York SW, Preston JF, Ingram LO (2000). Use of UV absorbance to monitor furans in dilute acid hydrolysate of biomass. Biotechnol. Prog., 16: 637-641.

Maye JP (2006). Use of hop acids in fuel ethanol production. US patent. Patent US2006263484.

Maziar SA (2010). A study on some efficient parameters in batch fermentation of ethanol using *S. cerevisea* SC1 extracted from fermented Siahe Sardasht Pomace, Afr. J. Biotechnol., 9: 2906-2912.

Miller RB (1999). Structure of wood. In*:* Wood handbook: Wood as an engineering material. Madison, WI: USDA Forest Service, Forest Products Laboratory, 1999. General Technical Report FPL. GTR, 113: 2.1-2.4.

Mitchell D (2008). A note of rising food prices. Policy research working paper 4682. Development Prospects Group. The World Bank, Washington D.C., USA., pp. 1-21.

Mosier N, Hendrickson R, Ho N, Sedlak M, Ladisch M (2005a). Optimization of pH controlled liquid hot water pretreatment of corn stover. Bioresour. Technol., 96: 1986-1993.

Mosier N, Wyman CE, Dale BD, Elander RT, Lee YY.Holtzapple M, Ladisch CM (2005b). Features of promising technologies for pretreatment of lignocellulosic biomass. Bioresour. Technol., 96: 673-686.

Nikolic S, Mojovic L, Pejin D, Rakin M, Vucurovic V (2009). Improvement of ethanol fermentation of corn Semolina hydrolyzates with immobilized yeast by medium supplementation. Food Technol. Biotechnol., 47: 83-89.

Palmqvist E, Hahn-Hagerdal A (2000a). Fermentation of lignocellulosic hydrolysates. I: Inhibitors and detoxification. Bioresour. Technol., 74: 17-24.

Palmqvist E, Hahn-Hagerdal A (2000b). Fermentation of lignocellulosic hydrolysates. II. Inhibitors and mechanisms of inhibition. Bioresour. Technol., 74: 25-33.

Pan XJ, Xie D, Gilkes N, Gregg DJ, Saddler JN (2005). Strategies to enhance the enzymatic hydrolysis of pretreated softwood with high residual lignin content. Appl. Biochem. Biotechnol., 121: 1069-1079.

Parekh S, Wayman M (1986). Fermentation of cellobiose and wood sugars to ethanol by *Candida shehatae* and *Pichia stipitis*. Biotechnol. Lett., 8: 597-600.

Perlack RD, Wright LL, Turhollow AF, Graham RL, Stokes BJ, Erbach DC (2005). Biomass as a feedstock for a bioenergy and bioproducts industry: The technical feasibility of a billion-ton annual supply. Oak

Ridge National laboratory. USA., pp. 1-78.

Pokharel S (2006). An econometric analysis of energy consumption in Nepal, Nayang Technological University, Singapore. J. Chem. Technol. Biotechnol., 82: 1039-1044.

Saha BC, Cotta MA (2007). Enzymatic saccharification and fermentation of alkaline peroxide pretreated rice hulls to ethanol. Enzyme Microbiol. Technol., 41: 528-532.

Shleser R (1994). Ethanol production in Hawaii: Processes, feedstock, and current economic feasibility of fuel grade ethanol production in Hawaii. Final Report. Department of Business,Economic Development and Tourism, State of Hawaii, USA., pp. 1-87.

Sternberg D (1976). Production of cellulase by Trichoderma. Biotechnol. Bioeng., Symp., pp. 35-53.

Sun FB, Cheng HZ (2007). Evaluation of enzymatic hydrolysis of wheat straw pretreated by atmospheric glycerol autocatalysis. J. Chem. Technol. Biotechnol., 82: 1039-1044.

Taherzadeh MJ, Karimi K (2008). Pretreatment of lignocellulosic wastes to improve ethanol and biogas production: A review. Int. J. Mol. Sci., 9: 1621-1651.

Teixeira LC, Linden JC, Schroeder HA (1999). Alkaline and peracetic acid pretreatments of biomass for ethanol production. Appl. Biochem. Biotechnol., 77: 19-34.

Teymouri F, Laureano-Perez L, Alizadeh H, Dale BE (2005). Optimization of the ammonia fiber explosion (AFEX) treatment parameters for enzymatic hydrolysis of corn stover. Bioresour. Technol., 96: 2014-2018.

Tolonen AC, Chilaka AC, Church GM (2009). Targeted gene inactivation in *Clostridium phytofermentans* shows that cellulose degradation requires the family 9 hydrolase Cphy3367. Mol. Microbiol., 74: 1300-1313.

Tu MB, Zhang X, Paice M, McFarlane P, Saddler JN (2009). Effect of surfactants on separate hydrolysis fermentation and simultaneous saccharification fermentation of pretreated lodge pole pine. Biotechnol. Prog., 25: 1122-1129.

United States Census Bureau (USCB) (2008). World population clock. United States Census Bureau. http://www.census.gov.

United States Department of Energy (USDOE) (2003). Advanced bioethanol technology-website www.ott.doe.gov/biofuels/ US Department of Energy, Office of Energy Efficiency and Renewable Energy and Office of Transportation Technologies, Washington DC., USA.

Van Walsum P, Lynd LR (1998). Allocation of ATP to synthesis of cells and hydrolytic enzymes in cellulolytic fermentative microorganisms: Bioenergetics, kinetics, and bioprocessing. Biotechnol. Bioeng., 58: 316-320.

Warnick TA, Methe BA, Leschine SB (2002). *Clostridium phytofermentans* sp. nov., a cellulolytic mesophile from forest soil. Int. J. Syst. Evol. Microbiol., 52: 1155-1160.

Weber C, Farwick A, Benisch F, Brat D, Dietz H, Subtil T, Boles E (2010). Trends and challenges in the microbial production of lignocellulosic bioalcohol fuels. Appl. Microbiol. Biotechnol., 87: 1303-1315.

WECS (2006). Energy Synopsis Report, Nepal, Water and Energy Commission Secretariat, Ministry of Water Resources, Nepal.

Wheals AE, Basso LC, Denise M, Alves G, Amorim H (1999). Fuel ethanol after 25 years. Trend. Biotechnol., 17: 482-487.

Wingren A, Galbe M, Zachhi G (2003). Techno-economic evaluation of producing ethanol from Softwood: Comparision of SSF and SHF and Identification of bottle necks. Biotechnol. Prog., 19: 1086-1093.

Wright JD (1988). Ethanol from lignocellulosics: An overview. Energy Prog., 84: 71-80.

Wright JD, Wyman CE, Grohmann K (1988). Simultaneous saccharification and fermentation of lignocellulose – Process evaluation. Appl. Biochem. Biotechnol., 18: 75-90.

Yamashita Y,Sasaki C, Nakamura Y (2010). Effective enzyme saccharification and ethanol production from Japanese cedar using various pretreatment methods. J. Biosci. Bioeng., 110: 79-86.

Yang B, Boussaid A, Mansfield SD (2002). Fast and efficient alkaline peroxide treatment to enhance the enzymatic digestibility of steam exploded soft wood substrate. Biotechnol. Bioeng., 77: 678-684.

Yang B, Wyman CE (2008). Pretreatment: The key to unlocking low-cost cellulosic ethanol. Biofuel. Bioprod. Biorefin., 2: 26-40.

Zhang X, Yu H, Huang H, Liu Y (2004). Evaluation of biological pretreatment with white-rot fungi for the enzymatic hydrolysis of bamboo culms. Int. Biodivers. Biodegrad., 60: 159-164.

Zhao XB, Peng F, Cheng KK, Liu DH (2009). Enhancement of the enzymatic digestibility of sugarcane bagasse by alkali–peracetic acid pretreatment. Enzyme Microbiol. Technol., 44: 156-164.

Zhu JY, Pan XJ, Wang GS, Gleisner R (2009). Sulfite pretreatment for robust enzymatic saccharification of spruce and red pine. Bioresour. Technol., 100: 2411-2418.

9

RNAi in medicine: current and future perspectives

L. Sudarsana Reddy, V. Sarojamma[$] and V. Ramakrishna*

Department of Biotechnology, Sri Krishnadevaraya University, Anantapur – 515 003.
[$]Department of Microbiology, Kurnool Medical College, Kurnool – 518 001, INDIA.

The discovery of 21 – 23 nucleotide RNA duplexes, called small interference RNA (siRNA) may well be one of the transforming events in biology in the past decade. RNAi can result in gene silencing or even in the expulsion of sequences from the genome. Efforts to understand its mode of action have revealed a central role in gene regulation and host defense. The specificity, efficiency and potency of RNAi makes it an attractive tool for analyzing the function of genes. RNA interference can be exploited artificially to inhibit the expression of any gene of interest. RNA interference systems could be used clinically to suppress gene expression as a therapeutic strategy in many diseases characterized by elevated gene function. Finally, as a therapeutic tool, it has shown enormous promise in the control of a large array of diseases. This review focuses on the potential therapeutic use of RNAi for various diseases, the current understanding of RNAi biology, and how RNAi has been utilized to study the role of different genes in the pathogenesis of cancer, HIV, infectious diseases, HBV, cardiovascular diseases, cerebral diseases, neurogenerative diseases, malaria, among others.

Key words: RNA interference – cancer – HIV – hepatitis B – neurogenerative diseases – malaria – infectious diseases.

TABLE OF CONTENTS

1. Introduction
2. RNA interference - Biology
3 Therapeutic applications of RNA interference
3.1 RNAi in Cancer: Advances, challenges and opportunities
3.2 Infectious diseases
3.3 RNAi in HIV treatment; Advances, challenges and opportunities
3.4 HBV (Hepatitis-B Virus) as an RNAi target
3.5 Cardiovascular and cerebral vascular diseases
3.6 Neurodegenerative disorders
3.7 Antiproliferative and proapoptotic RNAi
3.8 Malaria
3.9 RNAi and metabolic diseases
4 The future of RNA interference in Medicine
5 Challenges for RNAi as a therapy
6 Conclusion
7 Acknowledgement
8 References

*Corresponding author's E-mail: vrkrishna70@rediffmail.com.

1. INTRODUCTION

In the early days of molecular biology, genes were first defined through the description of their mutant phenotype. Forward genetics has the advantage that the phenotype of the mutant gives a clue to the function of the gene. But with the advent of large-scale genome sequencing, thousands of genes have been identified without of their function. Reverse genetics is now the most effective way to assess the function of a gene, but so far there has been no general method for reverse genetics other than gene targeting by homologous recombination, which is slow and costly. Antisense approaches, such as antisense oligonucleotide and ribozyme technologies, have been useful in reverse genetics, but only to a limited degree. By contrast, the promise of small interfering RNA (siRNA) technology to 'knock down' the expression of any gene in vertebrate cells is set to revolutionize reverse genetic approaches (Michael et al., 2002). In 1998, and at the time of the completion of the *Caenorhabditis elegans* genome project, Andrew Fire and Craig Mello described this new technology based on the silencing of specific genes by double stranded RNA (dsRNA); a technology they called RNA interference (RNAi).

2. RNA interference - Biology

The discovery of 21–23 nucleotide RNA duplexes (Elbashir et al., 2001; Dykxhoorn et al., 2003; Eugene Berezikov et al., 2006), called small interference RNA (siRNA) may well be one of the transforming events in biology in the past decade. Production of small interfering RNAs (siRNAs) that bind to and induce the degradation of specific endogenous mRNAs is now recognized as a mechanism that is widely employed by eukaryotic cells to inhibit protein production at a posttranscriptional level (Olivier Milhavet et al., 2003; Meister et al.,2004). In animals, micro RNAs are transcribed as long primary transcripts (pri-iRNAs) by RNA polymerase II enzyme Esquela-Kerscher and Slack, 2006). They are cropped into hairpin-shaped pre-iRNAs by the nuclear RNase III Drosha-DGCR8 complex. The processing intermediates, pre-iRNAs are exported out of the nucleus by exportin-5 (Exp-5) and subsequently cleaved by the cytoplasmic RNase III, Dicer, into ~22 nucleotides duplexes (Lee et al., 2002; Basyuk et al., 2003; Bryan R Cullen 2006). One strand of the duplex (the mature micro RNA) is incorporated into the RISC (RNA –induced Silencing Complex) or iRNP complex, a protein complex that, when bound to the micro RNA is responsible for the post-transcriptional regulation of the corresponding protein (Denli et al., 2003; Gregory et al., 2005; Matranga et al., 2005; Rand et al., 2005). This complex down regulates the gene expression in one of two ways. (i) Translation inhibition, in case where micro RNA only partially complements to its corresponding mRNA and (ii) Target mRNA cleavage, in case where there is a near perfect complementary between the iRNA and the mRNA.

Salient features of RNA interference are - double stranded RNA, high degree of specific gene silencing with less effort, highly potent and effective, silencing can be introduced in different developmental stages, systemic silencing, avoids problems with abnormalities caused by a knocked out gene in early stages and silencing effects passed through generations.

3. Therapeutic Applications of RNA Interference

RNAi was described by Jennifer Couzin in the journal *Science* as the "Break through of the Year" in 2002 having the potential to become a powerful therapeutic drug toward targeted and personal medicine. RNAi-based therapeutics has potentially significant advantages over traditional approaches to treating diseases, including broad applicability, therapeutic precision, and selectivity avoiding side effects associated with currently marketed pharmaceutical products. This widespread applicability, coupled with relative ease of synthesis and low cost of production make siRNAs an attractive new class of small-molecule drugs. RNAi-based drugs are designed to destroy the target RNA and therefore stop the associated undesirable protein production required for disease progression (Julian Downward, 2004; Peter Sandy, 2005; Ryther, 2005; Cristina Rondinone, 2006; Morris, 2006). Finally, as a therapeutic tool, it has shown enormous promise in the control of a large array of diseases. This review focuses on the potential therapeutic use of RNAi for various diseases, the current understanding of RNAi biology, and how RNAi has been utilized to study the role of different genes in the pathogenesis of cancer, HIV, infectious diseases, HBV, cardiovascular and cerebral vascular diseases, neurogenerative diseases, malaria, diabetes and obesity.

3.1 RNAi in Cancer: Advances, Challenges and opportunities

There are two general abnormalities in cancer cells- they exhibit dysregulation of the cell cycle resulting in uncontrolled growth and they are resistant to death as a result of abnormalities in one or more proteins that mediate apoptosis. In cancer cells, proto-oncogenes have frequently been activated by various mechanisms, producing oncogenes that act in a dominant fashion (Nam and Parang, 2003; Hwang and Mendell, 2006). In epithelial tumors, point mutations are predominant whereas hematological malignancies often show gene fusions that result from chromosomal translocations (Arndt Bork-Borkhardt, 2002). The goals for RNAi approaches for cancer therapy are therefore to knock out the expression of a cell cycle gene and/or an anti-apoptotic gene in cancer cells thereby stopping tumor growth and killing the cancer cells. To selectively eliminate cancer cells without damaging normal cells, RNAi would target a gene specifically involved in the growth or survival of the cancer cell, or the siRNAs would be selectively delivered into the cancer cells by transfection (Olivier Milhavet et al., 2003). Because interference RNA can function either as tumor suppressor or as oncogenes, they have been referred as "Oncomirs" (Auror Esquela-Koscher and Frank Slack, 2006). Recent evidence has shown that miRNA mutations or mis-expression correlate with vario-us human cancers (Auror Esquela-Koscher and Frank Slack, 2006), further asserting their oncogenic or tumor suppressor function. miRNAs have been shown to repress the expression of important cancer-related genes and might prove useful in the diagnosis and treatment of cancer (Calin et al., 2004 ; Lu et al., 2005; Auror Esquela-Koscher and Frank Slack, 2006). A recent study (Lu et al., 2005) showed that about 50% of annotated human miRNAs are located in areas of the genome, known as fragile sites that are associated with cancer and indicates a crucial role in cancer progression (Lin et al., 2005).

The differential expression of certain miRNAs in various tumours might become a powerful tool to aid in the diagnosis and treatment of cancer. Northern-blot analy-ses and miRNA microarrays have been useful in determining tissue-specific 'signatures' of miRNA genes in humans (Liu et al., 2004; Nelson et al., 2004; Miska et al., 2004; Monticelli et al., 2005). Researchers are now using miRNA-expression signatures to classify cancers and to define miRNA markers that might predict favorable prog-nosis (Iorio et al., 2005; Cimmino et al., 2005; Chan et al., 2005; He et al., 2005; O'Donnell et al., 2005). Gene therapies that use miRNAs might be an effective appro-ach to blocking tumour progression. miRNAs such as *let-7*, which has been shown to negatively regulate the Ras oncogenes, and miR-15 and miR-16, which negatively regulate *BCL2*, are promising candidates for cancer treatment. miRNAs have been shown to repress the expression of important cancer-related genes and might prove useful in the diagnosis and treatment of cancer (Cimmino et al., 2005; Calin et al., 2005).

Recent evidence has shown that microRNAs play a significant role in the development of different forms of cancer. Based on these important findings, Scientists can pursue a range of product development opportunities that will harness the power of microRNAs in the diagnosis and treatment of many different forms of cancer in the years ahead. The key to understanding the roles of iRNA in cancer will be identifying and validating the critical targets responsible for the phenotypic effects of miRNA loss- or gain-of-function. Another potential challenge for future studies relates to the probable tissue-specific functions of some iRNAs.

3.2 Infectious diseases

Diseases caused by viruses and bacteria continue to be major causes of death worldwide and are an increasing concern because of the emergence of resistant strains and the potential use of infectious pathogens by terrorists (Tan et al., 2000; Franz and Zajtchuk, 2002; Olivier Milhavet et al., 2003). The prominent viral infectious diseases include HIV, Influenza, Hepatitis, and West Nile virus and from bacterial infectious diseases pneumonia and sepsis are the prominent examples. The ability of RNAi to inhibit the replication or cellular uptake of viruses and other infectious agents has been clearly demonstrated in cell culture studies (Leonard and Schaffer, 2005). Transfection of human cells with siRNAs directed against different genes in the poliovirus genome resulted in resistance of the cells to infection with poliovirus (Gitlin et al., 2002). The ability of siRNAs targeting the gene encoding the death receptor Fas to protect mice from liver failure and fibrosis in two models of autoimmune hepatitis was tested by Song and colleagues (Song et al., 2003). Intravenous injection of Fas siRNA specifically reduced Fas protein levels in the livers of mice during a 10-day period. Fas siRNA treatment abrogated hepatocyte necrosis and infla-mmatory infiltration and markedly reduced serum conc-entrations of transaminases demonstrating a clear heap-topro-tective effect of the siRNA therapy. The obser-vations made in reference (Jopling, 2005; Krützfeldt et al., 2005) was the specific cellular miRNAs can facilitate virus replication, together with the finding that several pathogenic human viruses encode miRNAs that undoubt-edly are important in the virus life cycle *in vivo*, suggests that technologies that allow the specific inhibition of miRNA function, such as the recently described anta-gomir molecules, could have a future role in the treatment of virally induced diseases.

3.3 RNAi in HIV treatment: Advances, challenges and opportunities

HIV was the first infectious agent targeted by RNAi, perhaps because the lifecycle and pattern of gene expression of HIV is well understood. Synthetic siRNAs and expressed shRNAs have been used to target several early and late HIV-encoded RNAs in cell lines and in primary haematopoietic cells including the TAR element, tat, rev, gag, env, vif, nef, and reverse transcriptase (Jacque et al., 2002; Novina et al., 2002; Leonard and Schaffer, 2005; John J Rossi, 2006). Despite the success of RNAi-mediated inhibition of HIV-encoded RNAs in cell culture, targeting the virus directly represents a substantial challenge for clinical applications because the high viral mutation rate leads to mutants that can escape being targeted (Martinez et al., 2002). Therefore RNAi-mediated downregulation of the cellular cofactors required for HIV infection is an attractive alternative or complementary approach. Cellular cofactors such as NFk-B, the HIV receptor CD4, and the co-receptors CXCR4 and CCR5 (Capodici et al., 2003) have been successfully down regulated by RNAi, resulting in the inhibition of HIV replication in numerous human cell lines and in primary cells including T lymphocytes and haematopoietic-stem-cell-derived macrophages (Gitlin et al., 2002; Martinez et al., 2002; Boden et al., 2003; Capodici et al., 2003; John J. Rossi, 2006). There are nonetheless, drawbacks in targeting cellular HIV cofactors because non-infected cells will inevitably be targeted and as well, leading to toxicities that are similar to those observed with the current anti-retroviral drugs. Viral targets will need to be included in any successful strategy using RNAi and these targets should be sequences that are highly conserved throughout the various clades to ensure efficacy against all viral strains (Jacque et al., 2002).

An alternative approach to relying solely upon RNAi as an anti-HIV approach is mixing a single shRNA with other antiviral genes to provide a potent combinatorial approach. This has been successfully accomplished by co-expressing an anti-*tat/rev* shRNA, a nucleolar- localizing TAR decoy, and an anti-CCR5 ribozyme in a single vector backbone (Cordelier et al., 2003). A somewhat different combination used a shRNA with a dominant negative Rev M10 protein in a co-expression system (Li et al., 2005). Future investigations should focus on developing more potent combinations of iRNAs with mixtures of noniRNA antivirals for testing in preclinical settings.

The delivery of iRNAs to HIV-infected cells is also a challenge. The target cells are primarily T lymphocytes, monocytes and macrophages (Olivier Milhavet et al., 2003; Cordelier et al., 2003; Anderson and Akkina, 2005; Dmitriy Ovcharenko et al., 2006). Since synthetic siRNAs do not persist for long periods in cells (Julian Downward, 2004), they would have to be delivered repeatedly for years to effectively treat the infection. Systemic delivery of siRNAs to T lymphocytes is probably not feasible owing to the immense number of these cells. Using viral vectors to deliver anti-HIV-encoding short hairpin RNA (shRNA) genes is also problematic, and systemic delivery is not yet practicable because the immunogenicity of these vectors themselves precludes performing multiple injections (Olivier Milhavet et al., 2003; Gitlin et al., 2002; Michienzi et al., 2003). Therefore the preferred method is to isolate T cells from patients, which are then transduced, expanded and re-infused into the same patients. In clinical trial, T lymphocytes from HIV-infected individuals are transduced *ex vivo* with a lentiviral vector that encodes an anti-HIV antisense RNA (Morris and Rossi, 2004; Morris and Looney, 2005). The transduced cells are subsequently expanded and reinfused into patients. This type of therapeutic approach would also be applicable to vectors harboring genes that encode siRNAs. A different approach is to transduce isolated haematopoietic progenitor or stem cells with vectors harbouring the therapeutic genes. These cells give rise to all the haematopoietic cells capable of being infected by the virus. Haematopoietic stem cells are mobilized from the patient and transduced *ex vivo* before reinfusion (Esquela-Kerscher and Slack, 2006) and demonstrated it is more feasible approach (Amado et al., 2004; Banerjea et al., 2003).

Viral vector mediated delivery to hematopoietic cells, including stem cells, is a feasible approach for shRNA gene delivery. Clearly, the barriers that initially confronted therapeutic applications of RNAi for HIV infection are rapidly being broken down, and one can expect to see this powerful cellular process applied clinically to HIV-1 infected patients within the year.

3.4 HBV (Hepatitis-B Virus) as an RNAi target

Over 500 million people worldwide are infected with one or more different and unrelated types of human hepatitis virus. Such individuals are at a high risk of developing acute or chronic hepatic disease, and ultimately dying from sequelae. Although a vaccine is available for hepatitis A and B virus, treatment options for chronically infected patients are limited, and particularly ineffective in case of hepatitis C virus (HCV) infection (Mast et al., 1999; Bowen and Walker, 2005). A promising new avenue currently being explored is to harness the power of RNA interference for development of an antiviral therapy (Grimm and Kay, 2006). The timing to pursue this particular approach is excellent, with the first *in vivo* animal models for HCV infection becoming available (McCaffrey et al., 2003; Persengiev et al., 2004; Taylor and Naoumov, 2005; Locarnini, 2005), and the technology for liver-specific expression of short hairpin RNAs advancing at a rapid pace. Very recently Kusov *et al* (2006) achieved an efficient and sustained suppression of the viral infectivity after consecutive applications of an siRNA targeting a computer-predicted hairpin structure. This siRNA holds promise as a therapeutic tool for severe courses of HAV infection and in addition, these results provide new insight into the structural bases for sequence specific RNAi. HBV makes extensive use of overlapping reading frames within its DNA genome (Cheng et al., 2005), suggesting that while the viral DNA itself cannot be targeted; the multiple HBV RNAs will make the virus highly susceptible for RNAi. HBV is in fact an excellent candidate for therapeutic RNAi because its unusually compact genome lacks redundancy, resulting in very limited sequence plasticity thus preventing the virus from evading RNAi through mutation. Thus ideally, a single iRNA can potentially target multiple viral transcripts simultaneously, efficiently inhibiting not only viral gene expression, but also DNA replication (Wieland and Chisari, 2005).

Most noteworthy findings were obtained by McCaffrey et al. (2003), Shlomai and Shaul (2004), and Uprichard et al. (2005) who independently assessed anti-HBV shRNAs in different *in vivo* models and found high Efficiencies of their constructs, albeit for only relatively short periods (up to 26 days). An important consistent finding in these studies was that inhibition of viral gene expression did not require active viral replication (Cheng et al., 2005; Ren et al., 2005; Grimm and Kay, 2006), suggesting that RNAi strategies are excellent options as adjutants to conventional anti-HBV therapies, for example with inhibittors of the viral reverse transcriptase. The current understanding of RNAi-related mechanisms raises considerable hope that we will see the clinical evaluation of efficient, safe and specific antiviral RNAi therapeutics for treatment of virally induced human liver diseases in the not too-distant future.

3.5 Cardiovascular and Cerebral vascular Diseases

Cardiovascular diseases are the leading cause of death in the United States and many other industrialized countries. These diseases are commonly results from the progressive occlusion of arteries in a process called Atherosclerosis, which can ultimately culminate in a myocardial infarction or stroke (Forbes et al., 2004). Atherosclerosis involves damage to vascular endothelial cells, local production of inflammatory cytokines, and the recruitment of macrophages to the site of foam cell formation; in addition to these, apoptosis of foam cells and vascular smooth muscle cells occurs (Geng and Libby, 2002). The severe irritability that occurs in heart or brain cells during a myocardial infarction or stroke results in the death of cardiac muscle cells and neurons. Although some of the cells die rapidly by necrosis, many other cells die more slowly by apoptosis; data from animal studies by Zhao and Vinten-Johansen (2002) and Mattson *et al* (2000) shown that such cardiac myocytes and brain neurons are saved from die by apoptosis.

RNA interference (RNAi) utilizing small interfering RNAs (siRNAs) is a recent advance that provides the possibility of reducing gene expression at the post-transcriptional level in cultured mammalian cells (Elbashir et al., 2001). This technology exploited in the process of atherosclerosis or to reduce the damage to heart tissue and brain cells that patients suffer following a myocardial infarction or stroke. A key step in the process of atherosclerosis is the up-regulation of cell adhesion molecules in vascular endothelial cells, which play an essential role in the recruitment of macrophages to the site of endothelial damage. The production of cell adhesion molecules can be selectively suppressed in cultured cells (Jarad et al., 2002).

3.6 Neurodegenerative Disorders

Recent rapid progress in the applications of RNAi to mammalian cells, offers new approaches to drug target identification and validation for treatment of disease. Advances in targeted delivery of RNAi-inducing molecules has raised the possibility of using RNAi directly as a therapy for a variety of human genetic and other neural and neuromuscular disorders (Steven et al., 2004). RNA interference (RNAi) is a powerful new gene knockdown technique that permits tissue-specific, temporally controlled suppression of gene expression (Rodriguez-Lebron, and Paulson, 2006). Alzheimer's disease, Parkinson's disease, Huntington's disease, and Amyotrophic lateral sclerosis (ALS) are examples of relatively common age-related neurodegenerative disorders that are increasing as average life expectancy increases (Raoul et al., 2006). Each disorder is characterized by the dysfunction and death of specific populations of neurons: Hippocampal and cortical neurons involved in learning and memory processes in Alzheimer's disease (Link, 1995; Miller et al., 2004), dopamine-producing neurons in the substantia nigra that control body movements in Parkinson's disease (Feany and Bender, 2000), and spinal cord motor neurons in amyotrophic lateral sclerosis (Elia et al., 1999). Specific genetic mutations are responsible for a small percentage of cases of Alzheimer disease, Parkinson's disease and Amyotrophic lateral sclerosis (Hardy, 2001), whereas all cases of Huntington's disease result from mutations (polyglutamine expansions) in the Huntington protein (Rubinsztein, 2002). Studies of patients, and of animal and cell culture models of each disease, have revealed shared biochemical cascades that result in neuronal death. Those cascades include increased oxidative stress, dysregulation of cellular calcium homeostasis and apoptosis (Zhao and Vinten-Johansen, 2002). As a result, different strategies preventative and therapeutic intervention strategies have been employed in neurodegenerative disorders. One strategy is to block the disease-specific events that are believed to initiate the neurodegenerative process, whereas the second strategy targets downstream events in the neurodegenerative cascade (Olivier Milhavet et al., 2003). Recent studies have shown that cultured neurons can be efficiently transfected with siRNAs, leading to effective silencing of target genes (Rubinsztein, 2002). In one study it was shown that cultured neurons can be depleted of the p75 neurotrophin receptor, a protein in the TNF receptor family that has been implicated in neuronal apoptosis in certain settings (Higuchi et al., 2003). Pro-apoptotic members of the Bcl-2 family (Colussi et al., 2000) and caspases (Quinn et al., 2000) have been effectively targeted and neuronal death prevented using RNAi methods. Caplen and colleagues (2002) performed studies aimed at determining whether RNAi could be used to target the pathogenic process in inherited neuro-degenerative disorders caused by polyglu-tamine expansions and showed loss of ARGFP aggre-gates by 80% in cotransfected S2 cells. Therefore, RNA interference could have considerable therapeutic potential in poly (Q) neurodegenerative disorders. Thus, now it is possible, at least in cell culture, to selectively silence a transcript associated with an important group of genetic diseases by RNAi.

Despite the clear therapeutic potential of RNAi-based gene therapy (Harper et al., 2005; Raoul et al., 2005; Miller et al., 2005) several issues are challenging translation towards the clinic. The first concerns the RNAi-mediated depletion of the wild-type allele. Indeed, the dominant trait of inheritance of neurodegenerative disorders often implies that only one allele harbours the causative mutation. Therefore, silencing of the wild-type gene could have a deleterious effect, owing to the loss of normal physiological functions (Zeitlin, 1995). To circumvent this limitation, allele-specific siRNAs discriminating between wild type and mutant allele have been designed (Link, 1995; Miller et al., 2003). Nevertheless, for neuro-degenerative diseases associated with a large number of different point mutations in the same gene, such as ALS-linked mutations in SOD, a selective silencing may not be clinically appropriate. Indeed, design of allele-specific siRNA is not so straightforward. Moreover, the silencing efficiency of the allele-specific siRNA may vary with the nature and the position of the point mutation. Finally, all the safety concerns of each newly designed siRNA will have to be addressed before its translation into the clinical situation. An interesting alternative consists of a gene replacement technology that potentially allows for the knockdown of the vast majority of mutated genes and the synthesis of a wild-type protein refractory to RNAi-based silencing (Raoul et al., 2005; Rodriguez-Lebron and Paulson 2006). By retaining normal gene function and switching off a potential broad range of mutations, this system opens a therapeutic avenue to a wide variety of dominantly inherited neurodegenerative disorders.

3.7 Antiproliferative and Proapoptotic miRNAs

miRNAs play essential roles in many basic physiologic processes including proliferation, differentiation, and apoptosis. The function of a given miRNA is dictated by the milieu of targets that are expressed in a given cell type. Thus, an miRNA that regulates both proproliferative and antiproliferative targets may act as a tumour suppressor in some cancers and an oncogene in others, depending on which targets are driving tumorigenesis (Hwang and Mendell, 2006). MicroRNAs with proproliferative and antiapoptotic activity would likely promote oncogenesis and may be over expressed in cancer cells. The most comprehensively studied example of miRNA locus with these properties is the mir-17 cluster, consisting of six miRNAs: miRs-17-5p, -18, -19a, -19b, -20, and -92. This miRNA cluster is located on human chromosome 13q31, a region that is frequently amplified in several types of lymphoma and solid tumors (Ota et al., 2004; Hayashita et al., 2005; He et al., 2005). The other miRNA with antiproliferative and proapoptotic activity are likely to function as tumour suppressor genes and thus may be under-expressed in cancer cells (Hwang and Mendell, 2006). The family of let-7 miRNAs represents a clear example of this. The let-7 family consists of a group of highly conserved miRNAs in multiple species including *C. elegans*, Drosophila, and vertebrates. let-60/RAS was recently identified as a target of let-7 in *C. elegans* (Johnson et al., 2005). The regulation appears to be conserved in humans where three RAS genes, known to be potent oncogenes, have also been demonstrated to be directly regulated by human let-7. Since RAS dysregulation is a key oncogenic event in lung cancer, loss of let-7 may contribute to pathogenesis in this disorder. Indeed, let-7 is generally expressed at low levels in cancerous lung tissue compared to normal tissue and low expression of this miRNA correlates with shorter postoperative lung cancer survival (Takamizawa et al., 2004; Johnson et al., 2005). miR-15a and miR-16-1 represent additional miRNAs with likely tumour suppressing activity. These miRNAs are located on human chromosome 13 in a region frequently deleted in B-CLL. Cimmino *et al* (2005) identified a conserved target site for miR-15a and miR-16-1 in the 3' UTR of BCL2, a potent inhibitor of cell death. Consistent with a regulatory interaction between these miRNAs and this gene, the levels of miR-15a and miR-16-1 are inversely correlated with Bcl2 protein levels in samples from CLL patients. Furthermore, over expression of miR-15a and miR-16-1 in a leukemia cell line results in decreased Bcl2 protein expression, activation of the intrinsic apoptosis pathway, and ultimately cell death. Thus, loss of these miRNAs may contribute to elevated Bcl2 expression and pathologic cell survival in B-CLL (Hwang and Mendell, 2006).

Potential challenge for future studies relates to the probable tissue-specific functions of some miRNAs because in some cancer cells act as oncogenes and in others tumor suppressors. Although we are at an early stage in our understanding of the roles of miRNAs in cancer, the importance of these molecules in this disease process are clear. Undoubtedly, continued efforts to delineate miRNA function in physiologic and patho-physiologic states will reveal novel insights into normal cellular and developmental biology and the normal cellular and developmental biology and the mechanisms that are disrupted when these processes go awry.

3.8 Malaria

Despite intense efforts to eradicate, malaria remains a leading cause of morbidity and mortality worldwide. The World Health Organization (2004) estimates that nearly half a billion clinical cases of malaria occur each year, with over one million deaths. Recent evidence strongly suggests that RNAi can play a key role in identifying which genetic factors shape the vector parasite relationship may be crucial to identifying new genetic means of controlling mosquito-borne diseases (Vlachou et al., 2005). RNAi can, and indeed has been made inheritable in *Anopheles* mosquitoes by stably transforming the mosquito with a transgene that contains two copies of the target gene arranged in an inverted repeat configuration (Brown et al., 2003; Brown and Catteruccia, 2006). Hairpin RNA is expressed in vivo whenever the inverted repeat is transcribed from an upstream promoter. By placing dsRNA expression under the control of a tissue- and time-specific promoter, dsRNA expression can be tailored to coincide spatially and temporally with the journey of the parasite through the mosquito. The cell autonomous nature of transgenic RNAi in *Anopheles* (Ito et al., 2002), implies that the loss-of-function phenotype would only be observed in those cells where target gene and dsRNA are co-expressed, avoiding the pleiotropic effects of the loss of that gene in other non-target tissues. Conceptually, both parasite receptors and immune components protective of the parasite are putative targets for engineering parasite resistance through RNAi and, in principle, mosquito strains that have been rendered refractory to malaria transmission could be released in the field to replace wild-type, permissive populations and achieve malaria eradication. Only few publications to date have described characteristic RNAi-like silencing phenomena [loss-of-function-like phenotypes, reduction in cognate messenger RNA (mRNA) levels, and the presence of miRNA species] (Kumar et al., 2002; Malhotra et al., 2002; Mohammad et al., 2003; Li et al., 2004; Gissot et al., 2005). In one such in vivo study so far published, injection of *P. berghei*-infected mice with siRNAs targeting the *P. berghei* cysteine protease, berghepain, resulted in its specific down-regulation (Blandin et al., 2002; Mohammad et al., 2003; Tina et al., 2006; Volz et al., 2006). Injections resulted in only approximately 0.01% of the siRNA being internalized into the parasite, and the observed 40–50% reduction in berghepain mRNA levels did not alter the parasitemia of the siRNA-treated mice. More studies will be needed to elucidate the mechanisms of gene silencing observed in *Plasmodium* and to assess the therapeutic potential of RNAi. The urgent need to develop new therapeutics for malaria control warrants a thorough assessment of the possible application of this novel technology.

3.9 RNAi and Metabolic diseases

Over the past years RNA interference (RNAi) has exploded as a new approach to manipulate gene expression in mammalian systems. More recently, RNAi has acquired interest as a potential therapeutic strategy. RNAi can be used to silence endogenous genes involved in the cause or pathway of metabolic diseases and holds considerable promise as a therapeutic approach to silence disease-causing genes, particularly those that encode so-called "non-drugable" targets. In addition, the high potency, specificity, and chemical structure of siRNAs may eliminate the toxicity and adverse events commonly seen with small molecule and other oligo-nucleotide approaches (Soutschek et al., 2004; Rondinone, 2006).

The liver's central role in the control of glucose homeostasis is subject to complex regulation by substrates, insulin, and other hormones. Insulin resistance is a major hallmark in the development of type II diabetes, which is characterized by an impaired ability of insulin to inhibit glucose output from the liver and to promote glucose uptake in muscle. A few studies utilized RNAi technology to target key genes involved in the regulation of gluconeogenesis and provided in vivo proof-of-principle for the development of RNAi-based therapeutics for diabetes (Koo et al., 2004; Taniguchi et al., 2005). One group (Taniguchi et al., 2005) developed an adenovirus-mediated RNAi technique that utilizes shRNAs to substantially and stably knock down insulin receptor substrates; IRS-1 and IRS-2 expression, specifically in the livers of mice to better understand the roles of these proteins in hepatic insulin action. By knocking down IRS-1 and IRS-2 separately and together in liver, they showed that IRS-1 signaling may be more closely linked to the regulation of genes involved in glucose homeostasis, whereas IRS-2 signaling may have specific roles in the regulation of hepatic lipid metabolism. Vector-based RNAi approach (Hanson and Reshef 1997; Makimura et al., 2002) was used to induce posttranscriptional gene silencing of hepatic PEPCK using nonviral gene delivery. PEPCK is the rate-controlling enzyme in gluconeogenesis, and altered rates of gluconeogenesis are responsible for increased hepatic glucose output and sustained hyperglycemia. Treatment of diabetic mice with PEPCK siRNA caused a 50% decrease in hepatic PEPCK protein content and was sufficient to lower blood glucose and to improve glucose tolerance (Makimura et al., 2002). These data reinforce the significance of PEPCK in sustaining diabetes-induced hyperglycemia and validate liver-specific intervention at the level of PEPCK for diabetes gene therapy.

Type I diabetes originates from autoimmune/inflamematory destruction of insulin-producing β cells in the pancreatic islet (Castano and Eisenbarth, 1990). Islet transplantation offers a potential cure for type I diabetes. However, its success has been limited, due to loss of cells by apoptosis. Therefore, any process that would inhibit apoptosis or related insults should increase the yield of viable islets available for transplantation from one donor pancreas. Two different studies have described the successful introduction of siRNA directly into pancreatic islet cells, both during in situ perfusion and from intravenous tail vein injection (Bradley et al., 2005a,b; Brown and Catteruccia, 2006). In one of the studies, the insulin 2 (Ins2) gene was targeted with siRNA, and mice received siRNA via hydrodynamic tail vein injection. In vivo delivery of siRNA to pancreatic islets revealed a 33% reduction in Ins2 mRNA levels. These studies provided the first steps for the use of RNAi technology in pancreatic islets and may provide a solution to maintain viable cells for transplantation.

RNAi holds promise for the development of novel therapeutic strategies for disorders that are yet difficult to treat and might be beneficial for the treatment of other diseases such as obesity, neuropathic pain, and depresssion (Rondinone, 2006). After the *in vitro* success in down-regulating gene expression in neurons, chemically synthesized siRNAs were tested for their *in vivo* gene knockdown ability in the brain (Makimura et al., 2002; Shishkina et al., 2004) and observed the physiological functions of mammalian genes especially those expressed in brain. The companies like Sirna Therapeutics Inc., CytRx Corporation, Alnylan Pharmaceuticals developmed chemically modified siRNAs to deliver systematically, opening the possibility of RNAi therapeuptics for a range of different diseases, such as metabolic and cardiovascular diseases (Zinker et al., 2002; Powelka et al., 2006).

Several studies have demonstrated efficient *in vivo* delivery of siRNAs and therapeutic benefit in mice in earlier studies. The ongoing and future preclinical studies in animal models will hopefully help optimize RNAi therapeutics for applications in humans. Although development of RNAi-based therapeutics for diabetes is in its infancy, early clinical studies are soon to begin assessing the use of this new class of therapeutics to tackle metabolic diseases, including diabetes and obesity.

4. The Future of RNA Interference in Medicine

Inhibiting the expression of genetic information has long been the domain of classical genetics. The discovery of an endogenous mechanism for regulation of gene expression, RNA interference (RNAi), has created a small revolution in functional genomics, which is rapidly spreading into therapeutics. SiRNA-mediated gene silencing is generally believed to be highly sequence-specific. The demonstration of efficient and sequence-specific gene silencing by synthetic siRNAs and expressed shRNAs has led to the beginning of a revolution in mammalian functional genetics (Mittal, 2004). Advances in the design of targeting molecules and in their delivery methods have allowed the interrogation of several biochemical and signaling pathways relevant for basic physiological and complex disease processes (Peter Sandy et al., 2005) and shRNAs will likely establish RNAi as a standard novel therapeutic tool in the treatment of human diseases.

Understanding the role of micro RNAs in cell biology and human health represents a major advance in the growing trend towards the development of new diagnostic and therapeutic products which are fore front of persona--lized medicine and biotechnology, this focus on the earliest stages of disease progression, by controlling and diagnosing diseases at the genetic level. This approach makes it possible to target disease while avoiding many of the abundant side effects associated with pharmaceutical products. microRNAs play a significant role in the development of many serious diseases, including different forms of cancer. Specific microRNA-based agents have been shown to increase apoptosis (prog-rammed cell death) in certain cell lines, while other sets of microRNA based agents have been found to provoke cell growth. Based on these important findings, companies can pursue a range of product development opportunities that will harness the power of microRNAs in the diagnosis and treatment of many different forms of cancer in the years ahead. RNAi-mediated down-regulation of cellular targets that encode receptors required for viral entry proved to be effective in the HIV treatment. The power of RNAi as an anti-HIV agent has propelled development of RNAi-based gene therapy approaches for the treatment of HIV infection in humans. Viral vector mediated delivery to hematopoietic cells, including stem cells, is a feasible approach for shRNA gene delivery for treatment of HIV.

RNAi-based functional analysis of genes associated with human recessively inherited neurodegenerative diseases can be addressed through an impressive development of various viral platforms (Deglon and Hantraye, 2005; Rodriguez-Lebron, 2006). Through these methods, understanding of neuronal death mechanisms could allow the identification of potential targets for RNAi-based gene silencing and open therapeutic perspectives for sporadic cases of neurodegenerative disorders. RNAi can also play a key role in identifying the genetic factors which shape the vector parasite relationship and this may be crucial to identifying new genetic means of controlling mosquito-borne diseases like malaria. Injections resulted in only approximately 0.01% of the siRNA being internalized into the parasite, and the observed 40–50% reduction in berghepain mRNA levels did not alter the parasitemia of the siRNA-treated mice. More studies will be needed to elucidate the mechanisms of gene silencing observed in *Plasmodium* and to assess the therapeutic potential of RNAi. The urgent need to develop new therapeutics for malaria control warrants a thorough assessment of the possible application of this novel technology.

Ongoing and future preclinical studies in animal models will hopefully help optimize RNAi therapeutics for applications in humans. Although development of RNAi-based therapeutics for diabetes is in its infancy, initial clinical studies will assess the use of this new class of therapeutics to tackle metabolic and other diseases. Because of the potential of RNAi for therapeutic intervention, major efforts should now be placed on preclinical studies using this technology.

5. Challenges for RNAi as a therapy

Two key challenges in developing RNAi as a therapy are avoiding off-target effects and ensuring efficient delivery. One potential risk for side effects emerges from the feature that distinguishes RNAi from other antisense technologies - the use of cellular machinery for directing sequence-specific silencing. Using iRNAs to target specific cellular or viral transcripts in essence hijacks the endogenous RNAi machinery, and we know little about the potential for saturating the RNAi pathway in primary cells, although saturation of RISC is demonstrable in cultured cells (Hutvágner et al., 2004). Consequently, endogenous RNAi pathways could potentially be affected by siRNAs. It will be important to pay close attention to basic research studies on off-target effects of siRNAs and on the design of effective iRNAs (Saxena et al., 2003; Scacheri et al., 2004). A better understanding of the mechanisms that lead to nonspecific effects of short dsRNAs is essential before the use of siRNAs or taking shRNAs to human-based patient trials.

The issue of delivery has restricted the antisense field for almost two decades. Targeted delivery to specific cell or tissue types is still not a practical reality for oligonucleotide-based therapeutics. The alternative approach is viral-vector-mediated delivery of therapeutic shRNA genes. Because this type of delivery results in gene therapy, there are several associated safety concerns, and systemic delivery of viral vectors is still a major hurdle. Nevertheless, the potency and potential for general therapeutic utility of RNAi is prompting renewed vigour into delivery-related research. It remains to be determined whether backbone-modified, nuclease-resistant siRNAs will move to the clinic more quickly than synthetic deoxyoligonucleotides.

6. Conclusions

The ability to target a specific gene or genes using siRNAs or vector-mediated RNA expression methods, suggests the potential of RNAi to block the disease process or relieve symptoms of the disease. Depletion of proteins critical for the cell cycle, such as cyclins, cyclin dependent kinases, or telomerase, might be effective in treating cancers and some neurodegenerative disorders. Blocking the production of anti-apoptotic proteins such as Bcl-2, inhibitor of apoptosis proteins and antioxidant enzymes may be used to kill cancer cells. Conversely, RNAi-mediated suppression of expression of apoptotic proteins (Bax, Par-4, p53, AIF, and caspases, for example) may slow or stop the degenerative processes such as myocardial, neurological, and autoimmune disorders. Ligands, receptors, and downstream signal transduction proteins critical for a specific disease process may offer suitable targets. Similarly, genes that encode proteins involved in oxidative stress and inflamemation are potential targets in autoimmune and infectious or inflammatory diseases. The more obvious and widely studied viral and bacterial genes targets should offer biomedical scientists a critical springboard in moving RNAi-based therapeutic intervention in infectious diseases a notch higher.

7. Acknowledgements

The publication was conducted as part of the programme 'Biotechnology for Dry land Agriculture in Andhra Pradesh' with financial support for the Research and Communication Division, Ministry of Foreign Affairs, the government of Netherlands. Andhra Pradesh Netherlands Biotechnology Programme, IPE, Hyderabad, INDIA.

8. REFERENCES:

Arndt Borkhardt (2002). Blocking oncogenes in malignant cells by RNA interference—New hope for a highly specific cancer treatment? Cancer cell 23: 167.

Auror Esquela-Koscher, Slack FJ (2006). Oncomirs-microRNAs with a role in cancer. Nature Cancer 6: 259-269.

Anderson J, Akkina R (2005). CXCR4 and CCR5 shRNA transgenic CD34+ cell derived macrophages are functionally normal and resist HIV-1 infection. Retrovirology 2: 53.

Amado RG, Mitsuyasu RT, Rosenblatt JD, Frances K. Ngok, Steve Cole AB, Chorn N, Lin LS, Bristol G, Boyd MP, MacPherson JL, Fanning GC, Todd AV, Ely JA, Zack JA, Symonds GP (2004). Anti-Human Immunodeficiency Virus Hematopoietic Progenitor Cell-Delivered Ribozyme in a Phase I Study: Myeloid and Lymphoid Reconstitution in Human Immunodeficiency Virus Type-1–Infected Patients. Hum. Gene Ther. 15 (3): 251 -262.

Banerjea A, Li MJ, Bauer G, Remling L, Lee NS, Rossi J, Akkina R (2003) Inhibition of HIV-1 by lentiviral vector-transduced siRNAs in T lymphocytes differentiated in SCID-hu mice and CD34+ progenitor cell-derived macrophages. Mol. Ther. 8: 62-71.

.Brown AE, Catteruccia F (2006). Toward silencing the burden of malaria: progress and prospects for RNAi basedapproaches. BioTechniques 40: S38-S44.

Brown AE, Bugeon L, Crisanti A, Catteruccia F (2003). Stable and heritable gene silencing in the malaria vector Anopheles stephensi. Nucleic Acids Res. 31: 85.

Bradley SP, Kowalik TF, Rastellini C, da Costa MA, Bloomenthal AB, Cicalese L, Basadonna GP, Uknis ME (2005)a. Successful incorporation of short-interfering RNA into islet cells by in situ perfusion. Transplant Proc. 37: 233-236.

Bradley SP, Rastellini C, da Costa MA, Kowalik TF, Bloomenthal AB, Brown M, Cicalese L, Basadonna GP, Uknis ME (2005)b. Gene silencing in the endocrine pancreas mediated by short-interfering RNA. Pancreas 31: 373-379.

Basyuk E, Suavet F, Doglio A, Bordonne R, Bertrand E (2003). Human jet-7 stem loop precursor harbor features of RNase III Cleavage products. Nucleic Acid Res. 31: 6593-6597.

Boden D, Pusch O, Lee F, Tucker L, Ramratnam B (2003). Human immunodeficiency virus type-1 escape from RNA interference. J. Virol. 77: 11531–11535.

Bryan R Cullen (2006). Viruses and MicroRNAs. Nat. Genet. 38: 25 - 30.

Bowen DG, Walker CM (2005). Adaptive immune responses in acute and chronic hepatitis C virus infection. Nature 436: 946–952.

Blandin S, Moita LF, Kocher T, Wilm M, Kafatos FC, Levashina, EA (2002). Reverse genetics in the mosquito *Anopheles gambiae*: Targeted disruption of the *Defensin* gene. EMBO Rep. 3, 852-856.

Calin GA, Sevignani C, Dumitru CD, Hyslop T, Noch E, Yendamuri S, Shimizu M, Rattan S, Bullrich F, Negrini M, Croce CM (2004) Human microRNA genes are frequently located at fragile sites and genomic regions involved in cancers. Proc. Natl. Acad. Sci. USA 101: 2999–3004.

Cimmino A, Calin GA , Muller F, Iorio MV, Ferracin M, Shimizu M, Wojcik SE, Aqeilan RI, Zupo S, Dono M, Rassenti L, Alder H, Volinia S, Liu CG, Kipps TJ, Negrini M, Croce CM (2005). *miR-15* and *miR-16* induce apoptosis by targeting BCL2. PNAS 102 (39): 13944-13949.

Chan JA, Krichevsky AM, Kosik KS (2005). MicroRNA- 21 is an antiapoptotic factor in human glioblastoma cells. Cancer Res. 65: 6029–6033.

Cordelier P, Morse B, Strayer DS (2003). Targeting CCR5 with siRNAs: using recombinant SV40-derived vectors to protect macrophages and microglia from R5-tropic HIV. Oligonucleotides 13: 281-294.

Cheng TL, Chang WW, Su IJ, Lai MD, Huang W, Lei HY, Chang WT. (2005). Therapeutic inhibition of hepatitis B virus surface antigen expression by RNA interference. Biochem Biophys Res Commun 336: 820–830.

Colussi PA, Quinn LM, Huang DC, Coombe M, Read SH, Richardson H, Kumar S (2000). Debcl, a proapoptotic Bcl-2 homologue, is a component of the Drosophila melanogaster cell death machinery. J. Cell Biol. 148: 703–714.

Calin GA, Cimmino FM, Di Leva G, Shimizu M, Wojcik SE, Iorio MV, Visone R, Sever NI, Fabbri M, Iuliano R, Palumbo T, Pichiorri F, Roldo C, Garzon R, Sevignani C, Rassenti L, Alder H, Volinia S, Liu CG,. Kipps TJ, Negrini M, Croce CM (2005). A microRNA signature associated with prognosis and progression in chronic lymphocytic leukemia. N. Engl. J. Med. 353: 1793-180.

Caplen NJ, Taylor JP, Statham VS, Tanaka F, Fire A, Morgan RA (2004). Rescue of polyglutamine-mediated cytotoxicity by double-stranded RNA-mediated RNA interference. Hum. Mol. Genet. 11: 175–184.

Castano L, Eisenbarth GS (1990). Type I diabetes: a chronic autoimmune disease of human, mouse and rat. Annu. Rev. Immunol. 8: 647-679.

Capodici J, Kariko K, Weissman D (2003). Inhibition of HIV-1 infection by small interfering RNA-mediated RNA interference. J. Immunol. 169: 5196–5201.

Denli AM, Tops BB, Plaserk RH, Ketting RF, Honnon GJ (2003). Processing of primary micro RNA's by the micro processor complex. Nature 432: 231-235.

Dykxhoorn DM, Novina CD, Sharp PA (2003). Killing the messenger: short RNAs that silence gene expression. Nat. Rev. Mol. Cell. Biol. 4: 457-467.

Dmitriy Ovcharenko, Kevin Kelnar, Rich Jarvis (2006). Efficient Delivery of siRNA to Primary Cells and Hard-to-Transfect Cell Lines. Ambion paper (2006).

Deglon N, Hantraye P (2005). Viral vectors as tools to model and treat neurodegenerative disorders. J. Gene Med. 7: 530–539.

Elbashir SM, Lendeckel W, Tuschl T (2001). RNA interference is mediated by 21-and 22-nucleotide RNAs. Genes Dev. 15:188-200.

Elia AJ, Parkes TL, Kirby K, St George-Hyslop P, Boulianne GL, Phillips JP, Hilliker AJ (1999). Expression of human FALS SOD in motorneurons of Drosophila. Free Radic. Biol. Med. 26: 1332–1338.

Eugene Berezikov, Edwin Cuppen, Ronald HA Plasterk (2006). Approaches to microRNA discovery. Nature Genet. 38: S2 –S7.

Franz DR, Zajtchuk R (2002). Biological terrorism: understanding the threat, preparation and medical response. Dis. Mon. 48: 493–564 .

Fire A, Xu S, Montgomery MK, Kostas SA, Driver SE, Mello CC (1998) Potent and specific genetic interference by double-stranded RNA in Caenorhabditis elegans. Nature 391: 806-811.

Feany MB, Bender WW (2000). A Drosophila model of Parkinson's disease. Nature 404: 394–398.

Forbes Josephine M, Louis Teo Loon Yee, Vicki Thallas, Markus Lassila, Riccardo Candido, Karin A Jandeleit-Dahm, Merlin C Thomas, Wendy C Burns, Elizabeth K Deemer, Susan M Thorpe, Mark E Cooper, Terri J Allen (2004). Advanced Glycation End Product Interventions Reduce Diabetes-Accelerated Atherosclerosis. Diabetes 53.

Gissot M, Briquet S, Refour P, Boschet C, Vaquero C (2005). PfMyb1, a Plasmodium falciparum transcription factor, is required for intra-erythrocytic growth and controls key genes for cell cycle regulation. J. Mol. Biol. 346: 29-42.

Grimm D, Kay MA (2006). Therapeutic short hairpin RNA expression in the liver: viral targets and vectors. Gene Therapy 13: 563–575.

Gregory RL, Chendrimada TP, Cooch N, Shiekhattar R (2005). Human RISC couples microRNA biogenesis and posttranscriptional gene silencing. Cell 123: 631-640.

Geng YJ, Libby P (2002). Progression of atheroma: a struggle between death and procreation. Arterioscler Thromb Vasc. Biol. 22: 1370–1380.

Gitlin L, Karelsky S, Andino R (2002). Short interfering RNA confers intracellular antiviral immunity in human cells. Nature (London) 418: 430–434.

Higuchi H, Yamashita T, Yoshikawa H, Tohyama M (2003). Functional inhibition of the p75 receptor using a small interfering RNA. Biochem. Biophys. Res. Commun., 301: 804–809.

Harper SQ, Staber PD, He X, Eliason SL, Martins IH, Mao Q et al (2005). RNA interference improves motor and neuropathological abnormalities in a Huntington's disease mouse model. Proc Natl Acad Sci USA; 102: 5820–5825.

Hardy J (2001). The genetic causes of neurodegenerative diseases. J. Alzheimers Dis. 3: 109–116.

Hutvágner G, Simard MJ, Mello CC, Zamore PD (2004). Sequence-specific inhibition of small RNA function. PLoS Biol. 2: E98.

He L, Thomson JM, Hemann MT, Hernando-Monge E, Mu D, Goodson S, Powers S, Cordon-Cardo C, Lowe SW, Hannon GJ, and Hammond SM (2005). A microRNA polycistron as a potential human oncogene. Nature 435: 828–833.

Hayashita Y, Osada H, Tatematsu Y, Yamada H, Yanagisawa K, Tomida S, Yatabe Y, Kawahara K, Sekido Y, and Takahashi T (2005). A polycistronic microRNA cluster, miR-17-92, is over-expressed in human lung cancers and enhances cell proliferation. Cancer Res. 65: 9628–9632.

Hanson RW, Reshef L (1997). Regulation of phosphoenolpyruvate carboxykinase (GTP) gene expression. Annu. Rev. Biochem. 66: 581-611.

Hwang HW, Mendell JT (2006). MicroRNAs in cell roliferation, cell death, and tumorigenesis. British Journal of Cancer, 94: 776 – 780.

Ito J, Ghosh A, Moreira LA, Wimmer EA, Jacobs-Lorena M (2002).Transgenic anopheline mosquitoes impaired in transmission of a malaria parasite. Nature 417: 452-455.

Iorio MV, Ferracin M, Liu CG, Veronese A, Spizzo R, Sabbioni S, Magri E, Pedriali M, Fabbri M, Campiglio M, Ménard S, Palazzo JP, Rosenberg A, Musiani P, Volinia S, Nenci I, Calin GA, Querzoli P, Negrini M, Croce CM (2005). MicroRNA Gene Expression Deregulation in Human Breast Cancer. Cancer Res. 65: 7065-7070.

Jennifer Couzin. (2002). Breakthrough of the year: Small RNAs makes big splash. Science 298: 2296.

Johnson SM, Grosshans H, Shingara J, Byrom M, Jarvis R, Cheng A, Labourier E, Reinert KL, Brown D, Slack FJ (2005). RAS is regulated bythe let-7 microRNA family. Cell 120: 635–647.

Jopling, CL, Yi M, Lancaster AM, Lemon SM, Sarnow P (2005). Modulation of hepatitis C virus RNA abundance by a liver-specific microRNA. Science 309: 1577–1581.

Jacque JM, Triques K, Stevenson M (2002). Modulation of HIV-1 replication by RNA interference. Nature 418: 435–438.

John J Rossi (2006). RNAi as a treatment for HIV-1 infection. BioTechniques, 40: S25-S29.

Julian Downward (2004). RNA interference: Science, Medicine and the Future. BMJ 328: 1245-1248.

Jarad G, Wang B, Khan S, DeVore J, Miao H, Wu K, Nishimura SL, Wible BA, Konieczkowski M, Sedor JR Schelling JR (2002). Fas activation induces renal tubular epithelial cell beta 8 integrin expression and function in the absence of apoptosis. J. Biol. Chem. 277: 47826–47833.

Krützfeldt J et al (2005). Silencing of microRNAs in vivo with 'antagomirs'. Nature 438: 685–689.

Kusov Yuri, Tatsuo Kanda, Ann Palmenberg, Jean-Yves Sgro, Verena Gauss-Mu¨ller (2006). Silencing of Hepatitis A Virus Infection by Small Interfering RNAs. J. Virol. 5599–5610.

Koo SH, Satoh H, Herzig S, Lee CH, Hedrick S, Kulkarni R,. Evans RM, Olefsky J, Montiminy M (2004). PGC-1 promotes insulin resistance in liver through PPAR-alpha-dependent induction of TRB-3. Natl. Med. 10: 530-534.

Kumar R, Adams B, Oldenburg A, Musiyenko A, Barik S (2002). Characterisation and expression of a PP1 serine/threonine protein phosphatase (PfPP1) from the malaria parasite, Plasmodium falciparum: demonstration of its essential role using RNA interference. Malar. J. 1: 5.

Li H, Han Z, Lu Y, Lin Y, Zhang L, Wu Y, Wang H (2004). Isolation and functional characterization of a dynamin-like gene from Plasmodium falciparum. Biochem. Biophys. Res. Commun. 320: 664-671.

Locarnini S (2005). Therapies for hepatitis B: where to from here? Gastroenterology 128: 789–792.

Lu J, et al. (2005). MicroRNA expression profiles classify human cancers. Nature 435: 834–838.

Li MJ, Kim S, Li J, Zaia JK, Yee J, Anderson RA, Rossi JJ (2005). Long-term inhibition of HIV-1 infection in primary hematopoietic cells by lentiviral vector delivery of a triple combination of anti-HIV shRNA, anti-CCR5 ribozyme, and a nucleolar-localizing TAR decoy. Mol. Ther. 12: 900-909.

Lu J, Getz G, Miska EA, Alvarez-Saavedra E, Lamb J, Peck D, Sweet-Cordero A, Ebert BL, Mak RH, Ferrando AA, Downings JR, Jacks T, Horvitz R, Golub TR (2005). MicroRNA expression profiles classify human cancers. Nature 435: 834–838.

Liu CG, Calin GA, Meloon B, Gamliel N, Sevignani C, Ferracin M, Dumitru CD, Shimizu M, Zupo S, Dono M, Alder H, Bullrich F, Negrini M, Croce CM (2004). An oligonucleotide microchip for genome-wide microRNA profiling in human and mouse tissues. Proc. Natl. Acad. Sci. USA. 101: 9740–9744.

Lin He, J. Michael Thomson, Michael T. Hemann, Eva Hernando-Monge, David Mu, Summer Goodson, Scott Powers, Carlos Cordon-Cardo, Scott W. Lowe, Gregory J. Hannon, Scott M. Hammond (2005). A microRNA polycistron as a potential human Oncogene. Nature 435.

Link CD (1995). Expression of human b-amyloid peptide in transgenic Caenorhabditis elegans. Proc. Natl. Acad. Sci. USA 92: 9368–9372.

Leonard JN, Schaffer DV (2005). Antiviral RNAi therapy: emerging approaches for hitting a moving target. Gene Ther. 13: 532-540.

Lee Y, Jeon K, Lee JT, Kim S, Kim VN (2002). MicroRNA maturation:step wise processing and Subcellular localization. EMBO J. 21: 4663-4670.

Michael T McManus, Phillip A (2002). Sharp. Gene silencing in mammals by small interfering RNAs Nature Reviews/Genetics 3: 737 – 747.

Meister G, Tuschl T (2004). Mechanisms of gene silencing by double-stranded RNA. Nature 431: 343-349.

Miska EA, Alvarez-Saavedra E, Townsend M, Yoshii A, Šestan N, Rakic P, Constantine-Paton MH, Robert Horvitz (2004). Microarray analysis of microRNA expression in the developing mammalian brain. Genome Biol. 5: R68.

Malhotra P, Dasaradhi PV, Kumar A, Mohmmed A, Agrawal N, Bhatnagar RK, Chauhan VS (2002). Double-stranded RNA-mediated gene silencing of cysteine proteases (falcipain-1 and -2) of Plasmodium falciparum. Mol. Microbiol. 45: 1245-1254.

Mohmmed A, Dasaradhi PV, Bhatnagar RK, Chauhan VS, Malhotra P (2003). In vivo gene silencing in Plasmodium berghei—a mouse malaria model. Biochem. Biophys. Res. Commun. 309: 506-511.

Monticelli S, Ansel K M, Xiao C, Socci ND, Krichevsky AM, Thai TH, Rajewsky N, Debora SM, Sander C, Rajewsky K, Anjana Rao and Kosik KS (2005). MicroRNA profiling of the murine hematopoietic system. Genome Biol., 6: R71.

Martinez J, Patkaniowska A, Urlaub H, Luhrmann R, Tuschl T (2002). Singlestranded antisense siRNAs guide target RNA cleavage in RNAi. Cell 110: 563–574.

Matranga C, Tomari Y, Shin C, Bartel DP, Zamore PD (2005). Passenger-strand cleavage facilitates assembly of siRNA into Ago2-containing RNAi enzyme complexes. Cell 123: 607-620.

Michienzi A, Castanotto D, Lee N, Li S, Zaia JA, Rossi JJ (2003). RNA-mediated inhibition of HIV in a gene therapy setting. Ann. NY. Acad. Sci. 1002: 63-71.

Morris KV, Rossi JJ (2004). Anti-HIV-1 gene expressing lentiviral vectors as an adjunctive therapy for HIV-1 infection. Curr. HIV Res. 2: 79-92.

Morris KV, Looney DJ (2005). Characterization of human immunodeficiency virus (HIV)-2 vector mobilization by HIV-1. Hum. Gene Ther. 16: 1463-1472.

Morris KV (2006). Therapeutic potential of siRNA-mediated transcriptional gene silencing. BioTechniques 40: S7-S13.

Mast EE, Alter MJ, Margolis HS (1999). Strategies to prevent and control hepatitis B and C virus infections: a global perspective. Vaccine 17: 1730–1733.

McCaffrey AP, Nakai H, Pandey K, Huang Z, Salazar FH, Xu H et al. (2003). Inhibition of hepatitis B virus in mice by RNA interference. Nat. Biotechnol 21: 639–644.

Mattson MP (2000). Apoptosis in neurodegenerative disorders. Nat. Rev. Mol. Cell Biol., 1: 120–129.

Miller TM, Kaspar BK, Kops GJ, Yamanaka K, Christian LJ, Gage FH et al. (2005). Virus-delivered small RNA silencing sustains strength in amyotrophic lateral sclerosis. Ann. Neurol. 57: 773–776.

Miller VM, Gouvion CM, Davidson BL, Paulson HL. (2004). Targeting Alzheimer's disease genes with RNA interference: an efficient strategy for silencing mutant alleles. Nucleic Acids Res 32: 661–668.

Miller VM, Xia H, Marrs GL, Gouvion CM, Lee G, Davidson BL, Paulson HL (2003). Allele-specific silencing of dominant disease genes. Proc. Natl. Acad. Sci. USA. 100: 7195–7200.

Makimura H, Mizuno TM, Mastaitis JW, Agami R, Mobbs CV (2002). Reducing hypothalamic AGRP by RNA interference increases metabolic rate and decreases body weight without influencing food intake. BMC Neurosci. 3: 18.

Mittal V (2004). Improving the efficiency of RNA interference in mammals. Natl. Rev. Genet. 5: 355-365.

Nam NH, Parang K (2003). Current targets for anticancer drug discovery. Curr. Drug Targets 4: 159–179.

Nelson PT et al. (2004). Microarray-based, high-throughput gene expression profiling of microRNAs. Nature Methods 1: 155–161.

Novina CD, Murray MF, Dykxhoorn DM, Beresford PJ, Riess J, Lee SK, Collman RG, Lieberman J, Shankar P, Sharp PA (2002). siRNA-directed inhibition of HIV-1 infection. Natl. Med. 8: 681–686.

Olivier Milhavet, Devin S, Gary Mark P Mattson (2003). RNA Interference in Biology and Medicine. Pharmacol Rev. 55: 629–648.

O'Donnell KA, Wentzel EA, Zeller KI, Dang CV, Mendell JT (2005). c-Myc-regulated microRNAs modulate E2F1 expression. Nature 435: 839–843.

Ota A, Tagawa H, Karnan S, Tsuzuki S, Karpas A, Kira S, Yoshida Y, Seto M (2004). Identification and characterization of a novel gene, C13orf25, as a target for 13q31–q32 amplification in malignant

lymphoma. Cancer Res. 64: 3087–3095.
Persengiev SP, Zhu X, Green MR (2004). Nonspecific, concentration-dependent stimulation and repression of mammalian gene expression by small interfering RNAs (siRNAs). RNA 10: 12–18.
Peter Sandy, Andrea Ventura, Tyler Jacks (2005). Mammalian RNAi: Practical guide. BioTechniques 39: 215-224.
Powelka AM, Seth A, Virbasius JV, Kiskinis E, Nicoloro SM, Guilherme A, Tang X, Straubhaar J, Cherniack AD, Parker MG, Czech MP (2006). Suppression of oxidative metabolism and mitochondrial biogenesis by the transcriptional corepressor RIP140 in mouse adipocytes. J. Clin. Invest. 116: 125- 136.
Quinn LM, Dorstyn L, Mills K, Colussi PA, Chen P, Coombe M, Abrams J, Kumar S, Richardson H (2000). An essential role for the caspase dronc in developmentally programmed cell death in Drosophila. J. Biol. Chem. 275: 40416–40424.
Rand TA, Petersen S Du F, Wang X (2005). Argonaute2 cleaves the anti-guide strand of siRNA during RISC activation. Cell 123: 621-629.
Rondinone Cristina M (2006). Therapeutic potential of iRNA in metabolic diseases. BioTechniques 40: S31-S36.
Ryther RC, Flynt AS, Phillips JA, Patton JG (2005). siRNA therapeutics: big potential from small RNAs. Gene Ther. 12: 5-11.
Ren XR, Zhou LJ, Luo GB, Lin B, Xu A (2005). Inhibition of hepatitis B virus replication in 2.2.15 cells by expressed shRNA. J. Viral Hepat. 12: 236–242.
Rodriguez-Lebron E, Paulson HL (2006). Allele-specific RNA interference for neurological disease. Gene Ther. 13: 576-581.
Raoul C, Barker SD, Aebischer P (2006). Viral-based modelling and correction of neurodegenerative diseases by RNA interference. Gene Therapy 13: 487–495.
Rubinsztein DC (2002). Lessons from animal models of Huntington's disease. Trends Genet. 18: 202–209.
Raoul **C**, Abbas-Terki T, Bensadoun JC, Guillot S, Haase G, Szulc J, Henderson CE, Aebischer P (2005). Lentiviral-mediated silencing of SOD1 through RNA interference retards disease onset and rogression in a mouse model of ALS. Natl. Med. 11: 423–428.
Song E, Lee SK, Wang J, Ince N, Ouyang N, Min J, Chen J, Shankar P, Lieberman J (2003). RNA interference targeting Fas protects mice from fulminant hepatitis. Natl. Med. 9: 347–351.
Surabhi RM, Gaynor RB (2002). RNA interference directed against viral and cellular targets inhibits human immunodeficiency virus type-1 replication. J. Virol. 76: 12963–12973.
Shlomai A, Shaul Y (2004). RNA interference – small RNAs effectively fight viral hepatitis. Liver Int 2004; 24: 526–531. 32 Stevenson M. Therapeutic potential of RNA interference. N. Engl. J. Med. 351: 1772–1777.
Steven D. Buckingham, Behrooz Esmaeili, Matthew Wood, David B (2004). Sattelle. RNA interference: from model organisms towards therapy for neural and neuromuscular disorders. Hum. Mol. Genet. Vol. 13 (2): 275 – 288.
Soutschek J, Akinc A, Bramlage B, Charisse K, Constien R, Donoghue M, Elbashir S, Geick A, Hadwiger P, Harborth J, John M, Kesavan V, Lavine G, Pandey RK, Racie T, Rajeev KG, Rohl I, Toudjarska I, Wang G, Wuschko S, Bumcrot D, Koteliansky V, Limmer S, Manoharan M, Vornlocher HP (2004). Therapeutic silencing of an endogenous gene by systemic administration of modified siRNAs. Nature 432: 173-178.
Shishkina GT, Kalinina TS, Dygalo NN (2004). Attenuation of alpha2A-adrenergic receptor expression in neonatal rat brain by RNA interference or antisense oligonucleotide reduced anxiety in adulthood. Neuroscience 129: 521-528.
Saxena S, Jonsson ZO, Dutta A (2003). Small RNAs with imperfect match to endogenous mRNA repress translation. Implications for off-target activity of small inhibitory RNA in mammalian cells. J. Biol. Chem. 278: 44312–44319.
Scacheri PC, Rozenblatt-Rosen O, Caplen NJ, Wolfsberg TG, Umayam L, Lee JC, Hughes CM, Shanmugam KS, Bhattacharjee A, Meyerson M, Collins FS (2004). Short interfering RNAs can induce unexpected and divergent changes in the levels of untargeted proteins in mammalian cells. Proc. Natl Acad. Sci. USA. 101: 1892–1897
Taylor JA, Naoumov NV (2005). The potential of RNA interference as a tool in the management of viral hepatitis. J Hepatol, 42: 139–144.
Tan YT, Tillett DJ, McKay IA (2000). Molecular strategies for overcoming antibiotic resistance in bacteria. Mol. Med. Today 6: 309–314.
Takamizawa J, Konishi H, Yanagisawa K, Tomida S, Osada H, Endoh H, Harano T, Yatabe Y, Nagino M, Nimura Y, Mitsudomi T, Takahashi T (2004). Reduced expression of the let-7 microRNAs in human lung cancers in association with shortened postoperative survival. Cancer Res. 64: 3753–3756.
Tina ML, Peterson, Shirley Luckhart (2006). A mosquito 2 – cys peroxidoxin protects against nitrosative and oxidative stresses associated with malaria parasite infection. Free Radical Biol. Med. 40(6): 1067 – 1082.
Taniguchi CM, Ueki K, Kahn CR (2005). Complementary roles of IRS-1 and IRS-2 in the hepatic regulation of metabolism. J. Clin. Invest. 115: 718-727.
Unwalla HJ, Li HT, Bahner I, Li MJ, Kohn D, Rossi JJ (2006). Novel Pol II fusion promoter directs human immunodeficiency virus type 1-inducible over expression of a short hairpin RNA and protein. J. Virol. 80: 1863-1873.
Uprichard SL, Boyd B, Althage A, Chisari FV (2005). Clearance of hepatitis B virus from the liver of transgenic mice by short hairpin RNAs. Proc. Natl. Acad. Sci. USA. 102: 773–778.
Vlachou D, Kafatos FC (2005). The complex interplay between mosquito positive and negative regulators of Plasmodium development. Curr. Opin. Microbiol. 8: 415-421.
Volz J, Muller HM, Zdanowicz A, Kafatos FC, Osta MA (2006). A genetic module regulates the melanization response of Anopheles to Plasmodium. Cell. Microbiol. 8: 1392.
Wieland SF, Chisari FV (2005). Stealth and cunning: hepatitis B and hepatitis C viruses. J. Virol. 79: 9369–9380.
The World Health Organization. 2004. The World Health Report. Geneva.
Zhao ZQ, Vinten-Johansen J (2002). Myocardial apoptosis and ischemic preconditioning. Cardiovasc. Res. 55: 438–455.
Zeitlin S, Liu JP, Chapman DL, Papaioannou VE, Efstratiadis A (1995). Increased apoptosis and early embryonic lethality in mice nullizygous for the Huntington's disease gene homologue. Nat. Genet. 11: 155–163.
Zinker BA, Rondinone CM, Trevillyan JM, Gum RJ, Clampit JE, Waring JF, Xie N, Wilcox D, Jacobson P, Frost L, Kroeger PE, Reilly RM, Koterski S, Opgenorth TJ, Ulrich RG, Crosby S, Butler M, Murray SF, McKay RA, Bhanot S, Monia BP, Michael Jirousek R (2002). PTP1B antisense oligonucleotide lowers PTP1B protein, normalizes blood glucose, and improves insulin sensitivity in diabetic mice. Proc. Natl. Acad. Sci. USA. 99: 11357-11362.

10

Phenolic compounds and related enzymes as determinants of sorghum for food use

Mamoudou H. Dicko[1,2,3*], Harry Gruppen[2], Alfred S. Traoré[1], Alphons G. J. Voragen[2] and Willem J. H. van Berkel[3]

[1]Laboratoire de Biochimie, UFR-SVT, CRSBAN, Université de Ouagadougou, 03 BP. 7021, Ouagadougou 03, Burkina Faso,
[2]Laboratory of Food Chemistry, Department of Agrotechnology and Food Sciences, Wageningen University, PO Box 8129, 6700 EV Wageningen, The Netherlands
[3]Laboratory of Biochemistry, Department of Agrotechnology and Food Sciences, Wageningen University, PO Box 8128, 6700 ET Wageningen, The Netherlands

Phenolic compounds and related enzymes such as phenol biosynthesizing enzymes (phenylalanine ammonia lyase) and phenol catabolizing enzymes (polyphenol oxidase and peroxidase) are determinants for sorghum utilization as human food because they influence product properties during and after sorghum processing. Phenolic compounds are quality-grade markers for the preparation of several foods because of enzyme inhibitory activities, color, or antioxidant activities. Large inter-varietal differences in contents of phenolic compounds and their antioxidant activities among sorghum varieties exist. Moreover, some red sorghum varieties have higher antioxidant activities than the most important sources of natural antioxidants. Oxidation products of peroxidase and polyphenol oxidase (benzoquinones and polymeric compounds) affect food quality. This paper reviews the current advances in phenolic compounds and phenolic enzymes in sorghum as human food, with emphasis on nutritional and health aspects. This may provide some guidance for researchers in further Investigations and for industries in developing practical health agents and functional foods.

Key words: sorghum, phenolic compounds, antioxidants, 3-deoxyanthocyanidins, proanthocyanidins, peroxidase, polyphenol oxidase, phenylalanine ammonia lyase.

Table of content

I. INTRODUCTION 2

II. SORGHUM PHENOLIC COMPOUNDS BIOCHEMISTRY 2
II. 1. Classes of phenolic compounds in sorghum 2
II. 2. Simple phenols and phenolic acids 5
II. 3. Flavonoids 5
II. 4. Polyflavans 7
II. 5. Role of phenolic compounds in plants 7
II. 6. Health related properties of phenolic compounds 7
II. 7. Relevance of phenolic compounds for sorghum food quality 8

III. SORGHUM PHENOLIC ENZYMES BIOCHEMISTRY 9
III. 1. Role of phenylalanine ammonia lyase in the biosynthesis of phenolic compounds 9
III. 2. Polyphenol oxidases 10
III. 3. Peroxidases 10
III. 4. Role of phenolic enzymes in food 12
III. 5. Sorghum grain as potential source of bioactive components 13

IV. CONCLUSION 13
V. ACKNOWLEDGEMENTS 14
VI. REFERENCES 14

*Correspondence author. E-mail: mdicko@univ-ouaga.bf

INTRODUCTION

Sorghum bicolor (L.) Moench is the fifth most important cereal crop after wheat, rice, maize, and barley in terms of production (FAO, 2005). The total world annual sorghum production is over 60 million tons from a cultivated area of 46 millions ha. Sorghum is particularly adapted to drought prone areas: hot, semi-arid tropical environments with 400-600 mm rainfall-areas that are too dry for other cereals. The sorghum genome is currently sequenced (Paterson et al., 2003; http://fungen.org/Sorghum.htm).

Phenolic sorghum phytochemicals are important for human nutrition (Awika et al., 2004a). Phenolic compounds are the most widely distributed secondary metabolites, ubiquitously present in the plant kingdom. Among cereals, sorghum has the highest content of phenolic compounds reaching up to 6% (w/w) in some varieties (Deshpande et al., 1986; Beta et al., 1999, Doka et al., 2004; Awika and Rooney, 2004a, Dicko et al., 2005a). Almost all classes or phenolic compounds are found in sorghum (Chung et al., 1998; Krueger et al., 2003, Awika and Rooney, 2004a).

Proanthocyanidins (condensed tannins), originally classified as antinutritional factors, may have health benefits for humans (Waniska, 2000, Parr and Bolwell, 2000; Clifford, 2000; Awika, 2004a). Some flavanols, e.g. flavan-4-ols have particular therapeutic interest because of their antitumor activity (Ferreira and Slade, 2002). Procyanidins may inhibit the growth of several viruses including the human immunodeficiency virus 1 (HIV-1) (Chan and Kim, 1998; Okuda et al., 1991, Lu et al., 2004). Phenolic compounds, together with other natural compounds scavenge free radicals (antioxidant activity). Interestingly, independent of germination, sorghum grains display high antioxidant activities related to their phenolic content (Dicko et al., 2005a, Dykes et al., 2005). Moreover, some red sorghum varieties have higher antioxidant activities than the most important sources of natural antioxidants such as Vaccinium species, e.g. blackberries (Awika et al., 2004a; Dicko et al., 2005a). Since sorghum is a staple food in many African countries, it may be the main potential source of natural antioxidants. Epidemiological studies suggest that the consumption of whole cereal grains including sorghum lowers the mortality from cardiovascular disease, which is probably linked to their antioxidant properties (Kushi et al., 1999; Awika and Rooney, 2004a).

The biosynthesis of phenolic compounds in plants proceeds through the production of phenylalanine, which is subsequently deaminated by the enzyme phenylalanine ammonia lyase (PAL). PAL is indirectly associated with the synthesis of several phenolic constituents, including cell wall polymers (Figure 1) (Parr and Bolwell, 2000). PAL activity has been detected in half of the sorghum varieties before germination and in all screened varieties after germination (Dicko et al., 2006).

Polyphenol oxidases and peroxidases are the main enzymes involved in cereal browning. In sorghum, PPO is present in the leaves (Stafford and Dresler, 1973; Vaughn and Duke, 1981) and in the grain (Glennie, 1981; Dicko et al., 2002a, 2006a). Peroxidases are ubiquitously present and relatively abundant in sorghum grain both before and after germination (Dicko et al., 2002a, 2006a). Polyphenol oxidase and peroxidases being involved in the oxidative cross-linking of phenolic compounds are determinants of food quality by influencing product properties during and after processing (Tomas-Barberan and Espín, 2001).

SORGHUM PHENOLIC COMPOUNDS BIOCHEMISTRY

Classes of phenolic compounds in sorghum

Phenolic compounds, of which more than 8000 are known, embrace a wide range of plants secondary metabolites (Harborne, 1994; Pietta, 2000). Phenolic compounds are the most widely distributed secondary metabolites, ubiquitously present in the plant kingdom. Located in the vacuole, they are found in free form or linked to carbohydrates (glucose, galactose, rhamnose, mannose, rutinose etc.). Among cereals, sorghum has the highest content of phenolic compounds reaching up to 6% (w/w) in some varieties (Deshpande et al., 1986; Beta et al., 1999, Awika and Rooney, 2004a) (Tables 1 and 2). While all sorghums contain phenolic compounds, its genotype and the environment in which it is grown influence the amount present in any particular cultivar. The main classes of phenolic compounds are simple phenols, hydroxybenzoic acids, hydroxycinnamic acids, flavonoids (flavanols, flavones, flavanones, isoflavones and anthocyanins), chalcones, aurones (hispidol), hydroxycoumarins, lignans, hydroxystilbenes and polyflavans (proanthocyanidins and pro-deoxyanthocyanidins) (Chung et al., 1998; Krueger et al., 2003). These compounds are soluble in water or organic solvents (methanol, HCl-methanol, acetone, dimethylformamide, etc.). Sorghum does not contain tannic acid and hydrolysable tannins (Waniska, 2000, Awika et al., 2004a). Sorghums with a pigmented testa and spreader genes (B1B2S) or with purple/red plants and thick pericarp genes have the highest levels of phenolic compounds (Dykes et al., 2005). Sorghums with a black pericarp have higher levels of flavan-4-ols and anthocyanins than the other varieties. This suggests that genes for plant color, pericarp thickness, presence of a pigmented testa, and spreader genes increase phenolic levels (Dykes et al., 2005, Dicko et al., 2005a). Lignans and hydroxystilbenes have not been detected in sorghum grain (Awika et al., 2004a). Tables 1 and 2 give the approximate range and intervarietal contents of phenolic

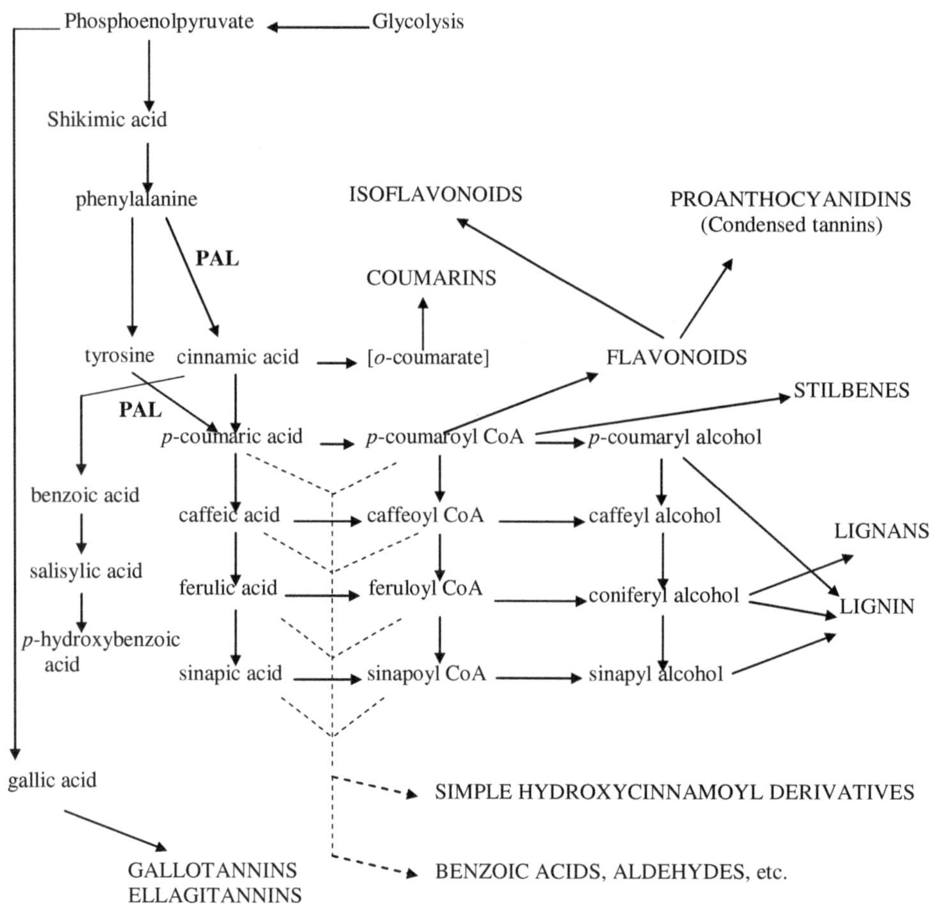

Figure 1. Schematic illustration of the biosynthesis of different phenolics from shikimic acid (Parr and Bolwell, 2000; Ryan and Robards, 2002). PAL= phenylalanine ammonia lyase.

Table 1. Phenolic contents in sorghum grain

Phenolic compounds	Content mg/g dry weight	References
Hydroxybenzoic acids		
p-Hydroxybenzoic	15-36	Hahn et al. (1983)
Gallic	26-46	Hahn et al. (1983)
Protocatechuic	24-141	Hahn et al. (1983)
Vanillic	8-50	Hahn et al. (1983)
Hydroxycinnamic acids		
p-Coumaric	100-200	Verbruggen et al. (1993)
Caffeic	25-52	Hahn et al. (1983)
Ferulic	300-500	Verbruggen et al. (1993)
Sinapic (sinapinic)	50-140	Hahn et al. (1983)
Flavonoids		
Anthocyanins	0-2800	Sérémé et al. (1992); Awika et al. (2003, 2004b)
3-deoxyanthocyanidins	0-4000	Dicko et al. (2005a)
Flavan-4-ols	0-1300	Bate-Smith (1969); Audilakshmi et al. (1999); Dicko et al. (2005a)
Proanthocyanidins	0-68000	Beta et al. (1999); Awika and Rooney (2004a); Dicko et al. (2005a)

Table 2. Comparison of total phenolic compounds and phenolic enzymes in sorghum varieties

Variety code	Total phenolics (%)[a]		Antioxidant activities[b]		PAL[c] (mU mg^{-1})		POX[d] (U mg^{-1})		mono-PPO[e] (mU mg^{-1})		di-PPO[f] (mU mg^{-1})	
	g-	g+	g-	g+	g-	g+	g-	g+	g-	g+	g-	g+
V1	0.58	0.96	43	54	nd	10.3	68.8	81.2	0.7	1.0	35.6	16.3
V2	0.85	0.61	59	21	3.9	1.7	11	49.9	0.7	0.2	39.7	18.7
V3	0.72	0.69	42	42	2.3	9.9	17.4	79.9	0.7	0.8	39.2	24.4
V4	0.66	0.41	45	12	4.1	1.8	15.1	106.7	0.7	0.9	44.2	14.6
V5	0.72	0.63	44	19	nd	5.8	93.3	175.9	0.7	0.9	39.9	11.3
V6	1.38	1.22	52	64	1.1	11.1	31.7	100.7	0.7	0.7	48.0	14.1
V7	0.59	0.87	33	50	nd	15.5	81.8	124.7	0.9	0.7	39.9	23.0
V8	0.71	0.67	40	35	2.4	8.7	59.9	76.3	0.9	0.5	42.4	15.7
V9	0.87	0.91	42	42	nd	2.5	56.5	75.8	1.0	0.1	48.1	14.0
V10	0.68	0.88	48	46	nd	10.1	60.8	71.4	1.3	0.2	57.0	14.5
V11	0.69	0.73	44	40	nd	7.1	50.9	81.6	0.7	0.2	36.4	20.7
V12	0.64	0.81	34	39	nd	3.2	73.2	99.9	1.0	0.5	40.5	17.3
V13	0.61	0.82	29	42	nd	8.8	44.2	95.4	0.6	0.5	34.6	15.1
V14	0.55	0.81	29	36	13.1	6.2	41.3	54.5	1.0	0.8	41.0	9.5
V15	0.76	1.22	37	51	9.0	11.2	59.7	144.3	1.0	0.8	40.7	0.9
V16	0.55	0.86	35	38	9.2	1.2	42.8	184.0	1.0	0.3	43.6	15.1
V17	1.28	1.47	66	53	3.1	2.8	20.9	89.6	0.7	0.4	45.6	14.7
V18	1.74	1.85	55	43	6.2	6.0	8.8	89.2	0.9	0.7	42.9	18.1
V19	3.01	2.95	80	70	2.9	6.6	15.2	109.7	0.9	1.0	41.4	20.3
V20	0.71	0.84	49	32	nd	0.9	22.6	105.3	1.0	1.4	50.3	9.7
V21	0.76	0.74	36	51	nd	16.5	62.3	75.8	0.8	1.9	43.3	19.8
V22	0.82	0.51	40	17	4.7	2.0	24.4	42.5	0.6	0.5	43.7	18.3
V23	0.64	0.78	37	19	5.1	0.6	23.2	83.0	0.6	1.2	40.5	20.5
V24	0.66	0.46	35	14	nd	1.4	25.7	56.5	0.5	0.8	40.6	27.0
V25	0.82	0.72	34	31	5.6	2.3	21	98.5	1.1	0.8	42.8	17.2
V26	0.92	0.75	42	25	3.9	1.2	24.6	75.3	1.0	0.3	47.1	21.0
V27	1.20	1.22	62	49	nd	3.8	12	15.6	1.0	0.6	42.8	15.3
V28	0.96	0.61	44	19	2.2	9.2	18.9	16.4	0.9	0.8	50.8	16.9
V29	0.87	0.79	48	22	4.1	1.9	21.4	49.1	1.1	0.6	47.9	13.0
V30	1.50	1.18	67	42	4.4	8.1	31.5	82.4	1.0	2.3	48.3	18.5
V31	0.65	0.63	35	37	nd	0.1	86.6	107.5	1.1	1.9	42.7	24.4
V32	0.73	0.83	36	48	nd	13.7	40.2	76.5	0.9	0.9	40.9	39.5
V33	0.60	0.97	30	43	2.7	13.6	37	92.6	1.1	1.4	42.3	40.7
V34	0.81	1.02	43	41	5.1	16.9	18.4	26.1	1.3	1.3	41.8	62.5
V35	1.28	0.97	52	44	3.6	19.7	36.3	45.7	1.3	1.9	43.4	54.6
V36	0.70	0.99	34	46	nd	30.1	71.7	86.6	0.7	1.6	46.0	43.5
V37	1.47	1.75	71	58	2.0	6.0	35.9	88.3	0.8	1.1	42.0	42.1
V38	0.66	1.16	40	45	nd	7.4	75.8	180.3	0.8	1.4	39.0	45.0
V39	0.82	0.76	51	49	nd	13.7	17.9	37.4	0.6	1.6	47.7	47.3
V40	0.72	1.01	40	47	nd	17.1	25.8	53.1	1.1	1.5	50.0	34.1
V41	0.73	0.7	30	52	7.7	8.7	31.7	39.5	1.1	2.1	52.7	46.3
V42	0.74	0.55	27	38	7.3	10.8	7.9	58.8	1.3	1.5	52.2	37.8
V43	0.83	1.05	26	53	1.0	12.7	23.9	51.7	1.0	1.4	49.6	11.7
V44	0.45	0.63	17	40	6.1	11.9	18.3	53.8	0.4	1.5	54.2	42.1
V45	0.98	0.77	41	46	nd	9.8	42.3	51.0	0.9	1.2	49.3	37.2
V46	0.63	0.71	16	38	nd	8.7	39.8	65.7	1.2	1.6	57.1	36.2
V47	1.10	0.96	36	46	nd	13.1	47.1	96.2	1.0	2.3	54.7	33.1
V48	0.58	0.61	20	40	nd	12.6	13.4	51.5	1.2	1.0	50.0	34.8
V49	0.46	0.81	25	46	nd	12.7	67.5	82.6	0.4	1.3	52.6	28.3
V50	1.71	1.01	67	52	1.7	7.1	32.3	58.8	1.0	1.4	59.2	39.2
Mean	0.88	0.92	42	41	2.49	8.49	38.21	79.90	0.90	1.05	45.3	25.5
Range	0.5-3	0.5-2.9	16-80	14-70	0-13	0.1-30	8-93	16-184	0.4-1.3	0.1-2	36-59	1-63
Std error	0.05		2		0.4		4		0.05		2	

[a]Gallic acid equivalents; [b]Antioxidant activities (μmol Trolox equivalents/g of fresh matter); [c]PAL = phenylalanine ammonia lyase, [d]POX = peroxidase, [e]mono-PPO = monophenolase activity of polyphenol oxidase (PPO), [f]*o*-diphenolase activity of PPO. Enzymes activities are expressed in terms of specific activities (Units/mg of protein). g- = ungerminated sorghum; g+ = germinated sorghum.
Data is from Dicko et al. (2005a, 2006a).

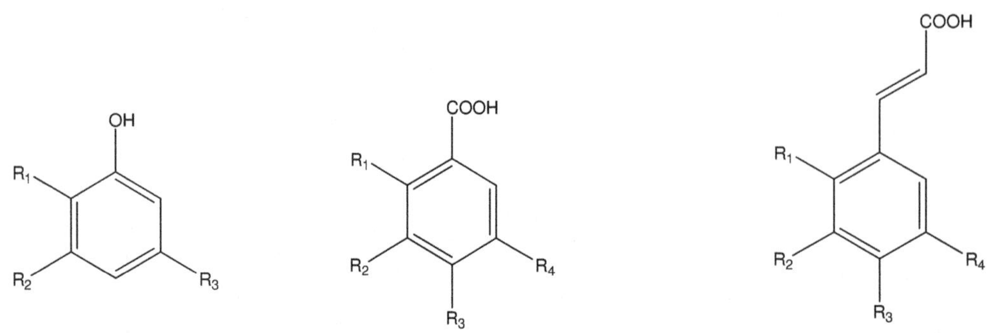

Simple phenols | **Hydroxybenzoic acids** | **Hydroxycinnamic acids**

Phenol: $R_1 = R_2 = R_3 = H$
Catechol: $R_1 = OH, R_2 = R_3 = H$
Resorcinol: $R_2 = OH, R_1 = R_3 = H$
Phloroglucinol: $R_2 = R_3 = OH, R_1 = H$

Gallic: $R_1 = H, R_2 = R_3 = R_4 = OH$
Gentisic: $R_1 = R_4 = OH, R_2 = R_3 = OH$
Salicylic: $R_1 = OH, R_2 = R_3 = R_4 = H$
p-(OH)-Benzoic: $R_1 = R_2 = R_4 = H, R_3 = OH$
Protocatechuic: $R_1 = R_4 = H, R_2 = R_3 = OH$
Syringic: $R_1 = H, R_2 = R4 = OCH_3, R_3 = OH$
Vanillic: $R_1 = R_4 = H, R_3 = OH; R_2 = OCH_3$

p-Coumaric: $R_1 = R_2 = R_4 = H, R_3 = OH$
Caffeic : $R_1 = R_4 = H, R_2 = R_3 = OH$
Ferulic: $R_1 = R_2 = H, R_4 = OCH_3, R_3 = OH$
Sinapic : $R_1 = H, R_2 = R_4 = OCH_3, R_3 = OH$

Figure 2. Structures of monophenols found in sorghum (Reed, 1995; Towo et al., 2003; Awika et al., 2004a).

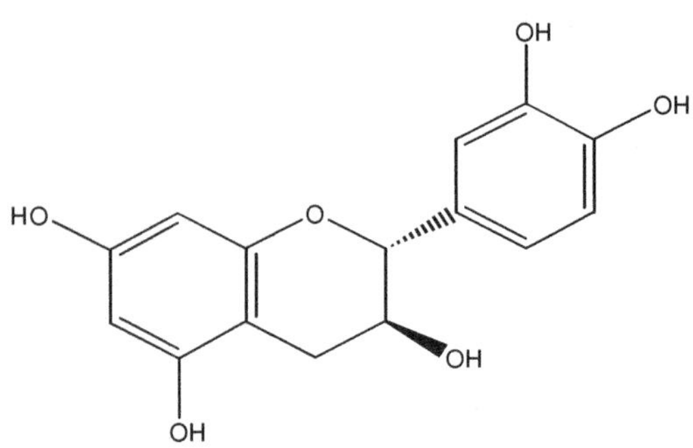

Flavan-3-ol
catechin

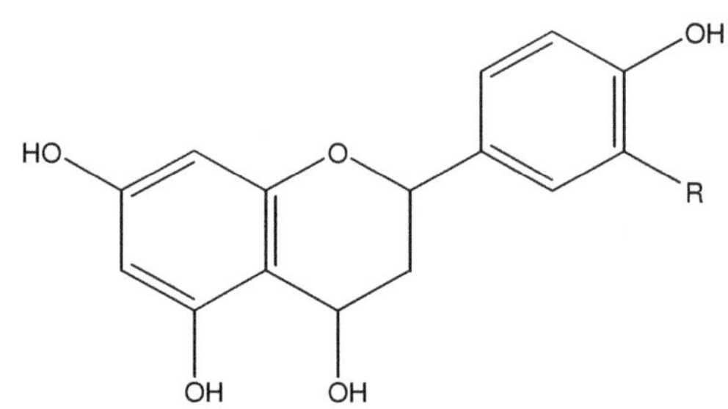

Flavan-4-ols
Apiforol (leucoapigeninidin) : R = H
Luteoforol (leucoluteolinidin) : R = OH

Figure 3. Structure of some sorghum flavanols (Melake-Berhan et al., 1996; Hagerman, 2005).

compounds in sorghum varieties.

Simple phenols and phenolic acids

Simple phenols are relatively rare in plants. Catechol and resorcinol were reported in sorghum grain; however their concentrations were not given (Watt and Breyer-Brandwijk, 1962; Czarnota et al., 2003; Towo et al., 2003). These compounds are undesired in food products because they are carcinogenic, hepatotoxic and goitrogenic (Gaitan et al., 1989, Reed, 1995). Interestingly, if present they may be removed by food processing like heating (Gaitan et al., 1989). Phenolic acids in sorghum include hydroxybenzoates and hydroxycinnamates (Figure 2). They are found in free form or bound as esters, and are concentrated in the outer layers of the grain (Waniska, 2000; Awika et al., 2004a). The most abundant phenolic acids in sorghum are ferulic acid and p-coumaric acid (Hahn et al., 1983; Verbruggen et al., 1993).

Flavonoids

Flavonoids sensu latu constitute the largest class of phenolic compounds with more than 3000 structures, possessing in common a flavylium unit (C_6-C_3-C_6) (Iacobucci and Sweeny, 1983). Sorghum contains flavonoids such as flavanols (flavan-3-ols, flavan-4-ols, etc., Figure 3), flavanones, flavones and anthocyanins (Haslam, 1998; Awika, 2004a). The flavan-4-ols apiforol

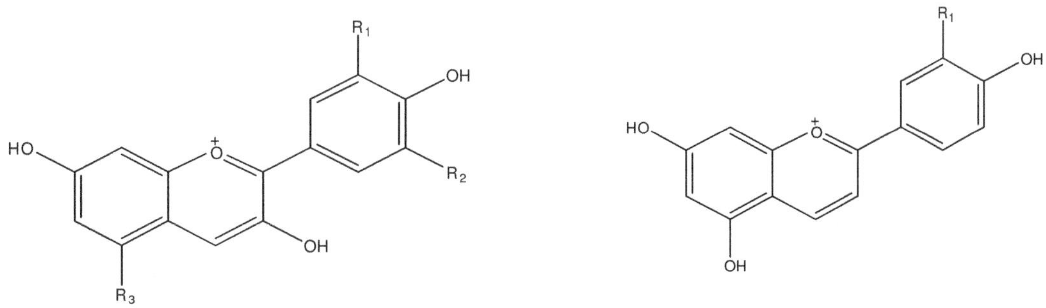

Anthocyanidins

Fisetinidin : $R_1 = R_2 = OH$, $R_3 = H$
Cyanidin : $R_1 = OH$; $R_2 = H$, $R_3 = OH$
Pelargonidin : $R_1 = R_2 = H$, $R_3 = OH$
Peonidin : $R_1 = OCH3$; $R_2 = H$, $R_3 = OH$
Malvidin : $R_1 = R_2 = OCH_3$, $R_3 = OH$
Delphinidin $R_1 = R_2 = R_3 = OH$
Petunidin $R_1 = OCH3$; $R_2 = R_3 = OH$

3-Deoxyanthocyanidins

Apigeninidin : $R_1 = H$
Luteolinidin : $R_1 = OH$

Figure 4. The main sorghum anthocyanidins and 3-deoxyanthocyanidins (Kouda-Bonafos et al., 1996; Palé et al., 1997; Awika et al., 2004a, 2004b).

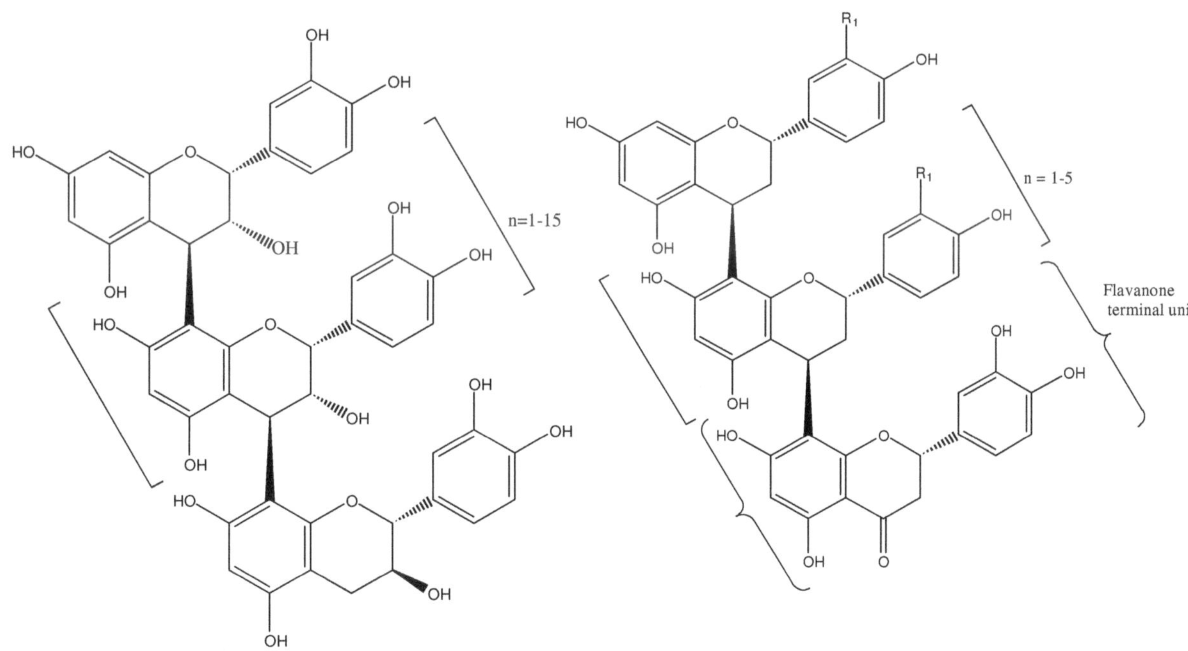

Proanthocyanidins

Pro-3-deoxyanthocyanidins
$R_1 = H$, pro-apigeninidins
$R_1 = OH$, pro-luteolinidins

Figure 5. Structure of sorghum polyflavans (Krueger et al. 2003; Chen and Hagerman, 2004; Awika and Rooney, 2004a).

(pro-apigeninidin or leuco-apigeninidin) and tuteoforol (proluteolinidin or leuco-luteolinidin) are abundant in sorghum (Dicko et al., 2005a), and precursors of apigeninidin and luteolinidin, respectively (Ferreira and Slade, 2002; Haslam, 1998; Hagerman, 2005). Sorghums with a black pericarp have higher levels of flavan-4-ols and anthocyanins than the other varieties (Dykes et al., 2005).

The most abundant anthocyanins in sorghum grain are 3-deoxyanthocyanidins, e.g. apigeninidin and luteolinidin (Figure 4) (Bate-Smith, 1969; Kouda et al., 1994, 1996; Palé et al., 1997; Awika et al., 2004b). The red color of the grain's pericarp is essentially due to the presence of 3-deoxyanthocyanidins (Bate-Smith, 1969). Like most anthocyanidins, these compounds are used as natural food colorants (Morazzoni and Magistretti, 1990; Coultate, 1996; Awika et al., 2004a). As a food (E163) or pharmaceutical (Myrtocyan) additive, these phenolic compounds represent a world market of 250 millions US dollars (Morazzoni and Magistretti, 1990; Coultate, 1996). 3-Deoxyanthocyanidins have interesting food applications because of their thermal and color stability (Iacobucu and Sweeny, 1983, Awika et al., 2003, 2004b). The 3-deoxyanthocyanidins (3-DAs), are particularly abundant in notably red sorghum grain (Dicko et al., 2005a), but rare or absent in other plants (Palé et al., 1997; Awika and Rooney, 2004a; Awika et al., 2004b). 3-DAs are of interest because they are more stable in organic solvents as well as in acidic solutions than anthocyanidins commonly found in fruits, vegetables and other cereals (Kouda-Bonafos et al., 1996; Palé et al., 1997; Awika et al., 2004b). This has been suggested as a potential advantage of sorghum as a viable commercial source of anthocyanins (Awika and Rooney, 2004a).

Polyflavans

The term polyflavan is referred to phenolic compounds formed by polymers of flavylium units in which, some hydrogen groups are substituted with hydroxyl groups (Krueger et al., 2003, Awika and Rooney, 2004a). Most polyflavans are often called condensed tannins, but this generic name is sometimes confusing because it does not give a structural definition of the compounds. The polyflavans found in sorghum are essentially polymers of flavan-3-ols (proanthocyanidins) and pro-3-deoxyanthocyanidins (Figure 5). Sorghum proanthocyanidins consist of flavan-3-ol units linked by C-C (type B proanthocyanidins) and occasionally C-O-C (type A proanthocyanidins) bonds ranging from one to fifteen (Krueger et al., 2003, Awika and Rooney, 2004a). The most abundant polyflavans in sorghum are homopolymers of catechin/epicatechin with uniform B-type interflavan bonds (Krueger et al., 2003). Not all sorghum varieties contain these polyflavans because their content is genetically governed by B1-B2 genes (Serna-Salvidor and Rooney, 1995; Butler, 1992; Waniska, 2000). In general, varieties with pigmented testa layers contain proanthocyanidins (Waniska, 2000, Dicko et al., 2002a, 2005a).

The main pro-deoxyanthocyanidins found in sorghum are pro-apigeninidins and pro-luteolinidins (Figure 5). Although present in sorghum, these polyflavans are very rare in other plants (Stafford, 1990). Hydrolysis of pro-apigeninidins and pro-luteolinidins yields apigeninidins and luteolinidins, respectively (Hagerman, 2005).

Role of phenolic compounds in plants

Phenolics play an important role in plant metabolism, but also protect the plant against stresses. For instance, it has been recently shown that flavonoids, such as cate chin, regulate the auxin transport in plants, and, therefore, play an important role in plant development (Brown et al., 2001). Several studies have shown that the plant resistance to both biotic (pathogens and predators) and abiotic (UV-radiation, drought, etc.) stresses is related to phenolic compounds (Parr, 2000, Dicko et al., 2005b). All classes of phenolic compounds (hydroxybenzoic acids, hydroxycinnamic acid derivatives, flavonoids, polyflavans, etc.) are involved in the resistance mechanisms. Sorghum 3-deoxyanthocyanidins are phytoalexins (plant-microbe interaction) or allelochemicals (plant-plant interaction), involved in plant resistance to biotic stresses such as fungi and parasitic plant invasion (Weiergang et al., 1996; Lo et al., 1999; Parr and Bolwell, 2000; Weir et al., 2004). Proanthocyanidins, 3-deoxyanthocyanidins and flavan-4-ols prevent losses from premature germination and damage due to mold (Waniska, 2000). Hydroxycinnamic acids are constituent of plant cell-wall polymers such as lignin, suberin and cutin. These polymers are physical barriers against invading predators, drought and several other stresses (Parr and Bolwell, 2000).

Health related properties of phenolic compounds

Condensed tannins, e.g. proanthocyanidins, may bind to proteins, carbohydrates and minerals, thereby affecting the nutritional and functional value of the bound constituents. Of major nutritional concern is the ability of proanthocyanidins to bind strongly to large proteins and to proline-rich proteins, thereby reducing their digestibility (Butler, 1992). Proanthocyanidins may be antinutritional through direct interference within the animal body of the digestive processes or inhibition of hydrolytic enzymes through formation of complexes (Nguz et al., 1998). One of the most undesired effects of phenolic compounds is their pro-oxidant activity, which can lead to mutagenicity and carcinogenicity (Stoewsand et al., 1984; Morton, 1992). This pro-oxidant activity depends strongly on the type of phenolic compounds (Awad et al., 2000; Rietjens et al., 2001; Awad, 2002; van der Woude et al., 2002).

The earlier idea of classification of proanthocyanidins as antinutritional factors is now questioned because they also are believed to have health benefits for humans (Hagerman et al., 1998; Waniska, 2000, Parr and Bolwell, 2000; Clifford, 2000; Awika, 2004a). Sorghum proanthocyanidins are unlikely to bind minerals (Waniska, 2000) and high molecular weight proanthocyanidins (DP >3) do not cross the gastrointestinal cell monolayer (Deprez et al., 2001). Bioavailability of iron in sorghum for human subjects was found to be affected more by phytic acid than by the proanthocyanidins content of the grains (Radhakrishnan and Sivaprasad, 1980). Gomez-Cordoves and co-workers (2001) demonstrated the effective therapeutic effect of sorghum proanthocyanidins against human melanoma. The binding of proanthocyanidins with proteins participates in their antibacterial activity (Murdiati and McSweeney, 1987; Scalbert, 1991). Proanthocyanidins have been shown to inhibit the growth of human immunodeficiency virus 1 (HIV-1), influenza virus, and herpes simplex virus by blocking their entry in the host cells (Chan and Kim, 1998; Okuda et al.; 1991, Lu et al., 2004; Hamauzu et al., 2005). The anti-HIV-1 activity is of high interest in Africa, and in Burkina Faso in particular, where HIV-1 prevalence is a major concern. The mechanism of proanthocyanidins toxicity against microbes is related to inhibition of hydrolytic enzymes, interactions to inactivate microbial adhesions and cell envelope transport proteins, and non-specific interaction with carbohydrates (Cowan, 1999). Among sorghum flavanols, the flavan-4-ols (Figure 5) have particular therapeutic interest because of their antitumor activity (Ferreira and Slade, 2002). Flavan-4-ols revealed strong host mediated antitumor activity, which is due to the enhancement of immune response of the host animals through the actions on tumor cells and some immunocytes (Okuda et al., 1991; Ferreira and Slade, 2002).

A number of highly reactive oxygen species such as singlet oxygen ($^1O_2$), superoxide anion radical ($O_2^{\bullet -}$), hydroxyl radical ($OH^\bullet$), nitric oxide radical ($NO^\bullet$), and alkyl peroxyl ($ROO^\bullet$) are regularly produced in the human body (Langseth, 1995). These radicals can damage lipids, proteins and DNA and participate in pathogenesis and ageing (Ryan and Robards, 1998; Santos-Buelga and Scalbert, 2000; Parr and Bolwell, 2000). Phenolic compounds, together with other natural compounds (vitamins C and E, and carotenoids), contribute to the defense by scavenging free radicals, by inhibiting oxidative enzymes such as lipooxygenase and cyclooxygenase and by chelating metal ions (Shi et al., 2001). The antioxidant activity of phenolic compounds is mainly due to their redox properties, which can play an important role in adsorbing and neutralizing free radicals (Bors et al., 2001). In general, phenolic compounds possessing ortho-hydroxyls, found for instance in caffeic acid and in the B-ring of some flavonoids (catechin, quercetin, luteolinidin, etc.), have higher antioxidant activities than the others (Natella et al., 1999; Shi et al., 2001). Antioxidant activities of phenolic compounds have been suggested to exert beneficial pharmacological effects on neurological disorders on the basis of in vitro observations (Moosmann and Behl, 1999). Epidemiological studies have shown that consumption of some phenolic compounds is associated with a reduced risk for developing chronic diseases, such as coronary heart disease, cancer, diabetes, and Alzheimers's disease, linked to their free radical scavenging activities (Ames et al., 1993; Block et al., 1992; Hertog et al., 1993; Temple, 2000; Joshipura et al., 2001; Willett, 2002; Khokhar and Magnusdottir, 2002; Yang et al., 2004). Another interesting property of phenolic compounds, notably hydroxyanthraquinones and hydroxynaptho-quinones, is their cathartic effect (Clifford, 2000). Cathartic compounds are believed to give a better feeling and help to deal with difficult emotions and eliminate them.

Relevance of phenolic compounds for sorghum food quality

Among phenolic compounds of particular interest in sorghum are proanthocyanidins (PAs), 3-deoxyanthocyanidins (3-DAs) and flavan-4-ols. This is because of their importance in human nutrition and because of their agronomic properties in grain preservation. For instance, the content of PAs is important in assessing the food grade quality of sorghum because of their possible antinutritional effect linked to the inhibition of hydrolytic enzymes (Butler, 1992). For the preparation of infant porridges, low PAs containing varieties, may be more convenient by avoiding enzyme-mediated oxidation of endogenous phenolic compounds into colored products or the inhibition of hydrolytic by PAs. On the other hand, food-based sorghum varieties rich in PAs can be suggested to obese people and diabetic patients by analogy with the 50% weight loss observed with animals (rabbits, pigs, etc.) fed with sorghums containing high levels of PAs (Ambula et al., 2001). This is supported by the observation that in certain cultures in Africa, people prefer to consume PAs containing sorghums because they have a longer "residence power" in the stomach (Awika and Rooney, 2004a). The low digestibility of high PAs containing sorghums through the inhibition of hydrolytic enzymes, together with their high antioxidant activities may be interesting from a nutritional standpoint for obese persons.

While most plants, including other cereals, are lacking 3-deoxyanthocyanidins, e.g. apigeninidins and luteolinidins, sorghum is unique in containing a relatively high level of these compounds (Awika et al., 2004b, Dicko et al., 2005a). Reports on the levels of phenolic compounds in sorghum varieties show a high inter-varietal difference of contents among varieties screened (Subramanian et al., 1992; FAO, 1995; Iwuoha and Aina, 1999; Bvochora et al., 1999, Dicko et al., 2002a, 2005a). Flavan-4-ols, present in only rare sorghum varieties (Audilakshmi et al., 1999, Dicko et al., 2005a), are of particular interest for grain storage (Melake-Berhan et al., 1996) and for treating various cancers (Ferreira and Slade, 2002). The contents of these compounds among sorghum varieties from Burkina Faso vary more than in other sorghums for instance from Nigeria, India and

Zimbabwe (Dicko et al., 2002a, 2005a). The majority of sorghum varieties do not contain high amounts of PAs, which is an advantage for weaning foods preparation.

Despite their importance for nutrition, the effect of germination on PAs, 3-DAs, and flavan-4-ols in sorghum varieties has been scarcely studied. Previous reports on 2-4 varieties indicated that the analyzable content of phenolic compounds may increase (Nwanguma and Eze, 1996) or decrease upon germination (Iwuoha and Aina, 1997). It was recently shown that, on average, germination does not affect the total phenolic compounds content of the fifty selected varieties (Dicko et al., 2005a). This could be related to the fact that during germination, both enzymes involved in phenolic compounds biosynthesis as well as enzymes involved in their degradation are activated (Dicko et al., 2006a). A clear decrease in PA, 3-DA and flavan-4-ol contents was observed. The decrease in extractable PAs and DAs upon germination could be due to leaching of water-soluble PAs and DAs, which are located in the pericarp and testa (Beta et al., 1999; Waniska et al., 2000) or due to formation of insoluble complexes between PAs and proteins (Hagerman et al., 2001).

Recently, Awika and co-workers (2003) reported methods to determine antioxidant activities of sorghum grains. It was also shown that independent of germination, there are large inter-varietal differences in antioxidant activities among sorghum varieties (Table 2) (Dicko et al., 2005a, Dykes et al., 2005). Phenolic compounds and antioxidant activities were well correlated before germination, but germination lowered this correlation (Dicko et al., 2005a). Although white sorghum varieties have lower antioxidant activities than red varieties (Dicko et al., 2005a), these activities are among the most important sources of natural antioxidants (Pellegrini et al., 2003). Since sorghum is a staple food in Africa, it can be tentatively inferred that it is the main potential source of natural antioxidants for people relying on it as a main source of energy. Cardiovascular disease (CVD) is one of the most important diseases in Western countries. Epidemiological studies suggest that the consumption of whole cereal grains, including sorghum lowers the mortality from CVD, linked probably to their antioxidant properties (Kushi et al., 1999; Awika and Rooney, 2004a). This implies that several sorghum varieties could be candidate to be processed into food not only for Africa, but also for other countries. For sorghum varieties displaying high levels of antioxidant activities, it would be interesting to perform a further analysis of the nature of phenolic compounds contained in these varieties and to study their antioxidant or pro-oxidant properties. A thorough analysis of the phenolic composition of these selected varieties may allow discovering new flavonoids displaying anti-HIV-1 activities (Li et al., 1998; Kitamura et al., 1998; Yao et al., 2004). The same analysis could allow finding lignans and other bioactive phenolic compounds in sorghum grains.

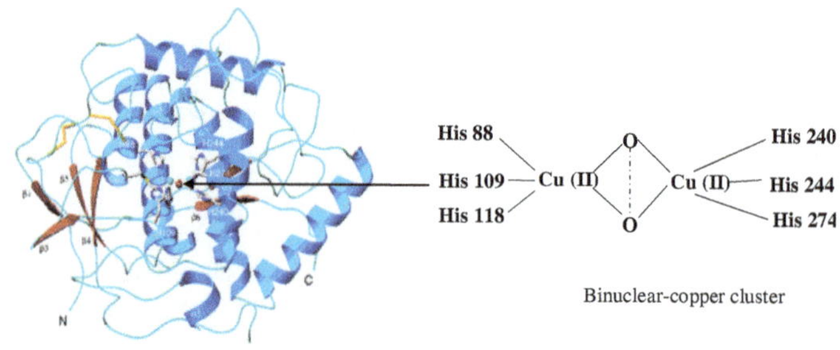

Figure 6. Structure of potato PPO and its binuclear copper-cluster (Klabunde et al., 1998).

SORGHUM PHENOLIC ENZYMES BIOCHEMISTRY

Role of phenylalanine ammonia lyase in the biosynthesis of phenolic compounds

The biosynthesis of phenolic compounds in plant (Figure 1) is initiated by the shikimic acid pathway (Tomas-Barberan and Espin, 2001; Heldt, 2005). This pathway continues with the production of phenylalanine, which is subsequently deaminated by the enzyme phenylalanine ammonia lyase [EC 4.3.1.5, PAL]. PAL can deaminate both L-phenylalanine and L-tyrosine into cinnamate derivatives (Rosler et al., 1997; Heldt, 2005). The released ammonia is refixed by glutamine synthetase [L-glutamate-ammonia ligase, EC 6.3.1.2] to produce glutamine (Singh et al., 1998; Heldt, 2005).

PAL is inhibited by its products, e.g. trans-cinnamates (Heldt, 2005). Apart from PAL, the main other important enzymes in phenolic synthesis are cinnamate-4-hydroxylase [EC 1.14.13.11]; 4-coumarate CoA ligase [EC 6.2.1.12], acetyl CoA carboxylase [EC 6.4.1.2]; chalcone synthase [EC 2.3.1.74); and chalcone-flavanone isomerase (EC 5.5.1.6] (Hrazdina, 1992, Haslam, 1998). PAL is indirectly associated with the synthesis of phenol polymers, including lignin and suberin (Parr and Bolwell, 2000; Heldt, 2005). In several fruits and vegetables, a high plant PAL activity has been associated with the accumulation of anthocyanins and other phenolic compounds (Tovar et al., 2002). In barley, PAL activity has been associated with the response to pathogen challenge (Shiraishi et al., 1995) and to light (Baztan and Torres, 1988). The inhibition of PAL activity in barley induces the susceptibility to fungal attack (Carver et al., 1994). PAL activity has been detected in the green shoots and leaves (Stafford, 1969, Mohan et al., 1988) of sorghum. In sorghum, the infection of the plant with pathogen involved a very rapid accumulation of PAL mRNA (Cui et al., 1996). The presence of PAL activity in sorghum grain and its activation upon germination was assessed (Dicko et al., 2006a). Although PAL is indirectly involved in the synthesis of almost all phenolic compounds, its activity was not correlated with the contents in phenolic compounds in both ungerminated and germinated varieties (Dicko et al., 2006a). This lack of correlation may be due to the presence of phenolic oxidizing enzymes in the grain.

Polyphenol oxidases

Polyphenol oxidases [monophenol, 3,4, L-dihydroxyphenylalanine: oxygen oxidoreductase, EC 1.14.18.1, PPOs] are type-3 copper containing oxidases that occur in plants as monomeric, dimeric and tetrameric structures, and as several isoforms (Kowalski et al., 1992; Martinez and Whitaker, 1995; Sheptovitsky and Bruvig, 1996; Jolivet et al., 1998; Klabunde et al., 1998; Timothy et al., 2001; Chazarra et al., 2001). In sorghum like in wheat, PPO isoenzymes displaying monophenolase activity were found exclusively in the endosperm while those having only o-diphenolase activity were localized in the pericarp (Mayer and Harel, 1979; Marsh and Galliard, 1986; Hatcher and Kruger, 1993, Dicko et al., 2005b). The crystal structure of a potato PPO has been solved (Klabunde et al., 1998). The active site of PPO is constituted of a binuclear copper-cluster (Figure 6). The copper atoms are linked to each other through oxygen atoms and bound to the polypeptide chain through three histidine residues. PPO may catalyze a regioselective aerobic two electrons transfer oxidation of monophenols (monophenolase or cresolase activity) to o-diphenols and their subsequent dehydrogenation to the corresponding o-quinones (diphenolase or catecholase activity) (Martinez and Whitaker, 1995; Timothy et al., 2001). The o-quinones produced may undergo non-enzymatic cyclization or polymerization reactions to yield melanin-like pigments (Martinez and Whitaker, 1995; Rodakiewicz-Nowak and Ito, 2003). Not all PPO isoenzymes can perform the hydroxylation step and even those possessing that activity may loose it during extraction or storage (Sanchez-Ferrer et al., 1995). The catecholase activity is a shared property of all PPOs.

PPOs are involved in various protection mechanisms, including human pigmentation and the browning of fruits and vegetables. The defense function of PPO in plants is attributed to the modification of endogenous phenolic compounds, notably allelochemicals or phytoalexins into o-quinones, which are toxic to the invading pathogens and pests (Mayer and Harel, 1979; Luthra et al., 1988; Kowalski et al., 1992; Weir et al., 2004). PPO is suggested to be indirectly involved in auxin biosynthesis because the o-quinones produced by PPO can react with tryptophan to form indole-3-acetic acid (Mayer and Harel, 1979). PPO was also found both in latent and active forms in the photosystem-II of the thylacoid membrane-protein complex, suggesting a function in photosynthesis (Sheptovitsky and Brudvig, 1996).

In sorghum, PPO is present in the leaves (Stafford and Dresler, 1972; Vaughn and Duke, 1981) and in the grain (Glennie, 1981, Dicko et al., 2002a, 2002b, 2006a). Three PPO isoforms with different molecular masses were found in the leaves, although no molecular masses were given (Stafford and Dresler, 1972). PPO activity in sorghum leaves has been associated in response to fungal invasion (Luthra et al., 1988; Gowda et al., 1989). However, until now no further purification and characterization of sorghum PPO has been performed.

Sorghum varieties display different monophenolase and o-diphenolase activities (Dicko et al., 2002a, 2002b, 2006a). However, the inter-varietal difference of PPO activity and the effect of germination on the occurrence of PPO isoenzymes in sorghum varieties are unknown. Like most other plant PPOs (Martinez and Whitaker, 1995), sorghum PPOs are more active with o-diphenols than with monophenols. Germination decreases the o-diphenolase activity and slightly increases the monophenolase activity. Zymography revealed that germination does not induce new PPO isoenzymes in sorghum grain (Dicko et al., 2006a). As reported for wheat (Hatcher and Kruger, 1993), a decrease of o-diphenolase activity of PPO was observed (Dicko et al., 2006a). This might be related to its localization in the pericarp, where it can leach out during germination or form complexes with phenolic compounds. Since both activities of PPO can be kinetically controlled for the synthesis of o-diphenols or o-quinones, sorghum varieties possessing these activities may find interesting applications in the food and chemical industries (Dubey et al., 1998; Espin et al., 2001).

Peroxidases

Peroxidases [hydrogen peroxide, oxidoreductase, EC 1.11.1.7, POXs] are ubiquitous enzymes found in bacteria, fungi, plants and animals (Krylov and Dunford, 1996, Dunford, 1999). They have in common that they accept hydrogen peroxide or hydroperoxide analogs as oxidant and form water as by-product. Most POXs studied to date contain ferric protoporphyrin IX as prosthetic group and act through a high-valence iron-oxo species. Besides these heme-containing POXs, selenium-, manganese-, vanadium- and flavin-containing POXs are known (Adam et al., 1999). Based on sequence similarity, numbers of calcium ions and origins, POXs have been divided in three major classes (Welinder, 1992). Class I includes yeast cytochrome c POXs, ascorbate POXs and bacterial catalase-POXs. Cytochrome c POXs are found in the mitochondrial electron transport chain, where they probably protect against toxic peroxides. Ascorbate POXs are the main enzyme responsible for hydrogen peroxide removal in chloroplasts and cytosol of higher plants (Dalton, 1993). Bacterial catalase-POXs have both classical POX and catalase activities (Welinder, 1991; Fraaije et al., 1996). Class II comprises secretory fungal POXs such as lignin POXs and manganese-dependent POXs (Reddy and

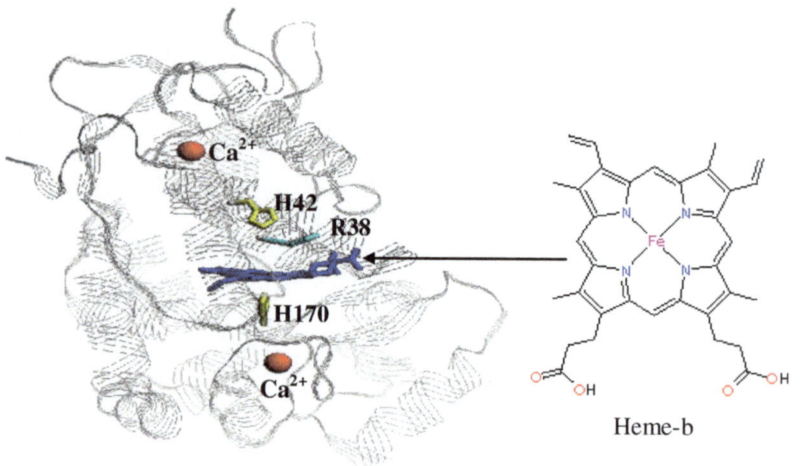

Figure 7. Three-dimensional structure of horseradish peroxidase. Shown are the heme, the structural calcium ions, and the three key amino-acid residues of the active site: Arg + 2 His. Source: Dunford (1999) and Silaghi-Dumitrescu (2005).

Souza, 1994). Class III consists of the secretory plant POXs found in e.g. horseradish, barley, and wheat.

Plant POXs have many different physiological functions including the removal of hydrogen peroxide from chloroplasts and cytosol; the oxidation of toxic compounds; the biosynthesis of cell walls (lignin and suberin); defense responses towards wounding and other stresses; indole-3-acetic acid regulation; ethylene biosynthesis; etc. (Dunford, 1999; Welinder et al., 2002; Duroux and Welinder, 2003). POXs occur in plants as several isoenzymes or glycoforms, with different cellular localization (Dunford, 1999; Duroux and Welinder, 2003). While in the five chromosomes of Arabidopsis 73 POX genes were found (Duroux and Welinder, 2003), 138 POX genes and 14 POX pseudogenes were annotated in the twelve chromosomes of rice (Oryga sativa japonica) (Passardi et al., 2004). Considering the phylogenetic linkage between rice and sorghum (Paterson, 2003), and the fact that the genome of sorghum (750 Mb) is twice as large as that of rice, even more POX genes are expected in sorghum. Moreover, the ongoing project of sorghum genome sequencing (http://peroxidase.isb-sib.ch/index.php) has allowed us to currently identify 160 stretches of sorghum peroxidase genes (dicko et al., 2006b). Plant secretory POXs generally have the following structural properties in common (Henriksen et al., 1998; Dunford, 1999; Welinder et al., 2002; Duroux and Welinder, 2003; Veitch, 2004):

monomeric glycoprotein structure of 300-310 amino-acids
putative N-glycosylation sites: Asn-X-Ser/Thr (X≠Pro)
non-covalently bound iron(III)-protoporphyrin-IX (type-b heme) key catalytic amino-acid residues around the heme prosthetic group: His^{42}, His^{170}, Arg^{38}, Asn^{70}, Asp^{238}, Phe^{41} and Pro^{139} (horseradish peroxidase numbering)four conserved disulphide bridges: Cys^{11-91}, Cys^{44-49}, Cys^{97-301}, and $Cys^{177-209}$ (horseradish peroxidase numbering)
two structural calcium ion binding sites
structural water molecules extending from the heme pocket to the distal calcium-binding site

The crystal structures of POXs from peanut (Schuller et al., 1996), horseradish (Gajhede et al., 1997); barley (Henriksen et al., 1998), and Arabidopsis (Mirza et al., 2000; Østergaard et al., 2000) have been elucidated. The cationic horseradish (HRP-C) isoenzyme is the most studied POX (Figure 7), because of its availability and its catalytic performance.

POXs oxidize reducing substrates (AH) either in presence of hydrogen peroxide (classical POX cycle) or molecular oxygen (oxidase cycle). Recently, it was shown (Berglund et al., 2002) that HRP-C is essentially in five oxidation states during catalysis (native enzyme, compound-I, compound-II, Compound-III, and ferro-peroxidase). In each of these states, the enzyme has a different conformation, especially at the heme environment (Berglund et al., 2002). Compound-I is obtained by two-electron oxidation of the resting enzyme by H_2O_2 and is stable (Hiner et al., 2002). In contrast to the generally assumed concept of irreversible formation of Compound–I, Rietjens and co-workers proposed that the formation of Compound I can be a reversible reaction via an uptake of a water molecule (van Haandel et al., 1998, van Haandel, 2000). The one-electron-reduction of Compound-I by AH produces Compound-II. Another electron from AH is required to reduce Compound-II to the native enzyme (Dunford, 1999). Formation of Compound-I (k_1) and the regeneration of the native enzyme (k_3) are generally rate-limiting steps in POX catalysis (Dunford, 1999). Compound-III is formed from Compound-II with excess of H_2O_2 or upon reaction of the ferrous-enzyme (oxyperoxidase) with O_2 (Dunford, 1999, Berglund et al., 2002). It is important to note that an excess of H_2O_2 can irreversibly inactivate POX because Compound-III is unstable (van Haandel, 2000). In addition, Compound-III can also react with an excess of H_2O_2 to produce verdohemoprotein called P-670; in this case H_2O_2 acts as a suicidal substrate (Dunford, 1999; Hernandez-Ruiz et al., 2001; Sakharov and Sakharova, 2002). The POX catalytic cycle yields free radicals ($A^{\bullet}$), which can undergo polymerization reactions, coupling with molecular oxygen, etc. In the oxidase cycle, POX transfers one electron to molecular oxygen, which in turn is transferred to a substrate; this is termed the monooxygenase activity (Dawson, 1988). The monooxygenase catalytic route of POX is performed via the formation of the ferrous form (Fe^{2+}) of the enzyme (ferro-POX) (Gazarian et al., 1998). Native POX or Compound III may also perform catalase or dismutase reactions (EC 1.11.1.6) by decomposition of H_2O_2 to water and oxygen (Hiner et al., 2001).

Some POXs are able to oxidize auxin in the absence of H_2O_2, thus display indole-3-acetic acid (IAA)-oxidase activity (Christensen et al., 1998, Gazarian et al., 1998; Dunford, 1999). POXs from non-plant sources are unable to oxidize IAA in the absence of H_2O_2, because they are lacking the IAA-binding site localized in the distal domain near the heme pocket (Savitsky et al., 1999). The IAA-oxidase activity of POX isoforms is related to their physiological role in auxin metabolism (Hiner et al., 2001; Marco et al., 1999).

POX activities have been detected in the leaves (Stafford and Bravinder-Bree, 1972; Vaughn and Duke, 1981; Luthra et al., 1988) and grains (Glennie, 1981) of sorghum. POX activities in sorghum varieties and the occurrence of isoenzymes have been reported (Ollitrault et al., 1989; Bvochora et al., 1999, Dicko et al., 2002a; Dicko et al., 2006a). There is a strong inter-varietal difference of POX activity among sorghum varieties, before and after germination (Dicko et al., 2002a, 2006a). Germination increased POX activity due to the activation of already present isoenzymes and/or de novo synthesis of POX isoenzymes (Dicko et al., 2006a). The cationic POX is the most abundant POX isoform in sorghum, and it is ubiquitously present in all varieties, in both ungerminated and germinated grains (Dicko et al., 2002a; 2006a). In other cereals such as barley, wheat and maize, the cationic isoenzymes are also the most abundant, accounting for 80-90% of total POX activity (Johansson et al., 1992; Converso and Fernandez, 1995; Billau et al., 1999).The major cationic POX in sorghum grain has been purified and characterized to the molecular level (Dicko et al., 2006b). Mass spectrometry analysis showed that the enzyme consists of two glycoforms with molecular masses of 34227 and 35629 Da and it contains a type-b heme. Chemical deglycosylation allowed the estimation of sugar contents of 3.0% and 6.7% (w/w) in glycoform I and II, respectively, and a mass of the apoprotein of 33246 Da. The enzyme is localized in the chromosome 1 of sorghum. The sequence of sorghum POX has a high sequence identity with barley BP1 (85%), rice Prx23 (90%), wheat WSP1 (82%) and maize POX (58%), indicative for a common ancestor. In spite of the conserved active sites, the sorghum cationic POX differs from the archetypically known barley POX BP1 in being active with aromatic compounds above pH 5. Another interesting POX isoenzyme in sorghum is an anionic POX (pI 3.1), which is expressed only in germinated grain (Dicko et al., 2006a). This germination-associated isoenzyme may play a role in cell wall synthesis and protect the seed against pathogens (Passardi et al., 2004). Purification and characterization of this anionic isoenzyme will give more insight in its specificity for phenolic compounds that are involved in plant growth.

Cationic POXs are generally more active on phenolic compounds than laccases and anionic POXs and are promising biocatalysts (Wallace and Fry, 1999). In view of this, the utilization of the sorghum cationic POX in biocatalysis could be a future challenge. Sorghum POX may be applied in food biotechnology for the modification of carbohydrate-containing hydroxycinnamates (Oudgenoeg et al., 2001, 2002; Boeriu et al., 2004; Regalado et al., 2004). In addition, POX catalyzed reactions are useful for (bio) chemical, clinical and biocatalytic applications (Adam et al., 1999; Regalado et al., 2004).

Role of phenolic enzymes in food

Contents of phenolic compounds and phenolic oxidizing enzymes are strongly associated with food quality (Deshpande et al., 1986; Hilhorst et al., 1999; 2002; Tomas-Barberan and Espín, 2001). Enzymes involved in the biosynthesis and oxidation of phenolic compounds have been shown to be determinants for the quality of plant-derived foods (Tomas-Barberan and Espin, 2001). The presence of PAL activity in sorghum grain and its activation upon germination was assessed (Dicko et al., 2006a). Although PAL is indirectly involved in the synthesis of almost all phenolic compounds, its activity was not correlated with the contents in phenolic compounds in both ungerminated and germinated varieties (Dicko et al., 2006a). This lack of correlation may be due to the presence of phenolic oxidizing enzymes in the grain.

PAL, POX and PPO were more activated during germination in red grains than in white ones, indicating that the phenolic biosynthesis and catabolism processes are relatively more important in red grains than in white grains (Dicko et al., 2006a). In wheat, the di-PPO activity was also higher in red grains than in white ones (Mayer and Harel, 1979; Hatcher and Kruger, 1993).

Phenolic compounds are potential substrates of POX

Table 3. Sorghum varieties as potential source of phenolic compounds or phenolic oxidizing enzymes.

(Bio)chemical component	Germination requirement	Utilizations	References
Total phenolic compounds	no	antioxidant, food additive, antimicrobial	1-3
Proanthocyanidins	no	antioxidant, antimicrobial and antiviral agent (including HIV1)	4-6
3-deoxyanthocyanidins	no	food colorant	7, 8
Flavan-4-ols	no	anti-cancer agent	9, 10
Polyphenol oxidase	no	synthesis of *o*-diphenols and *o*-quinones	11, 12
Peroxidase	yes	synthesis of various organic compounds, functional modification of carbohydrates and proteins	13, 14, 15-17

[1]Awika et al. (2003); [2]Pellegrini et al. (2003); [3]Tomas-Barberan and Espín (2001); [4]Scalbert (1991); [5]Chan and Kim (1998); [6]Lu et al. (2004); [7]Awika and Rooney (2004a), [8]Kouda-Bonafos et al. (1996); [9]Ferreira and Slade (2002); [10]Okuda et al. (1991); [11]Dubey et al., 1998; [12]Espin et al. (2001); [13]Guerra and Ferraz (2003); [14]Adam et al. (1999); [15, 16]Oudgenoeg et al. (2001, 2002); [17]Boeriu et al. (2004).

and PPO and their oxidation products (benzoquinones and polymeric compounds) affect food quality (Matheis and Whitaker, 1984, Martinez and Whitaker, 1995). PPO and POX work synergistically because PPO may generate H_2O_2 during the course of catalysis (Richard-Forget et al., 1997) and POX on other hand can generate O_2 (Matheis and Whitaker, 1984). Both PPO and POX generate benzoquinones, which are also new substrates of POXs. Because of their electrophilic nature, these quinones undergo secondary reactions, such as cross-linking with amino acid side groups in proteins (Anderson and Morris, 2001). PPO and POX are known to influence product properties during and after food processing (Matheis and Whitaker, 1984; Haslam, 1998, Feillet et al., 2000). They affect post harvest degradation of food by causing browning and the development of off-flavors in raw and unblanched cereals (March and Galliard, 1986; Hatcher and Kruger, 1993). In wheat the appearance of colored products in the flour is attributed to the oxidation of endogenous phenolic compounds by PPO (March and Galliard, 1986; Hatcher and Kruger, 1993). PPO and POX are generally admitted to be the most determinant enzymes for the preservation and organoleptic qualities of fruits and vegetables (Matheis and Whitaker, 1984; Parr, 2000, Tomas-Barberan and Espin, 2001). In analogy with what was found in wheat dough (Hilhorst et al., 1999), POX may mediate the thickness of sorghum flour during "tô" preparation by cross-linking carbohydrates containing endogenous hydroxycinnamate derivatives (Dicko et al., 2002a; 2006a). Low POX and PPO activities to avoid beer darkness and haze occurrence by the oxidation of endogenous phenolic compounds may be other criteria for selection of sorghum for industrial brewing (Clarkson et al., 1992).

For couscous preparation, the formation of a gel, mediated by POX via the cross-linking of macromolecules, is not desired. In most African countries, bakers do not use composite sorghum/wheat flour. However, acceptable bread can be produced with 30-50% sorghum substitution for wheat (Carson et al., 2000, Dicko et al., 2006c). Addition of sorghum flour possessing high POX activities (Hilhorst et al., 1999, Baik et al., 2003) could lead to the cross-linking of sorghum glucuronoarabinoxylans. These polysaccharides are known to contain ferulic acid (Verbruggen et al., 1993; 1996) and may make addition of a higher percentage of sorghum flour acceptable, in analogy with the action of POX in wheat bread (Hilhorst et al., 1999). These criteria may give directions for selecting sorghum varieties for bread making.

Sorghum grain as potential source of bioactive components

The investigation of several constituents within sorghum varieties has shown their wide (bio) chemical diversity. This large (bio) chemical diversity could be rationally exploited. Among the phenolic compounds and phenolic enzymes studied, some are of particular interest as bioactive constituents or biocatalysts. Therefore, it is useful to identify some varieties as sources of specific (bio) molecules. For instance, varieties containing a high level of total phenolic compounds could be novel sources of antioxidants. Varieties containing high levels of PAs could be candidate to isolate these compounds because they are currently gaining attention as antibacterial and anti-HIV1 agents (Chan and Kim, 1998). Varieties containing high levels of 3-DAs will be of interest for the isolation of apigeninidins and luteolinidins that are currently commercialized as food colorants and for various other industrial utilizations (Morazzoni and Magistretti, 1990; Coultate, 1996). As reviewed above, some sorghum varieties could be used as sources of specific biochemical (Table 3).

CONCLUSION

Sorghum is a tropical cereal that has great diversity in its content of phenolic compounds as well as phenolic related enzymes. Phenolic compounds, phenylalanine ammonia lyase, polyphenol oxidase and peroxidase are effective biochemical determinants for sorghum use as food or as source of bioactive components. Research on endogenous bioactive components such as phenolic compounds and phenolic enzymes, will contribute to unleash the capacity of sorghum to be the cornerstone of food security in Africa as well as in many developing countries.

ACKNOWLEDGEMENT

The organization for the prohibition of chemical weapons (OPCW) via the International Foundation for Science (Sweden) is acknowledged for supporting the research carried out by Dr. M. H. DICKO.

REFERENCES

Adam W, Lazarus M, Saha-Moller CR, Weichold O, Hoch U, Haring D, Schreider P (1999). Biotransformation with peroxidases. Adv. Biochem. Eng. Biotechnol. 63: 73-107.

Ambula MK, Oguho GW, Tuitoek JK (2001). Effects of sorghum tannins, a tannin binder (polyvinylpyrrolidone) and sorghum inclusion level on the performance of broiler chicks. Asian-Austr. J. Anim. Sci. 14: 1276–1281.

Ames BN, Shigenaga MK, Hagen TM (1993). Oxidants, antioxidants and the degenerative diseases of aging. Proc. Natl. Acad. Sci. U.S.A. 90: 7915-7922.

Anderson JV, Morris CF (2001). An improved whole-seed assay for screening wheat germplasm for polyphenol oxidase activity. Crop Sci. 41: 1697-1705.

Audilakshmi S, Stenhouse JW, Reddy TP, Prasad MVR (1999). Grain mould resistance and associated characters of sorghum genotypes. Euphytica 107: 91-103.

Awad HM (2002) Studies on the pro-oxidant chemistry of flavonoids. PhD Thesis, Wageningen University, Wageningen, The Netherlands,

Awad HM, Boersma MG, Vervoort J, Rietjens IM (2000). Peroxidase-catalyzed formation of quercetin quinine methide-glutathione adducts. Arch. Biochem. Biophys. 378: 224-233.

Awika JM, Rooney LW (2004a). Sorghum phytochemicals and their potential aspects on human health. Phytochemistry. 65:1199-1221.

Awika JM, Rooney LW, Waniska RD (2004b). Properties of 3-deoxyanthocyanins from sorghum. J. Agric. Food. Chem. 52: 4388-4394.

Awika JM, Rooney LW, Wu X, Prior RL, Cisneros-Zevallos L (2003). Screening methods to measure antioxidant activity of sorghum (Sorghum bicolor) and sorghum products. J. Agric. Food Chem. 51: 6657-6662.

Baik BK, Park CS, Paszczynska B, Konzak CF (2003). Characteristics of noodles and bread prepared from double-null partial waxy wheat. Cereal Chem. 80: 627-633.

Bate-Smith EC (1969). Luteoforol (3', 4, 4', 5, 7–pentahydroxyflavan) in Sorghum vulgare L. Phytochemistry 8: 1803-1810.

Baztan J, Torres M (1988). The effects of daily exposure to UV-A light on phenylalanine ammonia-lyase activity during the growth of barley plants. Rev. Esp. Fisiol. 44: 73-80.

Berglund GI, Carlsson GH, Smith AT, Szoeke H, Henriksen A, Hajdu J (2002). The catalytic pathway of horseradish peroxidase at high resolution. Nature 417: 463-468.

Beta T, Rooney LW, Marovatsanga LT, Taylor, JRN (1999). Phenolic compounds and kernel characteristics of Zimbabwean sorghums. J. Sci. Food Agric. 79: 1003-1010.

Billau C, Louarme L, Nicolas J (1999). Comparison of peroxidases from barley kernel (Hordeum vulgare L.) and wheat germ (*Triticum aestivum* L.): isolation and preliminary characterization. J. Food Biochem. 23: 145-172.

Block G, Patterson B, Subar A. (1992). Fruit, vegetables, and cancer prevention: a review of the epidemiological evidence. Nutr. Cancer 18: 1-29.

Boeriu CG, Oudgenoeg G, Spekking WTJ, Berendsen LBJM, Vancon L, Boumans H, Gruppen H, van Berkel WJH, Laane C, Voragen AGJ (2004). Horseradish peroxidase-catalyzed cross-linking of feruloylated arabinoxylans with β-Casein. J. Agric. Food Chem. 52: 6633-6639.

Bors W, Michel C, Stettmaier K (2001). Structure-activity relationships governing antioxidant capacities of plant polyphenols. Methods Enzymol. 335: 166-181.

Brown DE, Rashotte AM, Murphy AS, Normanly J, Tague BW, Peer WA, Taiz L, Muday GK (2001). Flavonoids act as negative regulators of auxin transport in vivo in Arabidopsis. Plant Physiol. 126: 524-535.

Butler LG (1992). Antinutritional effects of condensed and hydrolysable tannins. In Plant polyphenols: synthesis. properties and significance; Hemingway RW, Laks PE (Eds). Plenum press: New York. pp. 693-698.

Bvochora JM, Reed JD, Read JS, Zvauya R (999). Effect of fermentation processes on proanthocyanidins in sorghum during preparation of Mahewu: a non-alcoholic beverage. Process Biochem. 35: 21-25.

Carson L, Setser C, Sun XS (2000). Sensory characteristics of sorghum composite bread. Int. J. Food Sci. Technol. 35:465-471.

Carver TLW, Zeyen RJ, Bushnell WR, Robbins MP (1994). Inhibition of phenylalanine ammonia lyase and cinnamyl alcohol dehydrogenase increases quantitative susceptibility of barley to powdery mildew (Erysiphe graminis D.C.). Physiol. Mol. Plant Pathol. 44: 261-272.

Chaman ME, Copaja SV, Argandeña VH (2003). Relationships between salicylic acid content. phenylalanine ammonia-lyase (PAL) activity and resistance of barley to aphid infestation. J. Agric. Food Chem. 51: 2227-2231.

Chan DC, Kim PS (1998). HIV entry and its inhibition. Cell 3: 681-684.

Chazarra S, Garcia-Carmona F, Cabanes J (2001). Evidence for a tetrameric form of iceberg lettuce (Lactuca sativa L.) polyphenol oxidase: purification and characterization. J. Agric. Food Chem. 49: 4870-4875.

Chen Y, Hagerman AE (2004). Quantitative examination of oxidized polyphenol-protein complexes. J. Agric. Food Chem. 52: 6061–6067.

Christensen JH, Bauw G, Welinder KG, Montagu MV, Boerjan W (1998). Purification and characterization of peroxidases correlated with lignification in poplar xylem. Plant Physiol. 118: 125-135.

Chung T, Wong TY, Wei CI, Huang YW, Lin Y (1998). Tannins and human health: A review. Crit. Rev. Food Sci. Nutr. 38: 421–464.

Clarkson SP, Large PJ, Bamforth CW (1992). Oxygen-scavenging enzymes in barley and malt and their effects during mashing. Inst. Brew. 98: 111-115.

Clifford MN (2000). Miscellaneous phenols in foods and beverages-nature, occurrence and dietary burden. J. Sci. Food Agric. 80: 1126-1137.

Converso DA, Fernández ME (1996). Ca^{2+} activation of wheat peroxidase: a possible physiological mechanism of control. Arch. Biochem. Biophys. 333:59-65.

Coultate T (1996). Food: The chemistry of its compounds. 3rd Ed, Edition Lavoisier : Paris.

Cowan, MM (1999). Plants products as antimicrobial agents. Clin. Microbiol. Rev. 12: 564-582.

Cui Y, Frederiksen R, Magill C (1996). Chalcone synthase and phenylalanine ammonia-lyase mRNA levels following exposure of sorghum seedlings to three fungal pathogens. Physiol. Mol. Plant Pathol. 49: 187-199.

Czarnota MA, Rimando AM, Weston EL (2003). Evaluation of root exudates of seven sorghum accessions. J. Chem. Ecol. 9: 2073-2083.

Dalton DA (1991). Ascorbate peroxidase. Peroxidases in chemistry and biology, Vol. II, Everse, J, Everse, K.E, Grisham, M.B. Eds; CRC Press: Boca Raton, pp. 139-153.

Dawson J (1988). Probing structure-function relations in heme-containing oxygenases and peroxidases. Science 240: 433-439.

Deprez S, Mila I, Huneau JF, Tome D, Scalbert A (2001). Transport of proanthocyanidin dimer, trimer, and polymer across monolayers of

human intestinal epithelial Caco-2 cells. Antio. Redox Signal. 3: 957-967.
Deshpande SS, Cheryan M, Salunkhe DK (1986). Tannin analysis of food products. CRC Crit. Rev. Food Sci. Nutr. 24: 401-449.
Deu M, Gonzalez de Leon D, Glaszman JC, Degremont I, Chantereau J, Lanaud C, Hamon P (1994). RFPL diversity in cultivated sorghum in relation to racial differentiation. Theor. Appl. Genet. 88: 838-844.
Dicko MH, Gruppen H, Traoré AS, van Berkel WJH, Voragen AGJ (2005a). Evaluation of the effect of germination on content of phenolic compounds and antioxidant activities in sorghum varieties. J. Agric. Food Chem. 53: 2581-2588.
Dicko MH, Gruppen H, Barro C, van Berkel WJH, Voragen AGJ (2005b). Impact of phenolics and related enzymes in sorghum varieties for the resistance and susceptibility to biotic and abiotic stresses. J. Chem. Ecol. 31: 2671-2688.
Dicko MH, Gruppen H, Zouzouho OC, Traoré AS, van Berkel WJH, Voragen AGJ (2006a). Effects of germination on the activities of amylases and phenolic enzymes in sorghum varieties grouped according to food end-use properties. J. Sci. Food Agric. Vol. 86. DOI: 86: 953-963.
Dicko MH, Gruppen H, Hilhorst R, Voragen AGJ and van Berkel WJH (2006b). Biochemical characterization of the major sorghum grain peroxidase. FEBS Journal. DOI:10.1111/j.1742-4658.2006.05243.x.
Dicko MH, Gruppen H, Traoré AS, Voragen AGJ, van Berkel WJH (2006c). Sorghum as human food in Africa: relevance of content of starch and amylase activities. Afr. J. Biotechnol. Vol 5. 5: 384-395.
Dicko MH, Hilhorst R, Gruppen H, Traoré AS, Laane C, van Berkel WJH., Voragen AGJ (2002a). Comparison of content in phenolic compounds, polyphenol oxidases and peroxidases in grains of fifty sorghum varieties from Burkina Faso. J. Agric. Food Chem. 50: 3780-3788.
Dicko MH, Hilhorst R, Gruppen H, Laane C, van Berkel WJH, Voragen AGJ (2002b). Zymography of the monophenolase and o-diphenolase activities of polyphenol oxidase. Anal. Biochem. 360: 336-339.
Doka O, Bicanic DD, Dicko MH, Slingerland MA (2004). Photoacoustic approach to direct determination of the total phenolic content in red sorghum flours. J. Agric. Food Chem. 52: 2133-2136.
Dykes L, Rooney LD, Waniska RD, Rooney WL (2005). Phenolic compounds and antioxidant activity of sorghum grains of varying genotypes. J. Agric. Food Chem. 53: 68136818
Dubey S, Singh D, Misra RA (1998). Enzymatic synthesis and various properties of poly (catechol)-bilirubin oxidase. Sorghum malts for the production of a lager. Enz. Microb. Technol. 23:432-437.
Dunford HB (1999). Heme peroxidases. Dunford, H. B, Ed, Wiley-VCH, New York, USA,
Duroux L, Welinder KG (2003). The peroxidase gene family in plants: a phylogenetic overview. J. Mol. Evol. 57:397-407.
Espin JC, Soler-Rivas C, Cantos E, Tomas-Barberan FA, Wichers HJ (2001). Synthesis of the antioxidant hydroxytyrosol using tyrosinase as biocatalyst. J. Agric. Food Chem. 149: 1187-1193.
Feillet P, Autran JC, Icard-Verniere CJ (2000). Pasta brownness: an assessment. J. Cereal Sci. 32: 215-233.
Ferreira D, Slade D (2002). Oligomeric proanthocyanidins: naturally occurring O-heterocycles. Nat. Prod. Rep. 19:517-541.
Food and Agricultural Organization (FAO) (1995). Sorghum and millet in human nutrition. FAO Food and Nutrition Series, No. 27. ISBN 92-5-103381-1.
http://www.fao.org/DOCREP/T0818e/T0818E00.htm#Contents [accessed on 10-09-2005].
Food and Agricultural Organization (FAO) (2005). FAOSTAT Agricultural database 2005. http://faostat.fao.org/faostat/.
Fraaije MW, Roubroeks HP, Hagen WR, van Berkel WJH (1996). Purification and characterization of an Intracellular Catalase-Peroxidase from Penicillium simplicissimum. Eur. J. Biochem. 235: 192-198.
Gaitan E, Lindsay, RH, Reichert RD, Ingbar SH, Cooksey RC, Legan J, Meydrech EF, Hill J, Kubota K (1989). Antithyroid and goitrogenic effects of millets: role of c -glycosyl flavones. J. Clin. Endocrinol. Metab. 68: 707-714.
Gajhede M, Schuller DJ, Henriksen A, Smith AT, Poulos TL (1997). Crystal structure of horseradish peroxidase C at 2.15 Å resolution. Nature Struct. Biol. 4: 1032-1038.
Gazarian IG, Lagrimini LM, Mellon FA, Naldrett MJ, Ashby GA, Thorneley RNF (1998). Identification of skatolyl hydroperoxide and its role in the peroxidase-catalyzed oxidation of indol-3-yl acetic acid. Biochem. J. 333: 223-232.
Glennie CW (1981). Preharvest changes in polyphenols, peroxidase, and polyphenol oxidase in Sorghum grain. J. Agric. Food Chem. 29:33-36.
Gomez-Cordoves C, Bartolome B, Vieira W, Virador VM (2001). Effects of wine phenolics and sorghum tannins on tyrosinase activity and growth of melanoma cells. J. Agric. Food Chem. 49:1620-1624.
Gowda P, Bhavanishankara S, Bhat SG, Bhat S, Shankara (1989). Peroxidase and polyphenol oxidase activities in sorghum and Peronosclerospora sorghi interaction. Curr. Sci. 58:1037-1039.
Guerra A, Ferraz A (2003). Enzymatic treatment of lignin-containing effluents. In Wastewater treatment using enzymes (Akihiko Sakurai, Ed). ISBN 81-7736-172-4, Research Signpost, Trivandrum, India. pp. 73-91.
Hagerman AE (2005). Tannin chemistry; Hagerman, A. E. Ed, Department of Chemistry and Biochemistry. Miami University Oxford, OH 45056 USA. Consulted on 10-09-2005. at: http://www.users.muohio.edu/hagermae/
Hagerman AE, Riedl KM, Jones GA, Sovik KN, Ritchard NT, Hartzfeld PW, Riechel TL (1998). High molecular weight plant polyphenolics (tannins) as biological antioxidants. J. Agric. Food Chem. 46: 1887-1892.
Hahn DH, Rooney LW, Fabion JM (1983). Sorghum phenolic acids. and their HPLC separation and relation to fungal resistance. Cereal Chem. 60: 255-259.
Hamauzu Y, Yasui H, Inno T, Kume C, Omanyuda M (2005). Phenolic profile, antioxidant property, and anti-influenza viral activity of chinese quince (Pseudocydonia sinensis Schneid.), quince (Cydonia oblonga Mill.), and apple (Malus domestica Mill.) fruits. J. Agric. Food. Chem. 53: 928-934.
Harborne JB (1994). Do natural plant phenols play a role in ecology. Acta Hort. 381: 36-45.
Haslam E (1998). Practical polyphenolics: from structure to molecular recognition and physiological action. Haslam, E. Ed, Cambridge University Press. Cambridge. UK.
Hatcher DW, Kruger JE (1993). Distribution of polyphenol oxidase in flour millstreams of Canadian common wheat classes milled to three extraction rates. Cereal Chem. 70: 51-55.
Heldt HW (2005). Phenylalanine ammonia lyase catalyzes the initial reaction of phenylpropanoid metabolism. Plant Biochemistry; Elsevier: Amsterdam. pp. 437-454.
Henriksen A, Welinder KG, Gajhede M. (1998). The structures of the horseradish peroxidase C-ferulic acid complex and the ternary complex with cyanide suggest how peroxidases oxidize small phenolic substrates. J. Biol. Chem. 273: 2241-2248.
Hernandez-Ruiz J, Arnao MB, Hiner AN, Garcia-Canovas F, Acosta M (2001). Catalase-like activity of horseradish peroxidase: relationship to enzyme inactivation by H_2O_2. Biochem. J. 15: 107-114.
Hertog MGL, Hollman PCH, Katan MB, Kromhout D (1993). Intake of potentially anticarcinogenic flavonoids and their determinants in adults in The Netherlands. Nutr. Cancer 20: 21-29.
Hilhorst R, Dunnewind B, Orsel R, Stegeman P, van Vliet T, Gruppen H, Schols H (1999). Baking performance, rheology and chemical composition of wheat dough and gluten affected by xylanase and oxidative enzymes. J. Food Sci. 64: 808-813.
Hilhorst R, Gruppen H, Orsel R, Laane C, Schols HA, Voragen AGJ (2002). Effects of xylanase and peroxidase on soluble and insoluble arabinoxylans in wheat bread dough. J. Food Sci. 67: 497-506.
Hiner ANP, Hernandez-Ruiz J, Rodriguez-Lopez JN, Arnao MB, Varon R, Garcia-Canovas F, Acosta M (2001). The inactivation of horseradish peroxidase isoenzyme A2 by hydrogen peroxide: an example of partial resistance due to the formation of a stable enzyme intermediate. J. Biol. Inorg. Chem. 6: 504-516.
Hiner ANP, Hernandez-Ruiz J, Williams GA, Arnao MB, Garcia-Canovas F, Acosta M (2001). Catalase-like oxygen production by horseradish peroxidase must predominantly be an enzyme-catalyzed reaction. Archiv. Biochem. Biophys. 392: 295-302.
Hiner ANP, Raven EL, Thorneley RNF, Garcia-Canovas F, Rodriguez-

Lopez JN (2002). Mechanisms of compound I formation in heme peroxidases. J. Inorg. Biochem. 91: 27-34.
Hrazdina G. Biosynthesis of flavonoids. In Plant polyphenols: synthesis, properties, significance Vol 59, Hemingway R. W. and Laks. P. E. Ed, Plenum Press: New York. 1992, pp. 61-72.
Iacobucci GA, Sweeny JG (1983). The chemistry of anthocyanins. anthocyanidins and related flavylium salts. Tetrahedron 39: 3005-3015.
Iwuoha CI, Aina JO (1997). Effect of steeping condition and germination time on the alpha-amylase activity, phenolics content and malting loss of Nigerian local red and hybrid short Kaura sorghum malts. Food Chem. 58: 289-295.
Johansson SK, Rasmussen JE, Harthill JE, Welinder KG (1992). cDNA, amino acid and carbohydrate sequence of barley seed-specific peroxidase BP1. Plant Mol. Biol. 18: 1151-1161.
Jolivet S, Arpin N, Wichers HJ, Pellon G (1998). Agaricus bisporus browning: a review. Mycol. Res. 102: 1459-1483.
Joshipura KJ, Hu FB, Manson JE, Stampfer MJ, Rimm EB, Speizer FE, Colditz G, Ascherio A, Rosner B, Spiegelman D, Willett WC (2001). The effect of fruit and vegetable intake on risk for coronary heart disease. Ann. Intern. Med. 134: 1106-1114.
Khokhar S, Magnusdottir SGM (2002). Total phenol, catechin, and caffeine contents of teas commonly consumed in the United Kingdom. J. Agric. Food Chem. 50:565-570.
Kitamura K, Honda M, Yoshizaki H, Yamamoto S, Nakane H, Fukushima M, Ono K, Tokunaga T (1998). Baicalin, an inhibitor of HIV-1 production in vitro. Antiviral Res. 37: 131-140.
Klabunde T, Eicken C, Sacchettini JC, Krebs B (1998). Crystal structure of a plant catechol oxidase containing a dicopper center. Nat. Struct. Biol. 5: 1084-1090.
Kouda-Bonafos M, Czyzewska E, Nacro M, Oehschager AC (1994). Isolation of apigeninidin from leaf sheaths of Sorghum caudatum. J. Chem. Ecol. 20: 2123-2125.
Kouda-Bonafos M, Nacro M, Ancian B (1996). Total assignment of ^{1}H and ^{13}C NMR chemical shifts of a natural anthocyanidin, apigeninidin, using two-dimensional COLOC and HMBC techniques. Magn. Reson. Chem. 34: 389-394.
Kowalski SP, Eannetta NT, Hirzel AT, Steffens JC (1992). Purification and characterization of polyphenol oxidase from glandular trichomes of Solanum berthaultii. Plant Physiol. 100: 677-684.
Krueger CG, Vestling MA, Reed JD (2003). Matrix-assisted laser desorption/ionization time-of-flight mass spectrometry of heteropolyflavan-3-ols and glucosylated heteropolyflavans in sorghum (Sorghum bicolor (L.) Moench. J. Agric. Food Chem. 51: 538–543.
Krylov S, Dunford HB (1996). Reaction of horseradish peroxidase with indole-3-acetic acid. In Plant peroxidases: biochemistry and physiology. Obinger, C, Burner, U, Ebermann, R, Penel, C, Greppin, H, Eds; Geneva, University of Geneva, pp. 59-69.
Kushi LH, Meyer KA, Jacobs DR (1999). Cereals, legumes, and chronic disease risk reduction: evidence from epidemiologic studies. Amer. J. Clin. Nutr. 70: 451S–458S.
Langseth L (1995). Oxidants, antioxidants, and disease prevention; Bracco U. Ed, ILSI Press: Belgium, 24p.
Li BQ, Fu T, Ya, YD, Mikovits JA, Ruscetti FW, Wang JM (1998). Flavonoid baicalin inhibits HIV- 1 infection at the level of viral entry. Biochem. Biophys. Res. Commun. 276:534-538.
Lo SCC, De Verdier K, Nicholson RL (1999). Accumulation of 3-deoxyanthocyanidin phytoalexins and resistance to Colletotrichum sublineolum in sorghum. Physiol. Mol. Plant Pathol. 55: 263-273.
Lu L, Liu S, Jiang S, Wu S (2004). Tannins inhibits HIV-1 entry by targeting gp41. Acta Pharma. 25: 213-218.
Luthra YP, Ghandi SK, Joshi UN, Arora SK (1988). Total phenols and their oxidative enzymes in sorghum leaves resistant and susceptible to Ramulispora soghicola Harris. Acta Phytopathol. Entomol. Hung. 23: 393-400.
Mangan JL (1981). Nutritional effects of tannins in animal feeds. Nutr. Res. Rev. 1: 209-231.
Marco A, Guzzardi P, Jamet E (1999). Isolation of tobacco isoperoxidases accumulated in cell-suspension medium and characterization of activities related to cell wall metabolism. Plant Physiol. 120: 371-381.
Marsh DR, Galliard T (1986). Measurement of polyphenol oxidase activity in wheat milling fractions. J. Cereal Sci. 4: 241-248.
Martinez MV, Whitaker JR (1995). The biochemistry and control of enzymatic browning. Trends Food Sci. Technol. 6: 195-200.
Matheis G, Withaker JR (1984). Modification of proteins by polyphenol oxidase and peroxidase and their products. J. Food Biochem. 8: 137-162.
Mayer AM, Harel E (1979). Polyphenol oxidases in plants. Phytochemistry 18: 193-215.
Melake-Berhan A, Butler LG, Ejeta G, Menki A (1996). Grain mold resistance and polyphenol accumulation in sorghum. J. Agric. Food Chem. 44: 2428-2434.
Mirza O, Henriksen A, Ostergaard L, Welinder KG, Gajhede M (2000). Arabidopsis thaliana peroxidase N: structure of a novel neutral peroxidase. Acta Crystal. 56: 372-375.
Mohan S, Purushothaman D, Jayaraj S, Rangarajan AV (1988). Phenylalanine ammonia lyase activity in the roots of Sorghum bicolor (L.) inoculated with Azospirillum. Curr. Sci. 57: 492-493.
Moosmann B, Behl C (1999). The antioxidant neuroprotective effects of estrogens and phenolic compounds are independent from their estrogenic. Proc. Natl. Acad. Sci. USA. 96: 8867-8872.
Morazzoni P, Magistretti MJ (1990). Activity of Myrtocyan® an anthocyanoside complex from Vaccinium myrtillus (VMA) on platelet agregation and adhesiveness. Fitoterapia 61: 13-20.
Morton JF (1992). Astringency masticatories. In Plant polyphenols: synthesis, properties and significance; Hemingway, R. W, Laks, P. E, Eds, Plenum press, New York, pp. 739-765.
Murdiati TB, McSweeney CS (1987). Binding properties of hydrolysable tannins to proteins. Proceedings of the 2nd International Symposium on the Nutrition of Herbivores. Rose, M. Ed, University of Queensland, Brisbane, pp. 41-42.
Natella F, Nardini M, Di Felice M, Scaccini C (1999). Benzoic and cinnamic acid derivatives as antioxidants: structure-activity relation. J. Agric. Food Chem. 47: 1453-1459.
Nguz K, van Gaver D, Huyghebaert A (1998). In vitro inhibition of digestive enzymes by sorghum condensed tannins [Sorghum bicolor L. (Moench)]. Sci. Aliments 18: 507-514.
Nwanguma B, Eze MO (1995). Heat sensitivity, optimum pH and changes in activity of sorghum peroxidase during malting and mashing. J. Inst. Brew. 101: 275-276.
Okuda T, Yoshida T, Hatano T (1991). Chemistry and biological activity of tannins in medicinal plants. Economic and medicinal plant research, Vol 5, Wagner, H, Farnsworth, N. R. Eds, Academic Press: London, 129-165.
Ollitrault P, Arnaud M, Chantereau J (1989). Polymorphisme enzymatique des sorghos. Agron. Trop. 44: 211-222.
Østergaard L, Teilum K, Mirza O, Mattsson O, Petersen M, Welinder KG, Mundy J, Gajhede M, Henriksen A (2000). Arabidopsis ATP A2 peroxidase: expression and high-resolution structure of a plant peroxidase with implications for lignification. Plant Mol. Biol. 44: 231-243.
Oudgenoeg G, Dirksen E, Ingemann S, Hilhorst R, Gruppen H, Boeriu CG, Piersma SR, van Berkel WJ, Laane C, Voragen AGJ (2002). Horseradich peroxidase-catalyzed oligomerization of ferulic acid on a template of a tyrosine-containing tripeptide. J. Biol. Chem. 277: 21332-213340.
Oudgenoeg G, Hilhorst R, Piersma SR, Boeriu CG, Gruppen H, Hessing M, Voragen AGJ, Laane C (2001). Peroxidase-mediated cross-linking of a tyrosine-containing peptide with ferulic acid. J. Agric. Food Chem. 49: 2503-2510.
Palé E, Kouda-Bonafos M, Nacro M, Vanhaelen-Fastré R, Ottinger R (1997). 7-O-methylapigeninidin, an anthocyanidin from Sorghum caudatum. Phytochemistry 45: 1091-1093.
Parr AJ, Bolwell GP (2000). Phenols in the plant and in man. The potential for possible nutritional enhancement of the diet by modifying the phenols content or profile. J. Sci. Food Agric. 80: 985-1012.
Passardi F, Longet D, Penel C, Dunand C (2004). The class III peroxidase multigenic family in rice and its evolution in land plants. Phytochemistry 65: 1879-1893.
Paterson AH, Bowers JE, Peterson DG, Estill JC, Chapman B (2003). Structure evolution of cereal genomes. Curr. Opin. Gen. Dev. 13: 644-650.

Pellegrini N, Serafini M, Colombi B, Del Rio D, Salvatore S, Bianchi M, Brighenti F (2003). Total antioxidant capacity of plant foods, beverages and oils consumed in Italy assessed by three in vitro assays. J. Nutr. 133: 2812-2819.
Pietta PG (2000). Flavonoids as antioxidants. J. Nat. Prod. 63: 1035-1042.
Radhakrishnan MR, Sivaprasad J (1980). Tannin content of sorghum varieties and their role in iron bioavailability. J. Agric. Food Chem. 28: 55-57.
Reed JD (1995). Nutritional toxicology of tannins and related polyphenols in forage legumes. J. Ani. Sci. 73: 1516-1528.
Reddy CA, D'Souza TM (1994). Physiology and molecular biology of the lignin peroxidases of Phanerochaete chrysosporium. FEMS Microbiol. Rev. 13: 137-152.
Regalado C, Garcia-Almendárez BE, Duarte-Vázquez MA (2004). Biotechnological applications of peroxidases. Phytochemistry Rev. 3: 243-256.
Richard-Forget F, Gaulliard FA. (1997). Oxidation of chlorogenic acid, catechins and 4-methylcatechol in model solutions by combinations of pear (Pyrus communis cv Williams) polyphenol oxidase and peroxidase: a possible involvement of peroxidase in enzymatic browning. J. Agric. Food Chem. 45: 2472-2476.
Riedl KM, Hagerman AE (2001). Tannin-complexes as radical scavengers and radical sinks. J. Agric. Food Chem. 49: 4917-4923.
Rietjens IM, Awad HM, Boersma MG, van Iersel ML, Vervoort J, van Bladeren PJ (2001). Structure activity relationships for the chemical behaviour and toxicity of electrophilic quinones/quinone methides. Advances in experimental medicine and biology. Adv. Exp. Med. Biol. 500: 11-21.
Rodakiewicz-Nowak J, Ito M (2003). Effect of water-miscible solvents on the organic solvent-resistant tyrosinase from Streptomyces sp. REN-21. J. Chem. Technol. Biotechnol. 78: 809–816.
Rodriguez-Lopez JN, Lowe DJ, Hernandez-Ruiz J, Hiner AN, Garcia-Canovas F, Thorneley RN (2001). Mechanism of reaction of hydrogen peroxide with horseradish peroxidase: identification of intermediates in the catalytic cycle. J. Am. Chem. Soc. 123: 11838-11847.
Rösler J, Krekel F, Amrhein N, Schmid J (1997). Maize phenylalanine ammonia-lyase has tyrosine ammonia-lyase activity. Plant Physiol. 113: 175–179.
Ryan D, Robards K (1998). Phenolic compounds in olives. Analyst 123: 31-44.
Sachdev A, Johari RP, Mehta SL (1997). Development of starch granules in high lysine sorghum grains. Ind. J. Plant Physiol. 2:1-4.
Sae SW, Kadoum AM, Cunningham BA (1971). Purification and some properties of sorghum grain esterase and peroxidase. Phytochemistry 10:1-10.
Sakharov IY, Sakharova RV (2002). Extremely high stability of African oil palm tree peroxidase. Biochim. Biophys. Acta. 1598:108-114.
Sanchez-Ferrer A, Rodriguez-Lopez JN, Garcia-Canovas F, Garcia-Carmona F (1995). Tyrosinase: A comprehensive review of its mechanism. Biochim. Biophys. Acta 1247:1-11.
Santos-Buelga C, Scalbert A (2000). Proanthocyanidins and tannin-like compounds: nature, occurrence, dietary intake and effects on nutrition and health. J. Sci. Food Agric. 80:1094-1117.
Savitsky PA, Gazaryan IG, Tishkov VI, Lagrimini LM, Ruzgas T, Gorton L (1999). Oxidation of indole-3-acetic acid by dioxygen catalyzed by plant peroxidases: specificity for the enzyme structure. Biochem. J. 340: 579-583.
Scalbert A (1991). Antimicrobial properties of tannins. Phytochemistry 30: 3875-3883.
Schuller DJ, Ban N, van Huystee RB, McPherson A, Poulos TL (1996). The crystal structure of peanut peroxidase. Structure 4: 311-321.
Sérémé A, Kouda-Bonafos M, Nacro M (1994). Tannins in utilization of sorghum in Burkina-Faso. Plant Foods Hum. Nutr. 46: 331-334.
Serna-Saldivar S, Rooney LW (1995). Structure and chemistry of sorghum and millets. In Sorghum and Millets: Chemistry and Technology; D.A.V. Dendy, Ed, American Association of Cereal Chemists, St. Paul, USA. pp. 69-124.
Sheptovitsky YG, Brudvig GW (1996). Isolation and characterization of spinach photosystem II membrane-associated catalase and polyphenol oxidase. Biochemistry 35: 16255-16263.
Shi H, Noguchi N, Niki E (2001). Flavonoids and other polyphenols. Methods Enzymol. 335: 157-166.
Shiraishi T, Yamada T, Nicholson RL, Kunoh H (1995). Phenylalanine ammonia-lyase in barley: activity enhancement In response to Erysiphe graminis f.sp. hordei (race 1) a pathogen, and Erysiphe pisi, a nonpathogen. Physiol. Mol. Plant Pathol. 46: 153-162.
Silaghi-Dumitrescu RL (2005). HRP: A Summary of its Structure, Mechanism and Substrate Diversity. Department of Chemistry, University of Georgia in Athens, Georgia. Available at: http://www.unige.ch/LABPV/publications/silaghi/Dumitrescu.html
Singh S, Lewis NG, Towers GH (1998). Nitrogen recycling during phenylpropanoid metabolism in sweet potato tubers. J. Plant Physiol. 153: 316-23.
Stafford HA (1990). Flavone and flavonone pathway. In Flavonoids metabolism. CRC Press: Boca Raton, USA.
Stafford HA (1969). Changes in phenolic compounds and related enzymes in young plants of sorghum. Phytochemistry 8: 743-752.
Stafford HA, Bravinder-Bree S (1972). Peroxidase isozymes of first internodes of sorghum: tissue and intracellular localization and multiple peaks of activity isolated by gel filtration chromatography. Plant Physiol. 49: 950-956.
Stafford HA, Dresler S (1972). 4-hydroxycinnamic acid hydroxylase and polyphenolase activities in Sorghum vulgare. Plant Physiol. 49: 590-495.
Stoewsand GS, Anderson JL, Boyd JN, Hrazdina G, Babish JG, Walsh KM, Losco P (1984). Quercetin: a mutagen, not a carcinogen, in Fischer rats. J. Toxicol. Env. Health 14: 105-114.
Subramanian V, Murty DS, Rao NS, Jambunatan R (1992). Chemical changes and diastatic activity in grains of sorghum (Sorghum bicolor) cultivars during germination. J. Sci. Food Agric. 58: 35-40.
Temple NJ (2000). Antioxidants and disease: more questions than answers. Nutr. Res. 20: 449-459.
Timothy EM, Quintanar L, Palmer AE, Hasset R, Severance S, Kosman DJ, Solomon EI (2001). Spectroscopy and reactivity of the type 1 copper site in Fet3p from Saccharomyces cerevisae: correlation of structure with reactivity in the multicopper oxidases. J. Am. Chem. Soc. 123: 5507-5517.
Tomas-Barberan F, Espín JC (2001). Phenolic compounds and related enzymes as determinants of quality in fruits and vegetables. J. Sci. Food Agric. 81: 853-876.
Tovar MJ, Romero MP, Girona J, Motilva MJ (2002). L-phenylalanine ammonia-lyase and concentration of phenolics in developing olive (Olea europaea L cv Arbequina) fruit grown under different irrigation regimes. J Sci Food Agric. 82: 892-898.
Towo EE, Svanberg U, Ndossi GD (2003). Effect of grain pre-treatment on different extractable phenolic groups in cereals and legumes commonly consumed in Tanzania. J. Sci. Food Agric. 83: 980-986.
van Haandel, MJH, Primus JL, Teunis C, Boersma MG, Osman AM, Veeger C, Rietjens IMCM (1998). Reversible formation of high-valent-iron-oxo porphyrin intermediates in heme-based catalysis: revisiting the kinetic model for horseradish peroxidase. Inorg. Chim. Acta 1: 98-105.
van Haandel Marjon JH (2000). Structure, function and operational stability of peroxidases. PhD thesis, Wageningen University, Wageningen, The Netherlands.
Vaughn C, Duke SO (1981). Tissue localization of polyphenol oxidase in Sorghum Bicolor, Bundle sheath plastids, mesophyll plastids, guard cells. Protoplasma 108: 319-327.
Veitch NC (2004). Structural determinants of plant peroxidase function. Phytochemistry Rev. 3:3-18.
Verbruggen. M. A.; Beldman. G.;Voragen. A. G. J.; Hollemans. M. Water-unextractable cell wall material from sorghum: isolation and characterization. J. Cereal Sci. 1993, 17: 71-82.
Verbruggen MA, Beldman G, Voragen AGJ (1998). Structures of enzymically derived oligosaccharides from sorghum glucuronoarabinoxylans. Carbohydr. Res. 306: 275-282.
Wallace G, Fry SC (1999). Action of diverse peroxidases and laccases on six cell wall-related phenolic compounds. Phytochemistry 52: 769-773.
Waniska RD (2000). Technical and institutional options for sorghum grain mold management: proceeding of an International consultation. Chandrashekar, A, Bandyopadhyay, R, Hall, A. J. Eds, ICRISAT,

Patancheru, India. pp. 72-106. http://www.icrisat.org/text/research/grep/homepage/sgmm/chap4.pdf
Watt JM, Breyer-Brandwijk MG (1962). The medicinal and poisonous plants of southern and eastern Africa. 2nd ed. E.&S. Livingstone, Ltd, Edinburgh and London.
Weiergang I, Hipskind JD, Nicholson RL (1996). Synthesis of 3-deoxyanthocyanidin phytoalexins in sorghum occurs independent of light. Physiol. Mol. Plant Pathol. 49: 377-388.
Weir TL, Park SW, Vivanco JM (2004). Biochemical and physiological mechanisms mediated by allelochemicals. Curr. Opin. Plant Biol. 7: 472-479.
Welinder KG (1991). Bacterial catalase peroxidases are gene duplicated members of the plant peroxidase superfamily. Biochim. Biophys. Acta 1080: 215-220.
Welinder KG (1992). Superfamily of plant, fungal and bacterial peroxidase. Curr. Opin. Struct. Biol. 2: 388-393.
Welinder KG, Justesen AF, Kjaersgard IVH, Jensen RB, Rasmussen SK, Jespersen HM, Duroux L (2002). Structural diversity and transcription of class III peroxidases from Arabidopsis thaliana. Eur. J. Biochem. 269: 6063-6081.
Willett WC (2002). Balancing life-style and genomics research for disease prevention. Science 296: 695-698.
Woude van der H, Awad HM, Boersma MG, Boeren S, van Zanden J, Van Bladeren PJ, Vervoort J, Rietjens IMCM (2002). Quinones/quinone methides as reactive intermediates in the pro-oxidant chemistry and toxicity of dietary quercetin. Free Rad. Res. 36: 103-104.
Yang J, Meyers KJ, van der Heide J, Liu RH (2004). Varietal differences in phenolic content and antioxidant and antiproliferative activities of onions. J. Agric. Food Chem. 52: 6787-6793.
Yao LH, Jiang YM, Shi J, Tomas-Barbéran FA, Datta N, Singanusong R, Chen SS (2004). Flavonoids in food and their health benefits. Plant Foods Hum. Nutr. 59: 113-122.

11

Bioconversion of pentose sugars into ethanol: A review and future directions

Anuj Kumar Chandel[1,4], Gajula Chandrasekhar[1], Konakalla Radhika[1], Rudravaram Ravinder[2] and Pogaku Ravindra[3]*

[1]Centre for Biotechnology, Jawaharlal Nehru Technological University, Hyderabad-500 085, India.
[2]Celestial Labs Limited, Banjara Hills, Hyderabad -500 034 (A.P) India.
[3]School of Engineering and Information Technology, University Malaysia Sabah, Kotakinabalu, 88999 Sabah Malaysia.
[4]Department of Biotechnology, School of Engineering of Lorena, University of Sao Paulo, Lorena- 12.602.810, Brazil.

Hemicelluloses, the second most abundant polysaccharide in nature, are well suited for ethanol production due to their enormous availability, low cost and environmental benign process. The major fraction in hemeicelluloses is pentosans and the conversion of pentosans to ethanol is problematic. To get the process economized, the conversion of hemicellulose to ethanol with a satisfactory yield is necessary. In recent years, significant advances have been made towards the technology of pentosans to ethanol conversion. However, there are technical and economical impediments to the development of commercially viable processes utilizing hemicellulosic derived sugars. This article provides an overview of the new insights in pentose sugars conversion into ethanol, pentoses resources, microorganisms and the technology.

Key words: Hemicellulose, bioethanol, pentose sugars, yeasts, fermentation.

INTRODUCTION

Lignocelluloses are the most abundant organic mass in the biosphere, which accounts for approximately 50% of the biomass. In nature the annual production of biomass is estimated to 10 to 50 × 10^9 tons (Chandel et al., 2010a). They get generated in large concentration through activation of agricultural, forestry, fruits and vegetable processing operation as residual waste.

The plant cell wall is composed mostly of the polymers cellulose, hemicellulose, and lignin (Figure 1). Lignocellulosic raw materials, by virtue of their structures, are relatively refractory to direct degradation followed by microbial bioconversion. Hemicelluloses constitute an important fraction of lignocelluloses and the conversion of hemicelluloses into fuel ethanol with utmost yields is the deciding factor for the overall economization of the process (Chandel et al., 2010b). Hemicellulose unlike cellulose is heteropolymer comprised of xylose linked backbone branched with other sugars like arabinose,mannose, galactose and glucose etc (Figure 2).

For bioconversion, pretreatment of such material is necessary. Several approaches including physical, thermal, chemical, biological or combination of these are being explored in order to obtain polymeric components of the lignocellulosic materials in maximum yield and purity (Moiser et al., 2005; Chandel et al., 2007a,b). The resulting hydrolysate after pre-hydrolysis and hydrolysis contain varying amounts of monosaccharides, both pentoses and hexoses and broad range of substances as reaction by-products from sugar and lignin degradation. Many of these substances have an inhibitory effect on the microorganism in subsequent fermentation steps (Balat and Balat, 2009; Chandel et al., 2007c).

The pretreatment process can remove hemicellulose, reduce cellulose crystallinity and increase the porosity of the materials (Saha, 2003; Moiser et al., 2005). Among all the pretreatment methods, dilute acid pretreatment

*Corresponding author. E-mail: dr_ravindra@hotmail.com.

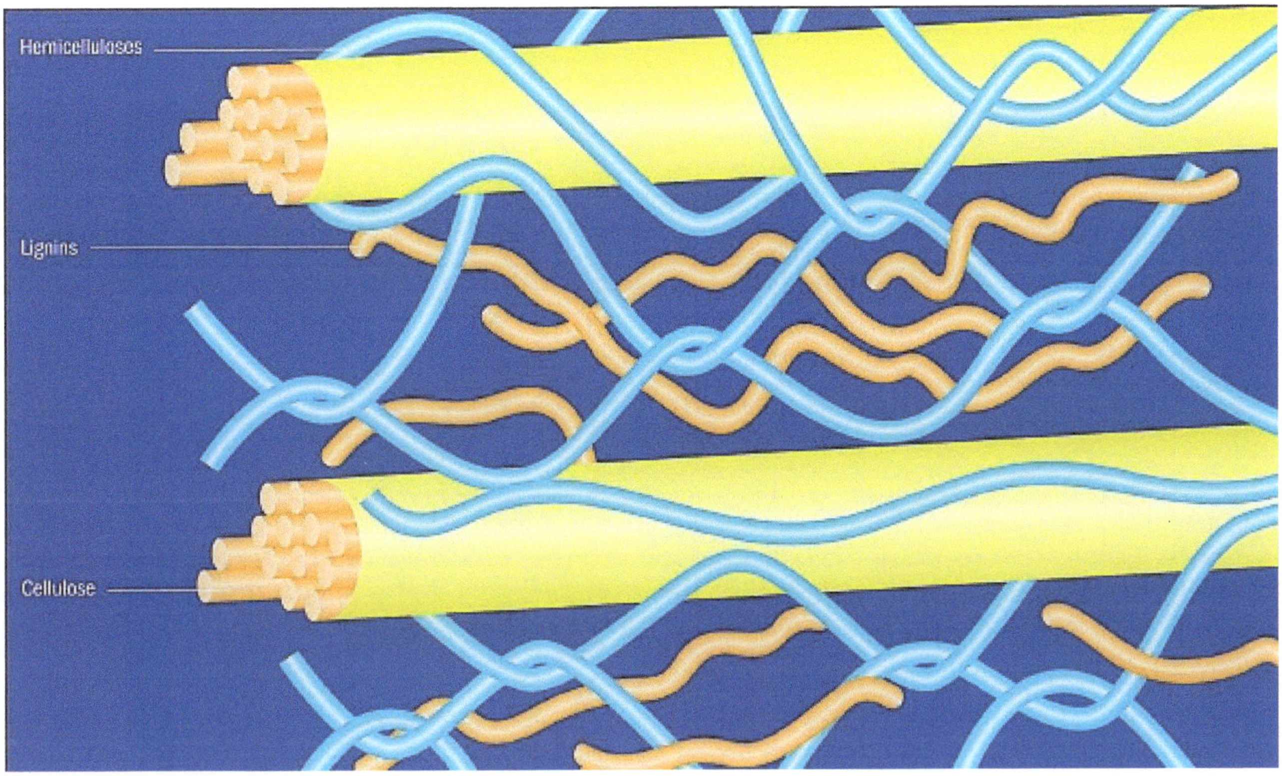

Figure 1. A schematic of a plant cell wall showing cellulose fibrils (brown) laminated with hemicellulose (turquoise) and lignin (orange) polymers (Source: www.scidacreview.org).

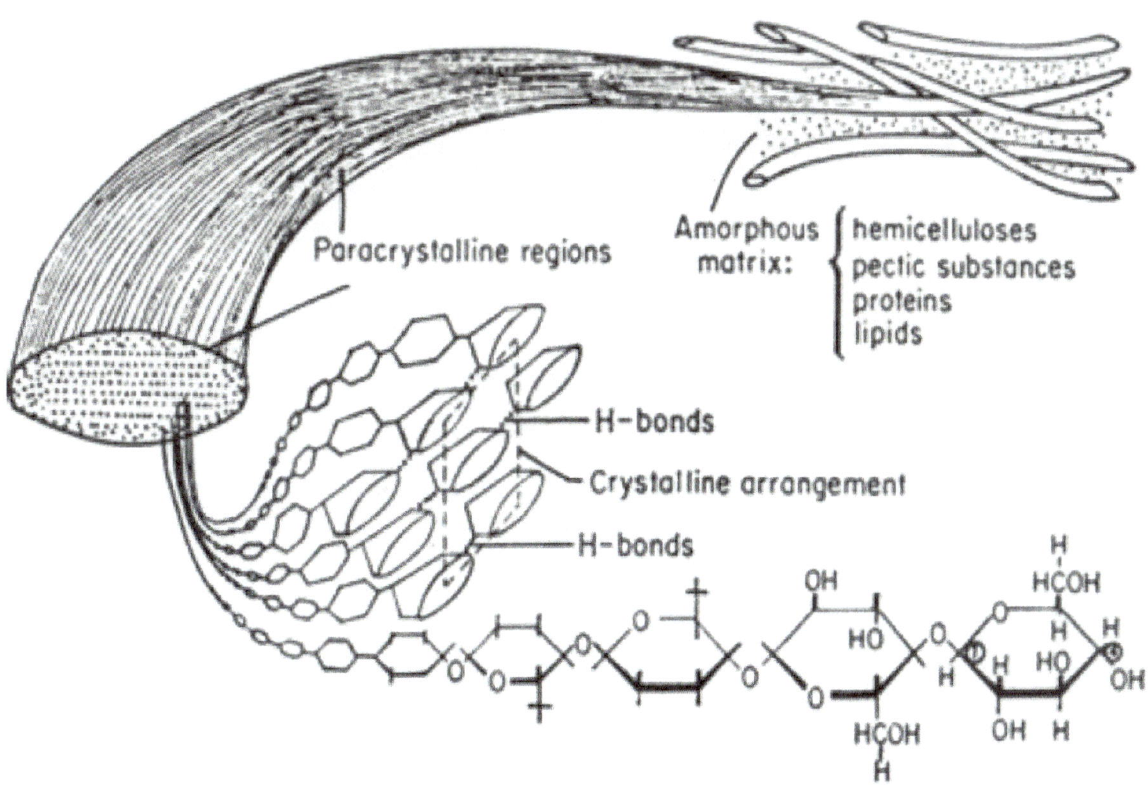

Figure 2. Schematic representation of basic plant cell wall structure showing hemicellulose arrangement (Source: www.uky.edu/~dhild/biochem).

Table 1. Cell wall composition among various lignocellulosic sources considered for biofuel (% of dry material) (Source: Chandel et al., 2010a).

Lignocellulosic source	Cellulose	Hemicellulose*				Lignin	Reference
	Glucan	Xylan	Arabinan	Mannan	Galactan		
Sugarcane bagasse	40.2	22.5	2.0	0.5	1.4	25.2	Neureiter et al., 2002
Wheat straw	32.1	19.5	2.8	0.6	1.1	20	Mani et al., 2006
Corn stover	37.5	21.7	2.7	0.6	1.6	18.9	Ohgren et al., 2006
Switch grass	34.2	22.8	3.1	0.3	1.4	19.1	Mani et al., 2006
Pine wood	44.8	6.0	2.0	11.4	1.4	29.5	Wayman and Parekh, 1990
Aspen wood	48.6	17.0	0.5	2.1	2.0	21.4	Wayman and Parekh, 1990
Spruce wood	41.9	6.1	1.2	14.3	1.0	27.1	Hayn et al., 1993
Saccharum spontaneum	36.81 ± 0.13	21.53 ± 0.04	2.16 ± 0.01	0.16 ± 0.04	0.72 ± 0.01	20.03 ± 0.12	Scordia et al. 2010
Salix wood	41.5	15.0	1.8	3.0	2.1	25.2	Wayman and Parekh, 1990
Douglas fir wood	46.1	3.9	1.1	14.0	2.7	27.3	Robinson et al., 2002

Source: Chandel et al., 2010a *Total hemicellulose amount present in lignocellulosics on the basis of percentage of dry material - Sugarcane bagasse, 27.5; Switch grass, 30; Corn stover, 26.8; Wheat straw, 50; Pine, 26; Aspen, 29; Spruce, 26; Birch wood, 23; Salix wood, 21.7; Douglas fir wood, 20.3

has been widely studied because it is highly active, inexpensive and easy to perform. The dilute acid pretreatment can effectively solubilize hemicellulose into monomeric sugars (arabinose, galactose, glucose, mannose and xylose) and soluble oligomers, thus improving cellulose conversion. The resulting pentose sugars can eventually be utilized for ethanol generation to improve the overall utilization efficiency of lignocelluloses (Moiser et al., 2005).

The bioconversion process would hold more promises of both hexoses and pentose sugars from lignocellulosic materials. The fermentation technology using bacteria and yeast to produce chemicals from hexoses is well known but the information of the ability of these organisms to ferment pentose sugars is somewhat scanty. The ability to ferment pentoses is not widespread among microorganism and most promising yeast species, identified so far, are *Candida shehatae Pichia stipitis* and *Pachysolen tannophilus* (Hahn-Hagerdal et al., 2007; Chandel et al., 2008a). To meet the suitability of these strains at industrial level, many efforts have been made to construct the choice of strain by cloning and further expression of pentose utilizing genes into common hosts such as *Saccharomyces cerevisiae*, *Zymomonas mobilis*, *Escherichia coli* etc (Jeffries, 2006; Hahn-Hagerdal et al., 2007). However it still remains a challenging issue to get the suitable strain fulfilling the requirements of ethanol production from lignocelluloses at industrial level (Zhang et al., 2010).

Rather than summarizing all of the existing literature on pentose sugar conversion into ethanol, the present article deals with the crucial parameters for developing an indigenous pentose sugars fermentation technology. Here we review the methods of hemicelluloses hydrolysis and the conversion of pentose sugars into ethanol, micro-organisms for ethanol production, methods used for strain improvement and the future needs for making this sustainable technology successful at large scale.

HEMICELLULOSES: SOURCES AND STRUCTURE

The name hemicelluloses are given to those polysaccharides, which would be extracted from plants by diluted mineral acids. Hemicelluloses are embedded in the cell walls of plants, sometimes in chains that form a 'ground' - they bind with pectin to cellulose to form a network of cross-linked fibres.

Unlike cellulose, hemicelluloses (also a polysaccharide) consist of shorter chains 500 - 3,000 sugar units as opposed to 7,000 - 15,000 glucose molecules per cellulose polymer (Saha, 2003; Chandel et al., 2010a). In addition, hemicellulose is a branched polymer, which is composed of both linear and branched heteropolymers of D-xylose, L-arabinose, D-mannose, D-glucose, D-galactose and D-glucuronic acid (Figure 2). These individual sugars may be acetylated or methylated. Most hemicelluloses from hardwoods have average degrees of polymerization (DP) of 150 - 200. Most hemicelluloses are based on 1,4-β-linkage of their backbone sugars. The amount of hemicelluloses in wood ranges from 20 to 30% (Abbi, 1996a; Saha, 2003). There are characteristic differences in the composition and structure of hemicelluloses in agricultural residues, soft wood and hard wood. Table 1 summarizes the composition of different hemicellulosic fraction of various lignocellulosics.

CONVERSION OF HEMICELLULOSE INTO FERMENTABLE SUGARS

Hemicellulose is a heteropolymer which has a random, amorphous structure with little strength. It can be depolymerized into simple form of fermentable sugars by dilute acid and enzymatic hydrolysis (Saha, 2003; Chandel et al., 2010a). A mild pretreatment with dilute acid or base prior to enzymatic hydrolysis could be effective for better hemicelluloses saccharification into fermentable sugars.

This reduces the crystallinity of the biomass and makes it more amenable to further coordinated enzymatic reactions (Howard et al., 2003; Taherzadeh et al., 2007b). In past, various pretreatment strategies with dilute acid, alkali, ammonia fiber explosion, hydrogen peroxide, steam explosion, wet oxidation, liquid hot water, sodium sulfite etc., have been reviewed (Moiser et al., 2005; Taherzadeh et al., 2007b).

Auto hydrolysis and dilute acidic hydrolysis

Auto hydrolysis is a simple method to break down the hemicellulosic backbone into monomeric sugar constituents with fewer by-products (Abbi et al., 1996b; Sun and Cheng, 2003). However, a significant fraction of hemicellulosic may remain with the substrate after the auto hydrolysis. To overcome from this problem, dilute acid hydrolysis is used for maximum degradation of hemicelluloses into monomers (Saha et al., 2005). Compared to other pretreatment methods, dilute acid hydrolysis is more useful for the conversion of maximum hemicellulosic fraction into xylose and other sugars, which can be fermented into ethanol by specialized microorganisms (Chandel et al., 2007a; Saha, 2003). Cell wall structure and components may be significantly different in different plants, which may influence the digestibility of the biomass (Hopkins, 1999). A broad discussion about dilute acid hydrolysis on a variety of lignocellulosics with respective sugar recovery has been reviewed by Taherzadeh et al. (2007a).

Formation of inhibitors during auto/acid hydrolysis and their removal

During acid hydrolysis of lignocellulosics, aliphatic acids (acetic, formic and levulinic acid), furan derivatives and phenolic compounds are formed in addition to the sugars. Furfural and 5-hydroxymethyl furfural (HMF) collectively known as furans are formed by decomposition of pentoses and hexoses respectively (Chandel et al., 2007c). Acetic acid or other weak acids are generated due to the hydrolysis of the acetyl groups or other linkage present in hemicellulosic backbone (Palmqvist and Hahn-Hagerdal, 2000).

To improve the efficiency of hydrolysate fermentation, these inhibitors are required to be removed prior to fermentation. Several detoxification methods such as neutralization, calcium hydroxide overliming, use of ion exchange resins, activated charcoal or tin oxides, enzymatic detoxification such as peroxidase and laccase have showed promising results (Chandel et al., 2007c; Palmqvist and Hahn-Hagerdal, 2000). To make the process more economic, it is necessary to either remove these fermentation inhibitors by less expensive methods or use less severe conditions for hemicellulose breakdown such as auto hydrolysis coupled with enzymatic degradation (Chandel et al., 2010b).

Enzymatic hydrolysis of hemicelluloses

Hemicellulases are multi-domain proteins generally containing structurally discrete catalytic and non-catalytic modules (Prates et al., 2001). The most important non-catalytic modules consist of carbohydrate binding domains (CBD), which facilitate the targeting of the enzyme to the polysaccharide, interdomain linkers and dockerin modules. The dockerin modules mediate the binding of the catalytic domain via cohesion-dockerin interactions, either to the microbial cell surface or to enzymatic complexes such as the cellulosome (Prates et al., 2001; Shallom and Shoham, 2003).

The coordinated action of hemicellulases is necessary to obtain the satisfactory yield of pentose sugars from lignocellulosic materials. Therefore, the development of low-cost and high effective hemicellulases is required for cellulosic ethanol production (Howard et al., 2003).

BIOCONVERSION OF PENTOSE SUGARS INTO ETHANOL

For the conversion of pentose sugars into ethanol, a lot of research reports are available in public domains which decipher the process well at laboratory scale. However, it is still challenging to get the desired yields of ethanol utilizing pentose sugars at large scale (Hahn-Hägerdal and Pamment, 2004). The reason is the unavailability of robust ethanol producing microorganisms from pentose sugars which could withstand against fermentation inhibitors and produce ethanol with high yields and productivities.

Figure 3 systematically shows the metabolic conversion

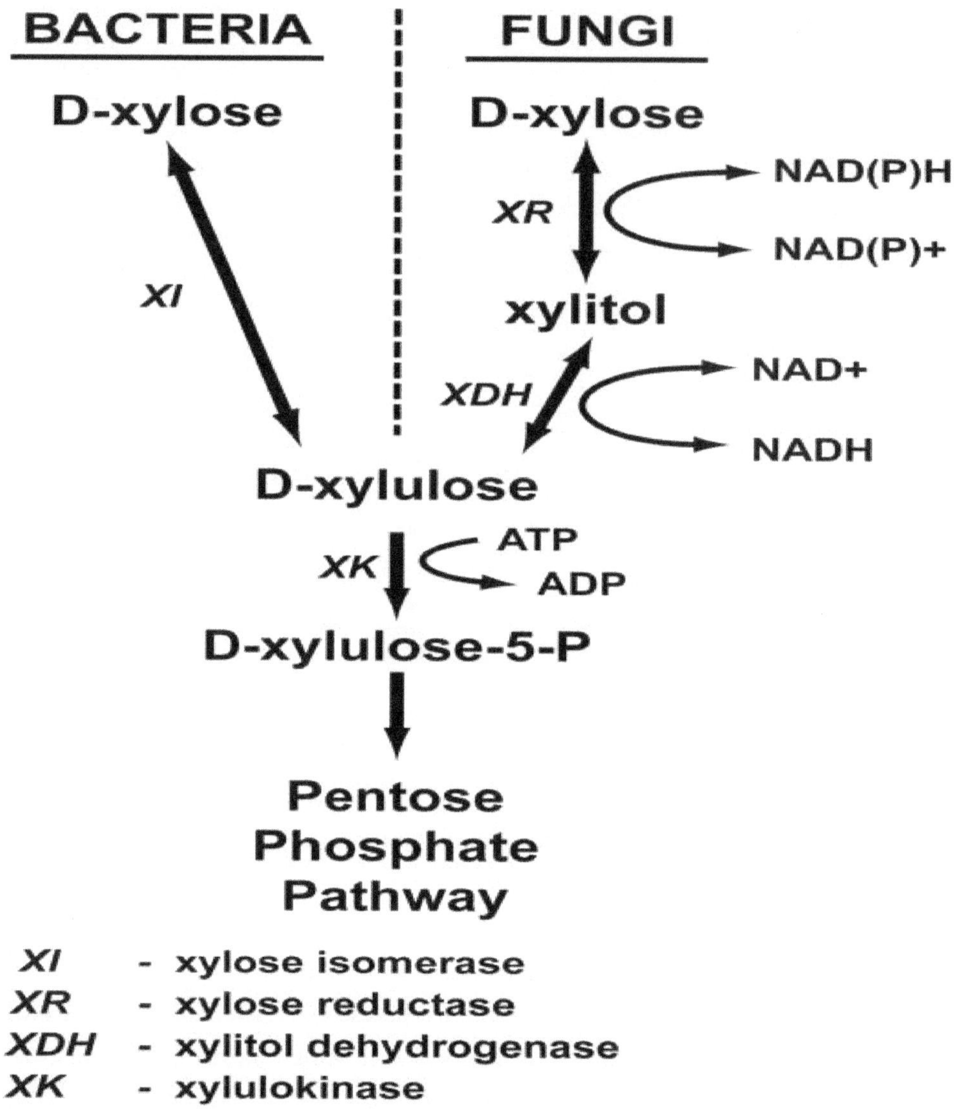

Figure 3. A depiction of xylose metabolic pathways (Source: Hector et al., 2008).

of pentose sugars in bacteria/fungi. D-xylose is first reduced to xylitol by D-xylose reductase (aldose reductase EC 1.1.1.21) and subsequently oxidized to D-xylose by xylitol dehydrogenase (D-xylose reductase, EC 1.1.1.9) to form D-xylose–5-phosphate. Ribolosephosphate-3-epimerase (5.1.3.1) transaldolase (EC 2.2.1.2) and transketolase (EC 2.2.1.1) sequentially convert alpha–xylose-5-phosphate into glyceraldehyde-3-phosphate and fructose-6-phoshate by non-oxidative rearrangement leading to ethanol formation by EMP (Emden-Meyorhoff Pathway). Alternatively, phosphoketolase (EC 4.1.2.9) can split D-xylose–5-phosphate into glycerealdehyde-3-phosphate and acetyl phosphate. Phosphoketolase is known to be important in lipid – producing yeast, especially when they are grown on D-xylose.

In some D-xylose fermenting yeasts, an oxidative pentose phosphate mechanism is present to metabolise D-xylose. Fructose–6-phosphate is oxidized to ribulose–5-phosphate. In this step, CO_2 is released in conjunction with the generation of NADPH. The pathway or its associated steps depend on the metabolic capacities of particular yeast and the growing conditions (Dien et al., 2003). In general, fermentative yeasts possess both aerobic and anaerobic pathways along with adaptive regulatory mechanisms. Some yeasts posses only aerobic pathways. However they can metabolize D-xylose anaerobically also.

D-xylose-fermenting yeasts require oxygen for growth, irrespectively the carbon source employed. This requirement for oxygen could indicate the use of fermentation pathway that does not give rise to sufficient ATP by substrate –level phosphorylation. In fermentation reaction, ethanol can be used as a carbon source after its

Table 2. Growth characteristics of natural pentose-fermenting microorganisms.

Microorganisms	Glu	Xyl	Ara	Man	Cel	Temperature range (°C)	pH range
			Filamentous fungi				
Fusarium oxysporum	+	+	+	+	+	28 - 32	5 - 6
Neurospora crassa	+	+	-	-	+	28 - 37	5 - 6
Monilia sp.	+	+	-	-	-	26	5
Mucor sp.	+	+	-	-	-	30	5.4
			Yeast				
Saccharomyces cerevisiae	+	-	-	+	-	30 - 35	3 - 7
Klyuvermyces marxians	+	+	+	+	-	30 - 35	3 - 7
Pachysolen tannophilus	+	+	+	-	-	28 - 32	2.5 - 7
Candida shehatae	+	+	+	+	-	28 - 32	3 - 7
Pichia stiptis	+	+	+	+	-	28 - 32	3 - 7
			Mesophilic bacteria				
Bacillus polymyxa	+	+	+	+	-	35 - 37	5.5 - 8
Aerobacter hydrophila	+	+	+	+	-	35 - 37	5.5 - 8
Klebsiella pneumonia	+	+	+	+	-	35 - 37	5 - 6
Clostridium acetobutylicum	+	+	+	+	+	35 - 37	4 - 8
			Thermophilic bacteria				
Clostridium thermocellum	+	+	+	-	+	65	4 - 8
C. thermohydrosulfuricum	+	+	+	-	-	65	4.7 - 8
C. thermosaccharolyticum	+	+	+	+	-	60	5 - 8
C. thermosulfurogenes	+	+	+	+	-	60	4.5 - 7.5
Thermoanerobacter ethanolicus	+	+	+	+	-	69	4.4 - 9.5

Glu-glucose Xyl-xylose Ara-arabinose Man-mannose cel-cellulose.(Source: Abbi, 1996a).

formation further to yield cell mass, CO_2 and acetate, even in the presence of substantial xylose concentrations (Abbi et al., 1996b). It has been observed that during xylose fermentation in aerobic conditions, ethanol production increases sharply followed by its declination and hike in biomass yields (Abbi et al., 1996b; Chandel et al., 2007c). However, ethanol uptake is not observed in anaerobic xylose fermentation, where the tricarboxylic acid (TCA) cycle is not operational.

PENTOSE FERMENTING MICROORGANISMS: AN ASSET IN BIO-REFINERY

During last decade, a number of laboratories have demonstrated the utilization of pentose sugars using different strains of yeast, molds and bacteria into ethanol alongside the production of acids, polyols and other by-products during bioconversion at diverse pH and temperature conditions (Table 2).

Ethanol production from different lignocellulosic sources depends upon many factors like the initial sugarconcentration in the hydrolysate, strain used for fermentation, presence of the inhibitory compounds and the employed cultivation conditions. Table 3 shows the ethanol production profile from different lignocellulosics using the different pentose utilizing yeasts.

Filamentous fungi

The ability of filamentous fungi to ferment pentose sugars has been known about 70 years. Several fungal species belonging to genera *Fusarium* (Singh and Kumar, 1991), *Rhizopus* (Perlman, 1950), *Monilia* (Gong et al., 1983), *Neurospora* (Deshpande et al., 1986) and *Paecilomyces* (Wu et al., 1986) were found to have potential for fermenting glucose as well as xylose.

In most fungi, the initial conversion of D-xylose and L-arabinose to D-xylulose 5-phosphate proceeds through a series of reduction and oxidation steps involving the cofactors $NAD(P)^+$/NAD(P)H (Hahn-Hagerdal et al., 2007). Among the ethanol producing fungi, *Fusarium oxysporum* has shown more ethanol production than *Neurospora crassa* and *Mucor sp* (Singh et al., 1992). Recently, Xiros and Christakopoulos (2009) explored *F. oxysporum* for ethanol production from brewer's spent grain (BG) under consolidated system. An ethanol yield of 109 g ethanol per kg of dry BG was

Table 3. Ethanol production from pentose sugars solution derived from lignocellulosics.

Lignocellulose material	Hydrolysis conditions	Released sugars (g/l)	Detoxification	Microbial strains	Ethanol yield (g/g)	References
Sugarcane bagasse	(2.5% (v/v) HCl, 140 °C, 30 min), # Solid: liquid = 1:10	30.29	Overliming + ion exchanger	*C. shehatae* NCIM3501	0.48	Chandel et al., 2007c
De-oiled rice bran	3.5% v/v H_2SO_4, 120 °C, 60 min	38.50±0.45	Overliming	*P. stipitis* NCIM 3499	0.42 ± 0.021	Chandel et al., 2009b
S. spontaneum	1.5% H_2SO_4, 15 min, 160ºC	32.15±0.22	Overliming	*P. stipitis* NCIM 3498	0.36 ± 0.011	Chandel et al., 2010c
Rice straw	Presoaking with 0.5% of H_2SO_4 for 18 h followed by steam heating at 15 bar pressure for 10 min, Solid: liquid = 600 g : 4 liter of 0.5% H_2SO_4	228 g sugar/ Kg of substrate	Overliming	*Mucor indicus*	0.24	Karimi et al., 2006
Wheat straw	(90 °C, 1.85% (w/v) H_2SO_4, 18 h), Solid: liquid= 1:20	17.10	Overliming	*P. stipitis* NRRL Y-7124	0.41 ± 0.01	Nigam, 2001
Corn cob	(0.3 M H_2SO_4, 98 °C , 1 h) Solid: liquid = 1:12	45.0	Overliming +ZSM-39 shaking	*P. stipitis*	0.44	Saracoglu-Eken and Arslan, 2000
Eicchornia crassipies	(1% v/v) H_2SO_4, 7 h), Solid: liquid = 1: 8	67.5	Overliming + sodium sulfite	*P. stipitis* NRRL Y-7124	0.35	Nigam, 2002
Pine	Sulfur dioxide (30 min, 160 °C + (225 °C, 30 s, HCl equivalent to 1% of dry weight), Solid: liquid = 1: 3	72	Overliming + sodium sulfite	*E. coli* K011	0.43	Barbosa et al., 1992
Willow	Steam [1 bar pressure, soaked with gaseous SO, (1 g SO_2/100 g willow) 6 min, 206 °C], Solid: gas = 100:1 g	9.0	Overliming + sodium sulfite	*E. coli* K011	0.51	Olsson et al., 1995
Mixed wood	# # Acid hydrolysis	70.4	Electrodialysis + Sodium hydroxide	*C. shehatae* FPL-Y-049	0.48	Sreenath and Jeffries, 2000

Table 3. Contd.

Paja brava	Pre-steamed, dilute sulfuric acid (0.5% or 1.0% by wt),at temperatures between 170-230℃ for 3 and 10 min. Solid: liquid = 1: 10	22.2	No Detoxification	*P. stipitis* CBS 6054	0.20	Sanchez et al., 2004

(Source: Chandel et al., 2010a) # Solid: liquid (Lignocellulose substrate: dilute acid solution) ## Details are not available.

obtained with alkali-pretreated BG under micro-aerobic conditions (0.01 vvm), corresponding to 60% of the theoretical yield based on total glucose and xylose content of BG.

Few fungal strains have been identified that ferment not only glucose and xylose but also other complex natural cellulosic substrates which is an advantage to get the maximum yield of ethanol (Singh et al., 1992). However, some critical factors such as their long generation and fermentation time, low tolerant to substrate and products and secretion of organic acids make them unattractive for ethanol production.

Bacteria

In bacteria, D-xylose and L-arabinose are isomerized to D-xylulose and L-ribulose, respectively (Chen, 1980) followed by their phosphorylation to D-xylulose 5-phosphate (Anderson and Wood, 1962) and to L-ribulose5-phosphate (Lee and Bendet, 1967), respectively. It is then epimerized to D-xylulose 5-phosphate (Lee et al., 1968), linking with the central metabolism through the pentose phosphate pathway (PPP). Watanabe et al. (2006) deciphered the alternative arabinose utilization pathway in bacteria. A review written by Senthilkumar and Gunasekaran (2005) describes ethanol production ability of wild and genetically engineered bacterial strains for ethanol production from cellulosic substrates under integrated process configurations.

Unlike yeast and filamentous fungi which do not ferment pentoses anaerobically, bacteria can readily convert xylose to ethanol under anaerobic conditions (Rosenberg, 1980; Sommer et al., 2003). These include *Bacillus macerans, Bacillus polymyxa, Kiebsiella pneumoniae, Clostridium acetobutylicum, Aeromonas hydrophila, Aerobacter* sp., *Erwinia* sp., *Leuconostoc sp* and *Lactobacillus* sp. (Gong et al., 1983; Detroy and Bolen, 1983; Abbi, 1996a). These microorganism convert pentose sugars into ethanol along with undesired by-products like acetic acid, lactic acid, 2, 3-butanediol and CO_2.

In addition to mesophillic bacteria, thermo-tolerant bacterial species such as *Clostridium thermocellum, Clostridim thermo hydrsulfurium, Clostridim thermosaccharolyticum, Clostridim thermosulfurogenes* and *Thermo anaerobacteretahnolicus* (Rosenberg, 1980; Sonneitner, 1983) can also produce ethanol. Sommer and colleagues (2003) carried out a mass screening to get thermophillic anaerobic bacteria for their ability to produce ethanol from paper pulp mills, hot springs and brewery waste water.

Due to low substrate and ethanol tolerance, pentose sugars utilizing ethanologenic bacteria are getting less attention. Though they grow fast relatively to fungi and yeast, but due to their low substrate specificity they are not considered right choice for ethanol production at large scale from pentose sugars (Senthilkumar and Gunasekaran, 2005).

Yeast

Yeasts are the most suitable microorganisms for ethanol production. They have been the center of attraction for bioethanol research since the dawn of biorefinery principle. They can grow on variety of sugars with high substrate and ethanol tolerance. However, they grow preferably on hexose sugars but few strains are also capable of utilizing pentose sugars and conversion into ethanol (Dien et al., 2003; Hahn-Hagerdal et al., 2007; Zhang et al., 2010). Among the pentose utilizing yeasts, microorganisms such as *P. tannophilus*, *C. shehatae*, *P. stipitis* and *Kluveromyces maxianus* have been discussed in detail for the ethanol production. Other yeast investigated for their xylose-fermenting ability include *Brettanomyces*, *Clavispora*, *Schizosaccharomyces*, other species of *Candida* viz., *Candida tenius*, *Candida tropicalis*, *Candida utilis*, *Candida blankii*, *Candida friedrichii*, *Candida solani* and *Candida parapsilosis* and species of *Debaromyces* viz., *Debaromyces nepalensis* and *Debaromyces polymorpha* (Abbi, 1996a). Gong et al. (1983) have tested 20 species of *Candida*, 21 strains of *Saccharomyces* and 8 strains of *Schizosaccharomyces* for their ability to ferment D-xylose. Xylitol was produced in the range of 10 - 15% by *Candida* strains with arabitol as the second major product. Ethanol was the major product with most of the *Candida* stains.

Schizosaccharomyces strains have also been reported to produce ethanol but the concentrations are low (1-5 gl). Maleszka and Schneider (1982) screened 15 yeast strains for their ability to utilize D-xylose, D-xylulose and xylitol for ethanol production under aerobic, microaerobic (low aeration) and anaerobic condition using rich undefined or defined media. In almost all cases, ethanol production by *P. tannophilus* and species belonging to *Candida* and *Pichia* was better on rich media under microaerobic conditions. Majority of yeasts are not able to ferment D-xylose directly. It has been observed that instead they utilize D-xylulose, an isomer of D-xylose, both oxidatively and fermentatively. The best xylulose-fermenting yeasts so far identified are species of *Btrettanomyces*, *Candida*, *Hansenula* and *Torulospora* (Skoog and Hahn-Hagerdal, 1988). Pentose utilizing yeasts also convert xylose into xylitol in addition to ethanol (Jeffries, 1982, 1984). Xylitol, for example, is generally a poor substrate, though it is an essential intermediate in D-xylose catabolic pathway. Pentiols are the major by-products of pentose fermentation with *C. shehatae*. However, important yeast, *P. stipitis* apparently produces no xylitol during pentose sugar fermentation (Nigam, 2001) which is a unique quality of this microorganism.

STRAIN IMPROVEMENT FOR PENTOSE FERMENTATION

Microorganisms have excelled at producing primary and secondary metabolites from a variety of raw carbohydrates for billions of years under varying cultivation processes. Todaythe results of studying the giant "microbial libraries" currently in vogue for microbial conversion of alternative carbohydrates into value-added products can be applied to ethanol production from cheap and surplus lignocelluloses. In addition, timely interventions, such as strain improvement through mutagenesis, gene cloning and expression and optimization of potential fermentation parameters can enhance the production of ethanol from pentose sugars (Chandel et al., 2008b).

Mutation

Although improving the yeasts strains by mutation is an old process. It has been quite successful at laboratory level. The major drawback with this method is to get the stable ethanol titers from mutagenic strains after every fermentation reaction. In past, multiple studies have shown progress toward strain improvement by mutation (Gong et al., 1983; Jeffries, 1984, 1985; Maleszka et al., 1983).

Candida sp. was mutated by UV mutagenesis that produced five times more ethanol than the native strain due to increased activity of xylulokinase and xylitol dehydrogenase (Gong et al., 1983). Later, Jeffries (1984) observed increased ethanol production from xylose by a mutant of *P. tannophilus* under aerobic conditions.

Based on single gene addition or deletion, various aneuploid and polyploid strains were developed to improve ethanol production from pentose sugars (James and Zahab, 1983; Maleszka et al., 1983a). Ingram et al. (1987) developed the mutants of *E. coli* that over-produced native alcohol dehydrogenase showing the high levels of ethanol.

Protoplast fusion

Protoplast fusion is a conventional but a successful technique for strain improvement. It enables a characteristic advantage of promoting high frequencies of genetic recombination between organisms which are genetically uncharacterized (Rajesh and Tauro, 1992). This technique is considered to be of exceptional importance in the construction of yeast strains, because sexual processes are frequently not available or not effective enough for efficient transfer of genetic information (Yoon et al., 1996).

In the past, protoplast fusion in bacteria and fungi has been reviewed by Peberdy (1980), Hopwood (1981). During the protoplast fusion, a hybrid state of microorganism develops showing the restoration of the genome on chromosomes which lead to further genetic recombination. Thermotolerant *S. cerevisiae* hybrids were developed by Kida et al. (1992) using protoplast fusion. Heluane et al. (1993) developed a protoplast fusant between *P. tannophilus* and S. *cerevisiae* which exhibited the improved utilization of xylose. Recently, protoplasts of thermotolerant *S. cerevisiae* VS_3 and mesophilic, xylose-utilizing *C. shehatae* were fused by electrofusion (Pasha et al., 2007). The fusants were selected based on their growth at 42°C and ability to utilize xylose. The mutant fusant CP11 was found to be stable and showed an ethanol yield of 0.459 ± 0.012 g/g, productivity of 0.67 ± 0.15 g /l/ h and fermentation efficiency of 90%.

Genetic engineering of pentose fermenting yeasts

A strong "degeneration" in selective strains can occur upon the storage of conidial material; therefore, molecular biology-based modern genetics and protein engineering have put forth new avenues to create genetically engineered microorganisms (GEMs) that can function as "booster biocatalysts" (Chandel et al., 2008b; Zhang et al., 2010). Despite the development of numerous industrial recombinant yeasts, GEMs have not been used commercially for ethanol production. Using GEM that can convert xylose and other pentose sugars along with cellulosics derived sugar constituents to ethanol can greatly improve the ethanol production efficiency and reduce the cost of the production.

Zhang et al. (1995) made the construct for xylose assimilation for the bacterium *Z. mobilis* for the effective ethanol yield. Further, Dien et al. (2000) developed a recombinant strain of *E. coli* with the genes from *Z. mobilis* for the conversion of pyruvate into ethanol. The recombinant plasmids with xylose reductase and xylitol dehydrogenase genes from *P. stipitis* and xylulokinase gene have been transformed into *S. cerevisiae* for the co-fermentation of glucose and xylose (Ho et al., 1998). Ingram and his colleagues has done extensive work on the development of efficient recombinant *E. coli* strains for ethanol production, eliminated the dependence on host of alcohol dehydrogenase (ADH) activity by combining *adh* B (coding for ADH II) and *pdc* genes of *Z. mobilis* to form a portable, plasmid borne operon for ethanol production (Ingram et al., 1987; Ingram and Conway, 1988; Beall et al., 1991). A genetically modified strain of *S. cerevisiae* containing the genes for pentose utilization and ligninolytic degradation could be the suitable strain for hemicellulosics derived sugars fermentation. Such strain can be directly used for hemicelluloses hydrolysate fermentation without detoxification by saving the cost of detoxification (Jonsson et al., 1998). Detailed information regarding the development of GEMs has been reviewed by Dien et al. (2003), Hahn-Hagerdal et al. (2007) and Matsushika et al. (2009).

Although bioethanol production from pentose sugars has been greatly improved by newly developed GEMs, there are still challenges that need further investigations. These challenges include maintaining a stable performance in commercial scale fermentation operations (Dipardo, 2000).

IMPROVED METHODS FOR PENTOSE FERMENTATION

To get the desired yields of ethanol from pentoses, essentially improved methods need to be used. Pentose fermentation is a tedious process unlike hexose or molasses fermentation. The choice of strain and other associated approach have an important role for the pentose sugars fermentation to get the desired yields and productivities. A suitable microorganism must be able to survive over a wide range of complex substrates at higher levels of carbohydrate with little or no toxicity of the formed product. Routinely applied methods in fermentation such as use of immobilized cells, adaptation of microorganisms with the lignocelluloses hydrolysates, recycling of free and immobilized cells in fermentation have shown promising results and can be used at industrial level also.

Immobilization of cells

In industry, application of immobilized cells during microbial fermentation is the most important parameter for obtaining the required yields and economizing the process (Williams and Munnecke, 1981; Kourkoutas et al., 2004). Immobilized cell systems are now used industrially for production of fructose syrups from glucose, ethanol and the production of stereo specific amino acids (Chibata and Tosa, 1977). A higher cell mass per unit fermentation volume can be achieved than with batch, continuous or cell recycle system, resulting in a corresponding increase in ethanol production. The risk of contamination is also reduced due to fast dilution rates and high cell densities (Norton and D'Amore, 1994; Chandel et al., 2009a).

A variety of synthetic and natural matrices are available for cell immobilization, synthetic materials such as polyacrylamide, poly urethane foam (PUF) may toxic to the cells, though they exhibit strong mechanical strength. Natural matrix such as alginates and carrageenan, yeast biocapsules and lignocellulosic materials (loofa sponge, pine apple cannary, sugarcane bagasse, *Saccharum spontaneum* pith etc) are non-toxic to cells and were used for yeast cell immobilization for ethanol production (Ogbonna et al., 2001; Peinado et al., 2006; Chandel et al., 2009a). However, their low mechanical strength and increased leaching of cell makes natural gels unsuitable for industrial application (Norton and D'Amore, 1994).

The immobilization of cells of xylose-fermenting yeasts has led to improved ethanol yields. *P. stipitis* cells entrapped in nylon net showed an ethanol productivity of 1 g/l/h compared with an average of 0.22 g/l/h obtained by free cells when grown on 50 g/l xylose (Linko et al., 1986). Immobilization of *P. tannophilus* in calcium alginate beads led to a high ethanol yield of 0.45 g/g initial xylose (Maleszka et al., 1982). Immobilized *C. shehatae* showed an ethanol yield of 0.5 g/g initial xylose from 1% xylose when compared with 0.48 g/g initial xylose obtained with free cells (Abbi, 1996b). Sliningher et al. (1985) has reported continuous conversion of D-xylose to ethanol by *P. tannophilus* immobilized in alginate beads.

Recycling of cells

Cell recycling and retention offer important advantages when the process micro-organisms are either slow growing or strongly affected by product inhibition (Nishiwaki and Dunn, 1977). Systems with cell recycling are finding increasing use in ethanol production to increase the productivity of the fermentation process. Cysewski and Wilke (1976) first proposed the use of yeast cell recycling to increase ethanol productivity. Recycling of cells can be performed by means of continuous ultra filtration and microfiltration (Hoffman et al., 1985).

Kargupta et al. (1998) used recycled cells in the case of a continuous membrane ferment or separator with continuous removal of product by per-evaporation. Abbi et al. (1996b) used the free cells of *C. shehatae* NCIM3501 for ethanol production from rice straw auto-hydrolysate and observed a constant ethanol production (0.44g/g) up to 5 batches.

Adaptation of microorganisms

Often, microorganisms can adapt to a variety of fermentation media. However, multiple essential nutrients along with suitable carbon and nitrogen sources are required for any fermentation reaction (Chandel et al., 2008b).

Adaptation of microorganisms to lignocelluloses hydrolysates is an important parameter to get the improved ethanol yields (Hahn-Hagerdal et al., 1991). It is evident that lignocelluloses hydrolysates contain a variety of inhibitory products (furans, furfurals and weak acids) along with the sugars. Adaptation of fermenting biocatalyst to these hydrolysates prior to using them in fermentation of lignocellulose hydrolysates will be yielding much improved ethanol productivities (Parawira and Tekere, 2010). Another important aspect of the adaptation of microorganisms is to get rid of detoxification step by saving the total cost incurred for ethanol production.

Adaptation of strains of *P. stipitis, C. shehatae* and *P. tanophillus* is expected to overcome such inhibition and improve ferment ability of xylose-containing hydrolyzates (Tran and Chambers, 1986; Parekh al., 1986). Parekh et al. (1987) used an adapted strain of *P. stipitis* on a steam-stripped hardwood hydrolysate which yielded 46% more ethanol, with over 90% xylose utilization than steam-unstripped hardwood hydrolysate. Nigam (2001) used *P. stipitis* NRRL Y-7124 adapted on wheat straw hemicellulosic hydrolysates which showed improved ethanol production (yield, 0.41 ± 0.01 g/g) equivalent to 80.4 ± 0.55% theoretical conversion efficiency.

CONCLUSIONS AND FUTURE PERSPECTIVES

The conversion of hemicelluloses into ethanol with utmost yields is crucial to get the best utilization of lignocelluloses in biorefinery. However, the lack of industrially suitable microorganisms for converting biomass into fuel ethanol has been a major issue for the success of bioethanol industries. The maximum recovery of pentoses from lignocelluloses with less inhibitor concentration and susequently their conversion into ethanol with desired yields and productivities are important issues to make bioethanol program successful at industrial scale. Both the issues have been addressed by scientists in past and research is going on currently in many laboratories in order to get the outcome of the said issues.

To get the maximum hydrolysis of hemicellulosics backbone with less inhibitors concentration, an optimized thermochemical pretreatment having less severity is required. A fine enzyme cocktail comprising hemicellulases, xylanases, pectinases and other ancillary enzymes is necessary to get the maximum sugars in quick time. The most significant issue in front of molecular biologists is to develop the robust strain having the ability to produce ethanol from all the sugars available in lignocellulose hydrolysates with maximum ethanol yields/productivities with minimum cultivation times.

The greatest successes have been in the engineering of pentose utilizing *S. cerevisiae, E. coli, Klebsiella oxytoca,* and *Z. mobilis.* Future development for these microorganisms will continue to emphasize increasing inhibitor tolerance, reducing growth factors and improving ethanol productivity by utilizing micro-array and robotic-aided directed evolution technologies.

REFERENCES

Abbi M (1996a): Bioconversion of rice straw into ethanol by Candida shehatae NCL 3501. M.Sc Thesis. University of Delhi, New Delhi, India. (in press)

Abbi M, Kuhad RC, Singh A (1996b). Bioconversion of pentose sugars to ethanol by free and immobilized cells of *Candida shehatae* (NCL-3501): fermentation behaviour. Proc. Biochem., 31: 555-560.

Anderson RL, Wood WA (1962). Purification and properties of L-xylulokinase. J. Biol. Chem., 237: 1029.

Balat M, Balat H (2009). Recent trends in global production and utilization of bio-ethanol fuel. Appl. Energy, 86: 2273-2282.

Barbosa MF, Beck MJ, Fein JE, Potts D, Ingram LO (1992). Efficient fermentation of *Pinus* sp. acid hydrolysates by an ethanologenic strain of *Escherichia coli.* Appl. Environ. Microbiol., 58: 1382-1384.

Beall DS, Ohta K, Ingram LO (1991). Parametric studies of ethanol production from xylose and other sugars by recombinant *Escherichia coli.* Biotechnol. Bioeng., 38: 296-303.

Chandel AK, Chan EC, Rudravaram R, Narasu ML, Rao LV, Ravindra P (2007a). Economics and Environmental Impact of Bioethanol Production Technologies: An Appraisal. Biotechnol. Mole. Biol. Rev., 2: 14-32.

Chandel AK, Kapoor RK, Narasu ML, Viswadevan V, Kumaran SSG, Ravinder R, Rao LV, Tripathi KK, Lal B, Kuhad RC (2007b). Economic evaluation and environmental benefits of biofuel: an Indian perspective. Int. J. Global Energy Issues, 28: 357-381.

Chandel AK, Kapoor RK, Singh AK, Kuhad RC (2007c). Detoxification of sugarcane bagasse hydrolysate improves ethanol production by *Candida shehatae* NCIM 3501. Biores. Technol., 98: 1947-1950.

Chandel AK, Narasu ML, Rudravaram R, Ravindra P, Narasu ML, Rao LV (2009b). Bioconversion of de-oiled rice bran (DORB) hemicellulosic hydrolysate into ethanol by *Pichia stipitis* NCIM3499 under optimized conditions. Int. J. Food Eng., 2(4): 1-12.

Chandel AK, Narasu ML, Chandrasekhar G, Manikeyam A, Rao LV (2009a). Use of *Saccharum spontaneum* (wild sugarcane) as biomaterial for cell immobilization and modulated ethanol production by thermotolerant *Saccharomyces cerevisiae* VS_3. Biores. Technol., 100: 2404-2410.

Chandel AK, Narasu ML, Ravinder R, Edula JR, Pasha C, Ravindra P, Rao LV (2008a). Forecasting bioethanol production from agro crop residues in Andhra Pradesh State: A case study. Technol. Spectrum, 1(2): 12-27.

Chandel AK, Singh OV, Chandrasekhar G, Rao LV, Narasu ML (2010b). Key-drivers influencing the commercialization of ethanol based biorefineries. J. Commercial Biotechnol., 16: 239-257.

Chandel AK, Singh OV, Chandrasekhar G, Rao LV, Narasu ML (2010c). Bioconversion of novel substrate, *Saccharum spontaneum,* a weedy material into ethanol by *Pichia stipitis* NCIM3498. Biores. Technol. (in press)

Chandel AK, Singh OV, Rao LV (2010a). Biotechnological applications of hemicellulosic derived sugars: State-of-the-art. In: O.V. Singh and S.P. Harvey (eds.) Sustainable Biotechnology: Renewable Resources and New Perspectives . Netherland: Springer Verlag, pp. 63-81.

Chandel AK, Venkateswar Rao L, Lakshmi Narasu M, Singh OV (2008b). The Realm of Penicillin G Acylase in Beta-lactam Antibiotics. Enz. Microb. Technol., 42: 199-207.

Chen WP (1980). Glucose isomerase (a review). Proc. Biochem., 15: 30-41.

Chibata I, Tosa T (1977). Transformation of organic compounds by immobilized microbial cells. Adv. Appl. Microbiol., 22: 27.

Cysewski GR, Wilke CR (1976). Utilization of cellulosic materials through enzymatic hydrolysis. I. Fermentation of hydrolysate to ethanol and single cell protein. Biotechnol. Bioeng., 18: 1297-1313.

Deshpande V, Keskar S, Mishra C, Rao M (1986). Direct conversion of cellulose/hemicellulose to ethanol by *Nuerospora* crassa. Enz. Microb. Technol., 8: 149-152.

Detroy RW, Bolen PL (1983). Pan-am. Sypsosium of fuels and Chemicals., 3rd, Guatemala, 24-26.

Dien BS, Cotta MA, Jeffries TW (2003). Bacteria engineered for fuel ethanol production current status. Appl. Microbiol. Biotechnol., 63: 258-266.

Dien BS, Nochols NN, O'Bryan PJ, Bothast RJ (2000). Development of new ethanologenic *Escherichia* coli strains for fermentation of lignocellulosic biomass. Appl. Biochem. Biotechnol., 84/86: 181-196.

Dipardo J (2000). Outlook for biomass ethanol production. Energy information. (in press)

Gong CS, Claypool TA, McCracken KD, Mann CM, Ueng PP, Tsao GT (1983). Conversion of pentoses by Yeast. Biotechnol. Bioeng., 25: 85-104.

Hahn-Hägerdal B, Karhumaa K, Fonseca C, Spencer-Martins I, Gorwa-Grauslund MF (2007). Towards industrial pentose-fermenting yeast strains. Appl. Microbiol. Biotechnol., 74: 937-953.

Hahn-Hagerdal B, Linden T, Senac T, Skoog K (1991). Ethanolic fermentation of pentoses in lignocellulose hydrolyzates. Appl. Biochem. Biotechnol., 28/29: 131-144.

Hahn-Hägerdal B, Pamment N. Microbial pentose metabolism (2004). Appl. Biochem. Biotech., 113-16: 1207-1209.

Hayn M, Steiner W, Klinger R, Steinmuller H, Sinner M, Esterbauer H (1993). Basic research and pilot studies on the enzymatic conversion of lignocellulosics. In: Bioconversion of Forest and Agricultural Plant Residues. Saddler, J. N., (Ed.) CAB International, Wallingford, pp. 33-72.

Heluane H, Spencer JFT, Spencer D, de Figueroa L, Callieri DA S (1993). Characterization of hybrids obtained by protoplast fusion between *Pachysolen tannophilus* and *Saccharomyces cerevisiae*. Appl. Microbiol. Biotechnol. 40:98-100.

Ho NWY, Chen Z, Brainard AP (1998). Genetically engineered Saccharomyces yeast capable of effective cofermentation of glucose and xylose. Appl. Environ. Microbiol., 64(5): 1852-1859.

Hoffman H, Kuhlmann W, Meyer HD, Schügerl K (1985). High productivity ethanol fermentations with cross flow membrane separation techniques for continuous cell recycling. J. Membrane Sci., 22: 235-243.

Hopkins WG (1999). Introduction to Plant Physiology, 2nd (ed.), JohnWiley and Sons Inc. pp. 12.

Hopwood DA (1981). Genetic studies with bacterial protoplasts. Ann. Rev. Microbiol. 35:237-272.

Howard RL, Abotsi E, Jansen van Rensburg EL, Howard S (2003). Lignocellulose biotechnology: Issues of bioconversion and enzyme production. Afr. J. Biotechnol., 2: 602-609.

Ingram LO, Conway J (1988). Expression of different level of ethanologenic enzymes from Zymomo*nas mobilis* in recombinant strains of *E. coli.* Appl. Environ. Microbiol., 54: 397-404.

Ingram LO, Conway T, Clark DP, Sewell GW, Preston III J F (1987). Genetic engineering of ethanol production in *Escherichia coli.* Appl. Environ. Microbiol., 53: 2420-2425.

James AP, Zahab DM (1983). The construction and genetic analysis of polyploids and aneuploids of the pentose-fermenting yeast *Pachysolen tannophilus*. J. Gen. Microbiol., 129: 2489-2494.

Jeffries TW (1982). Utilization of xylose by bacteria, yeasts, and fungi. Adv. Biochem. Eng. Biotechnol., 27: 1 -32.

Jeffries TW (1984). Unstable petite and grand variants of *Candida shehatae*. Biotechnol. Lett., 6(12): 777-782.

Jeffries TW (1985). Emerging technology for fermenting D-xylose. Trends Biotechnol., 3(8): 208-212.

Jeffries TW (2006). Engineering yeasts for xylose metabolism. Curr. Opin. Biotechnol., 17(3): 320-326

Jonsson LJ, Palmquist E, Nilvebrant NO, Hahn-Hagerdal B (1998). Detoxification of wood hydrolysates with laccase and peroxidase from the white-rot fungus *Trametes versicolor*. Appl. Microbiol. Biotechnol. 49(6): 691-697.

Kargupta K, Datta S, Sanyal SK (1998). Analysis of the performance of a continuous membrane bioreactor with cell recycling during ethanol fermentation. Biochem. Eng. J., 1: 31-37.

Karimi K, Emtiazi G, Taherzadeh MJ (2006). Production of ethanol and mycelia Biomass from rice straw hemicellulose hydrolyzate by *Mucor indicus*. Proc. Biochem., 41: 653-658.

Kida K, Morimura S, Sonoda Y (1992). Repeated batch fermentation process using a thermotolerant flocculating yeast constructed by protoplast fusion. J. Ferment. Bioeng., 74: 169-173.

Kourkoutas Y, Bekatorou A, Banat IM, Marchant R, Koutinas AA (2004). Immobilization technologies and support materials suitable in alcohol beverages production: a review. Food Microbiol., 21: 377-397.

Lee N, Bendet I (1967). Crystalline L-ribulokinase from *Escherichia coli.* J. Biol. Chem., 242(9): 2043-2050.

Lee N, Patrick JW, Masson M (1968). Crystalline L-ribulose 5-phosphate 4-epimerase from *Escherichia coli.* J. Biol. Chem., 243(18): 4700-4705.

Linko YY, Kautola H, Uotila S, Linko P (1986). Alcoholic fermentation of D-xylose by immobilized *Pichia stipitis* yeast. Biotechnol. Lett., 8: 47-52.

Maleszka R, Schneider H (1982). Concurrent production and consumption of ethanol by cultures of *Pachysolen tannophilus* bearing on D-xylose. Appl. Environ. Microbiol., 44: 909-912.

Maleszka R, Schneider H (1983). Concurrent production and consumption of ethanol by cultures of *Pachysolen tannophilus* growing on D-xylose. Appl. Environ. Microbiol., 44: 909-913.

Mani S, Tabil LG, Sokhansanj S (2006). Effects of compressive force, particle size and moisture content on mechanical properties of biomass pellets from grasses. Biomass Bioenergy, 30: 648-654.

Matsushika A, Inoue H, Watanabe S, Kodaki T, Makino K, Sawayama1 S (2009). Efficient Bioethanol Production by a Recombinant Flocculent *Saccharomyces cerevisiae* Strain with a Genome-Integrated $NADP^+$-Dependent Xylitol Dehydrogenase Gene. Appl. Environ. Microbiol., 75(11): 3818-3822.

Moiser N, Wyman C, Dale B, Elander R, Lee YY, Holtzapple M, Ladisch M (2005). Features of promising technologies for pretreatment of lignocellulosic biomass. Biores. Technol., 96: 673-686.

Neureiter M, Danner H, Thomasser C, Saidi B, Braun R (2002). Dilute-acid hydrolysis of sugarcane bagasse at varying conditions. Appl. Biochem. Biotechnol., 98: 49-58.

Nigam JN (2001). Ethanol production from wheat straw hemicellulose hydrolysate by *Pichia tipitis*. J. Biotechnol., 87(1): 17-27.

Nigam JN (2002). Bioconversion of water-hyacinth (*Eichhornia crassipes*) hemicellulose acid hydrolysate to motor fuel ethanol by xylose–fermenting yeast. J. Biotechnol., 97: 107-116.

Nishiwaki A, Dunn IJ (1977). Performance of a multistage fermentor with cell recycles for continuous ethanol production. Chem. Eng. Comm., 162: 179-198.

Norton S, D'amore T (1994). Physiological effects of yeast cell immobilization: Application for brewing. Enz. Microb. Technol., 16: 365-375.

Ogbonna JC, Mashima H, Tanaka H (2001). Scale up of fuel ethanol production from sugar beet juice using loofa sponge immobilized bioreactor. Biores. Technol. 76:1-8.

Ohgren K, Rudolf A, Galbe M, Zacchi G (2006). Fuel ethanol production from steampretreated corn stover using SSF at higher dry matter content. Biomass Bioenergy, 30: 863-869.

Olsson L, Hahn-Hagerdal B, Zacchi G (1995). Kinetics of ethanol production by recombinant *Escheichia coli* K011. Biotechnol. Bioeng., 45: 356-365.

Palmqvist E, Hahn-Hagerdal B (2000). Fermentation of lignocellulosic

hydrolysates. Iinhibition and detoxification. Biores. Technol., 74: 17-24.
Parawira W, Tekere M (2010). Biotechnological strategies to overcome inhibitors in lignocellulose hydrolysates for ethanol production: review. Crit. Rev. Biotechnol. (In Press).
Parekh SR, Parekh RS, Wayman M (1987). Fermentation of wood derived acid hydrolysates in a batch bioreactor and in a continuous dynamic immobilized cell bioreactor by *Pichia stipitis* R. Proc. Biochem., 22(3): 85-91.
Parekh SR, Yu S, Wayman M (1986). Adaptation of *Candida shehatae* and *Pichia stipitis* to wood hydrolyzates for increased ethanol production. Appl. Microbiol. Biotechnol., 25: 300-304.
Pasha C, Kuhad RC, Rao LV (2007). Strain improvement of thermotolerant *Saccharomyces cerevisiae* VS_3 strain for better utilization of lignocellulosic substrates. J. Appl. Microbiol., 103(5): 1480-1489.
Peberdy JF (1980). Protoplast fusion - a tool for genetic manipulation and breeding in industrial microorganims. Enz. Microb. Technol., 2: 23-29.
Peinado RA, Moreno JJ, Villalba JM, Gonzalez-Reyes JA, Ortega JM, Mauricio JC (2006). Yeast biocapsules: A new immobilization method and their applications. Enz. Microb. Technol., 40: 79-84.
Perlman D (1950). Observations on the production of ethanol by fungi and yeast. Am. J. Bot., 37: 237-244.
Prates JAM, Tarbouriech N, Charnock SJ, Fontes CM, Ferreira L, Davies GJ (2001). The structure of the feruloyl esterase module of xylanases 10B from *Clostridium thermocellum* provides insight into substrate recognition. Structure, 9: 1183-90.
Rajesh G, Tauro P (1992). Standardization of conditions for protoplast production from Endomycopsis fibuligera. Indian J. Microbiol., 32: 323-326.
Robinson J, Keating J, Boussaid A, Mansfield SD, Saddler JN (2002). The influence of bark on the fermenation of douglas-fir whitewood pre-hydrolysates. Appl. Microbiol. Biotechnol. 59:443-8.
Rosenberg SL (1980). Fermentation of pentose sugars to ethanol and other neutral products by microorganisms. Enz. Microb. Technol., 2: 185-193.
Saha BC (2003). Hemicellulose bioconversion. J. Ind. Microbiol. Biotechnol., 30: 279-91.
Saha BC, Iten LB, Cotta MA, Wu YV (2005). Dilute acid pretreatment, enzymatic saccharification and fermentation of wheat straw to ethanol. Proc. Biochem., 40: 3693–3700.
Sanchez G, Pilcher L, Roslander C, Modig T, Galbe M, Liden G (2004). Dilute-Acid hydrolysis for fermentation of the Bolivian straw material *Paja brava*. Biores. Technol., 93: 249-256.
Saracoglu-Eken N, Arslan Y (2000). Comparison of different pretreatments in Ethanol fermentation using corn cob hemicellulosic hydrolysate with *Pichia stipitis* and *Candida shehatae*. Biotechnol. Lett., 22: 855-858.
Scordia D, Cosentino SL, Jeffries TW (2010). Second generation bioethanol production from *Saccharum spontaneum* L. ssp. *aegyptiacum* (Willd.) Hack. Biores. Technol., 101(14): 5358-5365.
Senthilkumar V, Gunasekaran P (2005). Bioethanol production from cellulosic substrates: Engineered bacteria and process integration challanges. J. scientific Indus. Res., 64: 845-853.
Shallom D, Shoham Y (2003). Microbial hemicellulases. Curr. Opin. Microbiol., 6: 219–228.
Singh A, Kumar PK (1991). Fusarium oxysporum: status in bioethanol production. Crit. Rev. Biotechnol., 11(2): 129-147.
Singh A, Kumar PKR,Schugerl K (1992). Bioconversion of cellulosic materials to ethanol by filamentous fungi. Adv. Biochem. Biotechnol., 45: 29-55.
Skoog K, Hahn-Hägerdal B (1988). Xylose fermentation. Enz. Microb. Technol., 10: 66-80.
Sliningher PJ, Bothast RH, Okos MR, Landish MR (1985). Comparative evaluation of ethanol production by xylose fermenting yeast presented high xylose concentration. Biotechnol. Lett., 7: 431-436.
Sommer P, Georgieva T, Ahring BK (2004). Potential for using thermophilic anaerobic bacteria for bioethanol production from hemicellulose. Biochem. Soc. Trans., 32(Pt 2): 283-289.
Sonneitner B (1983). Biotechnology of Thermophilic bacteria –Growth, products and Application. Adv. Biochem. Eng. Biotechnol., 28: 69-94.
Sreenath HK, Jeffries TW (2000). Production of ethanol from wood hydrolyzate by yeasts. Biores. Technol., 72: 253-260.
Sun Y, Cheng J (2002). Hydrolysis of lignocellulosic materials for ethanol production: a review. Biores. Technol., 83: 1-11.
Taherzadeh MJ, Karimi K (2007b). Enzyme-based hydrolysis processes for ethanol from lignocellulosic materials: A Review. Bioresources., 2(4): 707-738.
Tran AV, Chambers RP (1986). Red oak wood derived inhibitors in the ethanol fermentation of xylose by *Pichia stipitis* CBS 5776. Biotechnol. Lett., 7: 841-846.
Watanabe S, Shimada N, Tajima K, Kodaki T, Makino K (2006). Identification and characterization of L-arabonate dehydratase, L- 2-keto-3-deoxyarabonate dehydratase and L-arabinolactonase involved in an alternative pathway of L-arabinose metabolism: novel evolutionary insight into sugar metabolism. J. Biol. Chem., 281: 33521-33536.
Wayman M, Parekh SR (1990). Biotechnology of Biomass Conversion; Fuel and Chemicals from Renewable Resources. Open University Press, Milton Keynes, pp. 181-232.
Williams D, Munnecke DM (1981). The production of ethanol by immobilized yeast cells. Biotechnol. Bioeng., 23: 1813-1825.
Wu JE, Lastik SM, Updegraff DM (1986). Ethanol production from sugars desired biomass by novel fungus. Nature (London), 321: 887-889.
Xiros H, Christakopoulos P (2009). Enhanced ethanol production from brewer's spent grain by a *Fusarium oxysporum* consolidated system. Biotechnol. Biofuels, 2: 4.
Yoon GS, Lee TS, Kim C, Seo JH, Ryu YW (1996). Characterization of alcohol fermentation and segregation of protoplast fusant of *Saccharomyces cerevisiae* and *Pichia stipitis*. J. Microbiol. Biotechnol., 6: 286-291.
Zhang M, Eddy C, Deanda K, Finkelstein M, Picataggio S (1995). Metabolic engineering of a pentose metabolism pathway in ethanologenic *Zymomonas mobilis*. Science, 267: 240-243.
Zhang Y, Zhu Y, Zhu Y, Li Y (2010). The importance of engineering physiological functionality into microbes. Trends Biotechnol. (In Press).

12

Immunomodulatory effects of carbohydrates and advanced glycation end products

Kankia, H. I.

Department of Biochemistry, Umaru Musa Yar'adua University, Katsina, Katsina State, Nigeria.

Carbohydrates which comprise of monosaccharides, disaccharides and polysaccharides are heterogeneous complex structure in living systems and are found to bind to other organic molecules such as proteins and lipids. Advanced glycation end products (AGEs) are also heterogeneous group of molecules that accumulate in plasma and other body fluids and in cell and tissues. This review focuses on the impact of polysaccharides and AGEs on immune function. A number of *in vitro* and *in vivo* studies have demonstrated that polysaccharides and AGEs modulate both innate and adaptive immune responses. Polysaccharides such as lectins are reported to activate innate immune cells and T-helper cells leading to the up-regulation of T-lymphocytes, B-lymphocytes and macrophages and release of different cytokine profiles such as tumour necrosis factor (TNF-α), interleukin (IL)-1, IL-6, IL-8, IL-10, IL-12, interferon gamma and beta (IFN-γ and IFN-β2). Both lectins which are non-enzymatic proteins present in plants and animals that preferentially bind to specific carbohydrate structures and AGE which triggers inflammatory response by binding to RAGE do so there by mediating inflammation known as a key underlying cause in the development of vascular complications leading to an increased expression of cytokines, growth factors and adhesion molecules with mediate and immune response.

Key words: Immunomodulation, carbohydrates, advanced glycation end products (AGEs), receptors, lectins, *in vitro* and *in vivo* studies.

INTRODUCTION

Immunomodulation through natural substances could be considered an alternative for the prevention and cure of various diseases. There is growing evidence suggesting that polysaccharides from natural plants can significantly enhance immune system responses. Plant polysaccharides are known to possess various biological and pharmacological activities hence are regarded as promising immunomodulatory agents which are relatively nontoxic and so have no significant side effects (Kim et al., 2011; Jin-Kyung et al., 2011; Tang et al., 2012; Xia et al., 2012). It has been demonstrated that most of the polysaccharides are activators of immune cells through their capacity to enhance production of immune mediators, such as reactive oxygen species (ROS), nitric oxide (NO), TNF-α and IL-10 and also their immunomodulatory characteristics are mainly dependent on their unique molecular structures which activate varied surface receptors (Jeurink et al., 2008; Weng et al.,

2011).

Macrophages and lymphocytes are important immune cells which play a pivotal role in immune responses by releasing cytokines that plays a critical role in mediating signal transduction and stimulating the immune defence system (Chen et al., 2008; Schepetkin and Quinn, 2006; Zha et al., 2007). Cytokines which are soluble glycol-proteins ought to critically involve in the immune response. The functions of these proteins are diverse and include roles in normal humoral and T- lymphocyte mediated immune response (Schepetkin and Quinn, 2006; Zha et al., 2007).

The killer cell immunoglobulin-like receptors (KIR) and C-type lectin-like receptors are the two major families of receptors expressed by human NK cells which recognize class I MHC molecules. The receptors within each family have similar features with highly homologous extracellular domains attached to various transmembrane regions and cytoplasmic tails. However, the inhibitory receptors belonging to both KIR and C-type lectin-like receptor families carry a pair of ITAMs in their cytoplasmic tails (Baba et al., 2000). Moreover, due to their early production of cytokines and chemokines and their ability to lyse target cells without prior sensitisation NK cells are crucial components of the innate immune system, providing a first line of defence against infectious agents. Hence, they can be activated non-specifically by several cytokines, including IL-2, IL-15 and IL-18 (Middleton et al., 2002)

Numbers of pathologic conditions are accompanied by changes in cytokine levels and by disturbances in the cytokine-mediated interplay between innate and acquired immune responses (Stanilova et al., 2005). IL-2 is found to be an essential cytokine for T lymphocyte growth. IFN-γ promotes the differentiation of Th1 cells and functions principally on macrophages to enhance antimicrobial properties. IL-4 is crucial for the differentiation of naive Th cells into the Th2 cells that promote humoral immunity and also has a central role in the pathogenesis of allergic inflammation (Chen et al., 2008). A study by Zha et al. (2007) using *Dendrobium huoshanense* in an *in vitro* study showed that the polysaccharide of HPS-1B23 was able to stimulate IFN-γ and TNF-α production in the culture medium of splenocytes and macrophages, respectively with no observed toxicity effect of the samples (Zha et al., 2007).

In the human body, the immune system is divided into the innate and adaptive immunities. Innate immunity refers to a non-specific immune response in encountering a particular antigen, while the adaptive immunity shows a high specificity to antigens invaded and is responsible for a memory response after the first exposure of the same antigen. These two systems do not work separately, they act closely together to help fight against the external threats by triggering a series of immune responses (Shao et al., 2004; Wong et al., 2011).

Carbohydrates are storage components in living system and comprise monosaccharides, disaccharides and polysaccharides (Khodse et al., 2007). Their role in biological processes makes them to be considered as vital biomolecules (Ghazarian et al., 2011; Hu and Xu, 2011). Carbohydrates decorate the surfaces of all cells (glycocalyx) in the form of polysaccharides, glycoproteins, glycolipids and other glycoconjugates. The saccharide binding proteins mediate cell-cell interactions in numerous biological processes, such as blood coagulation, immune response, viral infection, pathological processes, inflammation, embryogenesis and cellular signal transfer (Han et al., 2011; Hu and Xu, 2011), thus they are recognized as ligands for carbohydrate recognition proteins (for example, lectins) (Hu and Xu, 2011; Nakamura et al., 2013; Wu et al., 2006).

AGEs are heterogeneous group of molecules that accumulate in plasma and other body fluids and in cell and tissues (Kasper and Funk, 2001). They are main pathogenic agent of advanced age, diabetes, atherosclerosis, renal failure and other chronic diseases. Researches has shown that AGEs are risk factors of disease as are known to be generated by the non-enzymatic reaction of amino groups with different biomolecules such as proteins and lipids(Hirasawa et al., 2011; Kasper and Funk, 2001; Wu et al., 2011). AGEs accumulate in the living system due to an increased level of sugars and reactive dicarbonyl compounds such as glucose, fructose, deoxyglucose, glyoxal, methylglyoxal and triosephosphates and their formation is irreversible, causing protease-resistant cross-linking of peptides and proteins, which can lead to protein deposition and amyloidosis (Krautwald and Münch, 2010).

CARBOHYDRATE STRUCTURE AND FUNCTION

Organic molecules in living systems are classified into four major groups: proteins, lipids, nucleic acids and carbohydrates. However carbohydrates are by far the most abundant organic molecules found in nature, and nearly all organisms synthesize and metabolize carbohydrates. It has the empirical formula $C_nH_{2n}O_n$, where n is ≥3 and glucose is the common monosaccharide that may be linked to form a variety of other macromolecules. Carbohydrates appear to form a heterogeneous complex structure in living systems by binding to other organic molecules such as proteins and lipids by the action of glycosyltransferases and glycosidases. The combined activity levels of these enzymes in the endoplasmic reticulum (ER) and the Golgi apparatus, and perhaps on the cell surface, determine the glycosylation patterns (carbohydrate domains) of glycolipids and glycoproteins (Ghazarian et al., 2011).

The structural features of a polysaccharide are defined by molecular weight, monosaccharide composition, sequence of monosaccharide, configuration and position of glycosidic linkages, type and polymerization degree of

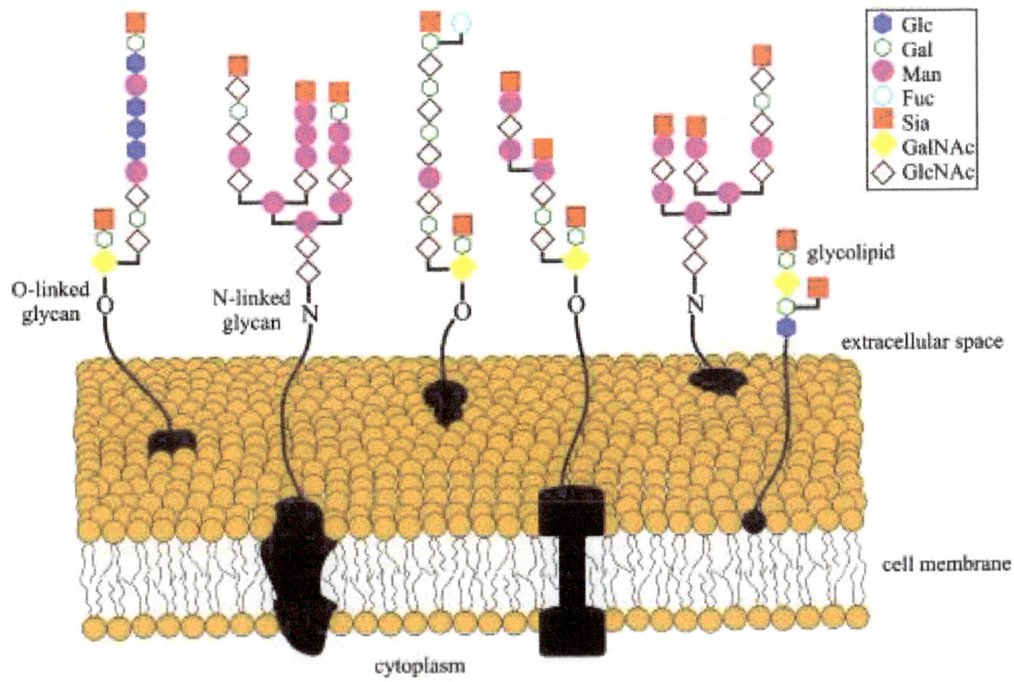

Figure 1. Showing carbohydrate heterogeneity found on cell surface glycoproteins and glycolipids. Sialic acids (Sia) are usually found at the terminal residue of O-linked and N-linked glycans of glycoproteins and glycolipids. Glc, glucose; Gal, galactose; Man, mannose; Fuc, fucose; GalNAc, N-acetylgalactosamine; GlcNAc, N-acetylglucosamine. Illustration prepared by HaikeGhazarian using Corel Draw X4 (Ghazarian et al., 2011).

branch, spatial configuration, etc (Wu et al., 2011). The extent to which polysaccharides exerts their bioactivities is closely related to their chemical composition, molecular weight, branching, chain conformation and water solubility. The basic understanding of both the primary and secondary structures for the polysaccharides is essential for the successful interpretation of their bioactivities, and therefore, the monosaccharide composition analysis of polysaccharides is the most important step for the further discovery of its physicochemical properties, structure and structure-bioactivity relationship (Huang et al., 2007; Xie et al., 2013a).

Possible structures are O- or N-linked glycans and composition varies, giving a wide variety of structures able to bind to many different ligands (Figure 1). However, it is the secondary structure that determines the biological activity.

Most of the polysaccharides, especially those found in the cell wall of the plant are mainly composed of monosaccharides such as xylose (xyl), arabinose (ara), mannose (man), sorbinose (sor), ribose (rib), Fuc, Glc, Gal, GlcNAc and GalNAc, along with small proportions of rhamnose (rha), fructose (fur), and 4-*O*-methyl galactosamine. The glycosyl linkages of these monosaccharide residues are those determined to give an insight into the structure of polysaccharides (Hsieh et al., 2008; Wang et al., 2012).

CELLULAR RECEPTORS FOR CARBOHYDRATES

Certain carbohydrates found on circulating cells, proteins and pathogens are recognized by specific lectin receptors involved in clearance or immunity (Nangia-Makker et al., 2002). Cell-surface carbohydrates, composed of several different kinds of monosaccharides, interact with many types of lectins as part of a communication signalling network between cells. Since the biogenesis of the cell-surface carbohydrates is tightly regulated, alterations in glycosylation patterns may play key roles in governing the invasive and metastasis potentials of many types of cancers. Viral virulence as well as bacterial and parasite infections are also mediated by host-cell surface carbohydrates, especially their terminal saccharides, which are recognized by invading pathogens (Hyun et al., 2007).

Complex carbohydrates (polysaccharides) are attached to the majority of secreted proteins and cover the surface of cells as a dense layer. However their importance, availability and structural variations reflect a key role in many biological functions. Lectins which are carbohydrate-binding proteins that are able to detect subtle differences between complex carbohydrate structures, decipher glycocodes recognizing certain sugars to carry out various functions like cell attachment, migration or invasion and that is why among all of the variety of

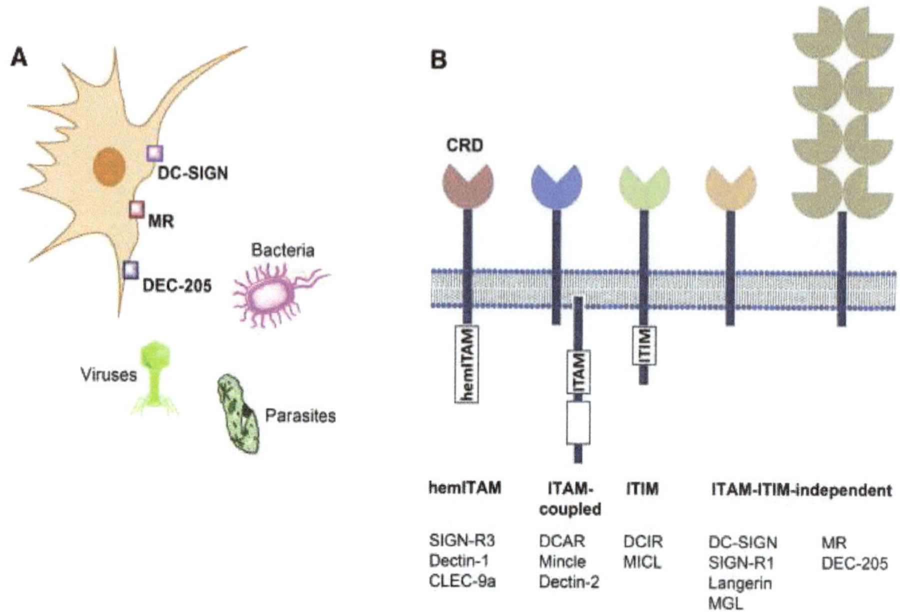

Figure 2. Myeloid C-type lectin receptors (CLRs) in innate immunity. A, In innate immunity, CLRs such as DC-SIGN, mannose receptor (MR), or DEC-205 serve as pattern-recognition receptors and bind to glycan structures present on pathogens such as bacteria, viruses, or parasites. B, Myeloid CLRs share one or more homologous carbohydrate-recognition domains (CRDs) that bind to carbohydrates often in a Ca^{2+}-dependent manner.

adhesion molecules, cell-surface carbohydrate structures have been a focus of many investigations (Nangia-Makker et al., 2002).

Lectins are found throughout nature in plants, bacteria and animals. The major types of animal lectins include galectins, C-type, I-type, P-type and R-type lectins. Most lectins contain a carbohydrate recognition domain (CRD) that binds with high specificity to the outermost sugars of *N*-linked and *O*-linked glycans as well as glycospingolipids (Sørensen et al., 2012).

Among other members, C-type lectin receptors (CLRs) are considered to be suitable target in innate immunity and cell specific drug delivery due to their endocytotic or cell activation properties. They represent a large receptor family including collectins, selectins, lymphocyte lectins and proteoglycan and share a structurally homologous carbohydrate-recognition domain (CRD) as well as often binding to carbohydrates in a Ca^{2+}-dependent pattern. Myeloid CLRs such as DC-SIGN, mannose receptor (MR), DEC-205 and others are crucial to initiate immune responses against various pathogens including bacteria, viruses, parasites and fungi (Figure 2a). However, some CLRs serve solely as phagocytic receptors whereas others activate signalling pathway by inducing the activation of tyrosine kinases such as spleen tyrosine kinase (SYK) through classical immunoreceptor tyrosine-based activation motifs (ITAMs) or hemITAMs (Figure 2b) (Lepenies et al., in press, accessed 2013).

The signaling pathway elicited upon CLR engagement depends on their cytoplasmic signaling motifs. Some CLRs contain an intracellular hemITAM or are associated with ITAM-bearing adaptor proteins leading to the activation of the tyrosine kinase Syk. In contrast, other CLRs contain an intracellular ITIM that induces the recruitment of phosphatases such as SHP-1 and SHP-2. Another group of myeloid CLRs including DC-SIGN or MR does not possess intracellular signaling motifs, thus signaling is independent of Syk or phosphatases (Lepenies et al., in press, accessed 2013).

CARBOHYDRATES IN CELLULAR FLUIDS

Cellular fluids, especially extracellular have been isolated from various bodily fluids and used as biological markers for various diseases among which are cancer and diabetes mellitus (Tzimagiorgis et al., 2011). Carbohydrates appeared to be basic to virtually every aspect of extracellular circulation, fulfilling roles from purely structural to mediating the highly specific recognition events that underlie cell-cell communication. Sugars are naturally chosen by nature to suitably mediate such an extraordinary variety of interactions. They are linked and similar to nucleotides and amino acids, also are able to form branched structures with stereospecific linkages, features that confer exponential capacity for structural diversity. Extracellular sugars are often elaborated with covalent modifications, such as sulfation, acetylation and

and phosphorylation that impart further structural variety (Bowman and Bertozzi, 1999).

Several conjugates including glycolipids and glycolproteins can adhere to each other by *trans* carbohydrate-carbohydrate interactions between apposed membranes. For example, myelin, the membranous sheath which is spirally wound around nerve axons is rich in two glycosphingolipids, GalC and its sulphated form, galactosylceramide I^3-sulfate. These interactions are often mediated by divalent cations. Water disrupts these interactions between single sugar molecules, but the interactions between polymeric multivalent glycoconjugates or membrane lipid domains, that present a multivalent array of carbohydrate head groups, are stronger and persist in water. These can be strong enough to cause specific adhesion of sponge cells into a multicellular organism. In the case of mammalian cells, they may cause transient adhesion of cells allowing up-regulation of proteins such as integrins to maintain adhesion and the interaction can be homotypic, similar to that for the lexganglioside (Boggs et al., 2008).

Immunomodulatory activity of polysaccharides

Polysaccharides isolated from various traditional medicinal plants, especially higher plants (Xu et al., 2009) have been shown to greatly affect the immune system both *in vivo* and *in vitro* by stimulating cytokine and chemokine production, ROS production and cell proliferation. They are reported to be non-toxic, with no significant side effects, which is a major problem associated with many other immunomodulatory agents, such as bacterial polysaccharides and synthetic compounds. These polysaccharides therefore have the potential as immunomodulators (Belska et al., 2010; Schepetkin and Quinn, 2006) as they have been shown to increase macrophage cytotoxic activity against tumour cells and microorganisms, activate phagocytic activity and enhance secretion of cytokines and chemokines (Chen et al., 2010; Schepetkin and Quinn, 2006).

Several studies have shown the immunomodulatory effects of certain polysaccharides and how they activate innate immune cells and T-helper cells leading to the up-regulation of T-lymphocytes, B-lymphocytes, and macrophages and release of different cytokine profiles (Chen et al., 2010; Cheng et al., 2008; Fan et al., 2012; Kouakou et al., 2013; Makino et al., 2006; Wang et al., 2013; Wong et al., 2011; Xia et al., 2012; Xu et al., 2009; Zhang et al., 2013). For example, an *in vivo* immunomodulatory study by Wong et al. (2011) using sclerotial polysaccharides isolated by hot water extraction from *Pleurotus tuber-regium* and *Pheidole rhinocerus* (PTW and PRW, respectively) on both healthy BALB/c mice and healthy athymic nude mice had shown that innate immune cells such as monocyte/macrophages and T-helper cells dendritic cells are activated by the mushroom are activated by the mushroom sclerotial polysaccharides upon detection of release cytokines like IL-1β, IFN-γ, TNF-α, IL-2 and IL-6 (Wong et al., 2011).

Also, in a study by Zhang et al. (2013) on hot water extracted polysaccharides from stems of *Taxillus chinensis* and *Uncaria rhyncophylla* in a dose-dependent manner (10 and 100 µg/ml) have shown an increase production of TNF and NO, especially on 100 µg/ml concentration in treated macrophages when compared with untreated ones. Hence it can be seen that these polysaccharide fractions display significant immunomodulatory activities as indicated by increased production of TNF-α as well as NO (Zhang et al., 2013).

Another example is on the recent study by Wang et al. (2013) on *Kadsura marmorata*. This plant has been used in traditional Chinese medicine as one of the major ingredients for the treatment of bronchial asthma, eczema and acute and chronic infection due to either viruses or bacteria. The study which was investigated *in vitro* on chicken lymphocytes and macrophages has shown that *K. rmorata* had up-regulate the production of chicken T-lymphocytes, B-lymphocytes and peritoneal macrophages which in turn could stimulate the cell proliferation. It also significantly enhances cytokine secretion. This was ascertained when different concentrations (dose dependent manner) of *K. marmorata* were used. Hence was suggested as having potential effects on regulating the immune system (Wang et al., 2013).

Protein glycation

During the lifetime of the cells, proteins are exposed to a multitude of modification process or non-enzymatic glycosylation reaction between reducing sugars and amine residues in proteins which usually occurred as a result of elevation of glucose concentrations (Shuvaev et al., 2001; Vlassopoulos et al., 2013). Oxidation and glycation are among the major non enzymatic mechanisms. Such kind of alterations can affect tissues and are accumulated in the cells with an indisputable effect on metabolism.

In the reaction known as Maillard, that involves three main stages: early, intermediate and late, glucose or other reducing sugars such as fructose, pentoses, galactose, mannose, ascorbate, xylulose, at the early stage, react with a free amino group of several molecules, including proteins, nucleic acids and lipids, to form an unstable aldimine compound called the Schiff base. However, through rearrangement, this base gives rise to a stable ketoamine, known as Amadori product (Figure 3) (Lapolla et al., 2005).

These molecular and cellular alterations, though not dangerous at the initial stage, can become damaging and pathogenic when sufficiently abundant. Among the various targets that are sensitive to alteration, blood plasma is sustainably and continually exposed to numerous

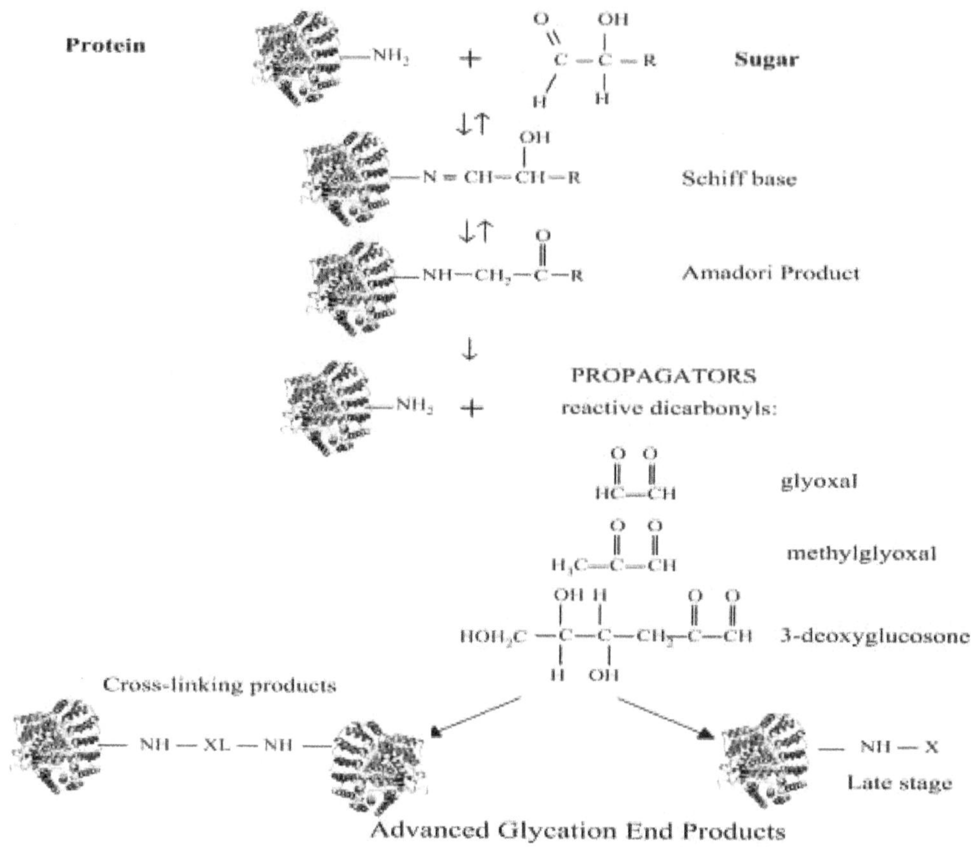

Figure 3. Maillard reaction showing the three steps of non-enzymatic glycation reaction (Lapolla et al., 2005).

metabolites, which in turn induce an oxidative stress (Rondeau and Bourdon, 2011).

Products of AGE are believed to be risk markers or risk factors of disease (Kasper and Funk, 2001; Vlassopoulos et al., 2013). Soluble AGE compounds in the circulation and extracellular fluids and another glycation product known as glycated hemoglobin (HbA1c) have been demonstrated to be the biomarkers for the diagnosis of diabetes and monitoring of glucose control in diabetic individuals as well as macrovascular diseases (Indurthi et al., 2012; Vlassopoulos et al., 2013).

Cellular receptors for advanced glycation end products

The AGEs that are in the circulation or extracellular space can interact with two types of cell surface receptors. Clearance and scavenger receptors are predominantly involved in AGE capture, removal and degradation. This group of receptors includes p60 (OST-48), p90 (80 K-H), galectin-3, type I and type II macrophage scavenger receptors, CD-36, FEEL-1 and -2, scavenger receptor proteins SR-BI and SR-BII and the lectin-like oxidized low-density lipoprotein receptor 1 (Lox-1). The other types of AGE-receptors initiate specific cellular signalling events in response to AGE exposure. Among these various AGE-receptors, a receptor for advanced glycation end-products (RAGE) is the receptor for AGEs which accumulate in various diseases and inflammations and is suggested to be involved in cellular migration and spreading over binding of amphoterin, a developmental-regulated heparin-binding protein (Indurthi et al., 2012; Kasper and Funk, 2001).

RAGE is a 35 kDa member of the immunoglobulin superfamily of cell surface proteins that interacts with a wide range of ligands, including AGE products, modified low density lipoproteins, amyloid fibrils, amphoterin (HMGB1) and various S100 proteins. It shares structural homology with other immunoglobulin like receptors (Leclerc et al., 2009) and is a specific binding protein for AGE moieties which have been detected on monocytes/macrophages, smooth muscle cells, hepatocytes, neurons, pericytes, T-lymphocytes, mesangial cells, type I pneumocytes, osteoblast-like cells and endothelial cells and its expression can be abnormally up or down-regulated in human disease (Kasper and Funk, 2001; Xie et al., 2013b).

Activation of the RAGE pathway has been shown to be important in wound healing, tumour growth and

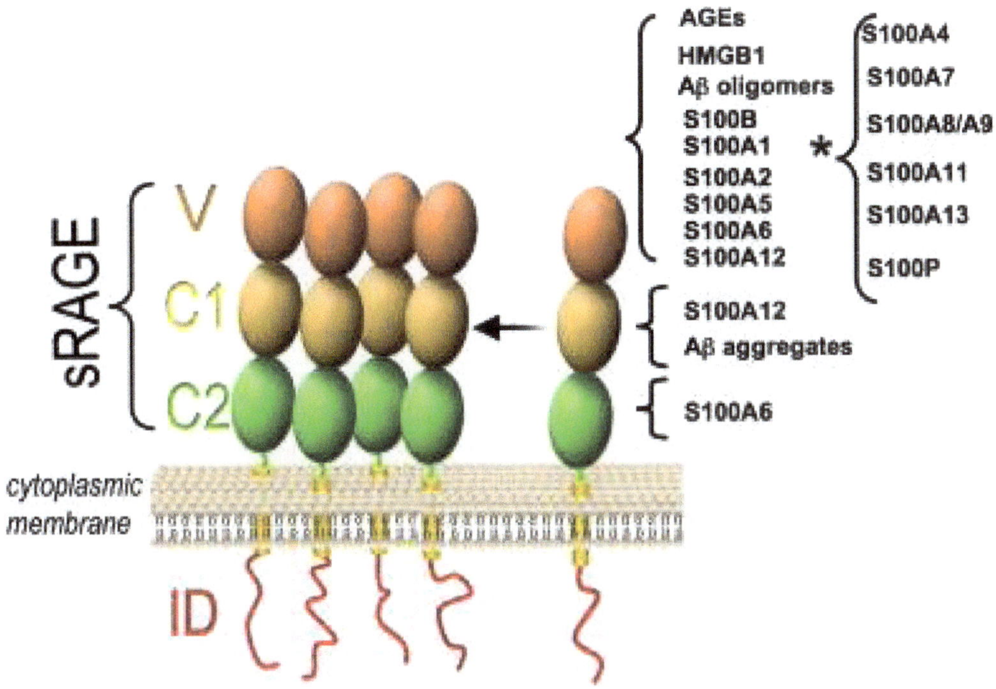

Figure 4. A schematic representation of RAGE showing the extracellular part comprising three immunoglobulin like domain, the V, C1 and C2 domain. A single transmembrane helix connects the extracellular domain with the short intracellular domain (ID). RAGE ligands interact with the extracellular domains as pointed out. (*) Indicates that the RAGE domain where S100 binds is not known (Leclerc et al., 2009).

metastasis, as well as in systemic amyloidosis (Germanová et al., 2010; Mahajan et al., 2010).

The mature 382 amino-acid long RAGE is composed of an extracellular part (314 aa), a single transmembrane spanning helix (27 aa) and a short cytosolic domain (41 aa) (Figure 4) (Leclerc et al., 2009). Complete length RAGE contains a V-type domain, which functions for ligand binding, two C-type domains, a transmembrane spinning helix, and a C-terminal cytosolic domain (ctRAGE) which is required for signal transduction. However RAGEs with the deletion of V-type domain (variant 2) are unable to interact with its ligands and therefore irresponsive to any stimuli, whereas RAGEs without ctRAGE (variant 1) have negative actions by competing with endogenous full-length RAGE for ligand binding, thereby blunting the RAGE-dependent signal transduction (Figure 5) (Xie et al., 2013).

RAGE has been documented in promoting inflammation and endothelial activation, which in turn also accelerates coronary atherosclerotic development. Its activation is enhanced by accumulation of its ligands in inflammatory compartments, initiating a vicious cycle causing further up-regulation of the receptor and sustained cell activation. These molecules has also been shown to be highly expressed in human atherosclerotic lesions where it co-localizes with pro-inflammatory and pro-oxidative mediators (Mahajan et al., 2010).

Advanced glycation end products in cellular fluids

Normal tissue remodelling is thought to be contributed by AGE modified proteins. However, under pathological conditions such as diabetes, rheumatoid arthritis (RA) and osteoarthritis, renal insufficiency and in smokers, AGEs can be found elevated in body fluids such as serum, synovial fluid and urine. Among them, serum and synovial fluid pentosidine is the superior indicator for so many diseases status (Kasper and Funk, 2001). Soluble receptor for advanced glycation end products (sRAGE) which is the alternative splicing of RAGE mRNA and the proteolytic cleavage of membrane-bound RAGE has the ability to bind with the same ligands, sRAGE acts as a competitive inhibitor of RAGE, and precludes the cell-bound RAGE signaling. RAGE adversely involves in several pathological processes of human diseases, such as diabetes, Alzheimer's disease and chronic inflammatory diseases such as RA (Chayanupatkul and Honsawek, 2010).

It has been reported recently that another extracellular newly identified RAGE-binding protein, known as EN-RAGE, has been found. It is a member of the S100 protein family, also called S100A12or CAAF1 or calgranulin C, which is comprised of closely related low molecular weight (10-14 kDa) acidic calcium-binding proteins with EF hands. These proteins were purified

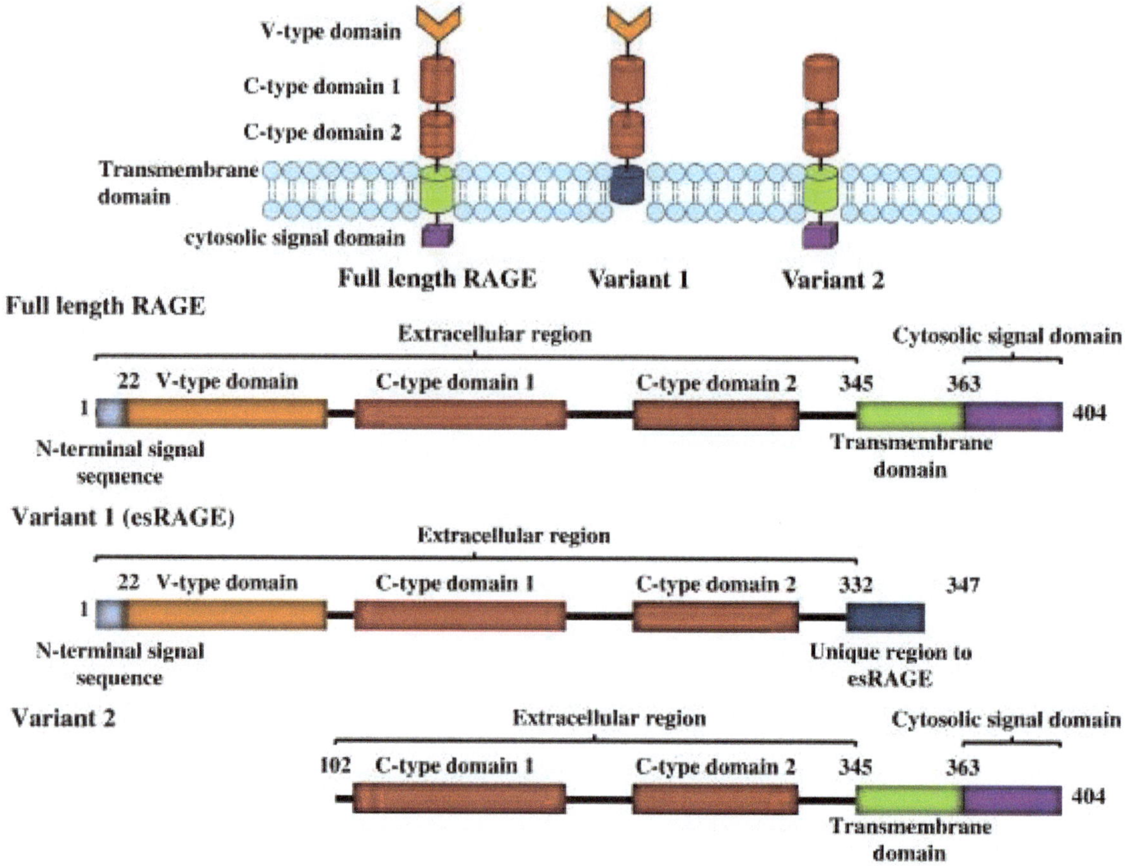

Figure 5. Structures of complete length RAGE and its variants which contains a first 22 amino acid signal recognition sequence, followed by a V-type domain, two C-type domains, a transmembrane domain and a cytosolic signal domain. RAGE variant 1 lacks the cytosolic domain whereas variant 2 lack the V-domain (Xie et al., 2013).

from bovine lung extract and expressed abundantly in the esophageal epithelium, neutrophils and monocytes/macrophages in humans. In the other hand to sRAGE, EN-RAGE is a natural pro-inflammatory ligand for RAGE.

EN-RAGE binding to RAGE activates intracellular signal cascades including MAP-kinase and NKκB which subsequently induces secretion of cytokines such as IL-1β and TNF-α, expression of adhesion molecules there by activating inflammatory responses and enhancing the activation and migration of monocytes/macrophages to damaged vessels leading to the pathogenesis of inflammatory diseases (Hasegawa et al., 2003; Kalousová et al., 2012; Kim et al., 2012).

Additionally, S100A12 is able to interact with scavenger receptors; it is also chemotactic for neutrophils and monocytes and by binding copper. It is able to generate ROS and so acts as a pro-oxidant agent. Increased level of EN-RAGE are found in various inflammatory compartments like synovial fluid and its serum concentrations are increased in inflammatory diseases and correlated with the disease activity. High concentrations of S100A12 are described in patients with neurodegenerative diseases and atherosclerosis. EN-RAGE interacting with RAGE could also be implicated in cancerogenesis particularly at sites with chronic and persistent inflammation (Kalousová et al., 2012).

Immunomodulatory effects of advanced glycation end-products

AGE triggers inflammatory response by binding to RAGE there by mediating inflammation known as a key underlying cause in the development of vascular complications leading to an increased expression of cytokines, growth factors and adhesion molecules with mediate an immune response (Choi et al., 2013; Hirose et al., 2011). Intracellularly, AGEs that are formed inside and outside of cells induces a more unspecific chemical reaction, the production of ROS increases which causes cross-linking with other cellular proteins and generation of lipid peroxidation product (Kasper and Funk, 2001). RAGE, found in most tissues, have potent immunomodulatory actions, promoting ROS production and inflammation (Vlassopoulos et al., 2013).

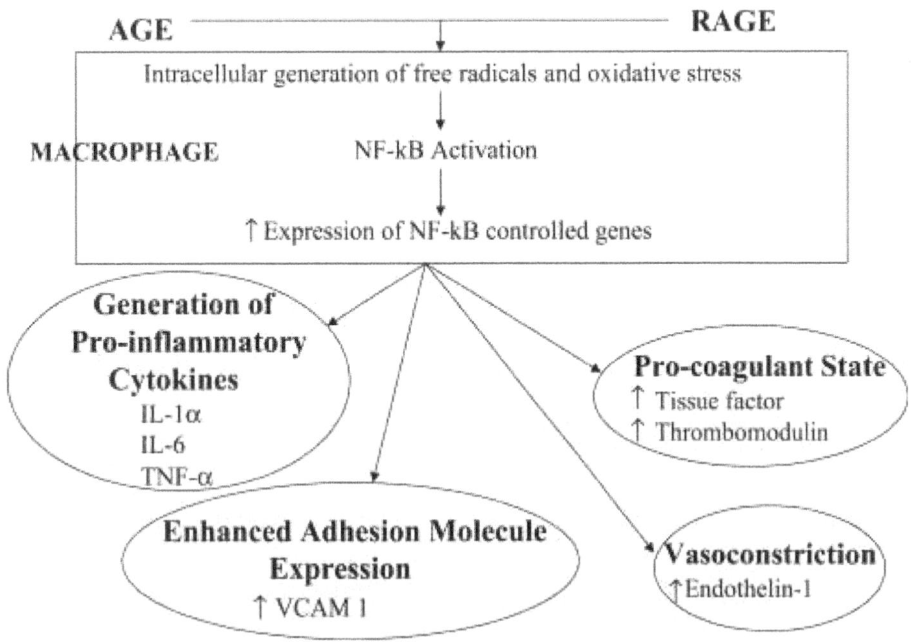

Figure 6. AGE-RAGE interaction and NF-kB activation leading to oxidant stress, vasoconstriction and a procoagulant state (Lapolla et al., 2005).

AGEs are known to be high in patients with poorly controlled disease, and its accumulation is a fundamental mechanism that causes the complications in diseases like diabetes. T lymphocytes contain a surface receptor for AGE-modified proteins that is inducible by the stimulation of the T-lymphocyte mitogen phytohemagglutinin. AGEs have also been found to act on macrophages through the AGE receptor on the plasma membrane, though it is not known whether the proliferation of activated T lymphocytes would be affected by AGEs (Liu et al., 2005).

Several targets of advanced glycation include structural proteins, such as collagen and aggrecan, plasma proteins, including immunoglobulins and albumin, and intracellular proteins, such as haemoglobin and lens crystallin. Protein glycation in particular has the potential to alter many cellular functions, and as a consequence AGE has been suggested to contribute towards the pathogenesis of many diseases. Moreover, AGE-modified proteins have been shown to generate free radicals such as superoxide and NO, and to up-regulate pro-inflammatory cytokines by monocyte and/or macrophages. Other responses elicited by AGE include induction of IL-1 and expression of tissue factor by monocytes and release of TNFα by THP-1 monocytic cells incubated with AGE-human serum albumin (HSA) (Pertyńska-Marczewska et al., 2004). Other consequences observed in AGE-RAGE interaction are an increase in the procoagulant response to TNF-α, a reduction in thrombomodulin expression, and an increase in endothelin-1 levels (Figure 6) (Lapolla et al., 2005).

CONCLUSION

Carbohydrates are favourable candidates for disease control, because they are present on the cell surface and act as an identification tag, through which they can interact with their surroundings, whether the extracellular matrix or homotypic or heterotypic cells. Interfering with the normal cell recognition phenomenon using a large or a small sugar molecule has been reported to block the progression of tumours by interfering with angiogenesis, cell-cell, cell-matrix interactions, tumour invasion and metastasis. Modification of a natural plant carbohydrate to facilitate its absorption by the body is a more cost-effective and easier way to prevent disease progression than synthesizing a competitive sugar molecule. The generation of AGEs is an inevitable process *in vivo* and their accumulation in tissues has been implicated in the process of aging and also in the pathogenesis of several pathological conditions. A number of *in vitro* and *in vivo* studies have demonstrated that carbohydrates and AGEs modulate various indices of immune function. Further analysis of monosaccharide composition of polysaccharides could be an essential parameter for the evaluation of quality of standard of polysaccharide materials and basic information on it before being applied for any investigation. Methods of isolating a particular polysaccharide is also paramount: their extraction by different methods might yield different chemical components, as such reliable and effective methods should be applied. Although, so far investigations of the receptor-mediated pathways for the immunomodulatory activity of certain polysaccharides have been reported, a more detailed mechanistic study is required to explain how the innate immune cells such as macrophages are activated by polysaccharides. Future studies could be directed not only to its ability to control cancer which is widely and commonly known it for, but also in other diseases. Further studies should be applied on the most effective compounds that could suppress AGE-mediated inflammatory signalling cascades, which involved the expression of pro-inflammatory intermediates and NF-κB transcriptional activity. Suppression of intracellular ROS production by scavenging ROS and reducing NADPH oxidase activity was a key regulatory mechanism for inhibiting AGE-mediated inflammatory response. Given the key role of AGE-RAGE signalling in the progression of diabetic complications, the suppressive effects of those compounds on AGE-mediated inflammation could be applied to prevent the development of diabetic complications.

Conflict of Interests

The author(s) have not declared any conflict of interests.

REFERENCES

Baba E, Erskine R, Boyson JE, Cohen GB, Davis DM, Malik P, Strominger JL (2000). N-linked carbohydrate on human leukocyte antigen-C and recognition by natural killer cell inhibitory receptors. Hum. Immunol. 61(12):1202-1218.

Belska NV, Guriev AM, Danilets MG, Trophimova ES, Uchasova EG, Ligatcheva AA, Belsky YP (2010). Water-soluble polysaccharide obtained from Acorus calamus L. classically activates macrophages and stimulates Th1 response. Int. Immunopharmacol. 10(8):933-942.

Boggs JM, Gao W, Hirahara Y (2008). Myelin glycosphingolipids, galactosylceramide and sulfatide, participate in carbohydrate-carbohydrate interactions between apposed membranes and may form glycosynapses between oligodendrocyte and/or myelin membranes. Biochim. Biophys. Acta 1780(3):445-455.

Bowman KG, Bertozzi CR (1999). Carbohydrate sulfotransferases: mediators of extracellular communication. Chem. Biol. 6(1):R9-R22.

Chayanupatkul M, Honsawek S (2010). Soluble receptor for advanced glycation end products (sRAGE) in plasma and synovial fluid is inversely associated with disease severity of knee osteoarthritis. Clin. Biochem. 43(13-14):1133-1137.

Chen J-R, Yang Z-Q, Hu T-J, Yan Z-T, Niu T-X, Wang L, Wang M (2010). Immunomodulatory activity *in vitro* and *in vivo* of polysaccharide from *Potentilla anserina*. Fitoterapia 81(8):1117-1124.

Chen Z, Kwong Huat Tan B, Chan SH (2008). Activation of T lymphocytes by polysaccharide-protein complex from Lycium barbarum L. Int. Immunopharmacol. 8(12):1663-1671.

Cheng A, Wan F, Wang J, Jin Z, Xu X (2008). Macrophage immunomodulatory activity of polysaccharides isolated from Glycyrrhiza uralensis fish. Int. Immunopharmacol. 8(1):43-50.

Choi H-J, Jang H-J, Chung T-W, Jeong S-I, Cha J, Choi J-Y, Ha K-T (2013). Catalpol suppresses advanced glycation end-products-induced inflammatory responses through inhibition of reactive oxygen species in human monocytic THP-1 cells. Fitoterapia 86:19-28.

Fan L, Ding S, Ai L, Deng K (2012). Antitumor and immunomodulatory activity of water-soluble polysaccharide from Inonotus obliquus. Carbohydr. Polym. 90(2):870-874.

Germanová A, Koucký M, Hájek Z, Pařízek A, Zima T, Kalousová M (2010). Soluble receptor for advanced glycation end products in physiological and pathological pregnancy. Clin. Biochem. 43(4-5): 442-446.

Ghazarian H, Idoni B, Oppenheimer SB (2011). A glycobiology review: Carbohydrates, lectins and implications in cancer therapeutics. Acta Histochem. 113(3):236-247.

Han E, Ding L, Jin S, Ju H (2011). Electrochemiluminescent biosensing of carbohydrate-functionalized CdS nanocomposites for in situ label-free analysis of cell surface carbohydrate. Biosensors Bioelectron. 26(5):2500-2505.

Hasegawa T, Kosaki A, Kimura T, Matsubara H, Mori Y, Okigaki M, Iwasaka T (2003). The regulation of EN-RAGE (*S100A12*) gene expression in human THP-1 macrophages. Atherosclerosis 171(2):211-218.

Hirasawa Y, Sakai T, Ito M, Yoshimura H, Feng Y, Nagamatsu T (2011). Advanced-glycation-end-product-cholesterol-aggregated-protein accelerates the proliferation of mesangial cells mediated by transforming-growth-factor-beta 1 receptors and the ERK-MAPK pathway. Eur. J. Pharmacol. 672(1-3):159-168.

Hirose J, Yamabe S, Takada K, Okamoto N, Nagai R, Mizuta H (2011). Immunohistochemical distribution of advanced glycation end products (AGEs) in human osteoarthritic cartilage. Acta Histochem. 113(6):613-618.

Hsieh YSY, Chien C, Liao SKS, Liao S-F, Hung W-T, Yang W-B, Wong C-H (2008). Structure and bioactivity of the polysaccharides in medicinal plant Dendrobium huoshanense. Bioorg. Med. Chem. 16(11):6054-6068.

Hu M-X, Xu Z-K (2011). Carbohydrate decoration of microporous polypropylene membranes for lectin affinity adsorption: comparison of mono- and disaccharides. Colloids Surf. B Biointerfaces 85(1):19-25.

Huang Q, Jin Y, Zhang L, Cheung PCK, Kennedy JF (2007). Structure, molecular size and antitumor activities of polysaccharides from Poria cocos mycelia produced in fermenter. Carbohydr. Polym. 70(3):324-333.

Hyun S, Kim J, Kwon M, Yu J (2007). Selection and syntheses of tentacle type peptides as 'artificial' lectins against various cell-surface carbohydrates. Bioorg. Med. Chem. 15(1):511-517.

Indurthi VSK, Leclerc E, Vetter SW (2012). Interaction between glycated serum albumin and AGE-receptors depends on structural changes and the glycation reagent. Arch. Biochem. Biophys. 528(2):185-196.

Jeurink PV, Noguera CL, Savelkoul HFJ, Wichers HJ (2008). Immunomodulatory capacity of fungal proteins on the cytokine production of human peripheral blood mononuclear cells. Int. Immunopharmacol. 8(8):1124-1133.

Kalousová M, Kuběna AA, Benáková H, Dusilová-Sulková S, Tesař V, Zima T (2012). EN-RAGE (extracellular newly identified receptor for advanced glycation end-products binding protein) and mortality of long-term hemodialysis patients: Aprospective observational cohort study. Clin. Biochem. 45(7-8):556-560.

Kasper M, Funk RHW (2001). Age-related changes in cells and tissues due to advanced glycation end products (AGEs). Arch. Gerontol. Geriatr. 32(3):233-243.

Khodse VB, Fernandes L, Gopalkrishna VV, Bhosle NB, Fernandes V, Matondkar SGP, Bhushan R (2007). Distribution and seasonal variation of concentrations of particulate carbohydrates and uronic acids in the northern Indian Ocean. Mar. Chem. 103(3-4):327-346.

Kim J-K, Cho ML, Karnjanapratum S, Shin I-S, You SG (2011). In vitro and in vivo immunomodulatory activity of sulfated polysaccharides from Enteromorpha prolifera. Int. J. Biol. Macromol. 49(5):1051-1058.

Kim J-K, Park S, Lee MJ, Song YR, Han SH, Kim SG, Yoo T-H (2012). Plasma levels of soluble receptor for advanced glycation end products (sRAGE) and proinflammatory ligand for RAGE (EN-RAGE) are associated with carotid atherosclerosis in patients with peritoneal dialysis. Atherosclerosis 220(1), 208-214.

Kouakou K, Schepetkin IA, Yapi A, Kirpotina LN, Jutila MA, Quinn MT (2013). Immunomodulatory activity of polysaccharides isolated from Alchornea cordifolia. J. Ethnopharmacol. 146(1):232-242.

Krautwald M, Münch G (2010). Advanced glycation end products as biomarkers and gerontotoxins - A basis to explore methylglyoxal-lowering agents for Alzheimer's disease? Exp. Gerontol. 45(10):744-751.

Lapolla A, Traldi P, Fedele, D. (2005). Importance of measuring products of non-enzymatic glycation of proteins. Clin. Biochem. 38(2):103-115.

Leclerc E, Fritz G, Vetter SW, Heizmann CW (2009). Binding of S100 proteins to RAGE: An update. Biochim. Biophys. Acta (BBA) Mol. Cell Res. 1793(6):993-1007.

Lepenies B, Lee J, Sonkaria S (in press, accessed 2013). Targeting C-type lectin receptors with multivalent carbohydrate ligands. Adv. Drug Deliv. Rev. 65(9):1271-1281.

Liu C-T, Chen K-M, Lee S-H, Tsai L-J (2005). Effect of supplemental l-arginine on the function of T lymphocytes and the formation of advanced glycosylated end products in rats with streptozotocin-induced diabetes. Nutrition 21(5):615-623.

Mahajan N, Bahl A, Dhawan V (2010). C-reactive protein (CRP) up-regulates expression of receptor for advanced glycation end products (RAGE) and its inflammatory ligand EN-RAGE in THP-1 cells: Inhibitory effects of atorvastatin. Int. J. Cardiol. 142(3):273-278.

Makino S, Ikegami S, Kano H, Sashihara T, Sugano H, Horiuchi H, Oda M (2006). Immunomodulatory Effects of Polysaccharides Produced by Lactobacillus delbrueckii ssp. bulgaricus OLL1073R-1. J. Dairy Sci. 89(8):2873-2881.

Middleton D, Curran M, Maxwell L (2002). Natural killer cells and their receptors. Transplant Immunol. 10(2-3):147-164.

Nakamura M, Tsutsumi M, Ishikawa Y, Umemiya H, Hasegawa T, Izawa K, Hasegawa T (2013). Glycosylated tris-bipyridine ferrous complexes to provide dynamic combinatorial libraries for probing carbohydrate-carbohydrate interactions. Tetrahedron 69(14):3019-3026.

Nangia-Makker P, Conklin J, Hogan V, Raz A (2002). Carbohydrate-binding proteins in cancer, and their ligands as therapeutic agents. Trends Mol. Med. 8(4):187-192.

Pertyńska-Marczewska M, Kiriakidis S, Wait R, Beech J, Feldmann M, Paleolog EM (2004). Advanced glycation end products upregulate angiogenic and pro-inflammatory cytokine production in human monocyte/macrophages. Cytokine 28(1):35-47.

Pesesse X, Backers K, Moreau C, Zhang J, Blero D, Paternotte N, Erneux C (2006). SHIP1/2 interaction with tyrosine phosphorylated peptides mimicking an immunoreceptor signalling motif. Adv. Enzyme Regul. 46(1):142-153.

Ponnusamy S, Zhang H, Kadam P, Lin Q, Lim TK, Sandhu JS, Choolani M (2012). Membrane proteins of human fetal primitive nucleated red blood cells. J. Proteom. 75(18):5762-5773.

Rondeau P, Bourdon E (2011). The glycation of albumin: Structural and functional impacts. Biochimie 93(4):645-658.

Schepetkin IA, Quinn MT (2006). Botanical polysaccharides: Macrophage immunomodulation and therapeutic potential. Int. Immunopharmacol. 6(3):317-333.

Shao B-M, Xu W, Dai H, Tu P, Li Z, Gao X-M (2004). A study on the immune receptors for polysaccharides from the roots of Astragalus membranaceus, a Chinese medicinal herb. Biochem. Biophys. Res. Commun. 320(4):1103-1111.

Shuvaev VV, Laffont I, Serot J-M, Fujii J, Taniguchi N, Siest G (2001). Increased protein glycation in cerebrospinal fluid of Alzheimer's disease. Neurobiol. Aging 22(3):397-402.

Sørensen ALT, Clausen H, Wandall HH (2012). Carbohydrate clearance receptors in transfusion medicine. Biochim. Biophys. Acta 1820(11):1797-1808.

Stanilova SA, Dobreva ZG, Slavov ES, Miteva LD (2005). C3 binding glycoprotein from Cuscuta europea induced different cytokine profiles from human PBMC compared to other plant and bacterial immunomodulators. Int. Immunopharmacol. 5(4):723-734.

Tang X-H, Yan L-F, Gao J, Yang X-L, Xu Y-X, Ge H-Y, Yang H-D (2012). Antitumor and immunomodulatory activity of polysaccharides from the root of Limonium sinense Kuntze. Int. J. Biol. Macromol. 51(5):1134-1139.

Tzimagiorgis G, Michailidou EZ, Kritis A, Markopoulos AK, Kouidou S (2011). Recovering circulating extracellular or cell-free RNA from bodily fluids. Cancer Epidemiol. 35(6):580-589.

Vlassopoulos A, Lean MEJ, Combet E (2013). Role of oxidative stress in physiological albumin glycation: A neglected interaction. Free Radic. Biol. Med. 60:318-324.

Wang H, Deng X, Zhou T, Wang C, Hou Y, Jiang H, Liu G (2013). The in vitro immunomodulatory activity of a polysaccharide isolated from Kadsura marmorata. Carbohydr. Polym. 97(2):710-715.

Wang X-M, Sun R-G, Zhang J, Chen Y-Y, Liu N-N (2012). Structure and antioxidant activity of polysaccharide POJ-U1a extracted by ultrasound from Ophiopogon japonicus. Fitoterapia 83(8):1576-1584.

Weng BB-C, Lin Y-C, Hu C-W, Kao M-Y, Wang S-H, Lo D-Y, Chiou R Y-Y (2011). Toxicological and immunomodulatory assessments of botryosphaeran (β-glucan) produced by Botryosphaeria rhodina RCYU 30101. Food Chem. Toxicol. 49(4):910-916.

Wong K-H, Lai CKM, Cheung PCK (2011). Immunomodulatory activities of mushroom sclerotial polysaccharides. Food Hydrocoll. 25(2):150-158.

Wu JH, Singh T, Herp A, Wu AM (2006). Carbohydrate recognition factors of the lectin domains present in the Ricinus communis toxic protein (ricin). Biochimie 88(2):201-217.

Wu L, Ma L, Nicholson LFB, Black PN (2011). Advanced glycation end products and its receptor (RAGE) are increased in patients with COPD. Respir. Med. 105(3):329-336.

Xia L, Liu X, Guo H, Zhang H, Zhu J, Ren F(2012). Partial characterization and immunomodulatory activity of polysaccharides from the stem of Dendrobium officinale (Tiepishihu) *in vitro*. J. Funct. Foods 4(1):294-301.

Xie J, Méndez JD, Méndez-Valenzuela V, Aguilar-Hernández MM (2013a). Cellular signalling of the receptor for advanced glycation end products (RAGE). Cell. Signal. 25(11):2185-2197.

Xie J-H, Shen M-Y, Nie S-P, Liu X, Zhang H, Xie M-Y (2013b). Analysis of monosaccharide composition of *Cyclocarya paliurus* polysaccharide with anion exchange chromatography. Carbohydr. Polym. 98(1):976-981.

Xu H, Yao L, Sun H, Wu Y (2009). Chemical composition and antitumor activity of different polysaccharides from the roots of Actinidia eriantha. Carbohydr. Polym. 78(2):316-322.

Zha X-Q, Luo J-P, Luo S-Z, Jiang S-T (2007). Structure identification of a new immunostimulating polysaccharide from the stems of Dendrobium huoshanense. Carbohydr. Polym. 69(1):86-93.

Zhang L, Koyyalamudi SR, Jeong SC, Reddy N, Bailey T, Longvah T (2013). Immunomodulatory activities of polysaccharides isolated from Taxillus chinensis and Uncaria rhyncophylla. Carbohydr. Polym. 98(2):1458-1465.

13

RNAi as a novel therapeutic platform technology for oncological solutions

Gowhar Shafi[1], Kaiser Jamil[1*], Atya Kapley[2], Hemant J. Purohit[2] and Mohana Ch Vamsy[1]

[1]Department of Genomics, Stem cells and Translational Medicine Indo-American Cancer Institute and Research Centre Road # 14, Banjara Hills, Hyderabad-500034, A.P. India.
[2]National Environmental Engineering Research Institute, Nehru Marg, Nagpur

RNA interference (RNAi) is a process in which double-stranded RNA triggers the degradation of a homologous messenger RNA (sharing sequence-specific homology). RNAi has been observed in all eukaryotes, from yeast to mammals. The power and utility of RNAi for specifically silencing the expression of any gene for which sequence is available has driven its incredibly rapid adoption as a tool for reverse genetics in eukaryotic systems. RNAi is remarkably potent i.e., only a few dsRNA molecules per cell (like catalytic reactions) are required to produce effective interference. The long dsRNAs enter a cellular pathway that is commonly referred to as the RNA interference (RNAi) pathway. RNAi is now recognized to be one of a larger set of sequence-specific cellular responses to RNA, collectively called RNA silencing. These responses have been shown to play a role not only in mRNA and dsRNA stability/degradation, but also in regulation of translation, transcription, chromatin structure, and genome integrity. RNAi has also been invaluable for unveiling critical pathway involved in cancer development, growth and metastasis and has been useful in identifying critical tumor type specific gene targets for cancer therapy. RNAi has been shown to inhibit gene expression and is considered as one of the most important recent discoveries in molecular oncology. It is already a widely used research tool in the analysis of molecular mechanisms for many diseases including cancer and this is because it allows researchers to silence the expression of specific gene targets, much like antisense technology, but with higher specificity and efficacy. Although RNAi is an evolutionarily conserved phenomenon for sequence specific gene silencing in mammalian cells, exogenous short interfering RNA (siRNA) and vector based short hairpin RNA (shRNA) can also invoke RNAi responses. Both are now not only experimental tools for analyzing gene functions but are also expected to be excellent avenues for drug target discovery and the emerging class of gene medicine for targeting incurable disease such as cancer. The success of therapeutic use of RNAi for cancer relies on the development of safe and efficacious delivery systems that introduce siRNA and shRNA expression vectors into target tumor cells. The present review provides a brief overview of RNAi based approaches for gene silencing as a potential therapy for cancer.

Key words: RNA interference (RNAi), Cancer, Posttranscriptional gene silencing, siRNA, miRNA.

TABLE OF CONTENTS

1. Abstract and Introduction
2. Evolution in understanding of RNA function
3. Mechanism of RNA Interference
4. RNAi and Single Nucleotide Polymorphism
5. RNAi interventions in various targets
6. RNAi and Single Nucleotide Polymorphism
7. Steps in RNAi therapeutic development
8. Selection of target gene(s)
9. siRNA Design
10. Method of siRNA Delivery system/ vehicles/ vectors
11. Delivery to live animals

12. MiRNA- a new class of RNA family
13. MiRNAs and Oncology
14. siRNA and miRNA
15. Interfacing micro array data with RNAi technology
16. Conclusion
17. Acknowledgement
18. References
19. Figures
20. Tables

INTRODUCTION

In recent years researchers have demonstrated the underlying mechanisms of how dsRNA results in the loss of the targeted homologous mRNA and the primary interference effects were post-transcriptional. In 1998, Craig Mello and Fire reported that only dsRNA targeting exon sequences were effective (promoter and intron sequences could not produce an RNAi effect). The small RNAs that provide target specificity to the silencing machinery, short interfering RNAs (siRNAs), repeat-associated siRNAs (rasiRNAs), and microRNAs (miRNAs)—can be distinguished by their origin. siRNAs are processed from dsRNA precursors made up of two distinct strands of perfectly base-paired RNA, while miRNAs originate from a single, long transcript that forms imperfectly base-paired hairpin structures. In plants and animals, RNA silencing has been adapted to play a critical role in regulation of cell growth and differentiation using a class of small RNAs called microRNAs (miRNAs). RNA interference is the process of mRNA degradation that is induced by double-stranded RNA in a sequence-specific manner.

In the last decade, the structural and functional analysis of a gene has acquired new dimensions. In parallel, it has also been assumed that the primary DNA sequence data is within the reach of available analytical tools (Venter et al., 2001). This collective consensus has revolutionized and remapped the road for defining a disease condition. Of which, biology of cancer has received tremendous attention. We are now much closer to genes that mediate cancer progression and many therapeutic gene targets that regulate apoptosis, proliferation and cell signaling. Long double-stranded RNAs (dsRNAs; typically > 200 nucleotide) can be used to silence the expression of target genes in a variety of organisms and cell types.

The initial evidence of RNAi in nature came from the work on putunia flowers in which over expression of the gene responsible for purple pigmentation actually caused the flowers to lose their endogenous colour (Napoli et al., 1990). RNAi was first described in animal cells by Fire and colleagues in the nematode *Caenorhabditis elegans* as a naturally occurring cellular mechanism that induces post transcriptional gene silencing, in which double stranded RNA (dsRNA) suppresses the expression of a target gene by triggering specific degradation of complementary messenger RNA (mRNA) sequence (Fire et al; 1998). The natural role of RNAi is thought to be that of cellular defense against viral infection or potentially harmful destabilizing genomic intruders such as transposons. The experimental designs were developed to influence cellular physiology by RNAi and following are few examples where different mode of RNAi introduction has been demonstrated. RNAi can be induced in mammalian cells by the introduction of synthetic siRNA 21 – 23 bp in length (Elbashir et al., 2001) or by plasmid (Brummelkamo et al., 2002; Lee et al., 2002; Paul et al., 2002) and viral vector systems (Brummelkamo et al., 2002) that express shRNA that are subsequently processed to siRNA by cellular machinery. Recently a nano particle based RNAi delivery has also been proposed and research on this is gaining momentum. Analysis of gene function in cultured somatic mammalian cells using siRNAs is now being described in a rapidly growing number of independent studies (Table 1).

Many diseases are rooted in the inappropriate activity or dysfunction of specific genes. RNAi has been heralded as a great therapeutic intervention for gene medicine against a wide range of human diseases. The pace of siRNA based drug development has been rapid, and some companies have already started clinical trials for RNAi therapeutics for age related macular degeneration (AMD) (Takeshita and Ochiya, 2006). AMD is caused by the abnormal growth of blood vessels behind the retina. Treatment strategies include inhibition of vascular endothelial growth factor pathways by siRNA. Such RNAi therapeutics was designed to be administered directly to the sites in the ocular diseases. However, like other forms of gene-based medicines, the clinical utility of systemic therapeutic siRNA will depend on the development of safe and efficacious delivery systems. Strategies for the inhibition of cellular proliferation by systemic treatment of tumors (in animals) with siRNA are mushroomming and are becoming suitable for evaluation of systemic delivery of siRNA as a means for personalized cancer treatment.

Evolution in understanding of RNA function

We have come long way in understanding the Ribonu-

*Corresponding author. E-mail: Kaiser.jamil@gmail.com.

Table 1. Human cell lines in which siRNA triggers silencing

Cell line	Tissue origin	Reference
A-431	human epidermoid carcinoma	Elbashir *et al.* (2002)
A549	human lung carcinoma	Bitko and Barik (2001)
BV173	human B-precursor leukemia	Tuschl and Borkhardt (2002)
C-33A	human papillomavirus-negative cervical carcinoma	Sui *et al.* (2002)
CA46	human Burkitt's lymphoma	Tuschl and Borkhardt (2002)
Caco2	human colon epithelial cells	Moskalenko *et al.* (2002)
CHO	Chinese hamster ovary	Elbashir *et al.* (2002)
H1299	human non small cell lung carcinoma	Sui *et al.* (2002)
HaCaT	human keratinocyte cell	Holen *et al.* (2002)
HEK 293	human embryonic kidney	Elbashir *et al.* (2001)
HeLa	human papillomavirus-positive cervical cervical carcinoma	Elbashir *et al.* (2001)
Hep3B	human hepatocellular carcinoma	Bakker *et al.* (2002)
HUVEC	human umbilical vein endothelial cells	Ancellin *et al.* (2001)
IMR-90	human diploid fibroblast	Paddison *et al.* (2002)
K562	human chronic myelogenous leukemiablast crisis	Tuschl and Borkhardt (2002)
Karpas 299	human T-cell lymphoma	Tuschl and Borkhardt (2002)
MCF-7	human breast cancer	Hirai and Wang (2002)
MDA-MB-468	human breast cancer	Hirai and Wang (2002)
MV-411	human acute monocytic leukemia	Tuschl and Borkhardt (2002)
SD1	human acute lymphoblastic leukemia	Tuschl and Borkhardt (2002)
SKBR3	human breast cancer	Elbashir *et al.* (2002)
U2OS	human osteogenic sarcoma cell	Martins *et al.* (2002)
EcR293	human embryonic kidney	Nham Tran *et al* (2003)
MDA MB 231	human breast cancer	Nham Tran *et al* (2003
SGC-7901	human gastric cancer	Wen-Hua Xu *et al* (2007)
MiaPaCa-2	human pancreatic cancer	Wei Wang *et al* (2005)
H1299	human non-small cell lung carcinoma cell	Sham S. Kakar (2006)

leic acid family (See table 2) and their role in cell regulation. Recent studies on the activities of RNA in the cell have revolutionized our understanding of the many roles played by this molecule. The role of RNA in protein synthesis had been suspected since 1939, based on experiments carried out by Torbjorn Caspersson, Jean Brachet and Jack Schultz. Hubert Chantrenne elucidated the messenger role played by RNA in the synthesis of proteins in ribosome. Robert W. Holley found the sequence of the 77 nucleotides of yeast RNA in 1964, winning the 1968 Nobel Prize for Medicine. In 1976, Walter Fiers and his team at the University of Ghent determined the complete nucleotide sequence of bacteriophage MS2-RNA (RNA world website-www.imb-jena.de/RNA.html). An "RNA world" is believed to have existed before DNA took over as the key genetic material and RNA was relegated to the role of a messenger. The dramatic new knowledge of gene regulation by RNAi

Table 2. RNA world listing the types of RNA species as described by the respective authors.

RNA types	Author	Year
siRNA	Tuschl et al.	1999
RNAi	Fire et al	1998
miRNA	Ambros et al	1993
RNAi	Fire et al	1998
tRNA	Holley et al	1964
mRNA	Jacob and Monod	1961
rRNA	Masayasu and Nomura	1960s

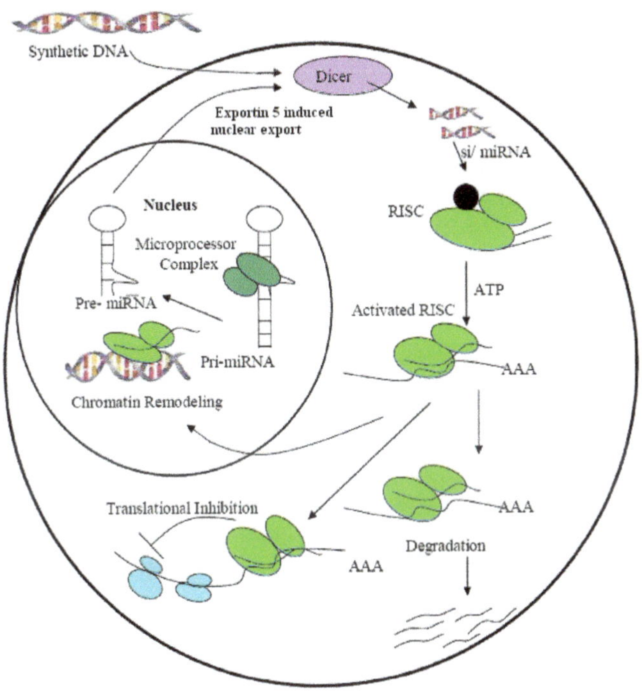

Figure 1. RNA interference: Processing and Activity: Depicts the formation of long primary micro RNA (pri-miRNA) in the nucleus which is processed by microprocessor complex [dark green] (Drosha- an RNase III enzyme and Pasha- double stranded RNA binding protein) into pre-microRNA(70 nucleotide stem loop structure) and transported to cytoplasm where Dicer [violet], an RNase II enzyme cleaves it to 20 – 21 nucleotide mature micro RNA that integrates into the RNA Inducing Silencing Complex (RISC)[light green]; thus, shares the same cellular machinery downstream of their initial processing with siRNA.

(siRNA and miRNA) has further widened our perspective of RNA's active role. The implications of the discovery of RNAi to the evolutionary process could, in fact, be revelatory. siRNAs were originally identified as intermediates in the RNAi pathway after induction by exogenous dsRNA; however, endogenous sources of siRNAs have now been recognized. Many of these endogenous siRNAs are derived from repetitive sequences within the genome, hence the term repeat-associated siRNAs, or rasiRNAs. MiRNAs were discovered through their critical roles in development and cellular regulation, and represent a large class of evolutionarily conserved RNAs. MiRNAs have always been recognized as being of endogenous origin. However, synthetic precursors and inhibitors of miRNAs are used to understand and exploit the various functions of this important class of small RNA.

Mechanism of RNA interference

RNAi is a conserved biological process among multi-protein synthesis and cleaving specifically targeted mRNA. The manner in which RNAi induced gene silencing functions is primarily dependent on the structure of the initiating RNA and thus RNA homology has been cellular organisms in, as diverse as plants, worms, yeast and humans, in which dsRNA suppresses the expression of target genes by triggering specific degradation of the complementary mRNA sequences (Fire et al., 1998). RNAi is a post transcriptional process that can effect gene silencing through chromatin remodeling, blocking exploited in several ways. Double stranded RNA that is homologous to the gene being silenced initiates the classical RNAi pathway.

It is now clear that higher eukaryotes contain a large number of genes that encode small RNAs referred to as micro RNAs (miRNAs) (Bartel et al., 2004). These miRNAs generally have only incomplete sequence homology to their targets, often recognizing sequences in the untranslated 3' end of a gene, and usually work by blocking the translation of mRNA into protein rather than destroying the mRNA transcript. The naturally occurring miRNAs are synthesized in the nucleus in large precursor forms (Pri-miRNA), which are processed within the nucleus by Drosha, and RNA III enzyme, into pre-miRNA (60 - 80 nucleotides long). Subsequently this is transported from nucleus to the cytoplasm by Exporting 5 and processed by Dicer, which generates duplexes of approximately 21 – 23 nucleotides with 3' overhangs of matured miRNA. These fragments activate a multisubunit ribonucleoprotein complex called RISC (RNA inducing silencing complex). RISC incorporates the antisense strand of the unwound miRNA and defines the target region of mRNA via a complementary sequence to promote its specific cleavage. These endogenous miRNA appear to regulate processes like proliferation, apoptosis and differentiation (Ambros et al., 2004). Dicer cleaves double stranded molecules, including those derived from vector based shRNA. The siRNA derived from vector based shRNA and synthetic siRNA are also incorporated into the RISC in a manner similar to naturally occurring miRNAs and are capable of inducing sequence specific and effective silencing of genes by mimicking the pathway. The silencing by siRNA is highly efficient, presumably because the guide strand RNA is protected from degradation by RISC and can repeat cleavage of many mRNA molecules (Figure 1).

RNAi interventions in various targets

Different experimental approaches have been used to compare tumor cells to normal cells. Single base pair mutations that alter the function of tumor suppressor genes and oncogenes occur frequently during oncogenesis. The guardian of the genome, p53, is inactivated by point mutations in more than 45 - 60% of human colorectal cancers. Synthetic siRNAs are highly sequence specific reagents and discriminate between single mismatched target RNA sequences, and may represent a new avenue for gene based medicines. Martines et al. (2002) demonstrated that a single base difference in siRNA discriminates between mutant and wild type p53 in cells expressing both forms, resulting in the restoration of wild type protein function (Martines et al., 2002). Therefore, it is suggested that siRNAs may be used to suppress expression of point mutated genes and provide the basis for selective and personalized cancer therapy.

The product of bcl - 2 genes are involved in regulation of apoptosis and proliferation and are associated with progression in several malignancies. It is reported that some researchers use mRNA – cDNA interference for silencing bcl - 2 expression in human prostrate cancer cell line, LNCap (Lin et al., 2001). c-Myc gene has a pivotal function in the development of breast cancer. It has been shown that decreasing the c – Myc protein level in MCF – 7 cells by RNAi could significantly inhibit tumor growth both *In vitro* and *in vivo* (Yi-hua Wang et al., 2005). Thereby, RNAi based treatment of breast cancer by targeting overexpression of oncogene such as c-Myc, might be a potential therapeutic target for human breast cancer treatment.

One of the major limitations of the current chemotherapy is the bone marrow toxicity associated with these drugs. However, it is well recognized that subpopulations of tumor cells are resistant to particular chemotherapeutic agents and continue to grow in a selective manner in the presence of such drugs. These cells contain specific genes which render them resistant to particular compounds. One such gene is the multiple drug resistance (MDR1) gene which confers resistance to vinca alkaloids (vinblastin, vincritine), anthracyclines (adriamycin, daunorubicin) etoposide and paclitaxel. For reversal of MDR1 gene dependent multidrug resistance, two siRNA constructs were designed to inhibit MDR1 expression by RNAi (Nieth et al., 2003). Researched data indicate that this approach will lead to a reversal of tumors with a P-glycoprotein dependent MDR phenotype back to drug sensitive ones. Using RNAi system, it has been shown that inducible knockdown of endogenous CXC chemokine receptor - 4 (CXCR4) gene expression in breast cancer cells resulted in significant inhibition of breast cancer cell migration in vitro (Yangchao chen et al., 2003).

VEGF is one of the archetypal angiogenic growth factor and has received considerable attention. Inhibition of VEGF activity or disabling the function of its receptors has been shown to inhibit both tumor growth and metastasis in a variety of animal tumor models (Haper et al., 2005; Jocelyn H et al -2002). Given the different isoforms and their various functions, the development of this RNAi technology and its ability to target specific VEGFs should facilitate both a greater understanding of this field and also the development of improved therapeutics.

Ligand targeted, stericall stabilized nanoparticles have been adapted for siRNA. Self assembling nanoparticles with siRNA have been constructed with polyethyleneeimine (PEI) that is PEGylated with an Arg-Gly-Asp peptide ligand attached at the distal end of the polyethylene glycol (GEG) as a means to target tumor neovasculature expressing integrins and are used to deliver siRNA Inhibiting VEGF-2 thereby resulting in tumor angiogenesis (Reymond et al., 2004).

Telomerase is another attractive molecular target toWard which to direct cancer therapeutic agents because telomerase activity is present in most malignant cells but undetectable in most normal somatic cells. Kosciolek et al., 2003, evaluated the ability of siRNA to inhibit telomerase activity in human cancer cells. In their research, human cancer cell lines were transfected with 21 nucleitode dsRNA homologous to either the catalytic subunit of telomerase (human telomerase reverse transcriptase) or its template RNA (human telomeric RNA). Both types of agents reduced telomerase activity in variety of human cancer cell lines representing both carcinomas and sarcomas and inhibition was dose dependent. Telomerase inhibition by siRNA is notable because telomerase is regarded as restricted to the nucleus, whereas, RNAi is commonly regarded as restricted to the cytoplasm.

As siRNA effects are extremely specific, initial in vitro studies have demonstrated effective silencing of a wide variety of mutated oncogenes such as K-Ras (Brummelkamp, 2002), mutated p53 (Martinez, 2002) Her2/neu (Chouldhury, 2004), and bcr-abl (Scherr, 2003), although RNAi approach has been successful on a relatively small scale to investigate cell death signaling by TRAIL (tumor necrosis factor related apoptosis inducing ligand), an agent that might have therapeutic potential against various cancers (Aza-Blance et al., 2003, Terwari M et al., 2003). In addition, the approach was used to identify familial cylindromatosis tumor suppressor gene (CYLD) as deubiquitinating enzyme in the nuclear factor-κB pathway (Brummelkamp, 2003). It has been recently used successfully to identify several novel components of the p53 tumor suppressor gene signaling pathways (Berns et al., 2004).

It has been identified that a single siRNA is capable of targeting both the major oncogenes of human papilloma virus. And it has been shown that the re-establishment of dormant tumor suppressor pathways via RNAi enhances the antitumor activity of cisplatin (Lisa NP et al 2005). The data suggest that a combination therapy involving cisplatin and siRNA could be advantageous in promoting

Homo sapiens Vimentin (VIM), mRNA, NM_003380

```
  1 GTCCCCGCGC CAGAGACGCA GCCGCGCTCC CACCACCCAC ACCCACCGCG CCCTCGTTCG
 61 CCTCTTCTCC GGGAGCCAGT CCGCGCCACC GCCGCCGCCC AGGCCATCGC CACCCTCCGC
121 AGCCATGTCC ACCAGGTCCG TGTCCTCGTC CTCCTACCGC AGGATGTTCG GCGGCCCGGG
181 CACCGCGAGC CGGCCGAGCT CCAGCCGGAG CTACGTGACT ACGTCCACCC GCACCTACAG
241 CCTGGGCAGC GCGCTGCGCC CCAGCACCAG CCGCAGCCTC TACGCCTCGT CCCCGGGCGG
301 CGTGTATGCC ACGCGCTCCT CTGCCGTGCG CCTGCGGAGC AGCGTGCCCG GGGTGCGGCT
361 CCTGCAGGAC TCGGTGGACT TCTCGCTGGC CGACGCCATC AACACCGAGT TCAAGAACAC
421 CCGCACCAAC GAGAAGGTGG AGCTGCAGGA GCTGAATGAC CGCTTCGCCA ACTACATCGA
481 CAAGGTGCGC TTCCTGGAGC AGCAGAATAA GATCCTGCTG GCCGAGCTCG AGCAGCTCAA
541 GGGCCAAGGC AAGTCGCGCC TGGGGGACCT CTACGAGGAG GAGATGCGGG AGCTGCGCCG
601 GCAGGTGGAC CAGCTAACCA ACGACAAAGC CCGCGTCGAG GTGGAGCGCG ACAACCTGGC
```

AA (N19) TT

```
     Vimentin cDNA: 5'AACTACATCGACAAGGTGCGC TT
                        | | || |||| | || | ||| |||| |
       Sense siRNA: 5'    CUACAUCGACAAGGUGCGC-dTdT
                        | ||| | ||| | || | ||| |||| |
Anti sense siRNA: 3'dTdT-GAUGUAGCUGUUCCACGCG 5'
```

Figure 2. Example of RNAi target selection.

better prognostic outcomes for patients and reducing the severe, toxic side effects associated with cisplatin therapy. The challenge now is to turn these findings into meaningful clinical applications.

RNAi was also used to target of BCR-ABL mRNA and this approach was compared to that of St1 571 (which is now used in the effective treatment of BCR-ABL positive leukemia) (Druker, 2002; Wilda et al., 2002) mediated cell killing in a cell culture model. The siRNA treatment reduced the expression of BCR-ABL mRNA, followed by a reduction of BCR-ABL oncoprotein leading to apoptosis in leukemic cells (Wilda et al., 2002).

RNAi and single nucleotide polymorphism

The use of single nucleotide–specific siRNA to reduce-expression of mutant, disease-causing alleles holds promise for the treatment of dominantly inherited human diseases caused by point mutations that lead to a gain of function. Moreover, human disease alleles often differ from their wild-type counterparts by SNPs that do not themselves cause the disease phenotype. By targeting the SNP isoform present in the disease allele, single nucleotide–specific siRNA might be used to reduce selectively expression of the disease-causing allele-without altering expression of the wild-type allele. Our earlier studies on SNPs of various genes like MTHFR, (Haranatha Reddy and Jamil, 2006) GST (M1 and T1) (Haranatha Reddy and Jamil, 2006), CYP3A4 (Suman and Jamil, 2006), DPD, (Kalyan and Jamil, 2007) relate to drug-gene interactions of cancer patients and susceptibility to the disease and their effect on drug-response. Patients with DPD spliced-site mutations were reported to be hypersensitive to certain drugs. These studies have helped us to establish a basic idea on mechanism of drug-gene interaction as well as to understand the how point mutations could affect patient's susceptibility to the disease. Hence, it is suggested that a study on cancer progression and gene silencing technique at posttranscriptional level i.e. RNAi will be more meaningful to elucidate the mechanism of reducing the cancer burden.

Ontogenesis refers to the sequence of events involved in the development of an individual organism from its birth to its death. The development of a complex organ relies on precise temporal and spatial gene expression patterns during ontogenesis. The unique adult phenotype is a result of a cascade of transcriptional events that finally trigger gene expression in an organ-specific fashion. Gene expression profiles can be assessed for human tumours, but from the pharmacological perspective, there is a problem: the associated treatment histories, if any, are generally complex, fragmentary and difficult to interpret (Uwe Scherf et al., 2000). The use of RNAi to analyze gene function, particularly when coupled with genome-wide analysis of gene expression, has enormous potential for elucidating biological pathways and the interaction between different pathways. Although many small RNAs appear to be extremely specific for their target gene, others have been found to induce off-target effects. As more is learned about small RNAs, improvements will be made to experimental methodologies. We believe that the rules for designing very specific siRNAs will soon be refined, and then siRNAs can be used to their fullest potential.

Steps in RNAi therapeutic development

Selection of target gene(s)

For the therapeutic interventions of siRNA in cancers, although the efficacy of siRNA has to be validated in animal models, evaluation in cultured cancer cells is also required before any *in vivo* applications. The first step is to identify the target gene for RNAi mediated knockdown. These genes could be selected from a variety of key oncogenes, antiapoptotic genes or tumor promoting genes, including growth and angiogenic factors or their recaptors. As a matter of course, cancer specific genes that are ideally mutated or translocated could be selected. When target genes are selected the second step would be to optimize the design of the siRNA sequences that would be required. For this purpose a computational algorithm that searches through a DNA or RNA sequence could be used. In this step the computational approach finds the regions in the sequence that meets the required criteria and displays the resulting RNAi candidates ranked in order of their fit to the user supplied rule set. Finally, it appears that specificity can be attained depending on the position and sequence of a given siRNA. After synthesis of two or three different sequences for siRNA target site, the most specific and effective siRNA sequence must be validated by measuring levels of target mRNA or protein *in vitro.*

Designing siRNA

While there are currently no reliable methods to identify the ideal sequence for a siRNA, a number of parameters have been suggested. These include: selection of a target cDNA region 50 - 100 nucleotides downstream of the start codon, selection of a 5'-AA (N19) UU target mRNA sequence where N is any nucleotide, 50% G/C content in the target sequence, avoidance of 5' or 3' untranslated regions and high G-rich areas, and confirmation of exclusive target-specific sequences (Figure 2) (Elbashir et al., 2002). Rational siRNA design schemes are being developed that are based on an understanding of RNAi biochemistry (Khvorova et al., 2003, Schwarz et al., 2003, Chiu and Rana, 2002, Reynolds et al., 2004, Ui-Tei et al., 2004) and on naturally occurring miRNA function. Presently, custom siRNA synthesis service is available through a number of companies, such as Sigma, Dharmacon, QIAGEN, and Ambion etc. Now that many siRNA companies have developed software for designing specific siRNA, it can be downloaded by researchers and can be used to design effective siRNA sequences more easily. Screening candidate siRNA for homology with available sequence databases can, in principle be predicted to avoid many off-target effects. An off-target effect is the silencing of an unintended target gene.

siRNA delivery systems

A number of different approaches have been used to introduce siRNAs into both cells and whole organisms (Figure 3). Successful approaches include: electroporation (McRobert and McConkey, 2002), soaking in siRNAs (Malhotra et al., 2002), feeding bacteria carrying dsRNA (Caplen et al., 2001), transfection with comercial reagents (Elbashir et al., 2001), and vector-based strategies (Lois et al., 2002; Paddison et al., 2002). Though transfection with commercial reagents is the most widely used RNAi technique at present (Kim, 2003), much focus has been placed on vector-based strategies due to their therapeutic potential. These vector-based strategies involve either DNA or viral vector-mediated RNAi. In the first approach, RNA Polymerase II or III promoters are incorporated into DNA vectors along with siRNA expression cassettes (Brummelkamp et al., 2002a; Paddison et al., 2002; Sui et al., 2002; Xia et al., 2002). These cassettes have included either sense and antisense siRNA strands expressed from tandem promoters or a short hairpin (shRNA) cassette whereby the two siRNA strands are separated by a short spacer (Dykxhoorn et al., 2003). Use of RNA Polymerase II promoters to generate shRNAs has the advantage that they permit easier adaptation of inducible/repressible tissue- or cell-specific siRNA expression (Wall and Shi, 2003).

Plasmid vectors

The ability of RNAi to silence disease associated genes in cell culture and animal models has spurred development of RNAi based reagents for clinical applications to treat diseases including cancer. However, their stability and delivery methods are challenges that must be solved for developing effective RNAi reagents for cancer therapy. More prolonged gene silencing has been

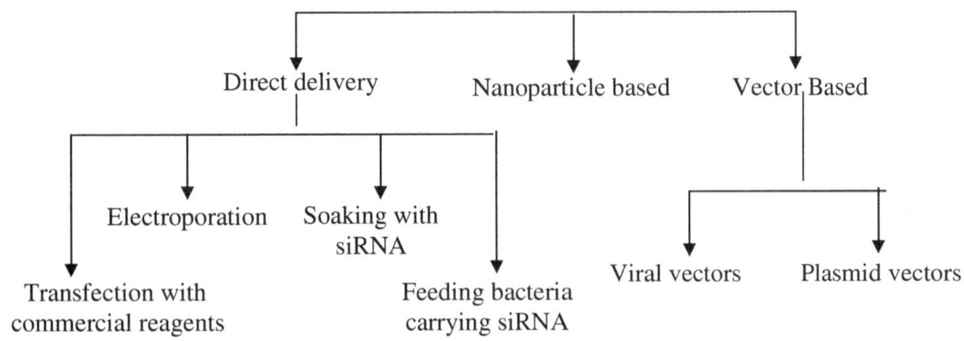

Figure 3. siRNA delivery methods.

achieved by expressing siRNAs from plasmid vectors that contain specific promoters (Lee et al., 2001; Paul et al., 2002; Sui et al., 2002). The most effective plasmid construct is the one that expresses the siRNA as a short hairpin (shRNA) precursor structure of around 70 nucleotides (Paddison et al., 2002). These shRNA precursors expressed from RNA polymerase promoters II or III are processed into a fully functional siRNA by the enzyme called Drosha. Plasmids are very useful for analyzing loss-of-function phenotypes that develop over extended periods of time. The shRNA is predicted to contain a perfectly double-stranded stem of 19 – 29 bp that is identical in sequence to the target mRNA; the 2 strandsof the stem are connected by a loop of 6 – 9 bases, which is removed *in vivo* by Dicer to generate effective siRNAs. The constitutive expression of plasmid-based shRNAs by RNA polymerase III (pol III) U6 and H1 snRNA promoters (Paddison et al., 2002; Sui et al., 2002,Paul et al., 2002; Yu et al., 2003; Miyagishi and Taira, 2002; McManus et al., 2002; Brummelkamp et al., 2002) tRNA promoters(Kawasaki and Taira, 2003) and RNA-pol-II-based CMV (cytomegalovirus) promoters (Xia et al., 2002, Shinagawa and Ishii. 2003) have been used successfully to obtain stable and efficient suppression of target genes (Dykxhoorn et al., 2003).

The pol III promoter is widely used for directing the expression of shRNAs, because it is active in all cell types and efficiently directs directs the synthesis of small, non-coding transcripts (which are structurally close to shRNAs) and bear well defined ends. By contrast, pol II promoters lack these properties and have therefore found limited use in mediating efficient gene silencing. Although the use of plasmid-based shRNA expression has quickly become the preferred method for targeted gene silencing, there are potential limitations. For example, in cells that are difficult to transfect, such as freshly isolated primary cells, neural cells, and stem cells because of inefficient transfection protocols, cell lines that stably express shRNA cannot be established.

The deli-very of siRNA by viral transduction is one way to over-come these shortcomings. Since plasmid-based siRNA expression has limitations in cases where transfection efficiency is concerned, which are relatively low, viral vectors have also been developed.

Viral vectors

Adenoviral vectors and various retroviral vectors (e.g. lentivirus-based) have proven to act as effective delivery systems in numerous cell types including non-cycling cells, stem cells, and zygotes (Abbas-Terki et al., 2002; Lois et al., 2002; Rubinson et al., 2003; Shen et al., 2003). In addition, viral vectors have also been developed for more stable and long term expression of shRNAs (Shen et al., 2003). Another natural process to achieve targeted delivery and release of siRNA inside cancer cells is that a specific bacteria infecting virus like phi29 uses a novel type of RNA (pRNA) to package its DNA into its protein shell (Guo S *et al*; 2006). pRNA, which acts as the delivery vehicle can be linked with siRNA, which acts as a therapeutic agent. To this folic acid molecule can be added which forms the targeting agent.

Nanoparticles as delivery vehicles

Nanoparticles, it turns out, are proving to be ideal carriers for siRNA molecules. With nanoparticles, we have the ability to load large number of these molecules into a protected environment, target them to cancer cells, and then have them taken up efficiently by those cells and release RNA molecules into interior of the cells. Not only can nanoparticles be engineered specific to the required RNA of interest, but also these can be designed more specific to the cell target, which means that there is almost no risk of targeting a wrong or non-tumor cell. More so with siRNA, this property is particularly attractive because there is some concern that siRNA delivered to wrong cells may have unintended consequences.

Another example is the successful delivery of a lead anticancer siRNA agent that has been achieved is by making use of a multicomponent polymer system that self-assembles into a rugged nanoparticle when mixed with siRNA molecules. A biocompatible polymer polyethyleneimine (PEI) is linked with its one end to a second biocompatible polymer-PEG (polyethylene glycol), to which the target agent (usually a three amino acid peptide ligand), arginine-glycine-aspartic acid (RGD) is attached, this can bind to a class of proteins known as integrins, and can be involved in a wide range of biological processes including angiogenesis, tumor cell growth, metastasis and inflammation (Raymond et al; 2004).

Another approach is to achieve delivery of siRNA to the target is to use Nanoparticles coated with folic acid molecules –a well studied nanoparticle targeting agent that binds to a receptor found in abundance on many types of cancers. In addition, the surfaces of these nanoparticles are decorated with sulfur containing chemical group-thiol, the sulfur molecule of which binds to the sulfur of gold nanoparticle. Target cells which contain large amounts of the glutathione, which is also a thiol results in the displacement of the thiols on the nanoparticles, triggering the release of RNA polyloads (Khaled A et al; 2005) and effect the delivery of the siRNA.

Delivery to live animals

The use of RNAi in live mice has been widely report-ed. The germline transmission of cells — embryonic stem cells in mice (Carmell et al., 2003) and fertilized eggs in rats (Hasuwa et al., 2002) that carry a shRNA transgene has been accomplished, and in some cases mice are phenotypically identical to those that carry a null mutation in the target gene (Kunath et al., 2003). In addition, the *in vivo* transfection of siRNA directly into the organs of postnatal mice inhibits the expression of co-transfected reporter plasmids (McCaffrey et al., 2002; Lewis et al., 2002) of the endogenous pro-apoptoic Fas receptor (Song et al., 2003), and of the *TH* (tyrosine hydroxylase) gene in the brain (Hommel et al., 2003). Although RNAi has been shown to down regulate the expression of various target genes effectively in animal models, it is still uncertain whether RNAi will be translated into effective therapy. More progress is required to improve RNAi-delivery systems and to evaluate the toxicity of exogenous RNA and its effects on endogenous RNAi processes.

MiRNA- a new class of RNA family

MicroRNAs (miRNAs) are a new class of non-protein-coding, endogenous, small RNAs. They are important regulatory molecules in animals and plants. MiRNA regulates gene expression by translational repression, mRNA cleavage, and mRNA decay initiated by miRNA-guided rapid deadenylation. A recently identified class of non-protein coding small RNAs, microRNAs (miRNAs), may provide new insight in cancer research. A recent study demonstrated that more than 50% of miRNA genes are located in cancer associated genomic regions or at fragile sites (Calin et al., 2004b), suggesting that miRNAs may play a more important role in the pathogenesis of a limited range of human cancers than previously thought.

Different experimental approaches have been used to compare tumor cells to normal cells, which have shown that defects in non-coding RNAs (ncRNAs) might be important. The first example of large ncRNA associated with tumors was the H19 gene (Hao et al., 1993) and there is now a growing list of ncRNA transcripts implicated in different types of cancer. NcRNAs that are implicated in cancer include both small and large RNAs and it has been reported that large ncRNAs can generate one or more microRNAs (miRNA). For example, the ncRNA BIC gene, which has been implicated in growth control and oncogenesis in cancer cells generates two miRNA, one of which miR155, which is over expressed in Burkitt and B cell lymphomas (Metzler et al., 2004; Eis et al., 2005).

MicroRNAs are an abundant class of endogenous small RNA molecules, 20 – 25 nucleotides in length (Ambros, 2001; Bartel, 2004; Ruvkun G et al 2004; Carrington and Ambros, 2003). All pre-miRNAs have hairpin secondary structure (Ambros, 2001; Bartel, 2004; Carrington and Ambros, 2003) with high minimal folding free energy index (MFEI) (Zhang et al., 2006). Some miRNAs are highly conserved from species to species in animals (Pasquinelli et al., 2000) and plants (Floyd and Bowman, 2004; Zhang et al., 2006).

At present, thousands of miRNAs have been predicted in animals, plants and viruses by different approaches (Zhang et al., 2006c), including experimental methods (Lee and Ambros, 2001), computational approaches (Brown and Sanseau, 2005), and expressed sequence tag (EST) and genomic survey sequence (GSS) analysis (Zhang et al., 2005, 2006a). However, only a few of them have been experimentally validated (Griffiths-Jones et al., 2006). Computational analysis indicates that the total number of miRNAs may be more than 1% of the total protein coding genes (Lai et al., 2003; Lim et al., 2003a,b); more than 30% of protein-coding genes may be targeted by miRNAs (Berezikov et al., 2005; Lewis et al., 2005).

MiRNAs and oncology

Differential expression of miRNAs has been reported in several human cancers including chronic lymphocytic leukemia (Calin et al., 2002), colorectal neoplasia (Michael et al., 2003) and Burkitt's lymphoma (Metzler et al., 2004) and human miRNAs are usually located at fragile sites at genomic loci involved in several cancers (Calin et al., 2004).

It has been reported that Prostate Specific Gene 1 (PCGEM1) ncRNA gene is over expressed in prostate tumors compared to normal and primary tumor specimens (Srikantan et al., 2000). In addition, over expression of PCGEM1 in LNCaP and in NIH3T3 cells enhance-ed cell proliferation and dramatically increased in colony formation suggesting a biological role for PCGEM1 in prostate tumorigenesis (Petrovics et al., 2004). Currently, almost all of the miRNA-related studies on cancers are based on the different expression profile of miRNAs in cancer cells vs. normal cells. Thus, methods used for detecting mRNA expression can also be used in studies on the potential role(s) of miRNAs in cancers. Recent findings suggest that miRNAs play an important role in regulation of gene activation by binding to the mRNAs of target genes and switching the genes on or off, or by fine-tuning the genes, to control key properties like, cell viability, cell cycle, proliferation or apoptosis thereby re-write the rules of molecular biology (Jamil, 2007). We are beginning to understand that miRNAs may act as oncogenes and/or tumor suppressor genes within the molecular architecture of gene regulatory networks, thereby contributing to the development of cancer. MiRNAs may provide useful diagnostic and prognostic markers for cancer diagnosis and treatment, as well as serving as potential therapeutic targets or tools.

When cells exhibit abnormal growth and loss of apoptosis function, it usually results in cancer formation. Several recent studies indicate that miRNA regulates cell growth and apoptosis (Cheng et al., 2005a; Tanno et al., 2005). For example, miR -15 and miR - 16 induce apoptosis by targeting antiapoptotic gene B cell lymphoma 2 (BCL2) mRNA (Cimmino et al., 2005), which is a key player in many types of human cancers, including leukemias, lymphomas, and carcinomas (Sanchez-Beato et al., 2003), Nairz et al., 2006) demonstrated that misexpression of miR - 278 in developing eyes causes massive overgrowth in Drosophila, partially due to inhibition of apoptosis by miR - 278 (Nairz et al., 2006). This suggests that miRNAs are involved in some cancer formation through regulation of cell growth and apoptosis. The initial evidence for the involvement of miRNAs in cancers came from a molecular study characterizing the 13q14 deletion in human chronic lymphocytic leukemia (CLL) (Calin et al., 2002), the most common form of adult leukemia in the Western world (Dohner et al., 2000). It was observed that two miRNAs, miR - 15a and miR - 16a, are located on chromosome 13q14, a region deleted in more than half of B cell chronic lymphocytic leukemia (B - CLL) cases. Detailed deletion analysis indicated that these two miRNAs are the only two genes within the small (30 kb) common region which are lost in CLL patients, and expression analysis indicated that miR - 15a and miR - 16a were either absent or down-regulated in the majority (68%) of CLL patients (Calin et al., 2002).

Recognition of miRNAs that are differentially expressed between tumor tissues and normal tissues may help to identify those miRNAs that are involved in human cancers and further establish the apparent pathogenic role of miRNAs in cancers (Iorio et al., 2005). Calin et al., (2004a, b) determined genome wide expression profiles of miRNAs in human B cell CLL using a microarray containing 368 probes corresponding to 245 human and mouse miRNA genes. This miRNA microarray analysis further confirmed that miR - 16 and miR - 15 are reduced in human CLL (Calin et al., 2004a). MicroRNA microarray analysis also indicated that miRNA expression patterns were related to the biological and clinical behavior of CLL (Calin et al., 2004a). A recent study indicated that BCL2 is one of the targets of miR-15a and miR-16-1. miR - 15a and miR -16 -1 expression was inversely correlated to BCL2 expression in CLL; both miRNAs negatively regulate BCL2 at the post-transcriptional level (Cimmino et al., 2005). This was also confirmed in a leukemic cell line model (Cimmino et al., 2005), suggesting that miR-15a and miR -16 -1 could be used therapeutically to cure tumors over-expressing BCL2.

Several experiments and clinical analysis suggest that miRNAs may function as a novel class of oncogenes or tumor suppressor genes. Those miRNAs whose expression is increased in tumors may be considered as oncogenes. These oncogene miRNAs, called “oncomirs”, usually promote tumor development by negatively inhibitting tumor suppressor genes and/or genes that control cell differentiation or apoptosis. Many miRNA genes have been found that are significantly over-expressed in different cancers. All of them appear to function as oncogenes; however, only a few of them have been well characterized.

In oncogenesis, some miRNAs expression is decreased in cancerous cells. These types of miRNAs are considered tumor suppressor genes. Tumor suppressor miRNAs usually prevent tumor development by negatively inhibiting oncogenes and/or genes that control cell differentiation or apoptosis. Currently, several miRNAs are considered as tumor suppressor genes, for example, miRNA let-7.

siRNA and miRNA

Although small interfering RNA and microRNA were initially discovered in unrelated studies, both types of small RNA are closely related in their biogenesis, assembly into RNA–protein complexes and ability to regulate gene transcripts negatively in diverse eukaryotes (Bartel, 2004; Meister and Tuschl, 2004; Baulcombe, 2004; Mello and Conte, 2004). Dicer, a multi domain enzyme of the RNase III family, generates both siRNAs and miRNAs. Dicer cuts long, double stranded RNA (dsRNA) into siRNAs and chops short precursor miRNAs with imperfect stem-loop structure into miRNAs. The nascent siRNAs and miRNAs are double-stranded duplexes. These duplexes need to be unwound before they can be assembled into an RNA-induced silencing complex (RISC). By comparing the thermodynamic stabilities at the two ends, siRNAs can be divided into two classes: symmetric siRNAs and asymmetric siRNAs. A symmetric siRNA has two equally stable ends; thus, both strands of the siRNA are assembled into the RISC with equivalent efficiency (Schwarz et al., 2003). By contrast, an asymmetric siRNA has one end that is less stable than the other. Because it is easier to unwind siRNA from the less stable end, one strand of the siRNA is preferentially incorporated into the RISC complex in a process referred to as the 'asymmetric assembly of RISCs' (Schwarz et al., 2003; Khvorova et al., 2003). Intriguingly, most miRNAs are highly asymmetric, ensuring efficient asymmetric assembly of miRISC in cells (Schwarz et al., 2003; Khvorova et al., 2003).

In vitro and in vivo biochemical studies have shown that siRISC can function as miRISC to repress translation of the target mRNA; similarly, miRISC can function as siRISC to cleave the target mRNA. This functional interchangeability between siRISC and miRISC argues that siRISCs and miRISCs are highly similar, if not identical [Hutvagner and Zamore, 2002; Tang. et al., 2003; Doench et al., 2003; Zeng et al., 2003]. Much evidence suggests, that siRISCs and miRISCs are distinct types of complexes. First, the biogenesis, maturation and subsequent assembly of siRNAs and miRNAs into silencing complexes are different (Bartel, 2004), which can result in RISCs with distinct functions. Second, Argonaute (AGO) proteins, which are principal components of RISCs, are encoded by a multigene family and can be divided into functionally distinct subgroups (Carmell et al., 2002; Hammond et al., 2001; Liu et al., 2004; Song et al., 2004; Meister et al., 2004). These functionally different AGOs endow their corresponding RISCs with distinct functions. Third, the complementarity between the small RNAs and their target mRNAs has been proposed to affect the functional mode of RISCs in terms of the regulation of mRNA stability and translation. Nevertheless, RISCs containing small RNAs that are extensively complementary to their target mRNAs do not always specify efficient cleavage of the targets as was previously predicted: some RISCs direct efficient target cleavage (Hutvagner and Zamore, 2002; Tang et al., 2003; Mallory et al., 2004; Llave et al., 2002), whereas others do not (Meister et al., 2004). Fourth, RISCs vary markedly in size, from the smallest 'coreRISC' of ~160 kDa to the largest 'holo RISC' of 80 Svedberg (80S) (Martinez and Tuschl, 2004; Martinez et al., 2002; Nykanen et al., 2001; Pham et al., 2004). Lastly, siRNA and miRNA programmed RISCs have distinct targeting functions in cells. Many endogenous miRNAs and their RISCs are genetically programmed to regulate gene expression and thus are important for the growth and development of an organism (Rhoades et al., 2002). By contrast, siRNAs are produced from dsRNAs that are often synthesized in vitro or in vivo from viruses or repetitive sequences introduced by genetic engineering. In addition, dsRNA can be produced from endogenously activated transposons. Thus, siRNAs have been proposed to function in: (i) antiviral defense (despite the fact that viruses develop counterdefense strategies as well (Pfeffer et al., 2004; Ding et al., 2004), (ii) silencing mRNAs that are overproduced or translationally aborted, and (iii) guarding the genome from disruption by transposons (Mello and Conte, 2004; Hannon, 2002; Tabara et al., 1999).

Interfacing microarray data with RNAi technology

RNAi that has progressed in the past few years from an obscure act on the international research stage to a major player, enjoying the spotlight as scientists race to study the previously unknown genomic regulatory molecules involved in this process. Even though genomic technologies are advanced, scientists continue to build upon them to make them even better. Two cutting-edge technologies in modern biology, microarrays and RNAi are in the lead to revolutionize biology and medicine. RNAi, a powerful new approach for achieving targeted gene silencing using double-stranded RNA as the triggering agent, offers remarkable specificity, scalability, potency, and reproducibility. On the other side, DNA microarrays provide a snapshot of gene expression of all the genes in a cell.

A microarray works by exploiting the ability of a given mRNA molecule to bind specifically to, or hybridize to, the DNA template from which it originated. By using an array containing many DNA samples, researchers can determine, in a single experiment, the expression levels of hundreds or thousands of genes within a cell by measuring the amount of RNA bound to each site on the array. With the aid of a computer and computational methods the amount of RNA bound to the spots on the microarray is precisely measured, generating a profile of gene expression in the cell. Using bioinformatics tools the data is interpreted for meaningful information. The process of microarray based microRNA profiling, initially developed by Liu et al. (2004) involves four main steps: target labeling, DNA-DNA hybridization, staining and signal detection.

Genomic sequencing and gene expression profiling have generated and continue to generate extensive information on the structure, sequence variation, and expression levels of human genes. However, an emerging challenge is the need to translate genomic data into meaningful insights for drug discovery. Identification of novel targets for the treatment of human diseases requires functional characterization of the products encoded by the tens of thousands of genes that have been identified by the Human Genome Project and EST (expressed sequence tag) sequencing efforts (Venter et al., 2001; Ewing and Green, 2000). Recent strides in genomics have promised to greatly facilitate the identification of targets for human diseases by making it feasible, in concept, to functionally screen all human gene products for those with roles in cellular processes associated with diseases.

One of the key advantages of the RNAi microarray for genome-scale screening is that it permits miniaturization far beyond what is physically possible in present well-based systems. At present, most cell-culture-based assays of gene function make use of 6-, 12-, 24-, or 96-well plate formats. Although RNAi experiments can be conducted in a 384 -well plate format, a "well-less" microarray platform will be significantly more effective. Even though the size and density of the spots are limited by the need to have at least several dozens if not hundreds of cells on each spot, it should be feasible to array and analyze as many as 5000 to 10,000 individual siRNAs spots per slide.

Another advantage of the array-based miniaturization is the reduction in reagent costs. In particular, chemically synthesized siRNAs are expensive and are produced in relatively small quantities. The RNAi microarray format will also reduce the cost of the phenotypic assays because only a single microscope slide needs to be subjected to analysis. The well-less nature of the platform also allows for improving the level of uniformity in the experimental culturing conditions and assay conditions, because all processes take place on the same surface.

Conclusion

An economic and novel way of ablating gene function holds out massive hope for improving our ability to untangle the complex regulatory pathways that control cellular behavior in health and disease. RNAi allows analysis of gene function at the cellular level and provides an effective design and delivery system. In vivo, siRNA technology has aroused entrepreneurship, as documented by Fortune magazine (2003) predicting siRNA as biotechnology's "next billion dollar breakthrough".

There is no golden principle to predict the length of sustaining time of RNAi inhibition, because each protein has its own turnover time. Normally, the transient effect of siRNA inhibition lasts less than one week. Multiple transfections may be needed to extend its effect for a longer time. Otherwise, vector based RNAi, either by polymerase II or III is required to achieve permanent knock down of target gene expression. The length of the target sequence could range from 21 to 23 nucleotide, which guarantees sequence specificity, without generating interferon response and mimic the product length of Dicer.

RNAi has become the method of choice for specific gene silencing in cell culture. However, the success of this approach depends on the effectiveness of siRNA, delivery and design of the siRNA sequence. Different biochemical, pharmacological and histological assays have been used to determine the effects of siRNA inhibittion of target genes and analyze the phenotypic changes in cells. Currently, the use of siRNA to characterize gene function and potential therapeutic drug targets is a highly promising application of this technology. Selected gene knockdown can be used to identify critical gene(s) and pathways that can be targeted by siRNA.

Thus, RNAi can be specifically designed to target anti-apoptotic genes which are over expressed in cancer cells and might explain the resistance of some tumors to drug induced apoptosis. At the same time, RNAi designed to target cell-proliferating protein encoding genes will help in curbing the rate of cell division and promote a balance between the rate of proliferation and cell death .Similarly, specific RNAi can be designed for other genes which are directly involved in cancer etiology such as RNAi for non-coding RNAs, cellular oncogenes (Ras gene family, bcl-2), mutated tumor suppressor gene, etc.

As miRNAs can function as oncogenes or tumor suppressors, it might be possible to regulate miRNA expression and/or inject miRNAs to regulate cancer formation, similar to the use of antisense mRNAs and RNAi which are widely used as tools for studying gene functions and in some case of gene therapy. Artificial miRNAs could be synthesized to down-regulate oncogenes and prevent the formation of cancer.

RNAi methods for both *in vitro* and *in vivo* target validation are being developed and are being utilized by pharmaceutical companies for prioritizing drug target candidates. SiRNAs can be readily synthesized with low production costs compared to protein or antibody therapies. In addition, siRNAs have favorable pharmacokinetic properties and can be delivered into a wide range of organs (Braasch et al., 2004). Several groups have been investigating the use of modified nucleotides to improve the clinical properties of these reagents. For example, by conjugating of the 3'-end of the sense strand of siRNA with cholesterol through a pyrrolidine linker, this was reported to markedly improve the pharmacological properties of siRNA molecules (Soutschek et al., 2004). The Cholesterol- conjugated siRNA are more resistant to nuclease degradation, exhibit increased stability in blood by increasing binding to human serum albumin and show increased uptake into liver. Another study has shown that boranophosphate modified siRNAs were 10 times more nuclease resistant than unmodified siRNAs (Hall et al., 2004). In addition, boranophosphate siRNAs were more potent than unmodified siRNAs and appeared to act through the standard RNAi pathway. Topical gels have also been used to deliver siRNAs to cells and could open the way for dermatological applications, as well as treatment for cervical cancer (Jiang et al., 2004). siRNAs are also rapidly catching up with ODNs (oligodeoxiribonucleotides) and ribosome for development as therapeutics owing to the establishment of siRNA-based biotechnology companies that focus on the development of clinical programs (Howard, K. 2003). Several proof-of-principle experiments have demonstrated the therapeutic potential of siRNAs: siRNAs protected mice from fulminant hepatitis (Zender et al., 2003; Song et al., 2003), viral infection (McCaffrey et al., 2003; Song et al., 2003), sepsis (Sorensen et al., 2003), tumour growth (Li K et al 2003; Filleur et al., 2003; Yang et al., 2003; Verma, 2003; Yoshinouchi et al., 2003) and ocular neovascularization causing macular degeneration (Reichet al 2003).

In an astonishingly short time, RNAi has gone from discovery to human clinical trials.' Acuity Pharmaceuticals in Philadelphia has completed safety tests on the first ever RNAi treatment for a health problem. Fifteen people have undergone injections into the eye to treat macular degeneration, the leading cause of blindness in the elderly. Other companies are also looking to harness RNAi to treat Alzheimer's disease, motor neuron disease, Parkinson's disease, hepatitis and even HIV. Benitec Inc, a company in Australia is working on a way of delivering a 'multiple warhead' that cripples several genes at once, which could aid the treatment of type 2 diabetes. In April 2007, Alnylam pharmaceuticals had developed a program for the treatment of liver cancers and potentially other solid tumors. Their RNAi therapeutic, ALN-VSP01, targets two separate genes involved in the growth and development of tumors: vascular endothelial growth factor (VEGF) and kinesin spindle protein (KSP). In addition, RNAi provides a new potential therapeutic tool applicable to a wide array of disease targets. However, many challenges exist in translating RNAi's potential anticancer activity into clinical benefit.

Cancers may provide distinctive advantages as proof of concept models for both the target validation and the anticancer therapeutic utility of RNAi. As current therapy is inadequate for many cancers, discoveries flowing from RNAi-based methods could make significant contributions.

ACKNOWLEDGEMENTS

We are grateful to the Chairman, and Scientific Advisers of Indo-American Cancer Institute and Research Centre for their encouragement and support for this work and to NEERI- Director for his support. We also greatly appreciate our reviewers for their thorough analysis and suggestions for improving the manuscript.

REFERENCES

Abbas-Terki T, Blanco-Bose W, Deglon N, Pralong W, Aebischer P (2002). Lentiviral-mediated RNA interference. Hum. Gene Ther. 13: 197-2201.

Ambros V (2001). MicroRNAs: tiny regulators with great potential. Cell 107: 23–826.

Ambros V (2004). The functions of animal microRNAs. Nature. 431: 350–355.

Ancellin N, Colmont C, Su J, Li Q, Mittereder N, Chae SS, Steffansson S, Liau G, Hla T (2001). Extracellular export of sphingosine kinase-1 enzyme: Sphingosine 1-phosphate generation and the induction of angiogenic vascular maturation. J. Biol. Chem. 227: 6667–6675.

Aza-Blanc P, Cooper CL, Wagner K, Batalov S, Deveraux QL, Cooke MP (2003). Identification of Modulators of TRAIL-Induced Apoptosis via RNAi-Based Phenotypic Screening. Mol. Cell 12: 627–637.

Bakker J, Lin X, Nelson WG (2002). Methyl-CpG binding domain protein 2 represses transcription from hypermethylated p-class glutathione S-transferase gene promoters in hepatocellular carcinoma cells. J. Biol. Chem. 17: 17.

Bartel DP (2004). MicroRNAs: genomics, biogenesis, mechanism, and function. Cell; 116: 281–297.

Bartel DP (2004). MicroRNAs: genomics, biogenesis, mechanism, and function. Cell. 116, 281–297

Baulcombe D (2004). RNA silencing in plants. Nature 431: 356–363 4

Berezikov E, Guryev V, van de Belt J, Wienholds E, Plasterk RHA, Cuppen E (2005). Phylogenetic shadowing and computational identification of human microRNA genes. Cell 120: 21–24.

Berns K, Hijmans EM, Mullenders J, Brummelkamp TR, Velds A, Heimerikx M, Kerkhoven RM, Madiredjo M, Nijkamp W, Weigelt B, Agami R, Ge W, Cavet G, Linsley PS, Beijersbergen RL, Bernards R (2004). A large-scale RNAi screen in human cells identifies new components of the pathway. Nature 428: 1-437.

Bitko V, Barik S (2001). Phenotypic silencing of cytoplasmic genes using sequence-specific double- stranded short interfering RNA and its application in the reverse genetics of wild type negative- strand RNA viruses. BMC Microbiol. 1: 34.

Braasch DA, Paroo Z, Constantinescu A, Ren G, Oz OK, Mason RP, Corey DR (2004). Biodistribution of phosphodiester and phosphorothioate siRNA. Bioorg Med. Chem. Lett. 14: 1139-1143.

Brown JR, Sanseau P (2005). A computational view of microRNAs and their targets. Drug Discovery Today. 10: 595–601.

Brummelkamp TR, Bernards R, Agami R (2002a). A system for stable expression of short interfering RNAs in mammalian cells. Sci. 296: 550-553.

Brummelkamp TR, Bernards R, Agami R (2002b). Stable suppression of tumorigenicity by virus-mediated RNA interference. Cancer Cell *2*: 243-247.

Calin GA, Dumitru CD, Shimizu M, Bichi R, Zupo S, Noch E, Aldler H, Rattan S, Keating M, Rai K, Dumitru CD, Shimizu M, Bichi R, Zupo S, Noch E, Aldler H, Rattan S, Keating M, Rai K, Rassenti L, Kipps T, Negrini M, Bullrich F, Croce CM (2002). Frequent deletions and down-regulation of micro-RNA genes miR15 and miR16 at 13q14 in chronic lymphocytic leukemia. Proc. Natl. Acad. Sci. U.S.A. 99: 15524–15529.

Calin GA, Sevignani C, Dan Dumitru C, Hyslop T, Noch E, Yendamuri S, Shimizu M, Rattan S, Bullrich F, Negrini M, Croce CM (2004b). Human microRNA genes are frequently located at fragile sites and genomic regions involved in cancers. Proc. Natl. Acad. Sci. U.S.A 101: 2999–3004.

Caplen NJ, Parrish S, Imani F, Fire A, Morgan RA (2001). Specific inhibition of gene expression by small double stranded RNAs in invertebrate and vertebrate systems. Proc. Natl. Acad. Sci. USA *98*: 9742-9747.

Carmell MA, Zhang L, Conklin DS, Hannon GJ, Rosenquist TA (2003). Germline transmission of RNAi in mice. Nature Struct. Biol. 10: 91–92.

Carmell MA. Xuan Z, Zhang MQ, Hannon GJ (2002) The Argonaute family: tentacles that reach into RNAi, developmental control, stem cell maintenance, and tumorigenesis. Genes Dev. 16: 2733–2742.

Carrington JC, Ambros V (2003). Role of microRNAs in plant and animal development. Sci. 30: 336–338.

Ch. Kalyana K, Sudha M, Kaiser J (2007). Possible Associations of Splice Site Mutation of Dihydropyrimidine Dehydrogenase (IVS14+1G>A) in Adverse Drug Reactions in Some Invasive Ductal Carcinoma Patients Int. J. Pharmacol. 3: 130-136.

Cheng AM, Byrom MW, Shelton J, Ford LP (2005a). Antisense inhibition of human miRNAs and indications for an involvement of miRNA in cell growth and apoptosis. Nucleic Acids Res. 33: 1290–1297.

Chiu YL, Rana TM (2002). RNAi in human cells: basic structural and functional features of small interfering RNA. Mol. Cell 10: 549–561.

Choudhury A, Charo J, Parapuram SK *et al.* (2004). Small interfering RNA (siRNA) inhibits the expression of the *Her2/neu* gene, upregulates HLA class I and induces apoptosis of *Her2/neu* positive tumor cell lines. Int J Cancer; 108: 71–77.

Cimmino A, Calin GA, Fabbri M, Iorio MV, Ferracin M, Shimizu M, Wojcik SE, Aqeilan RI, Zupo S, Dono M, Rassenti L, Alder H, Volinia S, Liu CG, Kipps TJ, Negrini M, Croce CM (2005). miR-15 and miR-16 induce apoptosis by targeting BCL2. Proc. Natl. Acad. Sci. U.S.A. 102: 13944–13949.

Ding S-W, Hongwei Li, Rui Lu, Feng Li, Wan-Xiang Li (2004) RNA silencing: a conserved antiviral immunity of plants and animals. Virus Res. 102: 109–115.

Doench JG, Christian PP, Phillip AS (2003) siRNAs can function as miRNAs. Genes Dev. 17: 438–442
Dohner H, Stilgenbauer S, Benner A, Leupolt E, Krober A, Bullinger L, Dohner K, Bentz M, Lichter P (2000). Genomic aberrations and survival in chronic lymphocytic leukemia. N. Engl. J. Med. 343: 1910–1916.
Dykxhoorn DM, Novina CD, Sharp PA (2003). Killing the messenger: short RNAs that silence gene expression. Nat. Rev. Mol. Cell Biol. *4*: 457-467.
Dykxhoorn DM, Novina CD, Sharp PA (2003). Killing the messenger: short RNAs that silence gene expression. Nature Rev. Mol. Cell Biol. 4: 457–467.
Eis PS, Tam W, Sun L, Wayne T, Liping S, Amy C, Zongdong Li, Mario FG, Elsebet L, James ED (2005). Accumulation of miR-155 and BIC RNA in human B cell lymphomas. Proc. Natl. Acad. Sci. USA, 102: 3627-3632.
Elbashir SM, Harborth J, Lendeckel W, Yalcin A, Weber K, Tuschl T (2001c). Duplexes of 21- nucleotide RNAs mediate RNA interference in mammalian cell culture. Nature 411: 494–498.
Elbashir SM, Harborth J, Lendeckel W, Yalcin A, Weber K, Tuschl T (2001a). Duplexes of 21-nucleotide RNAs mediate RNA interference in cultured mammalian cells. Nature 411: 494-498.
Elbashir SM, Harborth J, Lendeckel W, Yalcin A, Weber K, Tuschl T (2001) Duplexes of 21-nucleotide RNAs mediate RNA interference in cultured mammalian cells. Nature; 411: 494–498.
Elbashir SM, Harborth J, Weber K, Tuschl T (2001b). Analysis of gene function in somatic mammalian cells using small interfering RNAs. Methods 26: 199-213.
Elbashir SM, Harborth J, Weber K, Tuschl T (2002). Analysis of gene function in somatic mammalian cells using small interfering RNAs. Methods 26: 199–213.
Elbashir SM, Lendeckel W, Tuschl T (2001a). RNA interference is mediated by 21 and 22 nt RNAs. *Genes Dev.* 15: 188–200.
Elbashir SM, Martinez J, Patkaniowska A, Lendeckel W, Tuschl T (2001b). Functional anatomy of siRNAs for mediating efficient RNAi in *Drosophila melanogaster* embryo lysate. EMBO J. 20: 6877–6888.
Ewing, B, Green P (2000). Analysis of expressed sequence tags indicates 35,000 human genes. Nat. Genet. 25: 232–234.
Filleur S, Aurélie C, Slimane A-S-A, Julien G, Carole M, Annick H-B, Philippe C, Florence C (2003). SiRNA-mediated inhibition of vascular endothelial growth factor severely limits tumor resistance to antiangiogenic thrombospondin-1 and slows tumor vascularization and growth. Cancer Res. 63: 3919–3922.
Fire S, Xu, Montgomery MK, Kostas SA, Driver SE, Mello CC (1998). "Potent and specific genetic interference by double-stranded RNA in *Caenorhabditis elegans*". *Nature* 391: 806-811
Floyd SK, Bowman JL (2004). Gene regulation: ancient microRNA target sequences in plants. Nature 428: 485–486.
Griffiths-Jones, S., Grocock, R.J., van Dongen, S., Bateman, A., Enright, A.J., (2006). miRBase: microRNA sequences, targets and gene nomenclature. Nucleic Acids Res. 34: 140–144.
Guo S, Huang F, Guo P (2006). Construction of a folate-conjugated pRNA of bacteriaphage phi29 DNA packaging motor for delivery of chimeric siRNA to nasopharyngeal carcinoma cells. Gene Therapy 13: 814-820.
Hall AH, Wan J, Shaughnessy EE, Ramsay SB, Alexander KA (2004). RNA interference using boranophosphate siRNAs: structure-activity relationships. Nucleic Acids Res. 32: 5991-6000.
Hammond SM, Boettcher S, Caudy AA, Kobayashi R, Hannon GJ (2001) Argonaute2, a link between genetic and biochemical analyses of RNAi. Science 293, 1146–1150
Hannon GJ (2002). RNA interference. Nature 418: 244–251.
Hao Y, Crenshaw T, Moulton T, Newcomb E, Tycko B (1993). Tumour-suppressor activity of H19 RNA. Nature, 365: 764-767.
Haranatha R, Kaiser J (2006). "Polymorphism in the *MTHFR* gene and their possible association with susceptibility to childhood acute Lymphocyte Leukemia in an Indian population" Leukemia &Lumphoma. 47:1333-1339.
Haranatha R, Kaiser J (2006). Polymorphisms in the GST (M1 andT1) gene and their possible association with susceptibility to childhood acute *lymphocytic leukemia* in Indian population, Afr. J. Biotechnol. 5: 1454-1456, 1417.
Hasuwa H, Kaseda K, Einarsdottir T, Okabe M (2002). Small interfering RNA and gene silencing in transgenic mice and rats. FEBS Lett. 532: 227–230.
Hirai I,Wang HG (2002). A role of the C-terminal region of hRad9 in nuclear transport of the hRad9-hRad1-hHus1 checkpoint complex. J. Biol. Chem. 277: 25722–25727.
Holen T, Amarzguioui M, Wiiger MT, Babaie E, Prydz H (2002). Positional effects of short interfering RNAs targeting the human coagulation trigger Tissue Factor. Nucleic Acids Res. 30: 1757–1766.
Hommel JD, Sears RM, Georgescu D, Simmons DL, DiLeone RJ (2003). Local gene knockdown in the brain using viral-mediated RNA interference. Nature Med. 9: 1539–1544.
Howard K (2003). Unlocking the money-making potential of RNAi. Nature Biotechnol. 21: 1441–1446
Hutvagner G, Zamore PD (2002). A microRNA in a multipleturnover RNAi enzyme complex. Sci. 297: 2056–2060
Iorio MV, Ferracin M, Liu CG, Veronese A, Spizzo R, Sabbioni S, Magri E, Pedriali M, Fabbri M, Campiglio M, Menard S, Palazzo JP, Rosenberg A, Musiani P, Volinia S, Nenci I, Calin GA, Querzoli P, Negrini M, Croce CM (2005). MicroRNA gene expression deregulation in human breast cancer. Cancer Res. 65: 7065–7070.
Jiang M, Rubbi CP, Milner J (2004). Gel-based application of siRNA to human epithelial cancer cells induces RNAidependent apoptosis. Oligonucleotides, 14: 239-248.
Jocelyn H, Sam D, Nick P, Susan D, Croll LH, Michelle R, et al. ??? Provide all others name (2002). VEGF-Trap: A VEGF blocker with potent antitumor effects. PNAS, 99: 11393-11398.
Kaiser J (2007). MiRNAs Rewrite the Rules of Molecular Biology, Res. J. Biotechnol. 2: 3-5.
Kawasaki H,Taira K (2003). Short hairpin type of dsRNAs that are controlled by tRNA(Val) promoter significantly induce RNAimediated gene silencing in the cytoplasm of human cells. Nucleic Acids Res. 31: 700–707.
Khaled A, Guo S, Li F, Fuo P (2005). Controllable self-assembly of nanoparticles for specific delivery of multiple therapeutic molecules to cancer cells using RNA nanotechnology. Nano Letters. 5: 1797-1808.
Khvorova A, Reynolds A, Jayasena SD (2003). Functional siRNAs and miRNAs exhibit strand bias. Cell 115: 209–216.
Khvorova A, Reynolds A, Jayasena SD (2003). Functional siRNAs and miRNAs exhibit strand bias. Cell 115: 209–216.
Kim VN (2003). RNA interference in functional genomics and medicine. J. Korean Med. Sci. 18: 309-318.
Kunath T, Gish G, Lickert H, Jones N, Pawson T, Rossant J (2003).Transgenic RNA interference in ES cell derived embryos recapitulates a genetic null phenotype. Nature Biotechnol. 21: 559–561.
Lai EC, Tomancak P, Williams RW, Rubin GM (2003). Computational identification of Drosophila microRNA genes. Genome Biol. 4, R42.
Landthaler M, Yalcin A, Tuschl T (2004). The human DiGeorge syndrome critical region gene 8 and its D. melanogaster homolog are required for miRNA biogenesis. Curr. Biol. 14: 2162–2167.
Lee RC, Ambros V (2001). An extensive class of small RNAs in Caenorhabditis elegans. Science, 294: 862-864.
Lewis DL, Hagstrom JE, Loomis AG, Wolff JA, Herweijer H (2002). Efficient delivery of siRNA for inhibition of gene expression in postnatal mice. Nature Genet. 32: 107–108
Lewis DL, Hagstrom JE, Loomis AG, Wolff JA, Herweijer H.(2002) Efficient delivery of siRNA for inhibition of gene expression in postnatal mice. Nat Genet. 32: 107–108.
Li K, Lin SY, Brunicardi FC, Seu P (2003). Use of RNA interference to target cyclin E-overexpressing hepatocellular carcinoma. Cancer Res. 63: 3593–3597.
Lim LP, Glasner ME, Yekta S, Burge CB, Bartel DP (2003a). Vertebrate microRNA genes. Sci. 299: 1540.
Lim LP, Lau NC, Weinstein EG, Abdelhakim A, Yekta S, Rhoades MW, Burge CB, Bartel DP (2003b). The microRNAs of Caenorhabditis elegans. Genes Dev. 17: 991–1008.
Lin SL, Chuong CM, Ying SY (2001). A Novel mRNA-cDNA interference phenomenon for silencing bcl-2 expression in human LNCaP cells. Biochem. Biophys. Res. Commun. 281:639-644.
Lisa NP, Megan JB, Wenyi Gu, Nicholas AS, Brian GG, Graham RL, Nigel A, McMillan J (2005). RNA Interference against Human

apillomavirus Oncogenes in Cervical Cancer Cells Results in Increased Sensitivity to Cisplatin, Mol. Pharmacol. 68:1311–1319.

Liu J, Michelle AC, Fabiola VR, Carolyn GM, Michael JT, Ji-Joon S, Scott MH, Leemor J-T, Gregory JHm (2004). Argonaute2 is the catalytic engine of mammalian RNAi. Sci. 305: 1437–1441

Llave C, Zhixin X, Kristin DK, James CC (2002). Cleavage of Scarecrow-like mRNA targets directed by a class of Arabidopsis miRNA. Sci. 297: 2053–2056.

Lois C, Hong EJ, Pease S, Brown EJ, and Baltimore D (2002) Germline transmission and tissue-specific expression of transgenes delivered by lentiviral vectors. *Science* (Wash DC). 295: 868–872.

Malhotra P, Dasaradhi PVN, Kumar A, Mohmmed A, Agrawal N, Bhatnagar RK, Chauhan VS (2002). Double-stranded RNA-mediated gene silencing of cysteine proteases (falcipain -1 and -2) of plasmo*dium falciparum.* Mol. Microbiol. *45*: 1245-1254.

Mallory AC, Reinhart BJ, Jones-Rhoades MW, Tang G, Zamore PD, Barton MK, Bartel DP (2004). MicroRNA control of PHABULOSA in leaf development: importance of pairing to the microRNA 50 region. EMBO J. 23: 3356–3364

Martines LM, Iaccarino I, Tenev T, Gschmeissner S, Totty NF, Lemoine NR, Savopoulos J, Gray CW, Creasy CL, Dingwall C, Downward J (2002). The serine protease Omi/HtrA2 regulates apoptosis by binding XIAP through a reaper-like motif. J. Biol. Chem. 277: 439–444.

Martinez J, Agnieszka P, Henning U, Reinhard L, Thomas T (2002). Single-stranded antisense siRNAs guide target RNA cleavage in RNAi. Cell 110: 563–574.

Martinez J, Tuschl T (2004). RISC is a 50 phosphomonoesterproducing RNA endonuclease. Genes Dev. 18: 975–980

Martinez LA, Naguibneva I, Lehrmann H, Arlette V, Thierry T, Guillermina L, Annick H-B(2002). Synthetic small inhibiting RNAs: efficient tools to inactivate oncogenic mutations and restore p53 pathways. *Proc Natl Acad Sci USA*; 99: 849–854.

McCaffrey AP, Meuse L, Pham TT, Conklin DS, Hannon GJ, Kay MA (2002). RNA interference in adult mice. Nature. 418: 38–39

McCaffrey AP, Nakai H, Pandey K, Huang Z, Salazar FH, Xu H, Wieland SF, Marion PL, Kay MA (2003). Inhibition of hepatitis B virus in mice by RNA interference. Nature Biotechnol. 21: 639–644.

McManus MT, Petersen CP, Haines BB, Chen J, Sharp PA (2002). Gene silencing using micro-RNA designed hairpins. RNA 8: 842–850

McRobert L, McConkey GA (2002). RNA interference (RNAi) inhibits growth of *Plasmodium falciparum.* Mol. Biochem. Parasitol. 199: 273-278.

Meister G, Landthaler M, Patkaniowska A, Dorsett Y, Teng G, Tuschl T (2004) Human Argonaute2 mediates RNA cleavage targeted by miRNAs and siRNAs. Mol. Cell 15: 185–197

Meister G, Tuschl T (2004) .echanisms of gene silencing by double-stranded RNA. Nature 43: 343–349

Mello CC, Conte DJr. (2004).Revealing the world of RNA interference. Nature 43: 338–342

Metzler M, Wilda M, Busch K, Viehmann S, Borkhardt A (2004). High expression of precursor microRNA-155/BIC RNA in children with Burkitt lymphoma. Genes Chromosomes Cancer, 39: 167-169

Michael MZ, O'Connor SM, Pellekaan NGV, Young GP, James RJ (2003). Reduced accumulation of specific microRNAs in colorectal neoplasia. Mol. Cancer Res. 1:882–891.

Miyagishi M , Taira K (2002). U6 promoter-driven siRNAs with four uridine 3′ overhangs efficiently suppress targeted gene expression in mammalian cells. Nature Biotechnol. 20: 497–500.

Moskalenko S, Henry DO, Rosse C, Mirey G, Camonis JH, White MA (2002). The exocyst is a Ral effector complex. Nat. Cell Biol. 4: 66–72.

Muneesh T, Marc V (2003). RNAi on the Apoptosis TRAIL The Mammalian Cell Genetic Screen Comes of Age. 5: 534-535.

Nairz K, Rottig C, Rintelen F, Zdobnov E, Moser M, Hafen E (2006). Overgrowth caused by misexpression of a microRNAwith dispensable wildtype function. Dev. Biol. 291: 314–324.

Napoli C, Lemieux C, Jorgensen R (1990). Introduction of a chimeric chalcone synthase gene into petunia results in reversible co-suppression of homologous genes in trans. Plant Cell. 2: 279–89.

Nieth C, Priebsch A, Stege A, Lage H (2003). Modulation of the classical multidrug resistance (MDR) phenotype by RNA interference (RNAi). FEBS Lett. 545: 144–150.

Nykanen A, Haley B, Zamore PD (2001). ATP requirements and small interfering RNA structure in the RNA interference pathway. Cell 107: 309–321.

Paddison PJ, Caudy AA, Berstein E, Hannon GJ, Conklin DS (2002). Short hairpin RNAs (shRNAs) induce sequence-specific silencing in mammalian cells. Genes Dev. 16: 948-958.

Pasquinelli AE, Reinhart BJ, Slack F, Martindale MQ, Kuroda MI, Maller B, Hayward DC, Ball EE, Degnan B, Muller P, Spring J, Srinivasan A, Fishman M, Finnerty J, Corbo J, Levine M, Leahy P, Davidson E, Ruvkun G, (2000). Conservation of the sequence and temporal expression of let-7 heterochronic regulatory RNA. Nature 408: 86–89.

Paul CP, Good PD, Winer I, Engelke DR (2002). Effective expression of small interfering RNA in human cells. Nature Biotechnol. 20: 505–508.

Petrovics G, Zhang W, Makarem M, Street JP, Connelly RR, Sun L, Sesterhenn IA, Srikantan V, Moul JW, Srivastava S (2004). Elevated expression of PCGEM1, a prostate-specific gene with cell growth promoting function, is associated with high-risk prostate cancer patients. Oncogene, 23: 605-611.

Pfeffer S, Mihaela Z, Friedrich AG, Minchen C, James JR, Jingyue JBJ, Anton JE, Debora M, Chris S, Thomas T (2004). Identification of virus-encoded microRNAs. Science 304, 734–736.

Pham JW, Pellino JL, Lee YS, Carthew RW, Sontheimer EJ (2004). A Dicer-2-dependent 80s complex cleaves targeted mRNAs during RNAi in Drosophila. Cell 117: 83–94

Raymond MS, Aslam A, Jun X, Qin Z, Qingquan T, Gert S, Grietje M, Patrick YL, Puthupparampil VS, Martin CW (2004). Cancetr siRNA therapy by tumor selective delivery with lingand-targeted sterically stabilized nanoparticle. Nucleic Acids Res. 32: p. 19.

Reich SJ, Fosnot J, Kuroki A, Tang W, Yang X, Maguire AM, Bennett J, Tolentino MJ (2003). Small interfering RNA (siRNA) targeting VEGF effectively inhibits ocular neovascularization in a mouse model. Mol. Vis. 9: 210–216.

Reynolds A, Leake D, Boese Q, Scaringe S, Marshall WS, Khvorova A (2004).Rational siRNA design for RNA interference. Nature Biotechnol. 22: 326–330

Rhoades MW, Reinhart BJ, Lim LP, Burge CB, Bartel B, Bartel DP (2002). Prediction of plant microRNA targets. Cell 110: 513–520

Rubinson DA, Dillon CP, Kwiatkowski AV, Sievers C, Yang L, Kopinja J, Rooney DL, Ihrig MM, McManus MT, Gertler FB, Scott ML, Van PL (2003). A lentivirus-based system to functionally silence genes in primary mammalian cells, stem cells and transgenic mice by RNA interference. Nat. Genet. 33: 401-6.

Ruvkun G, Wightman B, Ha I (2004). The 20 years it took to recognize the importance of tiny RNAs. Cell. 116:S93-96.

Sanchez-Beato M, Sanchez-Aguilera A, Piris MA (2003). Cell cycle deregulation in B-cell lymphomas. Blood 101: 1220–1235.

Scherr M, Battmer K, Winkler T, Heidenreich O, Ganser A, Eder M (2003). Specific inhibition of *bcr-abl* gene expression by small interfering RNA. Blood, 101: 1566–1569.

Schwarz DS, Hutvágner G, Du T, Xu Z, Aronin N, Zamore PDm (2003). Asymmetry in the assembly of the RNAi enzyme complex. Cell 115: 199–208.

Sham SK, Mohammad TM (2006). Suppression of lung cancer with siRNA targeting PTTG: Int. j. oncol. 29: 387-395.

Shen C, Buck AK, Liu X, Winkler M, Reske SN (2003). Gene silencing by adenovirus-delivered siRNA. FEBS Lett. 539: 111-114.

Shinagawa T, Ishii S (2003). Generation of *Ski*-knockdown mice by expressing a long double-strand RNA from an RNA polymerase II promoter. Genes Dev. 17: 1340–1345.

Song E, Lee SK, Wang J, Ince N, Ouyang N, Min J, Chen J, Shankar P, Lieberman J (2003a). RNA interference targeting Fas protects mice from fulminant hepatitis. Nature Med. 9: 347–351.

Song E, Lee SK, Dykxhoorn DM, Novina C, Zhang D, Crawford K, Cerny J, Sharp PA, Lieberman J, Manjunath N, Shankar P (2003b). Sustained small interfering RNA-mediated human immunodeficiency virus type 1 inhibition in primary macrophages. J. Virol. 77: 7174–7181.

Song JJ, Smith SK, Hannon GJ, Joshua-TL (2004). Crystal structure of Argonaute and its implications for RISC slicer activity. Sci. 305: 1434 –1437

Sorensen DR, Leirdal M, Sioud M (2003). Gene silencing by systemic delivery of synthetic siRNAs in adult mice. J. Mol. Biol. 327: 761–766.
Soutschek J, Akinc A, Bramlage B, Akinc A, Bramlage B, Charisse K, Constien R, Donoghue M, Elbashir S, Geick A, Hadwiger P, Harborth J (2004). Therapeutic silencing of an endogenous gene by systemic administration of modified siRNAs. Nature, 432: 173-178.
Srikantan V, Zou Z, Petrovics G, Leirdal M, Sioud M (2000). PCGEM1, a prostate-specific gene, is overexpressed in prostate cancer. Proc. Natl. Acad. Sci. U.S.A, 97: 12216-12221.
Sui G, Soohoo C, Affar E, Gay F, Shi Y, Forrester WC, Shi Y (2002). A DNA vector-based RNAi technology to suppress gene expression in mammalian cells.Proc. Natl. Acad. Sci. USA. 99: 5515-5520.
Sui G, Soohoo C, Affar el B, Gay F, Shi Y, Forrester WC (2002). A DNA vector-based RNAi technology to suppress gene expression in mammalian cells. Proc. Natl. Acad. Sci. 99:5515–5520.
Sui, G. *et al.* (2002) A DNA vector-based RNAi technology to suppress gene expression in mammalian cells. *Proc. Natl Acad. Sci.* U.S.A. 99, 5515–5520.
Suman G, Kaiser J (2006). Novel CYP3A4 Gene Polymorphisms in Post Chemo Breast Cancer Patients: International J. Cancer Res. 2: 358-366.
Tabara H, Sarkissian M, Kelly WG, Fleenor J, Grishok A, Timmons L, Fire A, Mello CC (1999) The rde-1 gene, RNA interference, and transposons silencing in C. elegans. Cell 99: 123–132.
Tang G, Reinhart BJ, Bartel DP, Zamore PD (2003). A biochemical framework for RNA silencing in plants. Genes Dev. 17: 49–63
Tanno B, Cesi V, Vitali R, Sesti F, Giuffrida ML, Mancini C, Calabretta B, Raschella G (2005). Silencing of endogenous IGFBP-5 by micro RNA interference affects proliferation, apoptosis and differentiation of neuroblastoma cells. Cell Death Differ. 12: 213–223.
Tuschl T, Borkhardt A (2002). Small interfering RNAs—A revolutionary tool for analysis of gene function and gene therapy. Mol. Intervent. 2: 42–51.
Tuschl T, Zamore PD, Lehmann R, Bartel DP, Sharp PA (1999). Targeted mRNA degradation by double-stranded RNA in vitro. Genes & Dev. 13: 3191–3197.
Ui-Tei K, Naito Y, Takahashi F, Haraguchi T, Ohki-Hamazaki H, Juni A, Ueda R, Saigo K (2004). Guidelines for the selection of highly effective siRNA sequences for mammalian and chick RNA interference. Nucleic Acids Res. 32: 936–948.
Uwe S, Douglas T, Ross MW, Lawrence H, Smith JK, Lee LT, Kurt WK, William CR, Timothy GM, Darren TA1, Dominic AS, Michael BE, Edward AS, Yves P1, David B, Patrick OB, John NW (2000). A gene expression database for the molecular pharmacology of cancer, nature genetics, 24: 236-244.
Venter JC, Adams MD, Myers EW, Li PW, Mural RJ, Sutton GG, *et al.* (2001). The sequence of the human genome Sci. 291(5507):1304–1351.
Verma UN, Surabhi RM, Schmaltieg A, Becerra C, Gaynor RB (2003.) Small interfering RNAs directed against β-catenin inhibit the *in vitro* and *in vivo* growth of colon cancer cells. Clin. Cancer. Res. 9: 1291–1300.
Wall NR, Shi Y (2003). Small RNA: can RNA interference be exploited for therapy? Lancet 362: 1401-1403.
Wei W, Chun-You W, Ju-Hua D, Xiong C, Min Z, Gang Z (2005). Identification of effective siRNA against K-ras in human pancreatic cancer cell line MiaPaCa-2 by siRNA expression cassette: World J Gastroenterol. 11:2026-2031.
Wen-Hua Xu, Yin-Lin Ge, Quan Li, Xiao Z, Jian-Hua D (2007). Inhibitory effect of vascular endothelial growth factors-targeted small interfering RNA on proliferation of gastric cancer cells: World J. Gastroenterol. 13: 2044-2047.
Wilda M, Fuchs U, Wossmann W, Borkhardt A (2002). Killing of leukemic cells with a BCR/ABL fusion gene by RNA interference (RNAi). Oncogene 21: 5716–5724.
Xia H, Mao Q, Paulson HL, Davidson BL (2002). siRNA-mediated gene silencing *in vitro* and *in vivo*. Nature Biotechnol. 20: 1006–1010.
Yang G, Thompson JA, Fang B, Liu J (2003). Silencing of *H-ras* gene expression by retrovirus-mediated siRNA decreases transformation efficiency and tumor growth in a model of human ovarian cancer. Oncogene. 22: 5694–5701.
Yangchao C, George S, Chao-Zhan S (2003). Down regulation of CXCR4by inducible small interfering RNA breast cancer cell invasion in vitro, Cancer Res. 63: 4801 – 4804.
Yi-hua W, Shuang L, Guo Z, Cui-qi Z, Hang-xi Z, Xiao-bo Z, Lan-ping Q, Jin fleng B, Ning-Zhi X (2005). Knockdown of c-Myc expression by RNAi inhibits MCF-7 breast tumor cells growth in vitro and in vivo, breast cancer Res. 7: R220-R228.
Yoshinouchi M, Yamada T, Kizaki M, Fen J, Koseki T, Ikeda Y, Nishihara T, Yamato K (2003). *In vitro* and *in vivo* growth suppression of human papillomavirus 16-positive cervical cancer cells by E6 siRNA. Mol. Ther. 8: 762–768.
Yu JY, Taylor J, DeRuiter SL, Vojtek AB, Turner DL (2003). Simultaneous inhibition of GSK3α and GSK3β using hairpin siRNA expression vectors. Mol. Ther. 7: 228–236.
Zender L, Hutker S, Liedtke C, Tillmann HL, Zender S, Mundt B, Waltemathe M, Gosling T, Flemming P, Malek NP, Trautwein C, Manns MP, Kuhnel F, Kubicka S (2003). Caspase 8 small interfering RNA prevents acute liver failure in mice. Proc. Natl Acad. Sci. U.S.A.. 100: 7797–7802.
Zeng Y, Rui Yi, Bryan RC, Yan Z, Rui Yi, Bryan RC (2003). MicroRNAs and small interfering RNAs can inhibit mRNA expression by similar mechanisms. Proc. Natl. Acad. Sci. U.S.A. 100: 9779–9784.
Zhang BH, Pan XP, Cannon CH, Cobb GP, Anderson TA (2006a). Conservation and divergence of plant microRNA genes. Plant J. 46: 243–259.
Zhang BH, Pan XP, Cobb GP, Anderson TA (2006b). Plant microRNA: a small regulatory molecule with big impact. Dev. Biol. 289: 3–16.
Zhang BH, Pan XP, Wang QL, Cobb GP, Anderson TA (2005). Identification and characterization of new plant microRNAs using EST analysis. Cell Res. 15: 336–360.

The large-scale use of feruloyl esterases in industry

Ahmed E Fazary and Yi-Hsu Ju*

Department of Chemical Engineering, National Taiwan University of Science and Technology, Taipei 106-07, Taiwan, ROC.

Presently industrial enzyme companies sell enzymes for a wide variety of applications. The estimated value of world enzyme market is presently about US $ 4.3 billion and it has been forecasted to grow to almost US $ 5.1 billion by 2009. Detergents (37%), textiles (12%), starch (11%), baking (8%) and animal feed (6%) are the main industries; totally these industries use about 74% of industrially produced enzymes. Enzymes are also indirectly used in biocatalytic processes involving living or dead and permeabilized microorganisms. This review concentrates on the use of isolated feruloyl esterases enzyme preparations in large scale and speciality applications and chemical manufacturing.

Key words: Feruloyl esterases, industry, biotechnology.

Table of contents

1. Introduction
2. What is Feruloyl Esterases?
3. Feruloyl Esterases; Research and Industry
4. Feruloyl Esterases applications in Pulp and Paper Industry
5. Feruloyl Esterases in Food and Pharmaceutical Industry
6. Feruloyl Esterases in Bio-fuel Industry
7. Conclusion
8. References

INTRODUCTION

There is an increased interest in feruloyl esterases (FAEs) production which has considerable roles in biotechnological processes for various industrial and medicinal applications. As reported by us in 2007 (Fazary et al., 2007, 2008), the quantum of research carried out in FAEs has incre-ased dramatically since 1990. For example, during 1990 – 2000 and 2001–mid 2008, the average number of re-fereed papers involving FAEs as a major component was 24 and 60, respectively. It is interesting to note that the dramatic increase in publications concerning to FAEs between 2001 (four papers) and mid 2008 (20 papers) coincides with the most recent discoveries on isolation, purification, and characterization of FAEs including fungal and bacterial FAEs (Figure 1).

Results of search for feruloyl esterases in US Patent Collection data base, US Patent applications, European Parliament documents (EP) and in World Intellectual Property Organization-Patent Cooperation Treaty (WIPO-PCT) reveal 29, 50, 11, and 75 patents, respectively (Figure 2) (Table 1).

There was also remarkable increase in the number of publications concerned with ferulic acid (the related compound to FAEs). The average number of documents was 242 paper for the period (1990 – 2000), while for the period 2001–2008, the average number was 575 paper

*Corresponding author. E-mail: yhju@mail.ntust.edu.tw.

Table 1. Summary of Patents on feruloyl esterases

No.	Patent No.	Patent Title
1	US20080145903	Process of Producing Fermentation Products
2	US20080138872	Processes for Producing Fermentation Products
3	US20080138862	Enzymatic Hydrolysis of Biomasses Having a High Dry Matter (Dm) Content
4	US20080138856	Enzymatic Enantioselective Ester or Amide Hydrolysis or Synthesis
5	WO/2008/064025A2	Cloning and sequencing of the ferulate esterases gene from lactobacillus buchneriln4017
6	WO/2008/061982A1	Novel Method to Reduce Compounds Involved in Maillard Reactions in Thermally Processed Plant-Based Food Products
7	US20080118491	Xylanase gene sequences from the genomic DNA of unpurified rumen microorganisms
8	7374925	*Penicillium funiculosum* mutant strain
9	US20080115241	Cloning and Sequencing of the Ferulate Esterase Gene from Lactobacillus Buchneri LN4017
10	7371423	Method for preparing flour doughs and products made from such doughs using lipase
11	US20080096253	Methods of making nanotechnological and macromolecular biomimetic structures
12	WO/2008/039353A2	Novel pectate lyase and method of. use for bio-scouring
13	US20080076821	Intranasally administering curcumin prodrugs to the brain to treat alzheimer's disease
14	WO/2008/028300A1	Oral polymeric membrane feruloyl esterase producing bacteria formulation_
15	US20080050321	Formulations with Feruloyl Glycerides and Methods of Preparation
16	US20080026101	Food products
17	US20080025921	Conductance of Improperly Folded Proteins Through the Secretory Pathway And Related Methods For Treating Disease
18	WO/2008/008070A2	Construction of highly efficient cellulase compositions for enzymatic hydrolysis of cellulose
19	WO/2008/003090A2	Formulations with Feruloyl Glycerides and Methods of Preparation

Table 1 Continued

21	US20070254336	Transcription Factors
22	WO/2007/115723A2	Fusion fungal proteins between plant-wall degrading enzymes and a swollenin and their uses
23	US20070238155	Construction of highly efficient cellulase compositions for enzymatic hydrolysis of cellulose
24	US20070232495	Compositions and methods to add value to plant products, increasing the commercial quality, resistance to external factors and polyphenol content thereof
25	US20070224668	Process for producing 4-vinylguaiacol by biodecaroxylation of ferulic acid
26	US20070202566	Hydrolases, Nucleic Acids Encoding Them And Methods For Making And Using Them
27	WO/2007/092314A2	Esterases and Related Nucleic acids and methods
28	US20070184175	Continuous production of cereal flour and whole-cereal flour for grain-based foods, using a low-moisture precooking
29	US20070179076	Detergent composition
30	WO/2007/076436A2	Continuous production of cereal flour and whole-cereal flour for grain-based foods, using a low-moisture precooking
31	US20070154591	Chewing gum comprising biodegradable polymers and having accelerated degradability
32	EP1309674B1	Stereo-selective esterase from Aspergillus oryzae
33	7226770	Lipolytic enzyme variant
34	WO/2007/055735A2	Hydrolyses, nucleic acids encoding them and methods for making and using them
35	US20070077636	Methods of making compositions comprising a uv-absorbing chromophore
36	US20070072185	Carbohydrate-binding modules of a new family
37	WO/2007/019949A1	Fusion proteins between plant cell-wall degrading enzymes, and their uses
38	US20070037259	Integration of alternative feedstreams for biomass treatment and utilization
39	EP1752533A1	Fusion proteins between plant cell-wall degrading enzymes, and their uses
40	US20070031953	Treatment of biomass to obtain ethanol
41	US20070031919	Treatment of biomass to obtain a target chemical
42	US20070031918	Treatment of biomass to obtain

Table 1 Continued

		fermentable sugars
43	US20070029252	System and process for biomass treatment
44	7172997	Lipolytic enzyme variant
45	WO/2006/136161A2	Amylases for pharmaceutical use
46	WO/2006/136160A2	Proteases for pharmaceutical use
47	WO/2006/136159A2	Lipases for pharmaceutical use
48	EP1726213A1	Soluble coffee product
49	7132589	Manipulation of the phenolic acid content and digestibility of plant cell walls by targeted expression of genes encoding cell wall degrading enzymes
50	US20060229223	Lipolytic enzyme variant
51	7091023	Stereoselective esterase from Aspergillus oryzae
52	US20060035800	Detergent composition
53	WO/2006/003009A2	NEW ESTERASES FROM RUMEN
54	US20060005270	Manipulation of the phenolic acid content and digestibility of plant cell walls by targeted expression of genes encoding cell wall degrading enzymes
55	WO/2005/115445A1	ENZYMES FOR PHARMACEUTICAL USE
56	US20050227344	Mixture obtained from penicillium funiculosu
57	US20050181446	Protein variants having modified immunogenicity
58	EP1555322A1	Lipolytic enzyme variant
59	US20050118665	Methods for conducting assays for enzyme activity on protein microarrays
60	WO/2005/042735A1	A carbohydrate-binding module of a new family
61	WO/2005/040107A2	Methods for Making Simvastatin and Intermediate
62	US20040266883	Conductance of improperly folded proteins through the secretory pathway and related methods for treating disease
63	6828136	Esterase enzymes, DNA encoding esterase enzymes and vectors and host cells incorporating same
64	US20040231060	Methods to enhance the activity of lignocellulose-degrading enzymes
65	WO/2004/081185A2	Methods to enhance the activity of lignocellulose-degrading enzymes
67	US20040152180	Lipolytic enzyme variant
68	WO/2004/053039A2	Detergent composition comprisingEndoglucanase
69	6750051	Compositions and methods for enhancing fiber digestion
70	EP0904396B1	Production of vanillin

Table 1 Continued

71	WO/2004/009804A2	Feruloyl esterases and uses thereof
72	US20040005674	Methods for enzymatic hydrolysis of lignocellulose
73	US20040002136	Transformation system in the field of filamentous fungal hosts
74	US20030236300	Conductance of improperly folded proteins through the secretory pathway and related methods for treating disease
75	6664088	Production of vanillin
76	WO/2003/093420A2	Methods enzymatic hydrolysis of lignocellulose
77	US20030199048	Stereoselective estarase from aspergillus oryzae
78	US20030175384	Method for the extraction of aleurone from bran
79	US20030167511	Production of p-hydroxybenzoic acid
80	US20030149113	Conductance of improperly folded proteins through the secretory pathway and related methods for treating disease
81	6602700	Phenolic acid esterases, coding sequences and methods
82	US20030144165	Lipolytic enzyme variant
83	6589760	Methods of separating a corn fiber lipid fraction from corn fiber
84	6586212	Corn fiber for the production of advanced chemicals and materials: derivatizable cellulose and cellulose derivatives made therefrom
85	WO/2003/049717A2	Conductance of improperly folded proteins through the secretory pathway and related methods for treating disease
86	US20030108642	Penicillium funiculosum strain useful for the production of enzymes
87	WO/2003/046163A2	Multifunctional Caffeic Acid O-methyltransferase
88	6573086	Transformation system in the field of filamentous fungal hosts
89	WO/2003/043411A2	Manipulation of the phenolic acid content and digestibility of plant cell walls
90	6534286	Protein production in Aureobasidium pullulans
91	6534101	Enzymes mixture obtained from Penicillium funiculosum
92	US20030035822	Compositions and methods for enhancing fiber digestion
93	US20030032161	Esterase enzymes, DNA encoding esterase enzymes and vectors and host cells incorporating same

Table 1 Continued

94	US20030024009	Manipulation of the phenolic acid content and digestibility of plant cell walls by targeted expression of genes encoding cell wall degrading enzymes
95	WO/2003/000898A1	Plant genes involved in defence against pathogens
96	WO/2002/068666A1	Manipulation of the phenolic acid content and digestibility of plant cell walls by targeted expression of genes encoding cell wall degrading enzymes
97	WO/2002/055679A2	thermostable lipolytic enzyme variant
98	6388069	Corn fiber for the production of advanced chemicals and materials:arabinoxylan and arabinoxylan derivatives made therefrom
99	6368833	Esterases, DNA encoding therefor and vectors and host incorporating same
100	6365390	Phenolic acid esterases, coding sequences and methods
101	6352845	Corn fiber for the production of advanced chemicals and materials: separation of monosaccharides and methods thereof
102	WO/2002/012472A1	Stereoselective esterase from Aspergillus oryzae
103	WO/2002/002825A1	Method for the production of proteins and polypeptides using fungal cells, in particular Aspergillus vadensis.
104	WO/2001/096578A2	Protein production in Aureobasidium pullulans
105	6323011	Production of vanillin
106	WO/2001/083770A2	Lipolytic enzyme variant
107	WO/2001/083559A2	Production and Use of Protein Variants having modified immunogenicity
108	US20010014467	Production of vanillin
109	WO/2001/042433A2	Talaromyces emersonii Xylanses
110	6245363	Methods of treating plant materials with hydrolytic enzymes isolated from humicola species
111	6232101	Oxidase-promoted gelling of phenolic polymers
112	6137032	Xylanase obtained from an anaerobic fungus
113	WO/2000/020555A2	Transformation system in the field of filamentous fungal hosts
114	WO/2000/017368A1	Orange fruit pectinacetylesterase
115	WO/2000/014243A1	Phenolic acid esterases, coding sequences and methods
116	EP0976838A1	Enzymes mixture

Table 1 Continued

117	WO/2000/004053A1	Purification of Hemicelluloses materials
118	WO/1999/057325A2	Enzyme Miurext
119	5948667	Xylanase obtained from an anaerobic fungus
121	WO/1999/011672A1	Fractionation of Hemicelluloses materials
122	5869720	Transgenic cotton plants producing heterologous peroxidase
123	5824533	Orpinomyces xylanase proteins and coding sequences
124	WO/1998/014594A2	Esterases, DNA encoding therefor and vectors and host incorporating same
125	WO/1997/035999A2	Production of vanillin
126	5608148	Transgenic cotton plants producing heterologous peroxidase
127	5591619	Aureobasidium pullulans xylanase, gene and signal sequence
128	WO/1996/040182A1	Identification of a potent antioxidant from Aloe barbadensis
129	WO/1996/036701A1	Orpinomyces xylanase proteins and coding sequences
130	WO/1996/003440A1	Oxidize-Promoted Gelling of Phenolic Polymers
131	WO/2004/009804A3	Feruloyl Esterases and uses thereof

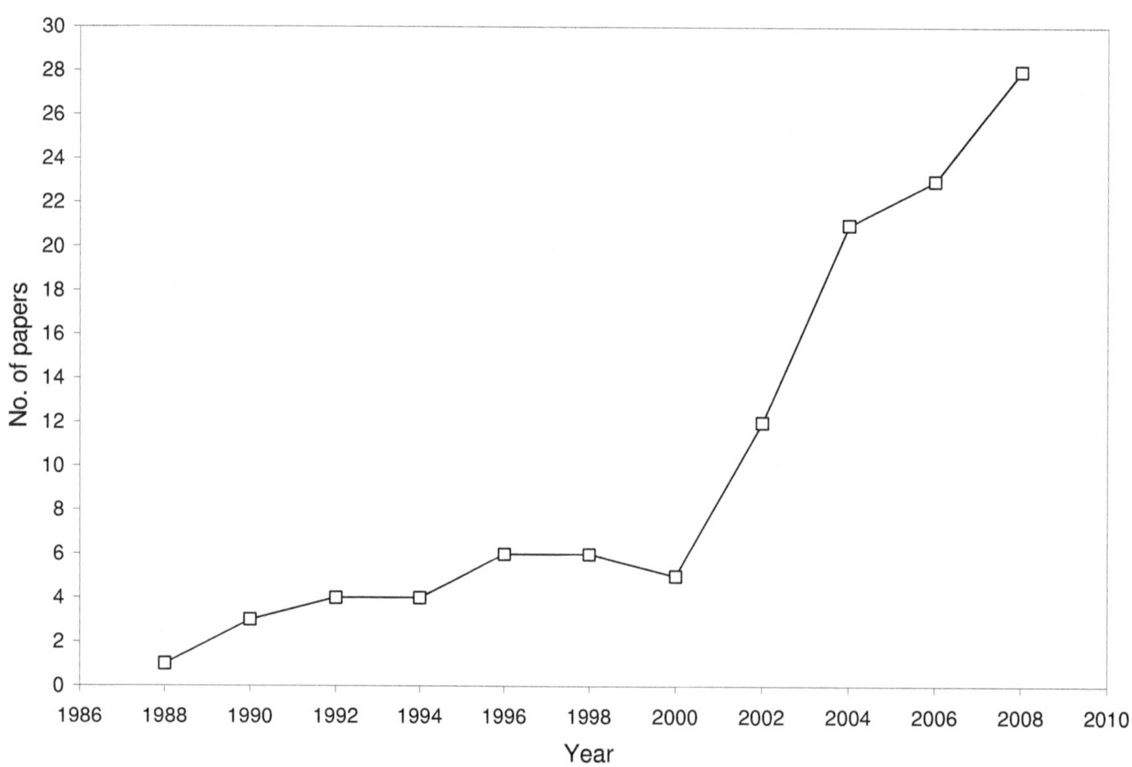

Figure 1. Annual publications on feruloyl esterases

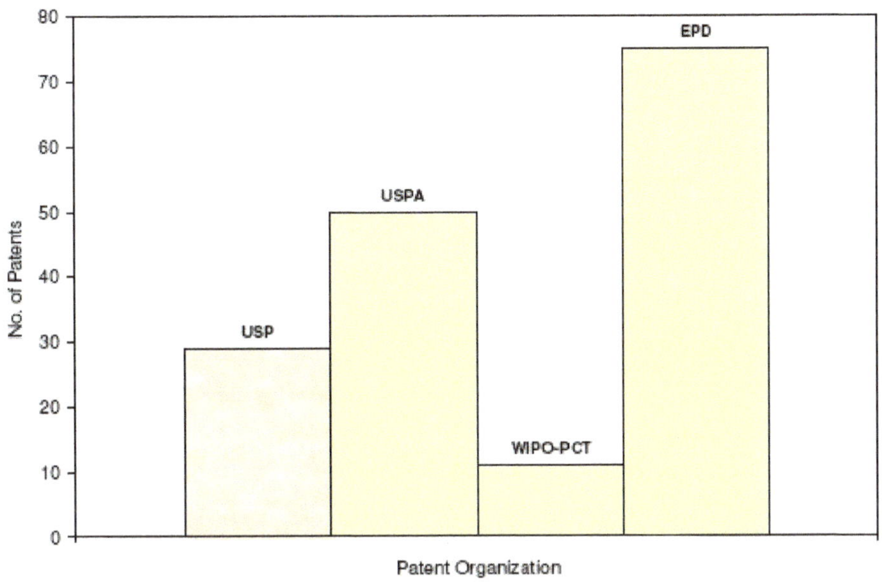

Figure 2. Patents recorded for feruloyl esterases. USP; United States Patent collection data base. USPA; United States Patent Applications data base. EPD; European Parliament documents. WIPO-PCT; World Intellectual Property Organization-Patent Cooperation Treaty.

using the same search engine (Elsevier SDOS) (Figure 3).

In this review, we will try to report concise and complete highlights of the large-scale usage of such kind of enzyme (feruloyl esterases) in both academic research and industry.

WHAT IS FERULOYL ESTERASES?

Recently, it was reported that ferulic acid had been shown to link hemicellulose and lignin. Cross-linking of ferulic acids within cell wall components influences wall properties such as extensibility, plasticity, and digestibility, and limits the access of polysaccharidaes to their substrates (Borneman et al., 1990).

The complete degradation of plant cell wall polymers requires an array of enzymes with different activities. The hydrolysis of hydroxycinnamate esters in plant cell walls is catalyzed by cinnamoyl esterases (CEs; such as cinnamoyl ester hydrolases, feruloyl/p-coumaroyl esterases, and ferulic/p-coumaric acid esterases). CEs are a subclass of the carboxylesterases (EC 3.1.1.1) and characterized by a relatively high activity on various hydroxycinnamate esters (Fazary et al., 2007).

Two enzymes, feruloyl and cinnamoyl esterases (FAE-B), purified from *Aspergillus niger* strains (Fazary et al., 2008), have different physicochemical characteristics and catalytic properties against cinnamoyl model substrates, such as methyl derivatives of hydroxycinnamic esters and soluble feruloylated oligosaccharides derived from plant cell wall (Ralet et al., 1994). These enzymes are exocellular and the corresponding expression is inducible. A new strain of *A. niger* I-1472 was shown to produce numerous polysac-charide-degrading enzymes as well as esterases that released ferulic acid from natural feruloylated oligosac-charides, when grown on sugar beet pulp and maize bran (Hutnan et al., 2000).

FAEs (EC 3.1.1.73), including cinnamoyl esterases and cinnamic acid hydrolases, are a subclass of the carboxylic acid esterases (EC 3.1.1) that play a key physiological role in the degradation of the intricate structure of plant cell wall by hydrolyzing ferulate ester groups involved in the cross-linking between hemicelluloses and between hemicellulose and lignin (Lee et al., 2005). The use of multiple alignments of sequences or domains that show FAE activity, as well as related sequences, helped to construct a neighborhood-joining phylogenic tree (Table 2). The outcome of this genetic comparison supported substrate specificity data and allowed FAEs to be sub-classified into four types, A, B, C, and D (Crepin, et al. 2003a, b, c, 2004; Donaghy et al., 1998, 1995; Faulds et al., 2002, 1997, 1993), based on their substrate specificities towards synthetic methyl esters of hydroxycinnamic acids (ferulic acid, diferulic acids, p-coumaric acid, sinapinic acid, and caffeic acid) for substitutions on the phenolic ring, and on their amino acid sequence identity (protein sequence), indicating an evolutionary relationship among FAEs, acetyl xylan esterases, and certain lipases.

By looking in the Comprehensive Enzyme Information System (http://www.brenda-enzymes.info/), we found that there are only 22 protein database entries in that type of enzyme. Considering the remarkable impact of FAEs in

Table 2. Gene Product Associations to feruloyl esterase activity

Symbol, Full Name	Primary Peptide Sequence	Species	GO Evidence*	Database	Reference
FAEA_ASPAW faeA: Feruloyl esterase A precursor	MKQFSAKHALAVVVTAGHALAASTQGISEDL YTRLVEMATISQAAYADLCNIPSTIIKGE KIYNSQTDINGWILRDDSSKEIITVFRGTGGD TNLQLDTNYTLTPFDTLPQCNGCEVHGG YYIGWVSVQDQVESLVKQQVSQYPDYALTV TGHSLGASLAALTAAQLSATYDNIRLYTFG EPRSGNQAFASYMNDAFQASSPDTTQYFR VTHANDGIPNLPPVEQGYAHGGVEYWSVDP Y SAQNTFVCTGDEVQCCEAQGGQGVNNAHT TYFGMTSGACTW	*Aspergillus awamori*	ISS	UniProtKB/Swiss-Prot Q9P979	de Vries st et al., 1997
FAEA_ASPNG faeA: Feruloyl esterase A precursor	MKQFSAKYALILLATAGQALAASTQGISEDLY NRLVEMATISQAAYADLCNIPSTIIKGE KIYNAQTDINGWILRDDTSKEIITVFRGTGSD TNLQLDTNYTLTPFDTLPQCNDCEVHGG YYIGWISVQDQVESLVKQQASQYPDYALTVT GHSLGASMAALTAAQLSATYDNVRLYTFG EPRSGNQAFASYMNDAFQVSSPETTQYFRV THSNDGIPNLPPADEGYAHGGVEYWSVDPY SAQNTFVCTGDEVQCCEAQGGQGVNDAHT TYFGMTSGACTW	*Aspergillus niger*	IDA	UniProtKB/Swiss-Prot O42807	de Vries st et al., 1997
FAEA_ASPTU faeA: Feruloyl esterase A precursor	MKQFSAKYAIAVVVTAGHALAASTQGISEDL YSRLVEMATISQAAYADLCNIPSTIIKGE KIYNSQTDINGWILRDDSSKEIITVFRGTGSD TNLQLDTNYTLTPFDTLPQCNSCEVHGG YYIGWISVQDQVESLVQQQVSQFPDYALTVT GHSLGASLAALTAAQLSATYDNIRLYTFG EPRSNQAFASYMNDAFQASSPDTTQYFRVT HANDGIPNLPPADEGYAHGVVEYWSVDPYS AQNTFVCTGDEVQCCEAQGGQGVNNAHTT YFGMTSGHCTW	*Aspergillus tubingensis*	ISS	UniProtKB/Swiss-Prot O42815	de Vries st et al., 1997
FAEA_ASPNG faeB: Feruloyl esterase B precursor	MKVASLLSLALPGAALAATDPFQSRCNEFQ NKIDIANVTVRSVAYVAAGQNISQAEVASV CKASVQASVDLCRVTMNISTSDRSHLWAEA WLPRNYTGRFVSTGNGGLAGCVQETDLNF A ANFGFATVGTNGGHDGDTAKYFLNNSEVLA DFAYRSVHEGTVVGKQLTQLFYDEGYNYSY YLGCSTGGRQGYQQVQRFPDDYDGVIAGS AAMNFINLISWGAFLWKATGLADDPDFISAN LWSVIHQEIVRQCDLVDGALDGIIEDPDFCAP VIERLICDGTTNGTSCITGAQAAKVNRA LSDFYGPDGTVYYPRLNYGGEADSASLYFT GSMYSRTEEWYKYVVYNDTNWNSSQWTLE S AKLALEQNPFNIQAFDPNITAFRDRGGKLLS YHGTQDPIISSTDSKLYYRRVANALNAAP SELDEFYRFFQISGMGHCGDGTGASYIGQG YGTYTSKAPQVNLLRTMVDWVENGKAPEY M PGNKLNANGSIEYMRKHCRYPKHNIHTGPG NYTDPNSWTCV	*Aspergillus niger*	IDA	UniProtKB/Swiss-Prot Q8WZI8	de Vries st et al., 2002

various industrial and medical applications, it is anticipated that these crystal structural studies (22 protein database) will provide an initial framework for the rational design of novel enzymes with improved biotechnological potential (Table 3) (Faulds et al., 1997).

Major extent of work on the microbial production of

FAEB_NEUCR fae-1, B22K18.040, NCU09491: Feruloyl esterase B precursor	MLPRTLLGLALTAATGLCASLQQVTNWGSN PTNIRMYTYVPDKLATKPAIIVALHGCGGT APSWYSGTRLPSYADQYGFILIYPGTPNMSN CWGVNDPASLTHGAGGDSLGIVAMVNYTI AKYNADASRVYVMGTSSGGMMTNVMAATY PEVFEAGAAYSGVAHACFAGAASATPFSPN Q TCARGLQHTPEEWGNFVRNSYPGYTGRRP RMQIYHGLADNLVYPRCAMEALKQWSNVLG V EFSRNVSGVPSQAYTQIVYGDGSKLVGYMG AGVGHVAPTNEQVMLKFFGLIN	*Neurospora crassa*	ISS	UniProtKB/Swiss-Prot Q9HGR3	Kroon PA et al., 2000
FAEB_PENFN faeB: Feruloyl esterase B precursor	MAIPLVLVLAWLLPVVLAASLTQVNNFGDNP GSLQMYIYVPNKLASKPAIIVAMHPCGGS ATEYYGMYDYHSPADQYGYILIYPSATRDYN CFDAYSSASLTHNGGSDSLSIVNMVKYVI STYGADSSKVYMTGSSSGAIMTNVLAGAYP DVFAAGSAFSGMPYACLYGAGAADPIMSNQ TCSQGQIQHTGQQWAAYVHNGYPGYTGQY PRLQMWHGTADNVISYADLGQEISQWTTIM G LSFTGNQTNTPLSGYTKMVYGDGSKFQAYS AAGVGHFVPTDVSVVLDWFGITSGTTTTTT PTTTPTTSTSPSSTGGCTAAHWAQCGGIGY SGCTACASPYTCQKANDYYSQCL	*Penicillium funiculosum*	IDA	UniProtKB/Swiss-Prot Q9HE18	Kroon PA et al., 2000
FAEB_PIREQ faeB: Feruloyl esterase B precursor	MKTSIVLSIVALFLTSKASADCWSERLGWPC CSDSNAEVIYVDDDGDWGVENNDWCGIQK EEENNNSWDMGDWNQGGNQGGGMPWG DFGGNQGGGMQWGDFGGNQGGGMPWG DFGGNQGG GMPWGDFGGNQGGNQGGGMPWGDFGG NQGGNQGGGMPWGDFGGNQGGGMQWG DFGGNQGG NQGGGMPWGDFGGNQGGGMQWGDFGG NQGGNQGGGMPWGDFGGNQGGGMQWG DFGGNQGG GMQWGDFGGNQGGNQDWGNQGGNSGPT VEYSTDVDCSGKTLKSNTNLNINGRKVIVKF PS GFTGDKAAPLLINYHPIMGSASQWESGSQT AKAALNDGAIVAFMDGAQGPMGQAWNVGP C CTDADDVQFTRNFIKEITSKACVDPKRIYAAG FSMGGGMSNYAGCQLADVIAAAAPSAFD LAKEIVDGGKCKPARPFPILNFRGTQDNVVM YNGGLSQVVQGKPITFMGAKNNFKEWAKM NGCTGEPKQNTPGNNCEMYENCKGGVKV GLCTINGGGHAEGDGKMGWDFVKQFSLP	*Piromyces equi*	IDA	UniProtKB/Swiss-Prot Q9Y871	Fillingham IJ et al., 1999

*Gene Ontology Evidence Codes: ISS; Inferred from Sequence or Structural Similarity. IDA; Inferred from Direct Assay

FAEs to date involves the isolation, (Topakas et al., 2006) purification, and characterization of FAEs derived from a wide range of microorganisms (fungi and bacteria), as well as the enzymetic release of the products from cell wall degradation (Figure 4). Generally, FAE enzymes are produced (purified and characterized) from several microorganisms (various bacteria and fungi).

FERULOYL ESTERASES, RESEARCH AND INDUSTRY

Generally, feruloyl esterases benefit microorganisms, industry, and biochemists (Figure 5). There has recently been considerable interest in a large number of potential applications of feruloyl esterases due to their roles in

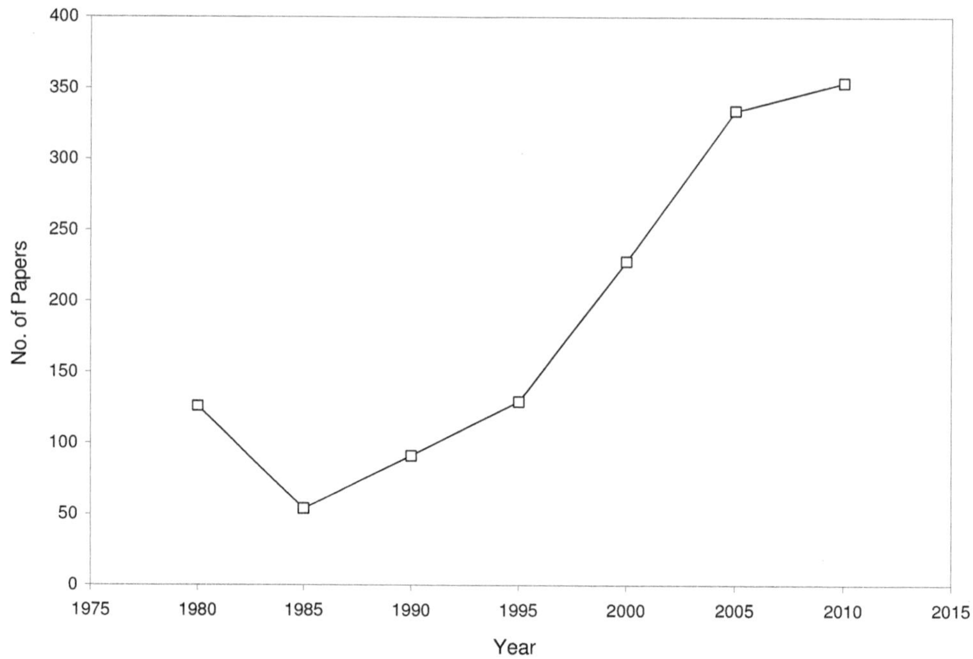

Figure 3. Annual publications on ferulic acid.

many biotechnological processes, in various industries (chemicals, fuel, animal feed, textile and laundry, pulp and paper, food and agriculture, and pharmaceutical), also in their potential applications in obtaining ferulic acid from agro-industrial waste materials such as those produced by milling, brewing, and sugar industries. The prospect of broad applications of FAEs has fueled much interest in these enzymes, as shown by the increasing number of FAEs discovered in microbial organisms in recent years.

FERULOYL ESTERASES APPLICATIONS IN PULP AND PAPER INDUSTRY

A great extent of work has been done during the last few decades to investigate many enzymes in pulp and paper industry. In spite of extensive research no viable techno-logy has so far been developed to utilize the oxidative enzymes for the pulp and paper industry. The major application is the use of xylanases in pulp bleac-hing. Xylanases liberate lignin fragments by hydrolysing residual xylan. This reduces considerably the need for chlorine based chemicals for bleaching process. Other minor enzyme applications in pulp production include the use of enzymes to remove fine particles from pulp. This facilitates water removal. FAEs could be used in pulp and paper processes (Record et al. 2003; Sigoillot et al., 2005). Hemicellulases and cellulases offer alternatives to augment chemical and mechanical paper-pulping methods, and there is a large extent of published work on this subject (Kroon et al. 1996; Williamson et al., 1998). Acetylxylan esterases and FAEs might enhance this process by removing substitutions and linkages between polymers during pulping, thus making the solubilization of lignin–carbohydrate complexes easier (Mathew et al., 2004; Topakas et al., 2004a, b; Tarbouriech et al., 2001; Faulds et al., 1993; Ralet et al., 1994; Kroon et al., 1999,1996a, b; Fillingham et al., 1999; Williamson et al., 1998). Pretreatment of lignocellulosic material by secreted fungal enzymes leads to de-esterification, which increa-sed the rate of *in vitro* digestion by ruminal micro-organisms by approximately 80% (Mathew et al., 2004; Topakas et al., 2004a, b; Tarbouriech et al., 2005; Faulds et al., 1993; Ralet et al., 1994; Kroon et al., 1999, 1996a, b; Fillingham et al., 1999; Williamson et al., 1998). In paper making, enzymes are used especially for modifica-tion of starch, which is used as an important additive. Starch improves the strength, stiffness and erasability of paper. The starch suspension must have a certain visco-sity, which is achieved by adding amylase enzymes in a controlled process. Enzymes used in paper industry should not contain cellulose-degrading activity as a side activity because this activity would damage the cellulose fibres (Nethaji et al., 1988; Mathew et al., 2004) to facilitate nutrient assimilation. Pitch is a sticky substance present mainly in softwoods. It is composed of lipids. It is a special problem when mechanical pulps of red pine are

Table 3. Some PDB entries of feruloyl esterases.

PDB Code	Title	Protein Source	Protein Structure	3D View	Reference
1usw	Crystal structure of ferulic acid esterase from aspergillus niger	Aspergillus niger. **Strain**: cbs 120.49/n400. **Expressed in**: pichia pastoris	Feruloyl esterase A. **Chain**: A. **Synonym**: ferulic acid esterase, FAE-III.	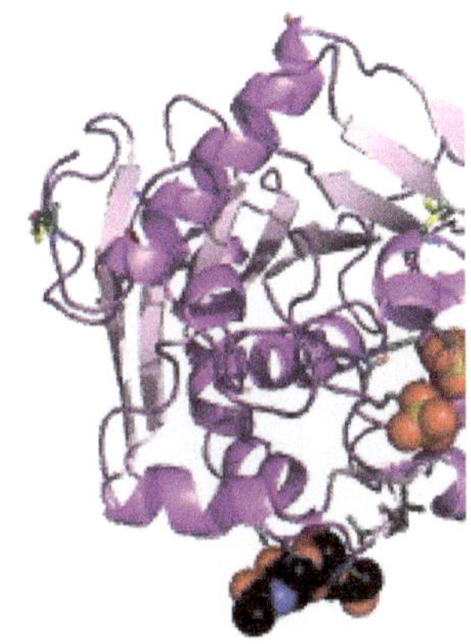	J.A.Hermoso et al. (2004)
1uwc	Feruloyl esterase a. Chain: a, b. Synonym: ferulic acid esterase a, fae-iii, cinnamoyl esterase. Engineered: yes. Other_details: n-acetylglucosamine at asn 79	Aspergillus niger. **Expressed in**: aspergillus oryzae	Feruloyl esterase a. **Chain**: a, b. **Synonym**: ferulic acid esterase a, FAE-III, cinnamoyl esterase.	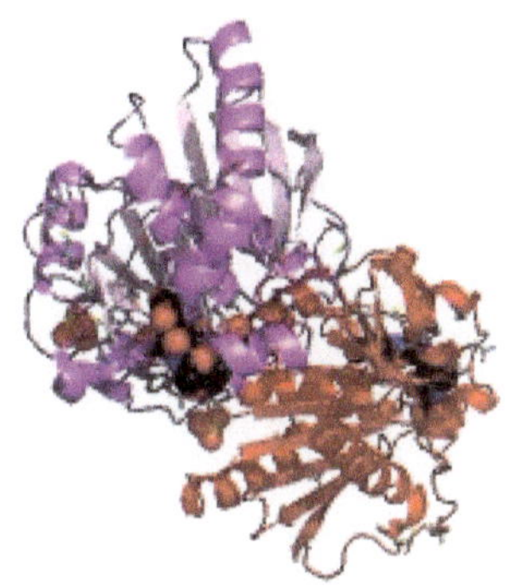	K.E.McAuley et al. (2004)
2bjh	Crystal structure of s133a anfaea-ferulic acid complex	Aspergillus niger. Strain: cbs 120.49-n400. **Expressed in**: pichia pastoris	Feruloyl esterase a. **Chain**: a, b, c. **Synonym**: ferulic acid esterase a, FAE-III, cinnamoyl esterase.	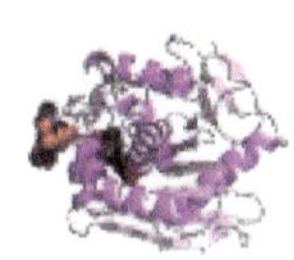 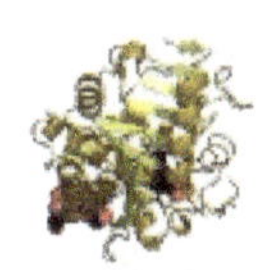 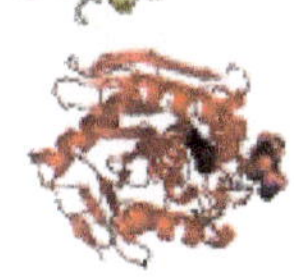	C.B.Faulds et al. (2005)
2hl6	Structure of homologously expressed ferrulate esterase of aspergillus niger in complex with caps	Aspergillus niger. Fungi. Gene: FAE-A. **Expressed in**: aspergillus niger.	Feruloyl esterase A. **Chain**: a, b. **Synonym**: ferulic acid esterase a, FAE-III, cinnamoyl esterase, ferrulate esterase A.	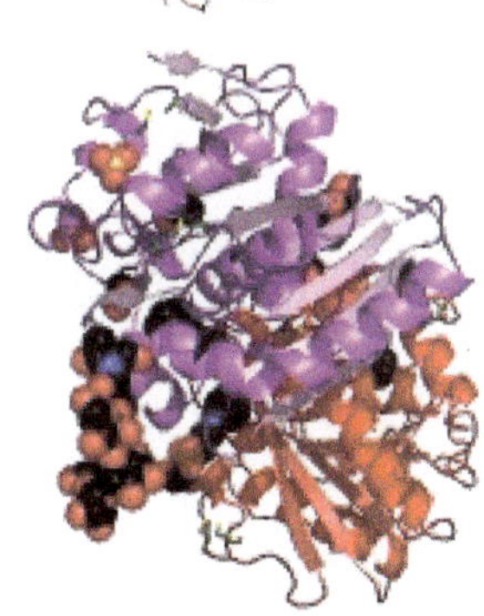	I.Benoit et al. (2006)

Table 3. Continued.

1uza	Crystallographic structure of a feruloyl esterase from aspergillus niger	Aspergillus niger. **Expressed in**: aspergillus oryzae	Feruloyl esterase a. **Chain**: a, b. **Synonym**: ferulic acid esterase A, FAE-III, cinnamoyl esterase.	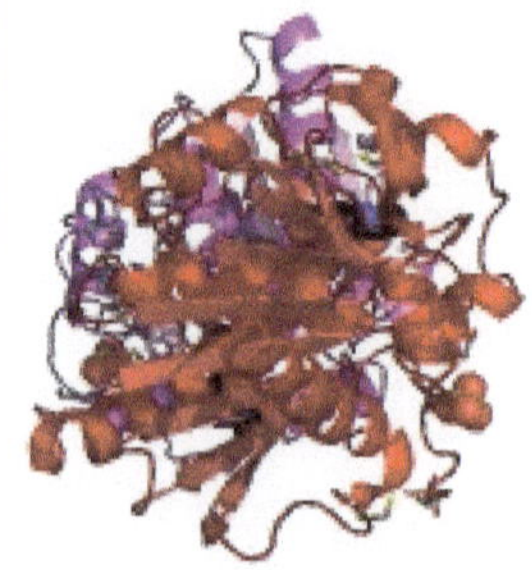	K.E.McAuley et al. (2004).
2ix9	Respective role of protein folding and glycosylation in the thermal stability of recombinant feruloyl esterase a	Aspergillus niger. Strain: d15pyrg. **Expressed in**: escherichia coli.	Feruloyl esterase a. **Chain**: a, b. Fragment: catalytic domain, residues 22-281. **Synonym**: feruloyl esterase types A, FAE-III, cinnamoyl esterase.	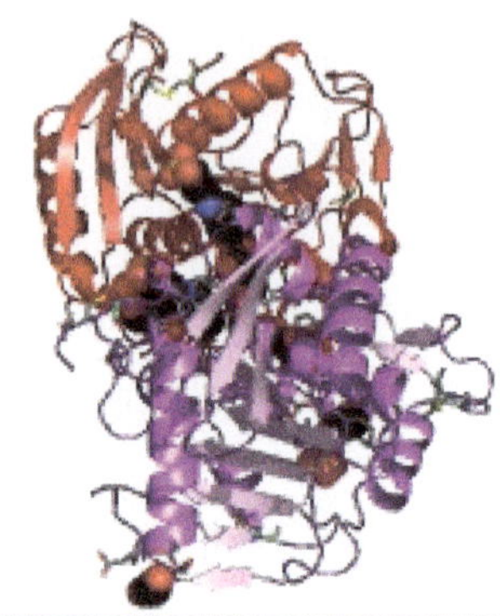	I.Benoit et al. (2006).

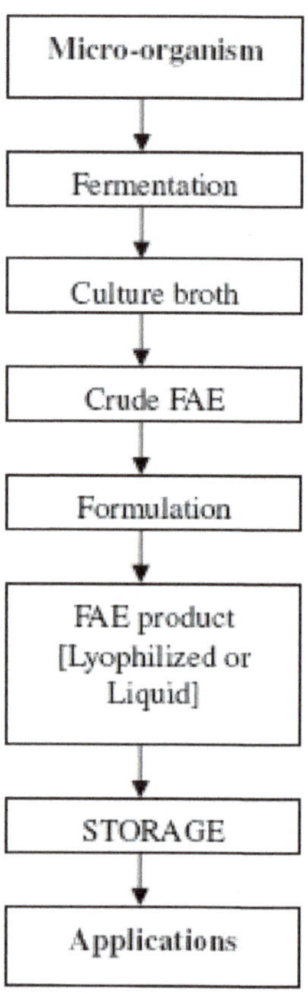

Figure 4. Flow diagram for the life cycle of feruloyl esterases from production to industry

used as a raw material. Pitch causes problems in paper machines and can be removed by lipases.

FERULOYL ESTERASES IN FOOD AND PHARMACEUTICAL INDUSTRY

FAEs derived from non-recombinant producing strains might be important for better acceptance by consumer's food related applications. In addition, ferulic acid, which is the most abundant cinnamic acid in plant cell walls, is a precursor for vanillin and its access through biotechnological methods is crucial in the quest for natural vanillin (Priefert et al., 2001). As well as being exploited as a hydrolase, FAE was shown to be a good catalyst in synthesizing sugar-phenolic esters (Mathew et al., 2004; Topakas et al., 2004a, b), and could also be used to functionalize sugar polymers by adding phenolic deriva-tives onto the natural biopolymers.

Some of the ester-linked substituents on plant cell wall polysaccharides retard or inhibit microbial infection. There are many examples in published reports concerning the antimicrobial nature of the phenolic compounds towards some microorganisms. Phenolic components of the plant cell wall, especially p-coumaric acid, ferulic acid, and p-hydroxybenzaldehyde, inhibit the growth of rumen micro-organisms (Williamson et al. 1998) and phenolic acids derived from plant cell walls have long been used as food preservatives (Garcia-Conesa et al., 1998a,b) to inhibit microbial growth. *Magnaporthe grisea*, a rice blast fungus,

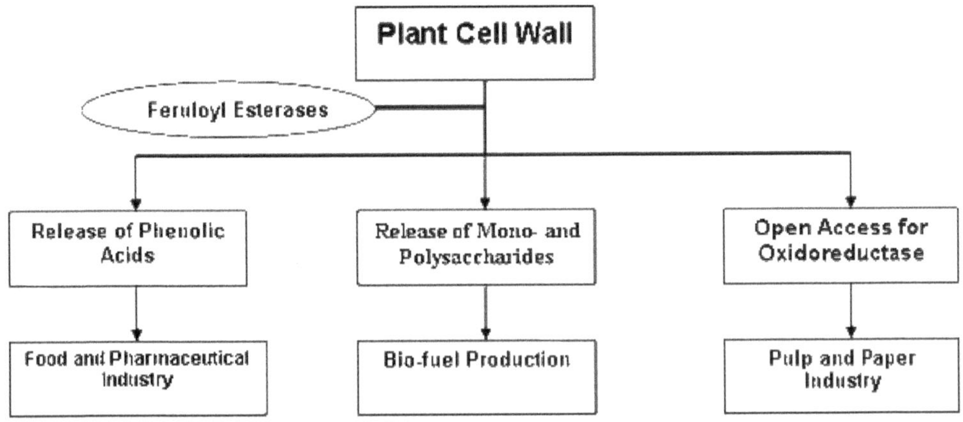

Figure 5. Diagrammatic representation of industrial and medicinal applications of feruloyl esterases

produces a xylanase and an arabinofuranosidase that act synergistically to release arabinoxylo-oligosaccharides from rice cell walls. These compounds contain esterified ferulic acid, and the release leads to the death (presumably programmed cell death) of surrounding rice cells. *In vitro* removal of the ferulic acid moiety destroyed more than 95% of this killing ability, which would enhance the chances of the colonization of the rice cell wall by the pathogen in the presence of a cinnamoyl esterase (Saija et al., 1999).

As phenolic acid sugar esters have clear antitumor activity and the potential to be used to formulate antimicrobial, antiviral, and/or anti-inflammatory agents (Omar et al., 1992; Garcia-Conesa et al., 2004), specific FAEs could be used in the tailored synthesis of such pharmaceuticals (Garcia-Conesa et al., 2004; Malherbo et al., 2002; Kondo et al., 1990). The FAE system of *T. stipitatus* has been studied and three discrete FAEs, including a type C esterase, with broad specificity against hydroxycinnamate esters (Kondo et al., 1990; Lam et al., 1990; Moore et al., 1996) have been found. The re-establishment of efficient use of cheap agricultural waste materials, with their synergistic action with other lignocellulose-degrading enzymes, is promising tools in various agro-industrial processes (Moore et al., 1996; Benoit et al., 2006). Other potential applications include production of important medicinal compounds, improvement of bread quality, pulp treatment, juice clarification, improvement of quality of animal feedstock, production of biofuel, and synthesis of oligosaccharides (Williamson et al., 1998). Therefore, effective FAEs production is a vital prerequisite for successful applications in various Industries. To achieve this goal, it is necessary to use FAEs, as well as defined polysaccharides and oligosaccharides from different agricultural raw materials.

FAEs are potential analytical aids in modern carbohydrate chemistry. In combination with other plant cell wall-degrading enzymes, the esterases will provide important tools in understanding the fine structure and linkage patterns that exist in the plant cell wall, but the science is at an early stage and ripe for exploitation (Mathew et al., 2004; Topakas et al., 2003a; Tarbouriech et al., 2005; Faulds et al., 1993; Ralet et al., 1994; Kroon et al., 1999, 1996a, b; Fillingham et al., 1999; Williamson et al., 1998). Arabinoxylans and α-glucans in the cell walls of barley have been shown to be associated either together or to a common component through an ester bond, as shown by specific hydrolysis by a pure cinnamoyl esterase (Campbell et al., 1998; Benoit et al., 2006). The exact nature of the covalent bond between lignin and carbohydrate polymers in the cell wall matrix of various plants has still to be determined, although evidence is beginning to accrue on these structures. FAEs could provide a useful tool in helping to determine this link (Campbell et al., 1998; Sun et al., 2002). FAEs are secreted by a number of bacterial and fungal organisms that exploit plants either to enter the plant cell or to use the cell wall material as a nutritional resource. The complete degradation of plant cell wall polymers requires multi-enzyme complex systems. Most FAEs have been shown to act synergistically with xylanases, cellulases, and pectinases to break down complex plant cell wall carbohydrates (Sun et al., 2002; Sahan et al., 2003; Sorensen et al., 2003).

By-products of the maize industry are ideal stock materials for biotechnology processes. An example is ferulic acid, an aromatic food anti-oxidant that can be isolated from maize fiber after wet milling and is converted to valuable compounds such as vanillin, an important flavoring agent used extensively for the foodstuffs (Luonteri et al., 1999; Sorensen et al., 2003). Apart from its use for flavoring, vanillin is also required for the synthesis of pharmaceutical drugs and is used extensively in the perfume and metal plating industries. In agriculture, it has herbicidal action, and can be used as a ripening agent to increase the yield of sucrose in sugar cane (Luonteri et al., 1999; Wood et al., 1996).

FERULOYL ESTERASES IN BIO-FUEL INDUSTRY

The demand for ethanol has been continuously increasing, where it is used either as a chemical feedstock or as an octane enhancer or petrol additive. Global crude oil production is predicted to decline from 25 billion barrels to approximately 5 billion barrels in 2050 (Benoitet et al., 2006). In the USA, fuel ethanol has been used in gasohol or oxygenated fuels since the 1980s. These gasoline fuels contain up to 10% ethanol by volume. It is estimated that 4.54 billion liters of ethanol is used by the American transportation sector and that this number will rise phenomenally as the American automobile manufacturers plan to manufacture a significant number of flexi-fueled engines which can use a blend of 85% ethanol and 15% gasoline by volume. The production of fuel ethanol from sugars or starch impacts negatively on the economics of the process, thus making ethanol more expensive compared with fossil fuels. Hence the technology development focus in the production of ethanol has shifted towards the use of residual lignocellulosic materials to lower production cost. An improved use of wheat endosperm by-products in this type of ethanol production would generate a fermentable hydrolysate based on the hemicellulose fraction. Complete enzymatic hydrolysis of arabinoxylan requires both depolymerizing and side-group cleaving enzyme activities such as FAEs. Any hemicellulose containing lignocellulose generates a mixture of sugars upon pretreatment alone or in combination with enzymatic hydrolysis. In Europe, potable alcohol manufacturing plants are based on wheat endosperm processing, with the hemicellulosic by-product remaining after fermentation consisting of approximately 66% (W/W) arabinoxylan (Benoit et al., 2006). Fermentable sugars from cellulose and hemicellulose will essentially be glucose and xylose, which can be released from lingocellulosics by single or two-stage hydrolysis, thereby leading to mixtures of glucose and xylose or separate glucose- and xyloserich streams. Conventional methods, applied for bioconversion of cellulose and hemicellulose to ethanol, involve acid or enzyme hydrolysis of biopolymers to soluble oligosaccharides followed by fermentation to ethanol. A synergistic action between cellulases, FAEs and xylanases may prove to be more effective when applied in a critical concentration in the saccharification of steam-exploded wheat straw (Borneman et al., 1993; Kennedy et al., 1999).

CONCLUSION

Feruloyl esterases are promising enzymes to have potential and significant contributions in research and industry. However, two of the problems to commercialize the use of feruloyl esterases are: to date there is no commercial full purified feruloyl esterases and the lack of sufficient enzyme stocks. Thus, efforts have to be made in order to achieve cheap overproduction of feruloyl esterases in heterologous hosts and also their modifica-tion by chemical means or protein engineering to obtain more robust and active purified feruloyl esterases enzymes.

REFERENCES

Benoit I, Navarro D, Marnet N, Rakotomanomana N, Lesage-Meessen L, Sigoillot J-C, Asther M (2006). Feruloyl esterases as a tool for the release of phenolic compounds from agro-industrial byproducts. Carbohydr. Res. 341: 1820–1827.

Borneman WS, Ljungdahl LG, Hartley RD, Akin DE (1993). Feruloyl and p-coumaroyl esterases from the anaerobic fungus *Neocallimastix* MC-2: Properties and functions in plant cell wall degradation. In: Coughlan MP, Hazlewood G, eds. Hemicellulose. Cambridge: Portland Press.

Campbell CJ, Laherrere JH. (1998). The end of cheap oil. Sci Am 3: 78–83.

Crepin VF, Faulds CB, Connerton IF (2003a). A non-modular type-B feruloyl esterase from *Neurospora crassa* exhibits concentration dependent substrate inhibition. Biochem. J. 370: 417–427.

Crepin VF, Faulds CB, Connerton IF (2003b). Production and characterization of the *Talaromyces stipitatus* feruloyl esterase FAEC in *Pichia pastoris*: Identification of the nucleophilic serine. Protein Expr. Purif. 29: 176–184.

Crepin VF, Faulds CB, Connerton IF. (2004). Functional classification of the microbial feruloyl esterases. Appl. Microbiol. Biotechnol. 63: 647–652

de Vries RP, Michelsen B, Poulsen CH, Kroon PA, van den Heuvel RH, Faulds CB, Williamson G, van den Hombergh JP, Visser J (1997). The faeA genes from Aspergillus niger and Aspergillus tubingensis encode ferulic acid esterases involved in degradation of complex cell wall polysaccharides.63: 4638-4644.

de Vries RP, vanKuyk PA, Kester HC, Visser J (2002). The Aspergillus niger faeB gene encodes a second feruloyl esterase involved in pectin and xylan degradation and is specifically induced in the presence of aromatic compounds. Biochem. J.15: 377-386.

Donaghy J, Kelly PF, McKay AM (1998). Detection of ferulic acid esterase production by *Bacillus* spp. and lactobacilli. Appl. Microbiol. Biotechnol. 50: 257–260.

Donaghy JA, McKay AM (1995). Production of feruloyl/p-coumaroyl esterase activity by *Penicillium expansum*, *Penicillium brevicompactum* and *Aspergillus niger*. J. Appl. Bacteriol. 79: 657–662.

Faulds C Hermoso J, Sanz-Aparicio J, Molina R (2005). Probing the determinants of substrate specificity of a feruloyl esterase, AnFaeA, from Aspergillus niger. FEBS J, 272: 4362-4371.

Faulds CB, deVries RP, Kroon PA, Visser J, Williamson G (1997). Influence of ferulic acid on the production of feruloyl esterases by *Aspergillus niger*. FEMS Microbiol. Lett. 157: 239–244

Faulds CB, Sancho AI, Bartolomé B (2002). Mono- and dimeric ferulic acid release from brewers' spent grain by fungal feruloyl esterases. Appl. Microbiol. Biotechnol. 60: 489–493.

Faulds CB, Williamson G (1993). Ferulic acid esterase from *Aspergillus niger:* Purification and partial characterization of two forms from a commercial source of pectinase. Biotechnol. Appl. Biochem. 17: 349–359

Faulds CB, Williamson G (1994) Purification and characterization of a ferulic acid esterase (FAE-III) from *Aspergillus niger*. Specificity for the phenolic moiety and binding to microcrystalline cellulose. Microbiol. 140: 779–787

Fazary AE, Ju YH (2007) Feruloyl Esterases as Biotechnological Tools: Current and Future Perspectives. Acta Biochimica et Biophysica Sinica 39: 811–828.

Fazary AE, Ju YH (2008). Production, Partial Purification and Characterization of Feruloyl Esterases by Aspergillus awamori in Submerged Fermentation. Biotechnol. J. DOI: 10.1002/biot.200800101.

Fillingham IJ, Kroon PA, Williamson G, Gilbert HJ, Hazlewood GP (1999). A modular cinnamoyl ester hydrolase from the anaerobic fungus *Piromyces equi* acts synergistically with xylanase and is part of a multiprotein cellulose-binding cellulase-hemicellulase complex. Biochem. J. 343: 215–224.

Garcia BL, Ball AS, Rodriguez J, Perez-Leblic MI, Arias ME, Copa-Patino JL (1998a). Production and characterization of ferulic acid esterase activity in crude extracts by *Streptomyces avermitilis* CECT 3339. Appl. Microbiol. Biotechnol. 50: 213–218.

Garcia BL, Ball AS, Rodriguez J, Perez-Leblic MI, Arias ME, Copa-Patino JL (1998b). Induction of ferulic acid esterase and xylanase activities in *Streptomyces avermitilis* UAH30. FEMS Microbiol. Lett. 158: 95–99.

Garcia-Conesa MT, Crepin VF, Goldson AJ, Williamson G, Cummings NJ, Connerton IF, Faulds CB (2004). The feruloyl esterase system of *Talaromyces stipitatus*: Production of three discrete feruloyl esterases, including a novel enzyme, TsFaeC, with a broad substrate specificity. J. Biotechnol. 108: 227–241.

Hermoso J, Sanz-Aparicio J, Molina R, Faulds C (2004) The crystal structure of feruloyl esterase A from Aspergillus niger suggests evolutive functional convergence in feruloyl esterase family. J. Mol. Biol. 338: 495-506.

Hutnan M, Drtil M, Mrafkova L (2000). Anaerobic biodegradation of sugar beet pulp. Biodegradation 11: 203–211.

Kennedy JF, Methacanon P, Lloyd LL (1999). The identification and quantification of the hydroxycinnamic acid substituents of a polysaccharide extracted from maize bran. J. Sci. Food Agric. 79: 464–470.

Kondo T, Mizuno K, Kato T (1990). Cell wall-bound p-coumaric and ferulic acids in Italian ryegrass. Can. J. Plant Sci. 71: 495–499.

Kroon PA, Faulds CB, Williamson G (1996b). Purification and characterization of a novel esterase induced by growth of *Aspergillus niger* on sugar-beet pulp. Biotechnol. Appl. Biochem. 23: 255–262.

Kroon PA, Garcia-Conesa MT, Fillingham IJ, Hazlewood GP, Williamson G (1999). Release of ferulic acid dehydrodimers from plant cell walls by feruloyl esterases. J. Sci. Food Agric. 79: 428–434.

Kroon PA, Williamson G (1996a). Release of ferulic acid from sugar-beet pulp by using arabinanase, arabinofuranosidase and an esterase from *Aspergillus niger.* Biotechno.l Appl. Biochem. 23: 263–267.

Kroon PA, Williamson G, Fish NM, Archer DB, Belshaw NJ (2000). A modular esterase from *Penicillium funiculosum* which releases ferulic acid from plant cell walls and binds crystalline cellulose contains a carbohydrate binding module. Eur. J. Biochem. 267: 6740–6752

Lam TBT, Liyama K, Stone BA (1990). Distribution of free and combined phenolic in wheat internodes. Phytochem. 29: 429–433.

Lee YS (2005). Role of NADPH oxidase-mediated generation of reactive oxygen species in the mechanism of apoptosis induced by phenolic acids in HepG2 human hepatoma cells. Arch. Pharm. Res. 28: 1183–1189.

Luonteri E, Kroon PA, Tenkanen M, Teleman A, Williamson G (1999). Activity of a *Aspergillus terreus* α -arabinofuranosidase on phenolic-substituted oligosaccharides. J. Biotechnol. 67: 41–48.

Malherbo S, Cloete TE (2002). Lignocellulose biodegradation: Fundamentals and applications. Rev. Environ. Sci. Biotechnol. 1: 105–114.

Mathew S, Abraham TE (2004). Ferulic acid: An antioxidant found naturally in plant cell walls and feruloyl esterases involved in its release and their applications. Crit. Rev. Biotechnol. 24: 59–83.

Mcauley KE, Svendsen A, Patkar S, Wilson KS (2004). Structure of a feruloyl esterase from Aspergillus niger.. Acta Crystallogr D. Biol. Crystallogr. 60: 878-887.

Moore J, Bamforth CW, Kroon PA, Bartolome B, Williamson G (1996). Ferulic acid esterase catalyses the solubilisation of □-glucans and pentosans from the starchy endosperm cell walls of barley. Biotechnol. Lett. 18: 1423–1426.

Nethaji M, Pattabhi V (1988). Structure of 3-(4-Hydroxy-3-Methoxy-phenyl)-2-propenoic acid (ferulic acid). Acta. Cryst. C44: 275–277.

Omar MM. (1992) In Phenolic compounds in food and their effect on health I. ACS Symposium Series 12: 154–168.

Priefert H, Rabenhorst J, Steinbuchel A (2001). Biotechnological production of vanillin. Appl. Microbiol. Biotechnol. 56: 296–314.

Ralet MC, Faulds CB, Williamson G, Thibault JF (1994). Degradation of feruloylated oligosaccharides from sugar-beet pulp and wheat bran by ferulic acid esterases from *Aspergillus niger.* Carbohydr. Res. 263: 257–269.

Record E, Asther M, Sigoillot C, Pages S, Punt PJ, Delattre M, Haon M van den Hondel CA , Sigoillot C A, Lesage-Meessen L, Asther M (2003). Overproduction of the *Aspergillus niger* feruloyl esterase for pulp bleaching application. Appl. Microbiol. Biotechnol. 62: 349–355.

Saha BC (2003). Hemicellulose bioconversion. J. Ind. Microbiol. Biotechnol. 30: 279–291

Saija A, Tomaino A, Cascio RL, Trombetta D, Proteggente A, De Pasquale A, Uccella N (1999). Ferulic and caffeic acids as potential protective agents against photooxidative skin damage. J. Sci. Food Agric. 79: 476–480.

Sigoillot C, Camarero S, Vidal T, Record E, Asther M, Perez-Boada M, Martinez MJ Maria MJ, Jean-Claude S, Marcel A, Colome JF, Angel TM (2005). Comparison of different fungal enzymes for bleaching high-quality paper pulps. J. Biotechnol. 115: 333–343.

Sorensen HR, Meyer AS, Pedersen S (2003). Enzymatic hydrolysis of watersoluble wheat arabinoxylan. 1. Synergy between α-*L*-arabinofuranosidases, endo-1,4- α -xylanases, and α -xylosidase activities. Biotechnol. Bioeng. 81: 726–731.

Sun Y, Cheng J (2002.Hydrolysis of lignocellulosic material from ethanol production. Biores Technol. 83: 1–11.

Tarbouriech N, Prates JA, Fontes CM, Davies GJ (2005). Molecular determinants of substrate specificity in the feruloyl esterase module of xylanase 10B from *Clostridium thermocellum.* Acta. Crystallogr. D. Biol. Crystallogr. 61: 194–197.

Topakas E, Christakopoulos P (2004b). Production and partial characterization of alkaline feruloyl esterases by *Fusarium oxysporum* during submerged batch cultivation. World J. Microb. Biot. 20: 245–250.

Topakas E, Stamatis H, Biely P, Christakopoulos P (2004a). Purification and characterization of a type B feruloyl esterase (StFAE-A) from the thermophilic fungus *Sporotrichum thermophile.* Appl. Microbiol. Biotechnol. 63: 686–690.

Topakas E, Stamatis H, Biely P, Christakopoulos P (2006). *Sporotrichum thermophile* type C feruloyl esterase (StFaeC): Purification, characterization, and its use for phenolic acid (sugar) ester synthesis. Enzyme Microb. Technol. 36: 729–736.

Topakas E, Stamatis H, Mastihubova M, Biely P, Kekos D, Macris BJ, Christakopoulos P (2003a). Purification and characterization of a *Fusarium oxysporum* feruloyl esterase (FoFAE-I) catalyzing transesterification of phenolic acid esters. Enzyme Microb. Technol. 33: 729–737.

Williamson G, Kroon PA, Faulds CB (1998). Hairy plant polysaccharides: A close shave with microbial esterases. Microbiol. 144: 2011–2023

Wood TM, McCrae SI (1996). Arabinoxylan-degrading enzyme system of the fungus *Aspergillus awamori*: Purification and properties of an - *L*-arabinofuranosidase. Appl. Microbiol. Biotechnol. 45: 538–545.

15

Emerging infections and bioterrorism emergencies: Where do we go from here?

I. O. Okonko[1]*, M. O. Ojezele[2], E. T. Babalola[3], J. C. Nwanze[4], O. K. Mejeha[5] and T. A. Amusan[6]

[1]Department of Biochemistry and Microbiology, Lead City University, Ibadan, Nigeria.
[2]Department of Nursing Science and Public Health, Lead City University, Ibadan, Nigeria.
[3]Department of Microbiology, Crawford University, Igbesa, Ogun State, Nigeria.
[4]Department of Pharmacology and Therapeutics, Igbinedion University, Okada, Edo State, Nigeria.
[5]Department of Microbiology, Federal University of Technology, Owerri, Imo State, Nigeria.
[6]Medical Laboratory Unit, Department of Health Services, University of Agriculture, Abeokuta, Ogun State, Nigeria.

Emerging infectious disease outbreaks and bioterrorism attacks warrant urgent public health and medical responses. Response plans for these events may include use of medications and vaccines for which the effects on pregnant women and fetuses are unknown. Recent experiences with outbreaks of severe acute respiratory syndrome, monkey pox and anthrax, as well as response planning for bioterrorism and pandemic influenza, illustrate the challenges of making recommendations about medical interventions for victims. Experience with bioterrorism attacks (anthrax), viroterrorism attacks (arena viruses) and emergency response preparedness (smallpox vaccination) has been gained. Understanding the physiology of the body, the factors that influence the teratogenic potential of medications and vaccines and the infection control measures that may stop an outbreak will aid planners in making recommendations for care during large-scale infectious disease emergencies.

Key words: Bioterrorism, emerging infections, emergence preparedness, pathogens, viroterrorism.

INTRODUCTION

Emerging infectious diseases defined as infectious diseases whose incidence in humans have increased during the past two decades or threatens to increase in the near future, are increasingly recognized by physicians as an important threat to pregnant women. Emerging infectious diseases include novel pathogens that have newly emerged, such as severe acute respiratory syndrome (SARS), as well as pathogens that could potentially be used as biologic weapons (Jamieson et al., 2006).

Bioterrorism and viroterrorism threats include plague, glanders, Q fever, filoviruses, anthrax, smallpox, brucellosis, botulism, and ricin. Indeed, the world is like a and the use of genomics for the agents of tularemia, brucellosis, and clostridial gas gangrene (Lindler et al., 2004; information on genetic fingerprinting for forensic studies). The threat of emerging infections and bioterrorist swarm with viral zoonoses. There are also genetically-engineered protein toxins. Some authors also provided attacks has heightened the need to make disease surveillance more sensitive, specific, and timely (Henderson, 1999; Fine and Layton, 2001). The primary purpose of reporting diseases is to trigger an appropriate public health response so that further illness can be prevented and public fears allayed (M'ikanatha et al., 2003). The concept of continual virus movement has prevailed over the past years in part because of the failure to identify a vertebrate reservoir (Monath, 1989),

*Corresponding author. E-mail: mac2finney@yahoo.com.

and in part because of the dearth of ecologic data to support or refute alternative modes of transmission (Carrion et al., 2007).

Challenges

The challenges facing issues of emerging infectious disease as bioterrorism agents is that unfortunately, information about how pregnant women are affected by many of these novel and emerging infections is limited (Jamieson et al., 2006). For example in the case of pregnant women during an outbreak; whether the use of a medication or vaccine is harmful to the embryo or fetus depends on multiple factors, including the nature of the agent e.g., live versus killed vaccine, its dose and route of administration, timing of use during gestation, concomitant use of other agents, nature of the infection being treated or prevented, and genetic susceptibility of the pregnant woman and of the embryo or fetus (Cono et al., 2006). Another challenge facing disease surveillance and emergency preparedness is to shorten the time of the entire process so; useful epidemiologic and viroterrorism response data can be rapidly obtained (Anderson et al., 2003). The problem of developing new antiviral agents, especially those specific for one or only a few viral diseases, is circular and is of great concern too. Without such treatments rapid agent identification is not necessary, but without such identification no pressing commercial justification for developing specific antiviral agents exists, except for HIV; because they will not be widely used. To be successful, diagnosis and therapy must be linked. Also, containment of infectious disease agents is also of great concern when dealing with radiologic hazards and to contain infectious agents (Gargan et al., 1988).

Identification of viral sources

To routinely detect new and potentially lethal viruses, researchers may need to create completely automated and contained laboratories that continually search for and sequence viruses from a wide variety of sources to hone skills, demonstrate efficiency and develop improved systems, methods and reagents (Anderson et al., 2003). Two approaches can be used, if necessary, to link viruses to patients. In the first approach, viruses would be tracked geographically, first in terms of large regions, and then, sequentially, in terms of smaller areas. Detecting a new agent in large pooled samples would thus be repeated in smaller, localized pools that had been combined hierarchically to generate the larger pool (Lee et al., 2001). According to Anderson et al. (2003) several key questions remain to be addressed in a practical project: (1) Can this process be carried out rapidly enough to support a timely therapeutic or prophylactic response to a new agent-natural or engineered? (2) Will the novel virions that originated from one or a very few infected persons be recovered and detected? (3) Can the affected persons be located? (4) Will the presence of antibodies in the starting samples against most known viruses affect the separations?

NOVEL PATHOGENS

Severe acute respiratory syndrome (SARS)

During the worldwide outbreak of SARS in 2003, several countries reported cases of SARS both in children and adults even in pregnant women. In a case-control study conducted in Hong Kong, pregnant women with SARS had more severe disease than nonpregnant women and an increased risk for admission to the intensive care unit, development of renal failure, development of disseminated intravascular coagulopathy and death (Lam et al., 2004). Of 8 cases of laboratory-confirmed SARS reported in the United States, 2 were in pregnant women; the small number of cases precludes definitive conclusions about the severity of the disease (Stockman et al., 2004).

Prions

Prions are believed to be the causative agents of a group of rapidly progressive neurodegenerative diseases called transmissible spongiform encephalopathies, or prion diseases. They are infectious isoforms of a host-encoded cellular protein known as the prion protein. Prion diseases affect humans and animals and are uniformly fatal. The most common prion disease in humans is Creutzfeldt-Jakob disease (CJD), which occurs as a sporadic disease in most patients and as a familial or iatrogenic disease in some patients (Claudio, 2007).

Avian influenza-H5N1

Avian influenza-H5N1 has caused the greatest number of human cases and death. It may develop the characteristics needed to start influenza pandemic by improving its transmissibility among humans through reassortment or through a more gradual process of adaptive mutation.

CONVENTIONAL INFECTIOUS DISEASE THREATS

Malaria

People in malaria-endemic regions are at risk of becoming infected with *Plasmodium falciparum*; 1 of 4 parasites that cause malaria in humans (Loebstein et al.,

1997).

Typhoid

Typhoid (enteric) fever is a global infection caused by *Salmonella typhi* with a fatality rate of 10%. The disease is a cause for concern and a major public health problem in developing countries (Asia and Africa), especially in Nigeria due to poor sanitary conditions and lack of or inadequate potable water. The World Health Organization (WHO) estimated an annual infectious rate of 21.6 million and approximate death rate of 600 000 with the highest percentage in Africa and Asia. An estimated 22 million cases of typhoid fever and 200,000 related deaths occur worldwide each year (Crump et al., 2004). Approximately, 400 cases of typhoid fever among persons with onset of illness in the United States, most of whom are recent travelers, are reported to CDC each year (Center for Disease Control and Prevention (CDC), 2008).

Toxoplasma gondii

T. gondii is a parasite that infects humans primarily through ingestion of infected raw or undercooked meat and, less frequently, by exposure to infected cat feces.

Hansen disease or leprosy

Hansen disease or leprosy is caused by *Mycobacterium leprae*, which can multiply and cause symptomatic disease, particularly in hosts with decreased immunity (Brent, 2004).

Listeriosis

In 2000, an outbreak of listeriosis among Hispanic persons in North Carolina was reported as a result of ingestion of contaminated homemade Mexican-style cheese (Doering et al., 2002). *Listeria monocytogenes*, a foodborne pathogen, is responsible for approximately 2,500 cases of serious illness in the United States each year (Lo and Friedman, 2002; Food and Drug Administration, 1979). There are still more of these conventional pathogens which have caused outbreaks in recent times.

POTENTIAL EFFECTS OF BIOTERRORISM

The working group on civilian biodefense has identified a limited number of biologic agents that are of particular concern (Borio et al., 2002). Evidence exist that infection with some of these pathogens, including smallpox virus and some of the hemorrhagic fever viruses, are very severe on their victims and may be more severe on women during pregnancy (Jamieson et al., 2006). Although, pregnant women are not immunosuppressed in the classic sense, immunologic changes of pregnancy may induce a state of increased susceptibility to certain intracellular pathogens, including viruses, intracellular bacteria and parasites (Cono et al., 2006).

Measles (Rubeola)

Evidence also indicates that measles (Rubeola) is more common and severe in pregnant women. Accounts of measles outbreaks before an effective vaccine was available indicate that pregnant women may be more severely affected.

Smallpox (Variola virus)

Clinical experience with smallpox (Variola virus) before vaccination and disease eradication indicates that pregnant women are more susceptible to variola infection and have more severe disease. Pregnant women are more likely than nonpregnant women to have hemorrhagic smallpox (purpura variolosa), a severe variety of the disease (Jamieson et al., 2006).

VIRAL HEMORRHAGIC FEVERS (VHFS)

The viral hemorrhagic fevers, including Lassa fever and Ebola, may be more severe during pregnancy.

Lassa virus

Lassa virus causes thousands of deaths annually in western Africa and is considered a potential biological weapon (Centers for Disease Control and Prevention (CDC), 2004). Lassa fever is a viral hemorrhagic fever caused by a rodentborne arenavirus that is endemic in West Africa. The first reported case of Lassa fever, caused by infection with an arenavirus, was described in a pregnant patient. In this initial outbreak, 11 patients and staff members who were exposed to the index patient died (White et al., 2002). One study found that after pregnancy ended, whether by abortion or normal delivery, women rapidly improved (Price et al., 1988). In 2004, the Centers for Disease Control (CDC) and prevention reported a fatal case of Lassa fever in New Jersey (Fisher-Hoch et al., 1995). Auditory or vestibular dysfunction may develop in patients with Lassa fever, and tinnitus, autophony, hearing loss, dizziness, vertigo, nystagmus and ataxia have been reported (Rybak, 1990). In their review of a 1989 nosocomial Lassa fever outbreak in a Nigerian hospital, Fisher-Hoch et al. (1995) noted a high fever in the index patient, who was taken to surgery on February 25. Solbrig and McCormick (1991) reported that neuropsychiatric sequelae of Lassa fever have included sleep disorders (insomnia), asthenia, multiple somatic complaints, psychosis, hallucinations, personality disorders, severe adjustment reactions, dementia, mania, depression and finally, induced damage to the brain stem with resultant autonomic dysfunction.

Ebola virus

Ebola virus, a member of the Filoviridae group, is transmitted by direct contact with blood, secretions, or contaminated objects and is associated with high case-fatality rates. Unlike risk for death from Lassa fever, which is highest during the third trimester of pregnancy, risk for death from Ebola is similar during all trimesters (Mupapa et al., 1999).

Yellow fever

Yellow fever (YF) is an important reemerging arboviral disease and a cause of severe illness and death in South America and Africa. The prevailing paradigm of yellow fever virus (YFV) ecology in South America is that of wandering epizootics. The virus is believed to move from place to place in epizootic waves involving monkeys and mosquitoes, rather than persistently circulating within particular locales (Monath, 1989).

OTHER EMERGING INFECTIONS

Pneumocystis jiroveci

P. jiroveci (formerly *P. carinii*) has long been identified as a cause of pneumonia in immunocompromised persons. *Pneumocytis* pneumonia was first identified in malnourished children in European orphanages during World War II and was later associated with severe immunosuppression in HIV-infected persons (McNally et al., 2005). However, this agent is increasingly causing infection among immunocompetent persons. Evidence also indicates that *Pneumocytis* pneumonia may be more severe during pregnancy and that *Pneumocystis* may be perinatally transmitted by HIV-infected women to their children (McNally et al., 2005).

Psittacosis

Psittacosis is primarily a flulike illness characterized by fever, headache, and atypical pneumonia. *Chlamydophila psittaci* (formerly *Chlamydia psittaci*), the causative agent, is transmitted by inhalation of material from infected birds or by exposure to infected amniotic fluid or placentas of sheep or goats. Illness during pregnancy can be quite severe, mimicking HELLP (hemolysis, elevated liver enzyme levels, and low platelet count) syndrome but without hypertension. Most women rapidly recover after pregnancy (McNally et al., 2005).

THE GROWING FIELD OF COUNTER-BIOTERRORISM

In recent years, the CDC and prevention have been funding several states to develop electronic laboratory reporting (Jorgensen, 1997). A more thorough understanding of the pitfalls of such existing systems can provide insights to improve the development and implementation of new media in infectious disease surveillance. With recent funding for activities to defend the public's health against terrorism and naturally occurring diseases, development of automated reporting systems has been accelerated (Centers for Disease Control and Prevention (CDC), 2002).

Counter bioterrorism measures include the following: Identification of viral sources, surveillance system, disease reporting system, early detection and management of a biological terrorism attack. As the popular saying to be forewarned is to be forearmed, given advance notice, even by weeks, of an impending viral outbreak, the hope exists that the tools and imaginations of molecular biology will find the means to prepare some effective biological defense (Centers for Disease Control and Prevention, 2001).

Automated surveillance and disease reporting system

Recent advances in provider and laboratory information management have facilitated one step towards the modernization of surveillance which is the development of automated reporting systems (Anderson et al., 2003). Evidence from deployed systems shows promise in the ability of electronic laboratory reporting to deliver more timely and complete notifications than paper-based methods (Henderson, 1999).

WHO global outbreak alert and response network (GOARN)

Beginning in March, 2003, after WHO recognized, through its GOARN, an outbreak of severe respiratory illness with high transmissibility in healthcare settings and international spread through airline travel, WHO issued a series of global alerts, travel advisories, and recommendations for diagnosis, clinical management, and prevention of transmission (Centers for Disease Control and Prevention (CDC), 1997).

Global emerging infections system (GEIS)

The GEIS of the US Department of Defense (DOD) as it applies to biodefense was reported by Bell et al. (2003). Surveillance systems used by US DOD-GEIS were described by Linder (2004), from the electronic surveillance system for the early notification of community-based epidemics to newer systems, along with ways to integrate DOD and civilian surveillance systems (Lindler et al., 2004).

The national electronic disease surveillance system (NEDSS)

The NEDSS and bioterrorism preparedness initiatives are expected to further enhance disease surveillance by supporting integration of electronic data from various sources (Lucey, 2006).

Aerosol pathogenesis and biological weapons defense

Effect levels are particularly relevant given by the US Cities Readiness Initiative that involves planning for an aerosol attack with anthrax or another agent in US metropolitan areas (Doyle et al., 2002). Anderson (2003) proposes a system for continuing surveillance of viral pathogens circulating in large human populations. Given this existing technology base, an integrated, automated, and contained system for surveillance of the human "virome" can be implemented within 1 to 2 years. While these newly available electronic transmission methods can increase timeliness and completeness of infectious disease reports, limitations of this technology may unintentionally compromise detection of, and response to, bioterrorism and other outbreaks (Anderson et al., 2003). Current federal funding for emergency preparedness surveillance and epidemiology capacity is expected to stimulate widespread use of automated systems in infectious disease reporting (Anderson et al., 2003).

Furthermore, the requirement that providers and laboratories report immediately by telephone when they detect organisms indicating an outbreak or an unusual occurrence of potential public health importance is expected to continue even when automated reporting systems are implemented. Modern technology can translate into better public health preparedness by enhancing and complementing existing reporting systems (Anderson et al., 2003). Such a system could monitor the levels of known viruses in human populations, rapidly detect outbreaks, and systematically discover novel or variant human viruses.

MEDICAL CONTRIBUTIONS OF GLOBAL SURVEILLANCE

There is a possibility of linking rapid detection to rapid responses, including vaccine and therapeutic antibody development, in an attempt to abort epidemics caused by new viruses while they are in progress. In the atmosphere of an actual terrorist attack with a biological agent, the consequences of false outcomes place an enormous strain on public services, as demonstrated by the recent anthrax episodes as reported by Anderson (2003).

Medical countermeasures for emerging infection and bioterrorism emergencies

Decisions about the treatment or prophylaxis of emerging infections must take into account the effect on the patients' health and the potential risks such as a mother's health and that of the embryo or fetus. In preparation for potential bioterrorism emergencies, the US government has stockpiled medications and vaccines, most of which are rated by FDA as one of the categories B through X, which indicates that they could pose a risk to the unborn fetus or that insufficient information exists to evaluate their potential fetal risk. Some of these products are commonly used in routine healthcare such as ciprofloxacin, gentamicin, and doxcycline, but others are reserved for emergency preparedness and response activities and for deployed military personnel such as smallpox and anthrax vaccines (Anderson et al., 2003).

Some emergency response medications and vaccines fall outside of the FDA labeling system because they are not licensed by FDA. Some are newly developed and still in pre-licensure clinical trials; others are no longer licensed and predate the classification system (Anderson et al., 2003). In an emergency setting with a high risk for life-threatening exposure to an infectious pathogen, recommendations likely will call for the use of vaccination and prophylactic medications, when they are available for pregnant women, despite unknown risks to the fetus. Other measures that can protect persons who are unable or choose not to receive vaccination or prophylactic medications include limiting exposure to persons who may be infectious, avoiding public gatherings and restricting travels to affected area.

Issues in medical countermeasures

In recent years, the public health and medical communities have faced several emerging infectious disease outbreaks, including SARS and monkey pox, and much consideration has been given to preparation for a future influenza pandemic. In addition, experience with bioterrorism attacks (anthrax) and emergency response preparedness (smallpox vaccination) has been gained. These events required careful consideration of recommendations for the care of patients, especially pregnant women (Cono et al., 2006).

The SARS outbreak of 2003, caused by a newly identified coronavirus, affected >8,000 persons worldwide (Jamieson et al., 2006). Reports suggest that the clinical course and outcomes of SARS might be more severe for pregnant than for non pregnant women (Ng et al., 2004). Identifying appropriate treatment modalities during the SARS outbreak was challenging, due to the lack of information about the newly identified disease (Lam et al., 2004). Ribavirin was initially chosen because of its broad antiviral spectrum. Corticosteroids were used in an attempt to limit the tissue damage caused by the inflammatory response (Anderson et al., 2003). However, issues regarding the teratogenicity of these medications have been raised, further complicating decisions about their use during pregnancy. Some animal studies have suggested that ribavirin is teratogenic, but limited experience is available regarding its effects on human pregnancies. On the basis of more recent data, the efficacy of ribavirin and corticosteroids in the treatment of patients with SARS has been questioned (Anderson et al., 2003). Other medications, such as interferons, have been proposed for use in future SARS outbreaks, but use of these medications in pregnant women may also be of concern.

In June 2003, the first outbreak of monkey pox in the Western hemisphere occurred in the United States (Lai, 2005). Because of the high death rate associated with monkey pox on the African continent and lack of experience with monkey pox in the United States, CDC recommended smallpox (vaccinia) vaccination approximately 85% effective against monkeypox (Centers for Disease Control and Prevention, 2003a). The outbreak was traced to importation of infected rodents that infected pet prairie dogs and other small mammals kept as pets. Smallpox vaccination during pregnancy poses a low risk for fetal vaccinia, which can lead to preterm birth, fetal and neonatal death (Centers for Disease Control and Prevention, 2003a).

Planning for a future influenza pandemic must include specific considerations for pregnant women (Centers for Disease Control and Prevention, 2003). Because pregnancy has been shown to increase the risk for influenza-associated complications, pregnant women are considered a high-risk group and are recommended to receive influenza vaccination during interpandemic years (Wharton et al., 2003). This vaccine is inactivated and is considered safe for pregnant women. It is reformulated each year to include the anticipated viral strains of the upcoming influenza season. During a pandemic, an effective vaccine may initially be unavailable or in limited supply. In such a situation, chemoprophylaxis will be an important option for pregnant women. Unfortunately, no information is available regarding the effects on the fetus of neuraminidase inhibitors (oseltamivir and zanamvir) and the medications likely to be useful in an H5N1 pandemic (Rasmussen and Hayes, 2005). Thus, weighing the risks associated with infectious exposure in a pregnant woman and risks associated with medication exposure to her unborn child is difficult.

The anthrax attacks of 2001 prompted the first, large-scale recommendations for use of prophylactic medications in response to bioterrorism. The recommended medication for initial antimicrobial drug prophylaxis of asymptomatic exposed adults was ciprofloxacin, doxycycline and amoxicillin as alternative therapies if the strain was susceptible (Laibl and Sheffield, 2005). Because of an observed association between fluoroquinolones and joint and cartilage toxicity in juvenile animals (Harper et al., 2005), ciprofloxacin was generally not recommended during pregnancy if efficacious alternatives are available. Although, information on the safety of ciprofloxacin in pregnant women was lacking, the available limited information suggested that its use during pregnancy was unlikely to be associated with a high risk for structural birth defects. In addition, maternal exposure to tetracyclines, which include doxycyline, carries the known risks of staining the primary teeth, concern about bone growth and abnormal tooth enamel in the fetus, and rare instances of hepatic necrosis in pregnant women (Dolin, 2005).

Although, penicillins are considered safe during pregnancy, the fact that *Bacillus anthracis* strains may have penicillinase activity led to the recommendation that amoxicillin be used for prophylaxis only if the specific strain was shown to be penicillin sensitive. On the basis of these considerations, CDC recommended that ciprofloxacin be the antimicrobial drug of choice for initial prophylactic therapy of asymptomatic pregnant women exposed to *B. anthracis* during the 2001 anthrax attacks (Inglesby et al., 2002). The American College of Obstetricians and Gynecologists Committee on Obstetric practice endorsed these recommendations and emphasized that prophylaxis be limited to women exposed to a confirmed environmental contamination or a high-risk source, as determined by local public health officials (Grady, 2003).

In 2003, the United States embarked on an effort to vaccinate public health and medical bioterrorism response teams against smallpox. In the absence of circulating smallpox virus, vaccination in pregnant women or women who desire to become pregnant within 28 days of the vaccination was contraindicated because of the risk for fetal vaccinia (Billings et al., 2004). However, after an intentional attack, pregnancy should not be considered an absolute contraindication to vaccination (Billings et al., 2004). In the event of exposure or high risk for exposure to smallpox, pregnant women are advised to receive the vaccine because the risk for death and serious illness from smallpox-particularly during pregnancy outweighs the risk for fetal vaccinia.

These examples demonstrate some of the challenges facing healthcare providers when considering prophylaxis and treatment in response to emerging infections or bioterrorism attacks. Decisions regarding appropriate prophylaxis and treatment must take into account the risks associated with specific medications or vaccines versus the risk for illness and death from a possible infectious exposure (Ng et al., 2004). Information on the effectiveness and safety of some medical countermeasures is limited for the general population, and even less information is available for pregnant women (Anderson et al., 2003).

SENTINELS OF BIOTERRORISM AGENTS

A number of biological terrorism agents have little potential for secondary spread in either animal or human populations, including *B. anthracis* and *Clostridium botulinum*. For other agents, however, evidence that their introduction into an animal population could cause an epizootic that would then place additional human populations at risk has been found. For example, studies of mosquitoes native to the United States have demonstrated their potential to spread a disease such as Rift Valley fever through livestock and other animal populations (White et al., 2002), even though person-to-person transmission does not occur. Agents such as *Coxiella burnetii* and *Brucella* spp. spread easily in animal populations through direct contact and can then pose a wider risk to humans, even though human-to-human transmission does not occur. Agents such as alphaviruses that are prevalent in wild bird populations can spread over a wide area in a short time (Gargan et al., 1988). For a numwetin ber of biological terrorism agents, there is evidence that animals could provide early warning of an acute attack. Most priority bioterrorism agents are zoonotic in origin. As a result, an attack on human populations with a bioterrorism agent would likely pose a health risk to animal populations in the target area; therefore, integrating veterinary and human public health surveillance efforts is essential. The potential use of animals as "sentinels" of a human bioterrorism attack can be differentiated from the possibility of a direct attack on animals of agricultural importance (agroterrorism) (Howard et al., 1996).

The CDC and prevention, in planning for the early detection and management of a biological terrorism attack, has recommended the "prompt diagnosis of unusual or suspicious health problems in animals," as well as establishing "criteria for investigating and evaluating suspicious clusters of human and animal disease or injury and triggers for notifying law enforcement of suspected acts of biological or chemical terrorism" (Rabinowitz et al., 2006). Similarly, an indicator of a biological terrorism attack would be "increased numbers of sick or dead animals, often of different species. Some biological warfare (BW) agents are capable of infecting/intoxicating a wide range of hosts" (Centers for Disease Control and Prevention, 2000). In part because of such recommendations, calls have been made for enhanced veterinary surveillance for outbreaks of animal disease caused by bioterrorism agents and better communication between animal health and human health professionals. For such efforts to succeed, the relevance to human health of disease events in animals must be established. The public health infrastructure must look beyond passive surveillance of acute animal disease events to build capacity for active surveillance and intervention efforts to detect and control ongoing outbreaks of disease in domestic and wild animal populations.

Animals could provide an early warning to humans if clinical signs could be detected before human illness emerged or soon enough to allow preventive measures to be initiated. This early detection could occur because an animal species had increased susceptibility to a particular agent, because the disease caused by the agent had a shorter incubation period, or because animals were exposed sooner (or at more intense and continuous levels) than the human population (Centers for Disease Control and Prevention, 2000). The simultaneous appearance of disease signs and symptoms in animals may contribute to the more rapid identification of a biological warfare agent that was produced nonspecific effects in nearby persons. If a released biological agent persists in the environment (such as soil, water, or air), active surveillance for sporadic illness in animals could help detect ongoing exposure risks. Additionally, the geographic pattern of sick or dead animals could indicate the persistence of a biological threat (Centers for Disease Control and Prevention (CDC), 2000). Animal populations such as wild birds, commercially shipped livestock, and animals involved in the local or international pet trade, could play a role in the maintenance and spread of an epidemic attributable to an intentional release of a biological agent.

STEPS IN PREPAREDNESS FOR BIOLOGICAL AGENT ATTACKS

The above findings suggest the need for certain steps related to preparedness for biological agent attacks and this include

1. Improved communication is needed between animal health and human health professionals, so that sentinel events could be rapidly detected. Such improvement would mean overcoming existing barriers to communication; a recent survey found that physicians and veterinarians communicate little about zoonotic issues (Cottrell and Morgan, 2003).
2. Also, an adequate surveillance network should be

developed to detect unusual health events in animal populations. Data on usual trends is missing for most animal species that could be potential sentinels. Whether public health resources can be committed to gathering such baseline data remains an open question.

3. Active surveillance of animal populations, including wildlife and companion animals, could fill a critical need in the aftermath of an attack involving certain bioterrorism agents by helping identify persistent sources of infection in the environment.

4. Better approaches for intervention are needed to be able to stem the propagation and amplification of an introduced biological warfare agent into a wild or domestic animal population. The US experience with West Nile virus reflects the difficulties of controlling an emerging zoonotic threat as it spreads through animal populations (Grant and Olsen, 1999).

5. There is need for additional research to fill knowledge gaps about animals as sentinels of human disease threats, including data on relative susceptibilities and exposure pathways for animal species living near human populations (Rabinowitz et al., 2006).

Concrete approaches

According to Rabinowitz (2006) concrete steps could include:

1. Establishment of surveillance veterinary clinics in strategic areas with incentives for practitioners to report unusual events.
2. Another approach would be to make greater use of electronic databases of animal diseases.
3. Common links or web-based interfaces should be developed to integrate human and animal disease surveillance information. Reporting systems for wildlife professionals and the public should be created, and their use should be encouraged to document unusual disease events and die-offs.
4. Another constructive step would be to improve the capacity of existing veterinary rapid-response teams, which exist in many states, to carry out active surveillance with animal populations as well as to improve the coordination of veterinary diagnostic laboratories.
5. Again, barriers to funding and cooperation between animal and human health agencies need to be addressed. In the past, these have hampered efforts to have a coordinated approach to the collection of animal surveillance data.
6. In addition, state-based efforts would need to be coordinated on a regional and national scale.

CONCLUSIONS AND RECOMMENDATIONS

Developing recommendations for prophylaxis and treatment of emerging and bioterrorism pathogens can be especially difficult. Data on the effects of some emergency response countermeasure treatments are limited especially on pregnant women and fetuses. Emergency response planners should include recommendations for treatment in pre-event response plans, rather than creating them during an emergency. Clinicians should become familiar with recommendations for prophylaxis and treatment of persons with emerging and bioterrorism pathogens so that they will be prepared to discuss risks and benefits of recommended treatments with their patients. Long-term goals should include evaluation of the effects of emergency response treatments, research and development of safer and effective medications when warranted.

Compared with what is known about conventional disease threats, knowledge about currently recognized emerging infectious diseases is quite limited. Soon we will likely be faced with novel pathogens about which little or nothing is known. Because the effects of emerging infections in some patients such as pregnant women might differ from those in the general population, it therefore become necessary that pregnancy must be considered a potential risk factor for disease susceptibility as well as for illness and death. Unfortunately, pregnancy issues are often not well addressed in outbreak investigations, ongoing prospective studies, or emergency preparedness planning. Future scientific inquiry and medical investigations must include pregnancy-related issues as a vital component.

The above listed steps in preparedness for biological agent attacks and bioterrorism emergencies would foster ongoing communication between community practitioners and regional public and private diagnostic laboratories (veterinary and medical inclusive) to establish baseline disease incidence trends and algorithms to identify outbreaks (Dauphin et al., 2004). The growing awareness that animal health and human health are inextricably linked, however, makes cooperation between human and animal health professionals imperative to strengthen the evidence base that will allow for rational use of animal data in public health decision-making. More so, detecting the agent in mobile animal populations could therefore signal the ongoing spread of the agent and provide an opportunity for interventions to prevent further spread. From all the issues discussed above it becomes imperative that the primary goal of public health response to emerging infections and bioterrorism attacks is to limit illness and death by providing the safest and most effective medical countermeasures in a timely manner to persons at greatest risk.

REFERENCES

Anderson NG, Gerin JL, Anderson NL (2003). Global screening for human viral pathogens. Emerging Infect. Dis. J., 9 (7): 768-773.

Bell D, Jenkins P, Hall J (2003). World Health Organization (WHO) Conference on Severe acute respiratory syndrome: where do we go from here? Emerging Infect. Dis. J., 9 (9):1091-1092.

Billings RJ, Berkowitz RJ, Watson G (2004). Teeth. Pediatrics, 113 (4): 1120-1127.

Borio L, Inglesby T, Peters CJ, Schmaljohn AL, Hughes JM, Jahrling PB, Ksiazek T, Johnson KM, Meyerhoff A, O'Toole T, Ascher MS, Bartlett J, Breman JG, Eitzen EM Jr, Hamburg M, Hauer J, Henderson DA, Johnson RT, Kwik G, Layton M, Lillibridge S, Nabel GJ, Osterholm MT, Perl TM, Russell P, Tonat K; Working Group on Civilian Biodefense. (2002). Hemorrhagic fever viruses as biological weapons: medical and public health management. J. Ame. Me. Assoc. (JAMA), 287: 2391-2405.

Brent RL (2004). Utilization of animal studies to determine the effects and human risks of environmental toxicants (drugs, chemicals, and physical agents). Pediatrics, 113: (4): 984-995.

Carrion Jr. R, Brasky K, Mansfield K, Johnson C, Gonzales M, Ticer A, Lukashevich I, Tardif S, Patterson J (2007). Lassa Virus Infection in Experimentally Infected Marmosets: Liver Pathology and Immunophenotypic Alterations in Target Tissues. J. Virolo., 81(12): 6482-6490

Center for Disease Control and Prevention (CDC, 2008). Prevention of Specific Infectious Diseases. CDC Health Information for International Travel 2008 available at: http://www.cdc.gov/travel/index.htm. (Accessed 20/06/2008)

Centers for Disease Control and Prevention (2001). Guidance for fiscal year 2001 supplemental funds for epidemiology and laboratory capacity for infectious diseases (ELC) cooperative agreement [ELC supplement A- NEDSSS FY2001: New activities]: National electronic disease surveillance system (NEDSS) activities. [Accessed June 20, 2008.] http://www.cdc.gov/nedss/About/NEDSS_RFA_2001.pdf

Centers for Disease Control and Prevention (CDC) (1997). Electronic reporting of laboratory data for public health. [Accessed June 12, 2008]. http://www.cdc.gov/od/hissb/docs/elr-1997.pdf

Centers for Disease Control and Prevention (CDC). (2000). Biological and chemical terrorism: strategic plan for preparedness and response. Recommendations of the CDC Strategic Planning Workgroup. *Morbidity and Mortality Weekly Report (MMWR) 49*:1–14.

Centers for Disease Control and Prevention (CDC). (2002). Electronic reporting of laboratory data for public health. 1997. (Accessed June 12, 2008). http://www.cdc.gov/od/hissb/docs/elr-1997.pdf

Centers for Disease Control and Prevention (CDC). 2004. Imported Lassa fever—New Jersey, 2004. *Morbidity and Mortality Weekly Report* (*MMWR*) *53*:894–897.

Centers for Disease Control and Prevention. (2003). Updated interim CDC guidance for use of smallpox vaccine, cidofovir, and vaccinia immune globulin (VIG) for prevention and treatment in the setting of an outbreak of monkeypox infections. http://www.cdc.gov/ncidod/monkeypox/treatmentguidelines.htm [Accessed July 13, 2008].

Centers for Disease Control and Prevention. (2003). Update: multistate outbreak of monkeypox—*Illinois, Indiana, Kansas, Missouri, Ohio, Wisconsin, Intials(2003). Morbidity and Mortality Weekly Report, (MMWR) *52*: 642-646.

Claudio Soto (2007). Prions: The New Biology of Proteins. CRC Press, Taylor and Francis Group, Boca Raton, Florida, USA, p.184.

Cono J, Cragan JD, Jamieson DJ, Rasmussen SA (2006). Prophylaxis and Treatment of Pregnant Women for Emerging Infections and Bioterrorism Emergencies. Emerging Infect. Dis. J. 12(11): 1631-1637

Cottrell TS, Morgan ER (2003). Animal surveillance in NBC defensive operations. J. Royal Army Med. Corps, 149: 225-230.

Crump JA, Luby SP, Mintz ED (2004). The global burden of typhoid fever. Bull. World Health Organ., 82(5): 346-353.

Dauphin G, Zientara S, Zeller H, Murgue B (2004). West Nile: worldwide current situation in animals and humans. Comprehensive Immunol. Microbiol. Infect. Dis., 27: 343-355.

Doering PL, Boothby LA, Cheok M (2002). Review of pregnancy labeling of prescription drugs: is the current system adequate to inform of risks? Am. J. Obstetrics and Gynecol, 187: 333-339.

Dolin R (2005). Influenza: interpandemic as well as pandemic disease. N. Engl. J. Med., *353*: 2535-2537.

Doyle TJ, Glynn KM, Groseclose SL (2002). Completeness of notifiable infectious disease reporting in the United States: an analytical literature review. Am. J. Epidemiol., 155: 866-874.

Fine A, Layton M (2001). Lessons from West Nile encephalitis outbreak in New York City, 1999: implications for bioterrorism preparedness. Clin. Infect. Dis., 32: 277-282.

Fisher-Hoch SP, Tomori O, Nasidi A, Perez-Oronoz GI, Fakile Y, Hutwagner L, McCormick JB. (1995). Review of cases of nosocomial Lassa fever in Nigeria: the high price of poor medical practice. British Med. J., 311: 857-859.

Food and Drug Administration (1979). 21 CFR 201.57 Use-in pregnancy ratingsystem.http://a257.g.akamaitech.net/7/257/2422/04nov20031500/edocket.access.gpo. gov/cfr_2001/aprqtr/21cfr201.57.htm (Accessed June 2008)

Gargan TP 2nd, Clark GG, Dohm DJ, Turell MJ, Bailey CL (1988). Vector potential of selected North American mosquito species for Rift Valley fever virus. Am. J. Trop. Med. Hygiene *38*:440–446.

Grady R (2003). Safety profile of quinolone antibiotics in the pediatric population. Pediatrics Infect. Dis. J., *22*:1128–1132.

Grant S, Olsen CW (1999). Preventing zoonotic diseases in immunocompromised persons: the role of physicians and veterinarians. Emerging Infect. Dis. J., *5*:159-163.

Harper SA, Fukuda K, Uyeki TM, Cox NJ, Bridges CB (2005). Prevention and control of influenza. Recommendations of the Advisory Committee on Immunization Practices (ACIP). Morbidity and Mortality Weekly Report (MMWR) Recommendation Report, 54: 1-40.

Henderson DA (1999). The looming threat of bioterrorism. Science, *283*: 1279-12782.

Howard JJ, Grayson MA, White DJ, Oliver J (1996). Evidence for multiple foci of eastern equine encephalitis virus (Togaviridae:Alphavirus) in central New York State. J. Med. Entomol., *33*: 421-432.

Inglesby TV, O'Toole T, Henderson DA, Bartlett JG, Ascher MS, Eitzen E, Friedlander AM, Gerberding J, Hauer J, Hughes J, McDade J, Osterholm MT, Parker G, Perl TM, Russell PK, Tonat K. (2002). Anthrax as a biological weapon, 2002: updated recommendations for management. JAMA, 287: 2236-2252.

Jamieson DJ, Theiler RN, Rasmussen SA (2006). Emerging Infections and Pregnancy. Emerging Infect. Dis. J., *12(11)*: 1638-1643

Jorgensen DM (1997). Gestational psittacosis in a Montana sheep rancher. Emerging Infect. Dis., 3: 191-194.

Lai ST (2005). Treatment of severe acute respiratory syndrome. Eur. J. Clin. Microbiol. Infect. Dis., 24: 583-591.

Laibl VR, Sheffield JS (2005). Influenza and pneumonia in pregnancy. Clin. Perinatol., 32: 727-738.

Lam CM, Wong SF, Leung TN, Chow KM, Yu WC, Wong TY, Lai ST, Ho LC (2004). A case-controlled study comparing clinical course and outcomes of pregnant and non-pregnant women with severe acute respiratory syndrome. Br. J. Gynecol (BJOG), 111: 771-774.

Lee TH, Montalvo L, Chrebtow V, Busch MP (2001). Quantitation of genomic DNA in plasma and serum samples: higher concentrations of genomic DNA found in serum than in plasma. Transfusion, 41: 276-282.

Lindler LE, Lebeda FJ, Korch GW (2004). Book Review on Biological Weapons Defense: Infectious Diseases and Counterbioterrorism. Humana Press, Totowa, New Jersey, pp. 597.

Lo WY, Friedman JM (2002). Teratogenicity of recently introduced medications in human pregnancy. Obstetrics and Gynecol, 100: 465-473.

Loebstein R, Lalkin A, Koren G (1997). Pharmacokinetic changes during pregnancy and their clinical relevance. *Clinical Pharmacokinetics 33*: 328–343.

Lucey DR (2006). Book Review on Biological Weapons Defense: Infectious Diseases and Counterbioterrorism. *Emerging Infectious Diseases J. 12(4)*: 713-714

M'ikanatha NM, Southwell B, Lautenbach E (2003). Automated laboratory reporting of infectious diseases in a climate of bioterrorism. *Emerging Infectious Diseases J. 9(9)*:1053-1057

McNally LM, Jeena PM, Lalloo U, Nyamande K, Gajee K, Sturm AW (2005). Probable mother to infant transmission of *Pneumocystis jiroveci* from an HIV-infected woman to her HIV-uninfected infant. AIDS, 19: 1548-1549.

Monath TP (1989). Yellow fever. Boca Raton (FL): CRC Press, pp. 139-231.

Mupapa K, Mukundu W, Bwaka MA, Kipasa M, De Roo A, Kuvula K, Kibadi K, Massamba M, Ndaberey D, Colebunders R, Muyembe-

Tamfum JJ (1999). Ebola hemorrhagic fever and pregnancy. J. Infect. Dis., 179(1): 11-12.
Ng PC, Leung CW, Chiu WK, Wong SF, Hon EK (2004). SARS in newborns and children. Biol. Neonate, 85: 293-298.
Price ME, Fisher-Hoch SP, Craven RB, McCormick JB (1988). A prospective study of maternal and fetal outcome in acute Lassa fever infection during pregnancy. Br. Med. J., 297: 584-587.
Rabinowitz P, Gordon Z, Chudnov D, Wilcox M, Odofin L, Liu A, Dein J (2006). Animals as sentinels of bioterrorism agents. Emerging Infect. Dis. J. 12(4): 647-652.
Rasmussen SA, Hayes EB (2005). Public health approach to emerging infections among pregnant women. Am. J. Public Health, 95: 1942-1944.
Rybak LP (1990). Deafness associated with Lassa fever. J. Am. Med. Assoc.(JAMA), 264: 2119.
Solbrig MV, McCormick JB (1991). Lassa fever: central nervous system manifestations. J. Trop. Geographical Neurol., 1: 23-30.
Stockman LJ, Lowther SA, Coy K, Saw J, Parashar UD (2004). SARS during pregnancy, United States. Emerging Infect. Dis. J., 10: 1689-1690.
Wharton M, Strikas RA, Harpaz R, Rotz LD, Schwartz B, Casey CG, PearsonML, Anderson LJ (2003). Recommendations for using smallpox vaccine in a pre-event vaccination program: supplemental recommendations of the Advisory Committee on Immunization Practices and the Healthcare Infection Control Practices Advisory Committee. Morbidity and Mortality Weekly Report (MMWR) Recommendation Report, 52: 1-16.
White SR, Henretig FM, Dukes RG (2002). Medical management of vulnerable populations and co-morbid conditions of victims of bioterrorism. Emerging Med. Clin. North Am., 20: 365-392.

16

Non-rhizobial nodulation in legumes

D. Balachandar*, P. Raja, K. Kumar and SP. Sundaram

Department of Agricultural Microbiology, Tamil Nadu Agricultural University, Coimbatore – 641 003, India

Legume - Rhizobium associations are undoubtedly form the most important N_2-fixing symbiosis and play a subtle role in contributing nitrogen and maintaining/improving soil fertility. A great diversity in the rhizobial species nodulating legumes has been recognized, which belongs to α subgroup of proteobacteria covering the genera, *Rhizobium, Sinorhizobium* (renamed as *Ensifer*), *Mesorhizobium, Bradyrhizobium* and *Azorhizobium*. Recently, several non-rhizobial species, belonging to α and β subgroup of Proteobacteria such as *Methylobacterium, Blastobacter, Devosia, Phyllobacterium, Ochrobactrum, Agrobacterium, Cupriavidus, Herbaspirillum, Burkholderia* and some γ-Proteobacteria have been reported to form nodules and fix nitrogen in legume roots. The phylogenetic relationship of these non-rhizobial species with the recognized rhizobial species and the diversity of their hosts are discussed in this review.

Key words: Legumes, Nodulation, Proteobacteria, Rhizobium

Table of content

1. Introduction
2. Non-rhizobial nodulation
2.1. ∝-Proteobacteria
2.1.1. *Methylobacterium*
2.1.2. *Blastobacter*
2.1.3. *Devosia*
2.1.4. *Phyllobacterium*
2.1.5. *Ochrobactrum*
2.1.6. *Agrobacterium*
2.2. β-Proteobacteria
2.2.1. *Cupriavidus*
2.2.2. *Herbaspirillum*
2.2.3. *Burkholderia*
2.3. γ-Proteobacteria
3. Conclusion
4. References

INTRODUCTION

Members of the leguminosae form the largest plant family on earth with around 19,000 species (Polhill et al., 1981). The success of legumes can largely be attributed to their ability to form a nitrogen-fixing symbiosis with specific bacteria known as rhizobia, manifested by the development of nodules on the plant roots in which the bacteria fix atmospheric nitrogen, a major contributor to the global nitrogen cycle. Symbiosis between legumes and rhizobia are of considerable environmental and agricultural importance since they are responsible for most of the atmospheric nitrogen fixed on land (Graham and Vance, 2003). Among the 19,000 species described so far, only a small proportion has been studied for their nodulation ability. The legume biodiversity is concentrated in tropical regions, while most studies are on cultivated leguminous plants from temperate region (Moulin et al., 2001).

Rhizobia are able to elicit symbiotic association on most of the species of leguminosae, forming nodules in roots in which they reduce atmospheric nitrogen to ammonia for the benefit of plant. Nodule formation is controlled by extracellular bacterial signal molecules, called

*Corresponding author. E-mail: dbalu2000@yahoo.com.

nod factors, which are recognized by the host plant (Lerouge et al., 1990; Schultze and Kondorosi, 1998). The rhizobial species described so far are very diverse and do not form an evolutionary homologous clade. They are ∝ subgroup Proteobacteria, belonging to three distinct branches which are phylogenetically intervened with many non-symbiotic bacteria (Young and Haukka, 1996; Zakhia and deLajudie, 2001). The first large branch groups the genera *Rhizobium, Sinorhizobium,* (renamed as *Ensifer*), *Mesorhizobium* with *Agrobacterium,* a pathogenic bacterium of plant. The second branch contains the genus *Bradyrhizobium* and the third branch includes the genus *Azorhizobium.* Each rhizobial species has a defined host range, varying from very narrow to very broad. Apart from rhizobia, very close relatives of rhizobia such as members of ∝ subgroup of proteobacteria and few β-Proteobacteria are also recognized to form nodules in legumes, which are the focus of this review.

Non-rhizobial nodulations

In the past five years, several symbionts capable of forming nodules and fixing nitrogen in legume roots have been documented and grouped under ∝ and β subclass of Proteobacteria, which include *Methylobacterium nodulans* (Sy et al., 2001), *Blastobacter denitrificans* (Van Berkum and Eardly, 2002), *Devosia* sp.(Rivas et al., 2002), *Ochrobactrum lupini* (Trujilo et al., 2005), *Agrobacterium* like strains (Mhamdi et al., 2005), *Phyllobacterium trifolii* (Valverde et al., 2005) , *Herbaspirillum lustianum* (Valverde et al., 2003), *Ralstonia taiwanensis* (renamed as *Cupriavidus taiwanensis*) (Chen et al., 2001), *Burkholderia tuberum, Burkholderia phymatum* (Vandamme et al., 2002), *Burkholderia cepacia* (Rasolomampianina et al., 2005) and few γ-proteobacteria (Benhizia et al., 2004). The legume host preferred by these non-rhizobial proteobacteria possesses high diversity.

∝ Proteobacteria

Methylobacterium

The genus *Methylobacterium* includes a variety of pink pigmented facultative methylotrophic bacteria (PPFMs) that are able to grow on C1 compounds such as formate, formaldehyde and methanol as sole carbon source, as well as on wide range of multi-carbon growth substances (Green, 1992). *Methylobacterium* belongs to ∝-Proteobacteria, comes under order Rhizobiales and family Methylobacteriaceae. Recently, Sy et al. (2001) reported *M. nodulans*, a facultative methylotrophic species forming nodules in the legume, *Crotalaria* spp., the only species of that genus able to form nodules and also fix nitrogen. It was also confirmed that this species harbors the nodulating genes, *nodABC* and genes encoding structural nitrogenase enzyme by PCR amplification (Sy et al., 2001; Jourand et al., 2004). Phylogenic analysis of *M. nodulans* shows it is closer relationship to other species of *Methylobacterium* (Sy et al., 2001), whereas, the phylogenic tree based on *nifH* gene shows that it is more related to *Gluconacetobacter diazotrophicus,* a sugarcane endophyte of ∝-Proteobacteria (Figure 1). *M. nodulans* type strain ORS2060 is able to form nitrogen fixing nodules in *Crotalaria glaucoides, Crotalaria perrottetii* and *Crotalaria podocarpa*. Similarly, another legume, *Lotonois bainesii* was also reported to be a host for nodulation and nitrogen fixation by pink pigmented *Methylobacterium* (Jaftha et al., 2002). Raja et al. (2006) isolated nodulating and non-nodulating *Methylobacterium* sp. from legumes with high nitrogenase activity. Similarly, Mathaiyan et al. (2006) isolated several nodulating *Methylobacterium* from tropical legumes such as field beans, cowpea, black gram, soybean, *Sesbania* with high nitrogenase activity. These isolates are able to form effective nodules in *Crotolaria juncea*. Their *nif* and *nod* gene diversity and mode of infection are to be examined.

Blastobacter

Blastobacter spp. is commonly aquatic (especially freshwater) budding bacteria that form rosette structures by cellular attachment to a common base. They belong to ∝-Proteobacteria, order Rhizobiales and family Brady-rhizobiaceae. The flood tolerant leguminous plant, *Aschynomene indica* is most frequently nodulated by *Bradyrhizobium japonicum* and *Bradyrhizobium elkanii* (Van Berkum et al., 1995). Van Berkem and Eardly (2002) isolated *B. denitrificans* capable of forming an effective nitrogen fixing symbiosis with *A. indica.* They are able to form effective nodules in the roots of *A. indica* and reduce the atmospheric nitrogen (11.5 μM of C_2H_4 produced per plant as ARA) (Table 1), which was again confirmed by presence of *nifHDK* genes by southern hybridization. The predicted results revealed that *B. denitrificans* could well be a symbiotic bacterium of *A. indica,* which has very close phylogenetic relation with *Bradyrhizobium* spp (Figure 2). Further confirmation of presence of *nif* gene diversity of this species is essential to reclassify this bacterium as *Blastobacter* or renaming the genera as *Bradyrhizobium.*

Devosia

The α-Proteobacterium, *Devosia,* currently includes a known species *Devosia riboflavina,* isolated from soil. This species is Gram-negative, aerobic, non-sporulated motile rod that contains long-chain 3-hydroxy fatty acids. Recently, isolates from nodules of aquatic legume *Neptunia natans* from India were reported to be phylogenetically very close to *D. riboflavina* (95.9 % similarity), but not the same species (Rivas et al., 2002).

Devosia is in the family Hyphomicrobiaceae within the order Rhizobiales of α-Proteobacteria. The other rhizobia of this family, *Azorhizobium caulinodans* does not show

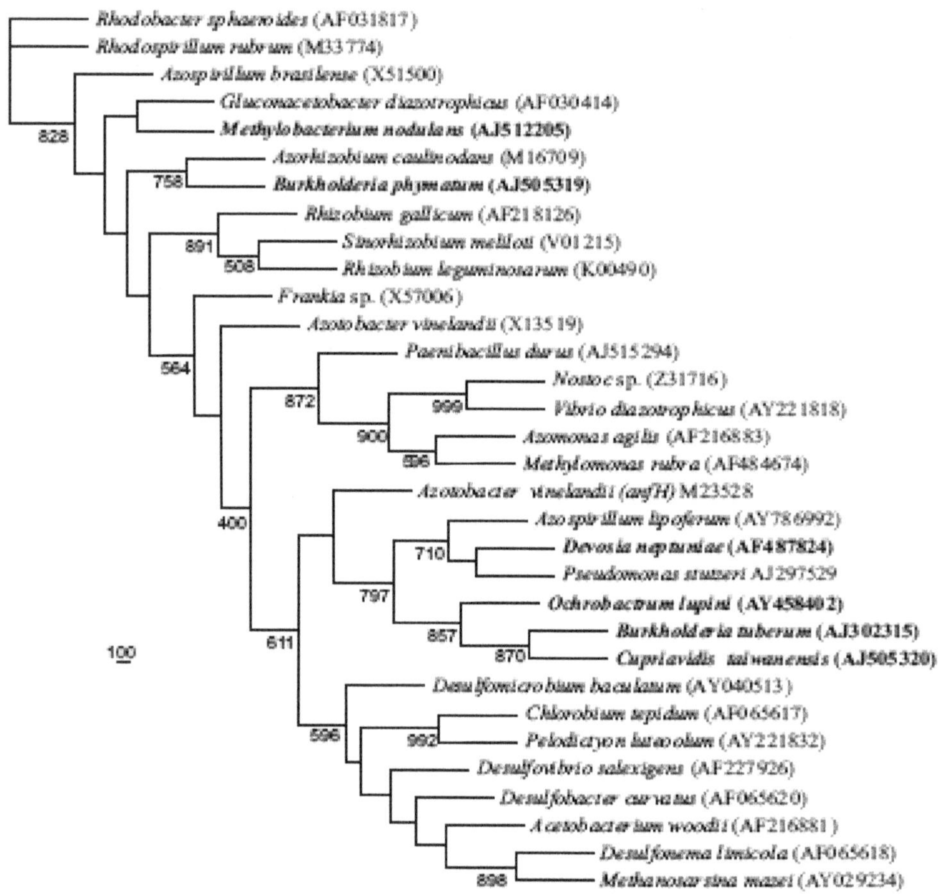

Figure 1. Phylogenetic tree based on a 368 bp fragment of *nifH* gene showing relationship of non-rhizobial nodulating proteobacteria with other nitrogen fixing bacteria. *nifH* sequences (accession number in parentheses) obtained from GenBank were used for phylogenetic tree construction. Phylogenetic tree was constructed by neighbor-joining method using DNADIST, SEQBOOT, NEIGHBOR and CONSENSE from PHYLIP v3.5c and the tree file was analysed using tree view. Bootstraps values of 500 or more (from 1000 replicates) are indicated at the nodes. The *nifH* of non-rhizobial proteobacteria are highlighted by bold letters.

Table 1. Nodulation and nitrogen fixation of non-rhizobial Proteobacteria in their respective legume roots.

Proteobacteria	No. of nodule /plant	Nitrogenase activity[a]	Reference
Blastobacter denitrificans (LMG8443[T])	NR	11.5	Vab Berkum and Eardly (2002)
Burkholderia sp. (Br3454[T])	19.0	2.91	Chen et al. (2005)
Ochrobactrum lupini (LUP21[T])	7.0	NR	Trujillo et al. (2005)

[a] Acetylene reduction assay in nmol of ethylene produced /h/plant, NR – Not reported

close relatedness, as far as 16S rDNA phylogeny is concerned. But, *nodD* gene of *Devosia* spp. nodulating *N. natans* is closely related to *Rhizobium leguminosarum* (Figure 3a). The presence of *nod* and *nif* genes were detected in *Devosia* isolates by PCR amplification, which are located in 170 kbp sized plasmids (known as *sym* plasmids) as that of most of the fast growing rhizobia. The species nodulating *N. natans* was named as *Devosia neptuniae* (Rivas et al., 2003).

Phyllobacterium

Phyllobacterium sp. is a soil inhabitating proteobacteria; belong to the family Phyllobacteriaceae of order Rhizobiales, having recognized species, *Phyllobacterium myrsinacearum* (Mergaert et al., 2002). *Phyllobacterium* species was originally proposed to accommodate bacteria isolated from leaf nodules of Rubiaceae and Myrsinaceae tropical plants (Knosel et al., 1984).

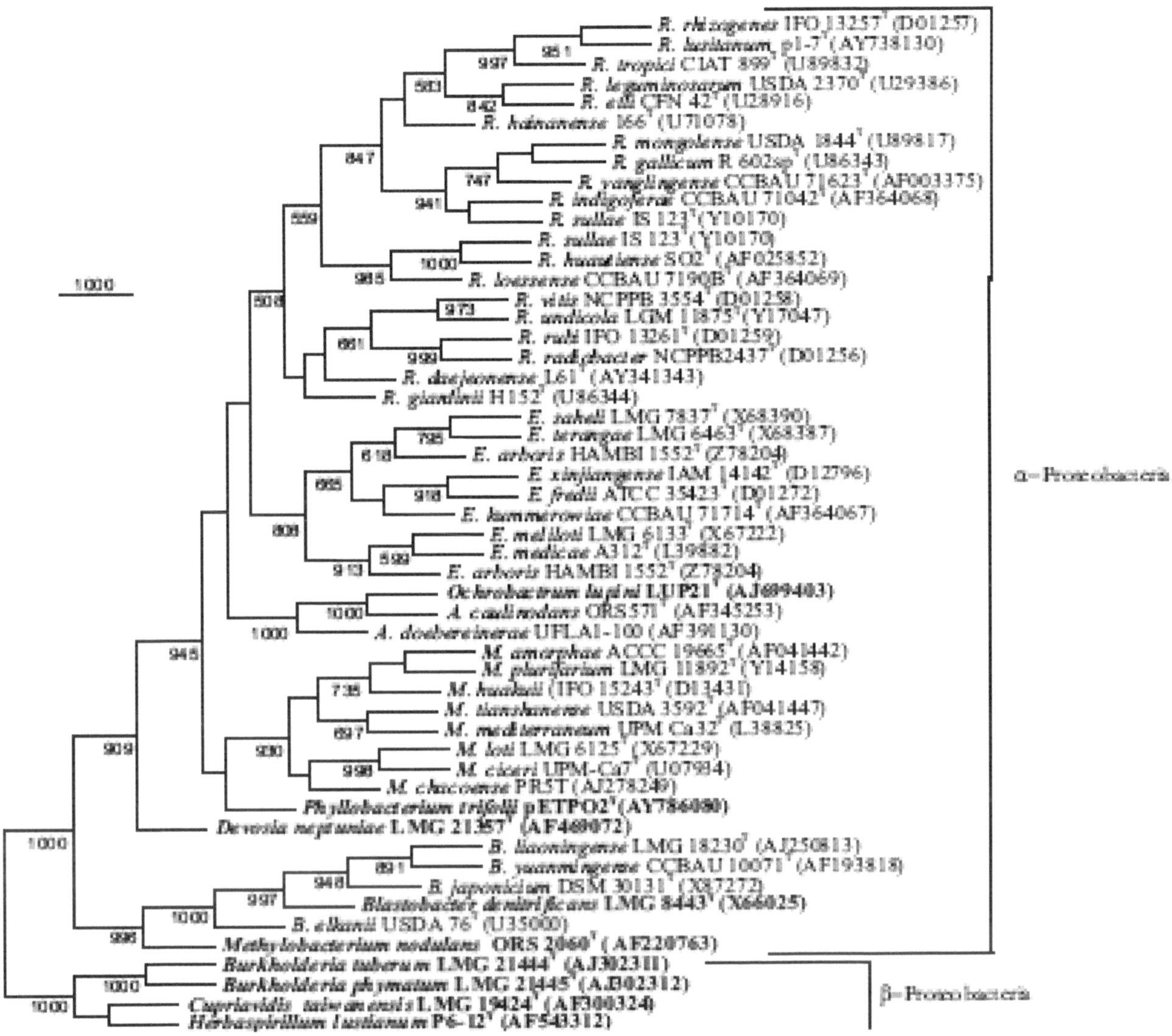

Figure 2. Phylogenetic relationship of rhizobia and other nodulating Proteobacteria based on 16S rRNA gene sequence comparisons. The sequence of type strain of representative rhizobial species and non-rhizobial species obtained from GenBank were used for phylogenetic tree construction. Sequence accession number of 16S rRNA of type strains are given in parentheses. Phylogenetic tree was constructed by neighbor-joining method using DNADIST, SEQBOOT, NEIGHBOR and CONSENSE from PHYLIP v3.5c and the tree file was analysed using tree view. Bootstraps values of 500 or more (from 1000 replicates) are indicated at the nodes.

Recently, another species of *Phyllobacterium*, *P. trifolii* was isolated from nodules of *Trifolium pratense* growing in Spanish soil (Valverde et al., 2005). Plant infection experiments with this species (Strain pETPO2^T) revealed that it forms effective nodules in the roots of *Trifolium repans* and *Lupinus albus*. The 16S ribosomal gene sequence comparative analysis revealed more closeness towards *P. myrsinacearum* (98 % similarity) and the closest rhizobia, *Mesorhizobium* spp (Figure 2). The species, *P. trifolii* differs with *P. myrsinacearum* both in cultural characters as well as the fatty acid methyl ester (FAME) profile. The sequence of *nodD* gene is more similar to other non-rhizobial nodule forming β-Proteobacterium, *Ochrobactrum*, (94.9 % similarity) which forms nodules in *Lupinus* than rhizobial species (Figure 3a). The plasmid profile revealed that *P. trifolii* harbors symbiotic plasmids in which nodulating and nitrogen fixing genes are located.

Ochrobactrum

Ochrobactrum, a α_2 subgroup of Proteobacteria belonging to the family Brucellaceae and order Rhizobiales, was first described by Holmes et al. (1988) with a single species, *Ochrobactrum anthropi.* Later four different species were recognized from different ecological origins, clinical sources, soil, wheat rhizosphere and chicken faeces.

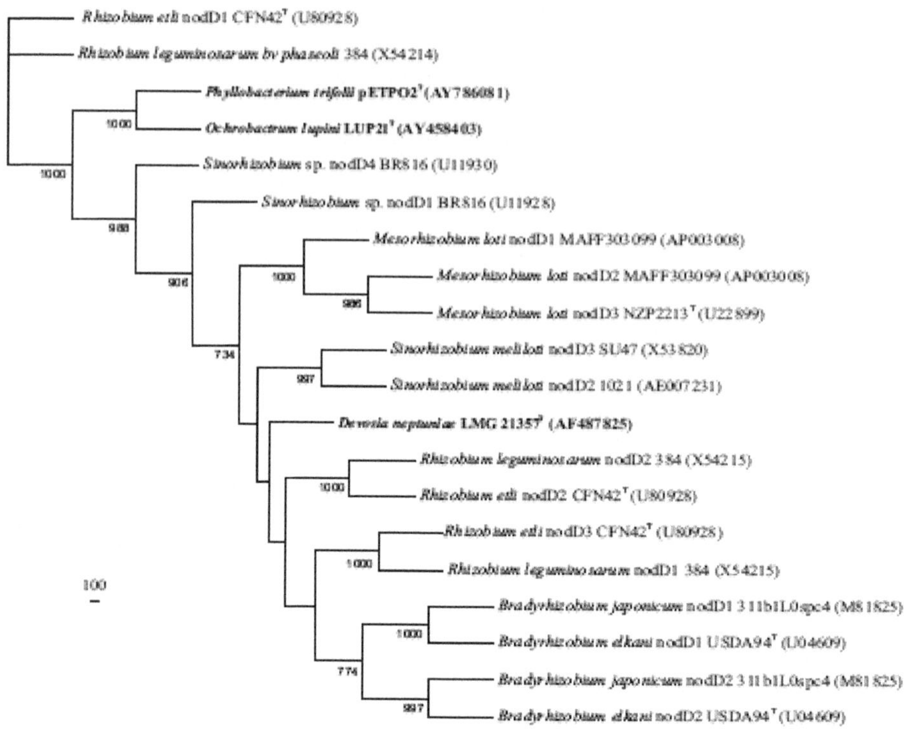

(a)

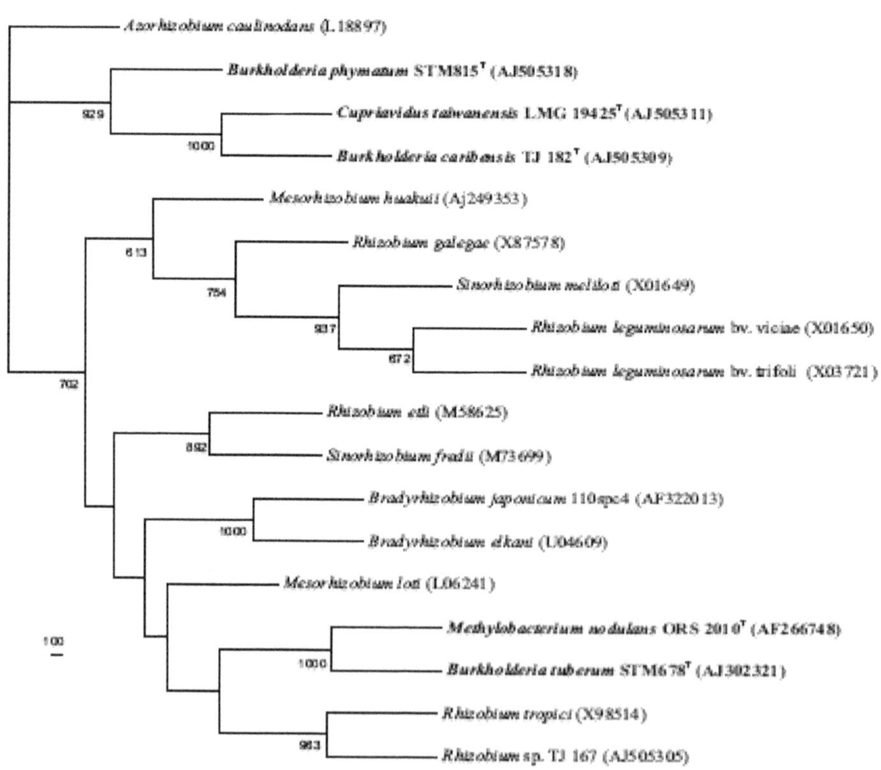

(b)

Figure 3. Comparative sequence analysis of *nodD* (a) and *nodA* (b) genes from non-rhizobial Proteobacteria and rhizobia. Sequence data obtained from GenBank were used for phylogenetic tree construction by neighbor-joining method using DNADIST, SEQBOOT, NEIGHBOR and CONSENSE from PHYLIP v3.5c and the tree file was analysed using tree view. Bootstraps values of 500 or more (from 1000 replicates) are indicated at the nodes. The non-rhizobial proteobacteria are highlighted by bold letters and sequence accession numbers are in parentheses.

Recently Ngom et al. (2004) isolated *Ochrobactrum* strain from root nodules of *Acacia mangium*. Trujilo et al. (2005) isolated a novel nodulating species, *O. lupini* from *Lupinus honoratum*. The 16S rRNA phylogeny showed that the strains of *O. lupini* do not belong to the rhizobial genera described until now. The nodulating species, *O. lupini* strain LUP23^{T} is more closely related to other species of *Ochrobactrum*, namely *O. anthropi, O. tritici* etc. *O. lupini* strain LUP21 and LUP23 harbor the mega plasmids of size 1500, 200 and 150 kbp. The nodulating and nitrogen fixing genes (*nod* and *nif* genes) were detected in all the *sym* plasmids using *nifH* and *nodD* probes. The isolate is able to form nitrogen fixing nodules in *L. albus* and this is the only species of Brucellaceae capable of nodule formation in legume, *Lupinus*.

Agrobacterium

Bacteria related to *Agrobacterium* were also identified among the root nodules of several tropical leguminous plants including *Acacia, Prosophis* and *Chamaecrista* from Africa (De Lajudie et al., 1999). Similarly, a significant portion of bacterial isolates of *Phaseolus vulgaris* grown in Tunisian soils has 16S rRNA genes similar to those of *Agrobacterium tumifaciens* and *Agrobacterium rubi* (Mhamdi et al., 2002). No symbiotic and nitrogen fixing genes were detected by PCR amplification of *nifH* and *nodC* genes and hybridization of total genomic DNA with nitrogen fixing gene probes. Mhamdi et al. (2002) confirmed that these *Agrobacterium* like strains entered the nodules by mixed infection with a rhizobial cell capable of nodule induction resulting mixed population within the nodule. When *Agrobacterium* cells mixed with rhizobia, (*R. gallicum*), it was able to occupy about 10-15 % of the total nodules in *P. vulgaris*, indicating that agrobacteria are likely to be confused as one of the non-rhizobial nodulations in legumes, while isolated from such broad hosts.

β Proteobacteria

Cupriavidus

Cupriavidus is a β-Proteobacteria, belonging to family Burkholderiaceae of order Burkholderiales. Several species of this genus were isolated from soil and human clinical specimens (Vandamme and Coeyne, 2004). Recently, *C. taiwanensis* (originally reported as *Ralstonia taiwanensis* or *Wautersia taiwanensis*) was isolated from nodules of *Mimosa pudica* and *Mimosa diplotricha* (Chen et al., 2001). Among many species of *Cupriavidus*, *C. taiwanensis* was the only species able to form root nodules in legumes. Phylogenetically, *C. taiwanensis* is distinct to all the rhizobial species (Figure 2) and are very close to other species of *Cupriavidus*. This species forms effective nodules in *M. pudica* and *M. diplotricha* and fixes atmospheric nitrogen. Further, two *sym* plasmids with approximate size of 3.5 and 2.4 Mb were identified in *C. taiwanensis*, in which *nod* and *nif* genes are presented (Chen et al., 2003).

Herbaspirillum

Herbaspirillum, a nitrogen fixing soil bacterium able to establish close associations with plants especially the grass plants, through epiphytic colonization in rhizoplanes and endophytic colonization in roots (Oliveres et al., 1997). It belongs to β-Proteobacteria, coming under the family Burkholderiaceae of order Burkholderiales. Recently a new species, *H. lustitanum* was isolated from root nodules of *P. vulgaris* from Portugal soils (Valverde et al., 2003). Complete 16S rDNA sequence based phylogenetic analysis of the isolate revealed that *H. lusitanum* is closely related to *Herbaspirillum seropedicae* and *Herbaspirillum rubrisubalbicans*. Further, this nodulating species is differentiated from other species by ARDRA (Amplified ribosomal DNA restriction analysis) and TP-RAPD (Two primer randomly amplified polymorphic DNA) patterns (Valverde et al., 2003). The presence of *sym* plasmids and host specificity of nodulation are to be evaluated for further establishment.

Burkholderia

Burkholderia, a β subclass Proteobacteria belonging to family Burkholderiaceae of order Burkholderiales, is a common soil habitant bacterium, often associated with plant roots. Few endophytic species were also recorded in maize. Moulin et al. (2001) isolated *Burkholderia* sp. from nodules of South African legume, *Aspalathus carnosa*. The 16S ribosomal DNA sequence homology revealed that it is very close to the species *Burkholderia kururiensis* (96.9 % identity). The isolate is able to form nodules in the roots of *Macroptilium atropurpureum,* a tropical legume, capable of establishing diverse rhizobial symbiosis. Further, the presence of *nodABC* through PCR confirmed that it is capable of producing nod factors to initiate the nodulation. Similarly, nodulating and nitrogen fixing *Burkholderia* species, namely *Burkholderia caribensis, B. phymatum* and *B. tuberum* were isolated from *Mimosa bimucronata* and *Mimosa pigra* (Vandamme et al., 2002; Chen et al., 2003; Chen et al., 2005). All these isolates are able to form effective nodules in *M. pudica, Mimosa diplotricha, M. pigra* and *Mimosa acutishpula* with nitrogenase activity (Table 1). Sequence similarity analysis of *nodA* showed that *B. phymatum* is closer to other non-rhizobial nodulating bacterium, *Cupriavidus taiwanensis,* whereas *B. tuberum* is close to *nodA* of *Methylobacterium nodulans* (Figure 3b).

γ-Proteobacteria

Benhizia et al. (2004) examined the nodules of three Mediterranean wild legume species, *Hedysarum carno-*

Table 2. The α- and β-Proteobacteria, nodulating legumes and their taxonomical position.

Non-rhizobial species	Host legume	Phylogenetic group	Taxonomical position		Reference
			Family	Order	
Methylobacterium nodulans	*Crotalaria glaucoides*	α-Proteobacteria	methylobacteriaceae	Rhizobiales	Sy et al. (2001)
Blastobacter denitrificans	*Aschynomene indica*	α-Proteobacteria	Bradyrhizobiaceae	Rhizobiales	Van Berkum and Eardly(2004)
Devosia neptuniae	*Neptunia natans*	α-Proteobacteria	Hyphomicrobiaceae	Rhizobiales	Rivas et al. (2002)
Phyllobacterium trifolii	*Trifolium repans*	α-Proteobacteria	Phyllobacteriaceae	Rhizobiales	Valverde et al. (2005)
Ochrobactrum lupini	*Lupinus albus*	α-Proteobacteria	Brucellaceae	Rhizobiales	Trujilo et al. (2005)
Agrobacterium like strains	*Phaseolus vulgaris*	α-Proteobacteria	Rhizobiaceae	Rhizobiales	Mhamdi et al. (2005)
Burkholderia tuberum, *B. phymatum*	*Mimosa*	β-Proteobacteria	Burkholderiaceae	Burkholderiales	Vandamme et al. (2002)
Cupriavidus taiwanensis	*Mimosa*	β-Proteobacteria	Burkholderiaceae	Burkholderiales	Chen et al. (2001)
Herbaspirillum lusitanum	*Phaseolus vulgaris*	β-Proteobacteria	Burkholderiaceae	Burkholderiales	Valverde et al. (2003)

sum, *Hedysarum spinosissimum* subsp. *capitatum* and *Hedysarum pallidum* and identified very high diversity among the microsymbionts nodulating these wild legumes. ARDRA profiles and RAPD fingerprinting analysis of these microsymbionts revealed that nodules of these wild legumes are not formed due to single species of rhizobia and surprisingly include many gamma Proteobacteria *viz., Enterobacter kobei, Enterobacte cloaceae, Leclercia adecarboxylata, Pantoea agglomerans, Escherichia vulneris* and *Pseudomonas* sp. No evidence of any rhizobial like sequence was found even upon amplifying from bulk of microbial cells obtained from squashed nodules, suggesting that exclusive occupants are members of the order Enterobacteriales or Pseudomonadales, which are belonging to gamma Proteobacteria (Bennizia et al., 2004). This is the first report of gamma Proteobacteria associated with legume nodules. Presence of *nifH, nodA, nodD* genes and their divergence in these γ-Proteobacteria need to be examined to include these bacteria as rhizobia.

Conclusion

Rhizobia are wide spread soil bacteria able to induce the formation of root nodules and to fix nitrogen on cultivated and wild legumes. These rhizobia are of economic importance in low input sustainable agriculture, agroforestry and land reclamation. Currently there are five genera of rhizobia in α-Proteobacteria recognized as *Azorhizobium, Bradyrhizobium, Rhizobium, Meso-rhizobium* and *Ensifer* (Zakhia and deLajudie, 2001). Besides these genera, several α- and β-Proteobacteria and several unclassified γ-Proteobacteria were recently isolated from nodules of diversified legume plants as summarized in Table 2. Several reports as summarized in this review are available about the occurrence of new species of Proteobacteria with the ability to nodulate specific host legumes. Most of these Proteobacteria nodulating legumes are characterized based on the genotypic, phenotypic and chemotaxonomic data and designated as new species. The *nifH* and *nodD* or *nodA* gene sequence divergence was also reported for few of these isolates. These approaches demonstrate the need for more mole-cular studies on:

i. Interaction between these Proteobacteria with their legume host, in terms of signal exchange.
ii. Diversity of flavonoids and nod factors which regulate the nodulation process.
iii. Mode of entry and infection process of these Proteobacteria in legumes.
iv. Interrelation between these Proteobacteria and with rhi-zobia in terms of *nif* and *nod* genes.
v. Source of *nif* and *nod* genes made these bacteria to form nodules (Perret et al., 2000).

These studies are to be addressed for better understanding the "origin of nodules" and for enhancing the nodulation and nitrogen fixing efficiency of these new group of Proteobacteria.

Nitrogen fixation, which is wide spread in Eubacteria and Archaea, is thought to be an ancestral function now lost by many bacteria. Conversely, nodulation is thought to have appeared in evolution, at the same time as the appearance of legumes on earth, about 70 -130 million years ago. At this period of history, α- and β-Proteobacteria and different rhizobial linkages already had diverged. The genes required for legume nodulation are thought to have been acquired subsequently by lateral transfer from undefined sources, thus converting soil saprophytes into symbionts (Hirsch et al., 2001). This hypothesis has been confirmed by many recent works (Sullivan and Ronson, 1998) and presence of very similar and phylogenetically related *nodABC* genes in nodulating α- and β-Proteobacteria strongly supports the hypothesis of a unique origin of *nod* genes. However, it is not clear whether a single transfer event was responsible for the spread of nodulation genes from one subclass to the other.

The wide spread character of nodulation by many Proteobacteria, dominated by rhizobia is also attested by very high diversity of legumes. Since many legumes and environments remain to be explored, it is likely that further characterization of rhizobia from new legumes will reveal an even greater diversity.

REFERENCES

Benhizia Y, Benhizia H, Benguedouar A, Muresu R, Giacomini A, Squartini A (2004). Gamma Proteobacteria can nodulate legumes of the genus *Hedysarum*. Syst. Appl. Microbiol. 27: 462 – 468.

Chen WM, deFaria SM, Straliotto R, Pitard RM, Simoes-Araujo JL, Chou JH, Chou, YJ, Barrios E, Prescott AR, Elliott GN, Sprent JI, Young JPW, James EK (2005). Proof that *Burkholderia* strains form effective symbioses with legumes: a study of novel *Mimosa* nodulating strains from South America. Appl. Environ. Microbiol. 71: 7461-7471.

Chen WM, Laevens S, Lee TM, Coenye T, Vos PD, Mergeay M, Vandamme P (2001). *Ralstonia taiwanensis* sp. nov., isolated from root nodules of *Mimosa* species and sputum of cystic fibrosis patient. Int. J. Syst. Evol. Microbiol. 51: 1729 – 1735.

Chen WM, Moulin L, Bontemps C, Vandamme P, Bena G, Masson CB (2003). Legume symbiotic nitrogen fixation by β Proteobacteria is wide spread nature. J. Bacteriol. 185: 7266-7272.

De Lajudie P, Willems A, Nick G, Mohamed SH, Torck U, Coopman R, Filali-Maltouf A, Kersters K, Dreyfus B, Lindstrom K, Gillis M (1999). *Agrobacterium* bv. 1 strains isolated from nodules of tropical legumes. Syst. Appl. Microbiol. 22: 119-132.

Graham PH, Vance CP (2003). Legumes: Importance and Constraints to Greater Use. Plant Physiol. 131: 872-877.

Green PN (1992). The genus *Methylobacterium* In A. Balows, H.G. Truper M, Dworkin M, Harder W, Schleifer KH (eds) The Prokaryotes, Springer – Verlag, New York. pp. 2342-2349.

Hirsch AM, Lum MR, Downie JA (2001). What makes the rhizobia-legume symbiosis so special? Plant Physiol. 127: 1484-1492.

Holmes B, Popoff M, Kiredjian M, Kersters K (1988). *Ochrobactrum anthropi* gen. nov., sp. nov., from human clinic specimens and previously known group Vd. Int. J. Syst. Bacteriol. 38: 406-416.

Jaftha JB, Strijdom BW, Steyn PL (2002). Characterization of pigmented methylotrophic bacteria which nodulate *Lotononis bainesii*. Syst. Appl. Microbiol. 25: 440-449.

Jourand P, Girand E, Bena G, Sy A, Willems A, Gillis M, Dreyfus B, deLajudie P (2004). *Methylobacterium nodulans* sp. nov., for a group of aerobic, facultatively methylotrophic, legume root-nodule-forming and nitrogen-fixing bacteria. Int. J. Syst. Evol. Microbiol. 54: 2269-2273.

Knosel DH (1984). Genus IV. *Phyllobacterium* nom. rev. In: Krieg NR, Holt JG, (eds) Bergey's Manual of systematic Bacteriology Vol. 1, Williams and Wilkins, Baltimore, pp. 254-256.

Lerouge P, Roche P, Faucher C, Maillet F, Truchet G, Prome JC, Denarie J (1990). Symbiotic host-specificity of *Rhzobium meliloti* is determined by a sulphated and acylated glucosamine oligosaccharide signal. Nature 344: 781-784.

Madhaiyan M, Poonguzhali S, Senthilkumar M, Sundaram SP, Sa T (2006). Nodulation and plant growth promotion by methylotrophic bacteria isolated from tropical legumes. Microbiol. Res. (On line Publ.)

Mergaert J, Cnockaert MC, Swings J (2002). *Phyllobacterium myrsinaecearum* (Subjective synonym *Phyllobacterium rubiacearum*) emend. Int. J. Syst. Evol. Microbiol. 52: 1821-1823.

Mhamdi R, Laguerre G, Aouani ME, Mars M, Amarger N (2002). Different species and symbiotic genotypes of field rhizobia can nodulate *Phaseolus vulgaris* in Tunisian soil. FEMS Microbiol. Ecol. 41: 77-84.

Mhamdi R, Mrabet M, Laguerre G, Tiwari R, Aouani ME (2005). Colonization of *Phaseolus vulgaris* nodules by *Agrobacterium* like strains. Can. J. Microbiol.51:105-111.

Moulin L, Munive A, Dreyfus B, Boivin-Masson C (2001). Nodulation of legumes by members of the β subclass of Proteobacteria. Nature 411: 948-950.

Ngom A, Nakagawa Y, Sawada H, Tsukahara J, Wakabayashi S, Uchiumi T, Nuntagij A, Kotepong S, Suzuki A, Higashi S, Abe M (2004). A novel symbiotic nitrogen-fixing members of the *Ochrobactrum* clade isolated form root nodules of *Acacia mangium.* J. Gen. Appl. Microbiol. 50: 17-27.

Oliveres FL, James EK, Baldani JI, Dobereiner J (1997). Infection of mottled stripe disease susceptible and resistant sugarcane varieties by the endophytic diazotroph *Herbaspirillum.* New Phytol. 135: 723-737.

Perret X, Staehelin C, Broughton WJ (2000). Molecular basis of symbiotic promiscuity. Microbiol. Mol. Biol. Rev. 64: 180-201.

Polhill RM, Raven PH, Stirton CH (1981). Evolution and systematics of the Leguminosae. In: Polhill RM, P.H. Raven (eds) Advances in Legume Systematics Part 1, Royal Botanic Gardens, Kew, UK, pp. 1-26

Raja P, Uma S, Sundaram SP (2006). Non-nodulating pink-pigmented facultative *Methylobacterium* sp. with a functional *nifH* gene. World J. Microbiol. Biotechnol. 22: 1381-1384

Rasolomampianina R, Bailly X, Fetiarison R, Rabevohitra R, Bena G, Ramaroson L, Raherimandimy M, Moulin L, DeLajudie P, Dreyfus B (2005). Nitrogen fixing nodules from rose wood legume trees (*Dalbergia* spp) endemic to Madagascar host seven different genera belongs to α and β Proteobacteria. Mol. Ecol. 14: 4135-4146.

Rivas R, Velazquez E, Willems A, Vizcaino N, Subbarao NS, Mateos PF, Gillis M, Dazzo FB Molina EM (2002). A new species of *Devosia* that forms a unique nitrogen fixing root symbiosis with aquatic legume *Neptunia natans* (l.f) Druce. Appl. Environ. Microbiol. 68: 5217-5222.

Rivas R, Willems A, Subbarao NS, Mateos PF, Dazzo FB, Kroppenstedt RM, Martinez-Molina E, Gillis M, Velazquez E (2003). Description of *Devosia neptuniae* sp. nov. that nodulates and fixes nitrogen in symbiosis with *Neptunia natans,* an aquatic legume from India. Syst. Appl. Microbiol.26: 47-53.

Schultze M, Kondorosi A (1998). Regulation of symbiotic root nodule development. Annu. Rev. Genet. 32: 33-57.

Sullivan JT, Ronson CW (1998). Evolution of rhizobia by acquisition of a 500 kb symbiosis island that integrates into a phe-tRNA gene. Proc. Natl. Acad. Sci. USA 95: 5145-5149.

Sy A, Giraud E, Jourand P, Garcia N, Willems A, DeLajudie P, Prin Y, Neyra M, Gillis M, Boivin-Masson C Dreyfus B (2001). Methylotrophic *Methylobacterium* bacteria nodulate and fix nitrogen in symbiosis with legumes. J. Bacteriol. 183: 214-220.

Trujillo ME, Willems A, Abril A, Planchuelo AM, Rivas R, Ludena D, Mateos PF, Molina EM Velazquez E (2005). Nodulation of *Lupinus albus* by strains of *Ochrobactrum lupini* sp. Nov. Appl. Environ. Microbiol.71: 1318-1327.

Valverde A, Velazquez E, Gutierrez C, Cervantes E, Ventosa A, Igual JM (2003). *Herbaspirillum lusitanum* sp. nov., a novel nitrogen fixing bacterium associated with root nodules of *Phaseolus vulgaris.* Int. J. Syst. Evol. Microbiol. 53: 1979-1983.

Valverde A, Velazquez E, Santos FF, Vizcaino N, Rivas R, Mateos PF, Molina EM, Igual JM Willems A (2005). *Phyllobacterium trifolii* sp. nov., nodulating *Trifolium* and *Lupinus* in Spanish soils. Int. J. Syst. Evol. Microbiol. 55: 1985-1989.

Van Berkum P, Eardly BD (2002). The aquatic budding bacterium *Blastobacter denitrificans* is a nitrogen fixing symbiont of *Aeschynomene indica.* Appl. Environ. Microbiol. 68: 1132-1136.

Van Berkum P, Tully RE, Keister DL (1995). Non-pigmented and bacteriochlorophyll containing bradyrhizobia isolated from *Aeschynomene indica.* Appl. Environ. Microbiol. 1: 623-629.

Vandamme P, Coeyne T (2004). Taxonomy of the genus *Cupriavidus*: a tale of lost and found. Int. J. Syst. Evol. Microbiol. 54: 2285-2289.

Vandamme P, Goris J, Chen WM, de Vos P, Willems A (2002) *Burkholderia tuberum* sp. nov. and *Burkholderia phymatum* sp. nov. nodulate the roots of tropical legumes. Syst. Appl. Microbiol. 25: 507-512.

Young JPW, Haukka KE (1996). Diversity and phylogeny of Rhizobia. New Phytol. 133: 87-94.

Zakhia F, deLajudie P (2001).Taxonomy of rhizobia. Agronomie 21: 569-576.

Transgenic plants: Successes and controversies

Alfred O. Ubalua

Microbiology/Tissue Culture laboratories, National Root Crops Research Institute (NRCRI), Umudike, P. M. B 7006 Umuahia, Abia State, Nigeria. E-mail: alfreduba@yahoo.com.

Applications of herbicides, pesticides, and fertilizers have greatly increased crop yields in the past. More recently, however, crop yields are barely keeping up with world growth, hence the need for new approaches. Agricultural biotechnology may likely play a key role in the race to feed the world's expanding population with fewer inputs and on less and less available land. Genetic engineering of plants represents the next stage of evolution in our continuing efforts to improve plants used for the production of food and animal feed. Agricultural biotechnology is a powerful technique offering great potential for agricultural sustainability and safe production of foods with increased nutritive value, improved flavour, prolonged freshness, and even disease-fighting properties, but it is not without controversy. The potential for the transgenic crops to alleviate human hunger, and the controversies which are invariably based on visions of the new technology from widely different ethical perspective which have divided both the public and the scientific communities are discussed. But, critical to its adoption and acceptance is by providing choice and accurate information to consumers from scientists, policy makers, industry and the press. The present review addresses the prospects of the technology, and the polemics concerning its adoption.

Key words: Transgene, pharmaceutical crops, risks, prospects and controversies.

INTRODUCTION

Biotechnology is not new. Cheese, bread, alcoholic beverages and yoghurt are products of (traditional) biotechnology and have been known for centuries. But recent developments in molecular biology have given biotechnology new meaning, new prominence, and new potential. It is (modern) biotechnology that has captured the attention of the public and can have a dramatic effect on the world economy and the society. Genetic engineering which is an example of modern biotechnology is the process of transferring individual genes between organisms or modifying the genes in an organism to remove or add a desired trait or characteristic. Through genetic engineering, genetically modified (GM) crops or organisms are formed. These GM crops or GMOs are used to produce biotech-derived foods. It is this specific type of modern biotechnology that seems to generate the most attention and concern by consumers and consumer groups. What is interesting is that modern biotechnology is far more precise than traditional forms of biotechnology. It allows for the transfer of only one or a few desirable genes, thereby permitting scientists to develop crops with specific beneficial traits and reduced undesirable traits. Traditional biotechnology such as cross-pollination in corn produces numerous, non-selective changes.

Actually, it can more accurately be said to originate with the publication of the Cohen and Boyer paper in 1973. Since then, the field of agricultural biotechnology (Ag Biotech) has grown rapidly over the last half-century. Modern agricultural biotechnology has offered opportunities to produce more nutritious and better tasting foods, higher crop yields and plants that are naturally protected from diseases and insects. Genetic modifications have produced fruits that can ripen on the vine for better taste, yet have longer shelf lives through delayed pectin degradation. Similarly, introducing genes that increase available iron levels in rice three-fold is a potential remedy for iron deficiency, a condition that affects more than two billion people and causes anemia in about half that number. Most of the today's hard cheese products are made with a biotech enzyme called chymosin, produced by genetically engineered bacteria which are considered more pure and plentiful than its naturally occurring counterpart, rennet, which is derived from calf stomach tissue (Kass, 2005). More so, the starch content of Russet Burbank potato has been successfully increased in 1992, by Monsanto Company through gene transfer. Such a gene from a bacterium increased the starch content of the potato. The advantage is that the higher starch content reduces oil absorption during frying , thereby lowering the cost of processing French fries and chips, with a concomitant reduction of the fat content in the finished product.

Sub-Saharan Africa is considered one of the world's food insecure regions. The available statistics indicates worsening scenario. Africa's overall food production capacity is said to be increasing at the rate of 1.4% while its population is expanding at about 2.4% per year (FAO, 2000). The continuing decline in food production will have to be reversed if massive food insecurity, poverty, social and political instability are to be averted. Green revolution type technologies requiring increased land, water and fertilizer use may not be appropriate for sub-Saharan Africa due to resource limitations and population pressure. Biotechnology application may be considered to be more appropriate to solving our agricultural and poverty reduction problems, so as it minimizes inputs while increasing yields (Alhassan, 2002). Reduced pesticide use can mean more abundant and diverse animal and beneficial insect populations, cleaner water, fewer containers for disposal, and reduced fuel use for pesticide applications. Improved weed control enhances conditions for no-till farming, which greatly reduces soil erosion and pre-emergent pesticide use. Improved crop yields mean that the acreage available for production is used in a more efficient way (Bridges et al., 2003). Biotechnology is currently changing the way plants and animals are grown, boosting their value to growers, processors, and consumers. The most common transgenic varieties available today are those that tolerate proprietary brands of herbicides, and those that contain insecticide genes.

In the developing nations, small and peasant farmers are the primary producers of staple foods. This sector, which is so important for food production, is itself characterized by poverty and hunger. Thus rural areas in the developing nations are today characterized by extreme inequalities in access to land, insecurity of land tenure and in the quality of land farmed. Furthermore, by keeping wages and living standards low, the elite guarantees that healthy domestic markets will never emerge. The result is a downward spiral into deeper poverty and marginalization. One irony of our world is that food and other farm products flow from areas of hunger and need to areas where money is concentrated.

APPLICATIONS OF BIOTECHNOLOGY IN FOOD PROCESSING

Historically, the food processing industry has had to accept and adapt to heterogeneous raw materials.

Biotechnology can be used to better tailor food crops to meet food processing and consumer needs. Tissue-culture techniques are being used to select or construct crop varieties with improved functional, processing, or nutritional characteristics. Plant tissue culture techniques can be used to produce food flavour and colouring ingredients. These methods could potentially replace production and extraction of these ingredients from plants (Fraley, 1991; Harlander, 1991).

Genetic engineering is also a means of altering food characteristics. Genes coding for enzymes involved in starch and lipid biosynthesis are being isolated and cloned, enhancing the prospects of engineering plants with specific composition of starch and oil. Genes coding for floral pigment pathways are also being isolated. Plants potentially can be engineered to produce pharmaceuticals such as blood clotting factors and growth hormones. For example, oilseed rapeseed has been genetically engineered to produce encephalin (Vandekerckhove, 1989). In addition antisense technology is being used to eliminate toxins allergenic compounds, or off-flavour components in plants, and to delay ripening of tomatoes (Fraley, 1991). An elegant study by Brookes and Barfoot, (2005) revealed that GM technology has had a very positive impact on farm income derived from a combination of enhanced productivity and efficiency gains (Table 1). In 2004, the direct global farm income benefit from GM crops was $4.8 billion. If the additional income arising from second crop soybeans in Argentina is considered, this income gain rises to $6.5 billion. This is equivalent to adding between 3.1 and 4.2% to the value of global production of the four main crops of soybeans, maize, canola, and cotton - a substantial impact. Since 1996, farm incomes have increased by over $19 billion or $27 billion inclusive of second-crop soybean gains in Argentina.

Biotechnology is also being used to improve microorganisms used as vegetable starter cultures and in brewing and baking (that is, organisms used in making sauerkraut, pickles, olives, soy sauce, wine, beer, and bread) such that these organisms tolerate different temperature and pH ranges. Similar work is being conducted with microorganisms used to produce food ingredients such as acetic acid, citric acid, ethanol (Ubalua, 2007), niacin, vitamin B 12, xantham gum, and monosodium glutamate. In addition, genetically engineered enzymes are being developed to treat food wastes (Harlander, 1991). Application of biotechnology in the development of methods to assay levels of pathogens toxins, and chemical contaminants in raw ingredients and final products have being perfected. DNA probes and poly and monoclonal antibody kits are beginning to replace traditional bioassay methods; an example is the detection of pesticide residues in food with monoclonal antibody kits. Molecular breeding by applying genetic engineering in the production of medicines, dyes, insecticides, flavours and fragrances is in this respect a promising approach (Verpoorte and Memelink, (2002).

Table 1. Global farm income benefits from growing GM crops, 1996-2004 (US$ million).

Trait	2004 increase in farm income	1996-2004 increase in farm income	2004 farm income benefit as % of total value of production of these crops in GM adopting Countries	2004 farm income benefit as % of total value of global production of these crops
GM HT Soybeans	2,440 (4,141)	9, 300 (17, 351)	5.6 (9.5)	4.0 (6.7)
GM HT maize	152	579	0.6	Less than 0.5
GM HT cotton	145	750	1.4	0.53
GM HT canola	135	713	8.3	1.34
GM IR maize	415	1, 932	1.4	0.8
GM IR cotton	1, 472	5, 726	10.5	5.3
Others	20	37	N/a	N/a
Totals	4, 779 (6, 480)	19, 037 (27, 088)	5.3 (7.2)	3.1 (4.2)

Note: HT = Herbicide tolerant, IR = insect resistant, others = Virus resistance papaya and squash, rootworm-resistant maize. Figures in parentheses include second-crop benefits in Argentina. Totals for the value shares exclude "other crops" (that is, relate to the four main crops of soybeans, maize, canola, and cotton. Source: Brookes and Barfoot, (2005).

Genetic engineering of a secondary metabolic pathway aims to either increase or decrease the quantity of a certain compound or group of compounds (Dixon, 2001; Facchini, 2001; Verpoorte et al., 2000; DellaPenna, 2001). To decrease the production of a certain unwanted (group of) compound (s) several approaches are possi-ble. An enzymatic step in the pathway can be knocked out, for example, by reducing the level of the corresponding mRNA via antisense, cosuppression or RNA interference technologies, or by over-expressing an antibody against the enzyme. The antisense gene approach has been successfully used for changing flower colours. Other approaches include diversion of the flux into a competitive pathway or an increase in the catabolism of the target compound (Verpoorte and Memelink, (2002).

PHARMACEUTICAL CROPS

The benefits of gene manipulations in agricultural production has obvious overwhelming potentials but with unconfirmed risks and controversies. Crops are being genetically engineered to produce a wide variety of drugs, vaccines, and other pharmaceutical proteins. Although they may open the door to less expensive and more-readily available drugs, there is concern regarding the potential for contamination of human food and livestock feed, as well as environmental harm (CSA, 2007). In light of the many pros and cons, three major approaches: the precautionary approach, risk analysis, and cost-benefit analysis could be used to move the debate about pharmaceutical crops forward. Today, however, crops are now being turned into factories producing not just food, but drugs, insulin, vaccines, enzymes, and antibodies. The first step in using crops to produce pharmaceutically active proteins is the synthesis or isolation of genes that code pharmaceutical proteins, followed by the transfer of those genes into the DNA of crop plants. These transferred genes, or "transgenes", can potentially come from a different plant species, an animal, or a bacterium. The genetically modified crops are then cultivated and harvested. In most cases, the crop-produced pharmaceutical protein is extracted, purified, and possibly modified further before it is administered to humans or livestock. In some instances, however, crops are being engineered so that a vaccine can be delivered through the direct consumption of leaves, fruits, or other plant parts, without the cost and the inconvenience of extracting the proteins and delivering them via pills or injections (Sala et al., 2003).

Benefits of pharmaceutical crops

The driving force behind pharmaceutical crops is the possibility of quick production of large quantities of drugs and vaccines, with the hope of reducing costs and increases the much needed pharmaceuticals (Giddings et al., 2000; Fischer et al., 2004; Horn et al., 2004; Ma et al., 2003; Ma et al., 2005). The potential products of transgenic plants include blood thinners, haemoglobin, insulin, growth hormones, cancer treatments, and contraceptives. Products already available include plant-produced vaccines for hepatitis-B, cholera, rabies, malaria, and influenza. Presently, genetically modified maize (corn) to produce lipase, a digestive enzyme used to treat patients with cystic fibrosis is under development. Others are arthritis and other autoimmune diseases. The use of maize, banana, tomatoes, carrots, and lettuce as possible oral-delivery mechanisms for such vaccines was reported by Sala et al. (2003), because these foods can be eaten raw, thereby avoiding the protein denature-tion that typically occurs during cooking. Such eliminates the need for refrigeration, which limits the usefulness of which limits the usefulness of certain vaccines in many parts of the world.

Currently, the method of choice for manufacturing many drugs is by laboratory cell culture of bacteria, yeast, or animal cells. There are many draw backs to this technique of mass producing drugs such as outlined below:

- Cell cultures requires constant monitoring and sampling, as it tends to require precise parameters in order to produce large amounts of protein
- Expansion is very costly and laborious, as new equipment is required and also space to store the equipment
- In order to retain biological activity, many proteins require modifications (addition of sugars, for example), some of which are only performed by mammalian cells and
- Most cells need to be ruptured (not a trivial procedure if you consider the size of modern bioreactors) to isolate and purify the protein of interest, which is much more difficult than purifying proteins from the blood or milk of an animal (Jamie, 2005).

Presumably, plants are an attractive alternative because they could potentially produce greater yields. This is especially important for monoclonal antibodies (such as entercept, which is used to treat rheumatoid arthritis) because current production methods cannot keep up with increasing demand (Elbehri, 2005). Moreover, faster and less expensive production could reduce prices for consumers. Another major benefit of utilizing plants is the reduced risk of disease transmission, in view of the concern that drugs produced via mammalian cell cultures or animal milk could facilitate the movement of certain viruses to humans (CSA, 2007). Generally, grain crops are favoured because protein yields from the large seeds of maize, rice and barley are typically much higher than those obtained from leaves and other vegetative parts. In addition, pharmaceutical proteins can remain stable in dried grain for several years, compared with the much-reduced stability of these same proteins in leaf tissues. Maize is generally recognized as safe for ingestion by the U.S. Food and Drug Administration (FDA) and therefore can be used as an inactive carrier, suitable for drug delivery. Fischer et al. (2004) opined that a growing number of companies are focusing on tobacco, or even mosses, algae, and duckweed, as platforms for pharmaceutical production. However, these plants, pose risks of their own that must be considered, though algae and duckweed, if cultivated would have greater potential than highly domesticated crop species to escape from cultivation.

CONTRIBUTIONS OF RECOMBINANT DNA TECHNOLOGY

Biotechnology is offering innovative possibilities for increasing crop and livestock production and for the protection of the environment by the reduced use of chemicals. The major thrust is presently directed towards medicine, industry, and agriculture in the industrialized countries, with significant investments by trans-national companies. The specificity with which genes can be examined and manipulated opens up real opportunities to tailor-made new plants, and new livestock for specific environment. While traditional plant breeding involves the transfer of large numbers of new genes into a crop followed by cycles of backcrossing and de-selection of undesired genes and traits, genetic manipulation can now be done in the laboratory. Biotechnology is thus comprised of a continuum of technologies, ranging from the long-established and widely used technologies, which are based on the commercial use of microbes and other living organisms, through to the more strategic research on genetic engineering of plants and animals. Genetic engineering evolved from an understanding of how cells function naturally, particularly how the genetic material (DNA) codes for the production of proteins essential for the life of the cell. Based on this understanding, other scientists then devised a series of new techniques, collectively called recombinant DNA technology, to allow the manipulation of these processes in the cell. The major limitation is to identify genes which, when transferred with appropriate molecular controls, will confer agriculturally useful traits on the recipient microorganism, plant or animal. Biotechnology is said to be "the first business with enough glamour to persuade eminent scientists that the entrepreneurial spirit and academic respectability are not mutually exclusive. It has been used in a number of crops for several years, and more genetically enhanced products are expected to be on the market in the coming years. By increasing a crop's ability to withstand environmental factors, growers will be able to farm in parts of the world currently unsuitable for crop production. Along with additional food, this could also provide the economies of developing nations with much-needed jobs and greater productivity.

Biotechnology will also enable growers to produce further enhancements in plant varieties. This would allow for the possibility of increasing the agricultural gene pool that billions of people rely on for basic foodstuffs.

IMPACT OF TRANSGENIC PLANTS ON SOIL AND MICROBIAL COMMUNITIES

Transgenic plants are those plants whose heredity DNA has been augmented by the addition of DNA from a source other than the parental germplasm, using recombinant DNA techniques. They possess novel genes that impart beneficial characteristics such as increased nutritive value, improved flavour, prolonged freshness and even disease-fighting properties. The debate surrounding the use and commercialization of genetically modified crops is emotive and presently unabating. The “perceived” risks include plant invasiveness or dispersal of the plant itself into the native ecosystem causing indirect impacts on the diversity of crops, gene flow through pollen transfer or through horizontal gene transfer with associated microorganisms, development of resistance in target organisms, and non-target effects on native flora and fauna including effects on the biodiversity of beneficial and antagonistic microorganisms (Eastham and Sweet, 2002; Nielsen et al., 2001; Wolfenbarger and Phifer, 2000; Riba et al., 2000).

One of the primary concerns about genetically modified crops is the presence of clinically important antibiotic resistance gene products in transgenic plants that could inactivate oral doses of the antibiotic. Another concern is that the antibiotic resistance genes could be transferred to pathogenic microbes in the gastrointestinal tract or soil, rendering them resistant to treatment with such antibiotics (Daniell et al., 2001). Though evidence for the persistence of transgenic plant DNA exists, the transformation of plant DNA to native soil microorganisms has not been found. Several studies attempted to assess natural transformation from plant DNA to soil microorganisms under field conditions and determined that while free DNA persisted in the soil, no proof of a plant gene being transferred to soil bacteria was found (Widmer et al., 1997; Paget et al., 1998; Gebhard and Smalla, 1999). Oger et al. (1997) demonstrated that genetically engineered plants might alter their biological environment, more precisely the root-associated bacterial populations. A response in the composition of the microbial population was observed after the introduction of a single genetic trait into the plant genome. According to Dunfield and Germida in 2004, the effect of plant variety on the microbial community at one field site was sometimes entirely different at another field site, suggesting that the environment will play a major role in determining the potential ecological significance of growing genetically modified plants. Furthermore, a time course study examining genetically modified plants over an entire field season suggests that changes to the microbial community structure associated with genetically modified plants are not permanent. Collectively, these results seem to indicate that microbial diversity can sometimes be altered when associated with transgenic plants; however, these effects are minor in comparison with environmental factors such as sampling date and field site.

HEALTH AND ENVIRONMENTAL RISKS

Microorganisms are ubiquitous in nature. Many are simply harmless commensals, and majorities are beneficial symbionts that protect us from pathogen colonization, synthesize vitamins, and contribute to proper functioning of the intestine. They are known as microbiota or the body’s normal flora. Regardless of the mode of entry, once inside the body, these pathogenic microorganisms may overcome our powerful defense mechanisms, multiply and release their enzymes, toxins and molecules, thereby causing diseases. Microorganisms have always been employed to improve the quality of life, to the extent that biotechnology can be considered the world’s second oldest profession. Humans first used microorganisms empirically, without really knowing what was responsible for processes such as the production of bread, wine, and beer, which dates back to biblical times. Currently as a consequence of advances in molecular biology, along with a better understanding of nucleic acids and enzymes, microorganisms have become the main tool for genetic engineering - the basis of modern biotechnology - since they constitute the best known source of genetic novelty, in addition to their versatility, simple genetics, and ease of use (Berraquero, 2006).

It is widely accepted that the potential applications of microorganisms through biotechnology offer immense benefits in medicine, food production and other fields. Recombinant DNA technique has been the key to developing improved varieties of plants. The genome modifications made possible by this process has made the modifications of plants virtually limitless. Instead of confining breeding to varieties within one species, plant reproduction can now be expanded to combining plant DNA with that of other species of plants, animals and bacteria.

The types of potential hazards posed by GMOs vary according to the type of organism being modified and its intended application. Most of the concern surrounding GMOs relates to their potential for negative effects on the environment and human health. There are several types of potential health effects that could result from the insertion of a novel gene into an organism. Health effects of primary concern to safety assessors are production of new allergens, increased toxicity, decreased nutrition and antibiotic resistance (Bernstein et al., 2003). There is concern that inserting an exotic gene into a plant could cause it to produce toxins at higher levels that could be dangerous to humans. This could happen through the process of inserting the gene into the plant. If other genes in the plant become damaged during the insertion process it could cause the plant to alter its production of toxins. Alternatively, the new gene could interfere with a metabolic pathway causing a stressed plant to produce more toxins in response. Although these effects have not been observed in GM plants, they have been observed through conventional breeding methods thereby creating a safety concern for GM plants.

Despite many reassuring words by companies, researchers and some governments, many concerns about the implications of GM crops remain. Major concerns relate to the consequences for the ecological systems into which they are being introduced. For instance, the insertion of Bt (*Bacillus thuringiensis*) genes was thought to be a silver bullet, a permanent solution to insect problems. But the model of "one-pest-one solution" does not work forever, as is the case with pesticides; sooner or later resistance builds up. Similarly, building of herbicide resistance in plants is headed for trouble as it unleashes basic ecological reactions. Excessive use of herbicides as a major or only tool of weed management will eventually reduce the sensitivity of weeds to herbicides and create an even worse weed problem. It is "to a large extent a victim of its own success". Yield decline in GM soybean, for instance, is being traced to reduced root development, nodulation and nitrogen fixation. Another effect is related to the unexpected impact of gene transferred and its consequences. One example from USA tells how genes from one bacterium, *Xanthomonas* were transferred to another soil bacterium, *Klebsiella planticola*. The new organism was meant to ferment stubble into alcohol, thus providing farmers with an extra source of income instead of burning the stubble. However, a test by the authorities found that wheat planted in the soil containing the new organism was killed by it (LEISA, 2001).

One thing that makes the development of GE unique in the history of agriculture is that it is almost fully controlled by private companies. It appears that GE technologies are not being developed because of their problem-solving capacity, but because of the patent and thus profit it can bring to the companies. For instance, in the 1980s, Monsanto was not interested in genetically engineering virus resistance into plants, as it would bring minimal profits. Terminator gene technology takes the issue further. This technology, in which genes are manipulated to be able to switch seeds on and off by treatment with chemicals provided by one and the same GM Seed Company, effectively prevents farmers from keeping their seeds for replanting. Strong public opposition has forced the companies to give up this line of research, but they still hold the patents to the technology. The question is: do we really need GM technology to combat malnutrition, to improve local production and to make agriculture more productive. Has the introduction of GM crops contributed to the reduction of poverty? The FAO (United Nations Food and Agricultural Organization), in a recent report, indicated that "for the world as a whole there is enough, or more than enough food production potential to meet the growth of effective demand, that is the demand for food of those who can afford to pay farmers to produce it". The underlying factor is the purchasing power in the developing countries of the world. A wealth of agro-ecological, low-external-input alternatives to agricultural production is far from exhausted.

The case of natural crop protection indicates that there are many plants in nature which can provide us with clues for better disease and pest management. Many ecological principles that are still being overlooked, underestimated or sidelined deserve more attention as they provide relatively cheap, controllable and low external input solutions to many problems that farmers face. Moreover, these approaches are not accompanied by the many risks - both economic and ecological - that GM crops may be posing. In agriculture, the fight against harmful organisms is essential, and it is therefore necessary to develop appropriate technologies to regulate and control pests. One of the justifications for the development of GM crops is to build-in resistance to insects and diseases, which in turn is expected to reduce the negative impact of agrochemicals on health and the environment. Environmentally-friendly biological options that do not make use of genetic modification do exist. These options are based on natural crop protection approaches that make use of the diversity found in nature itself, providing excellent and competitive alternatives for pest management. The challenge in natural crop protection is to have simple and low-cost technologies that are able to regulate pests and diseases and to reduce or completely avoid the problem of contamination by agrochemicals. One such natural crop protection approach is based on the use of plants with biological control properties (Ubalua and Oti, 2008).

ETHICS AND CONTROVERSIES

Biotechnology is more than just a scientific issue. It is capable of engendering disagreement and controversy, and highlighting moral and ethical concerns which are difficult to resolve. These concerns include or arise from uneasiness over the fact that biotechnology is seen by some to "interfere with the workings of nature and creation", and that it might involve risk-taking for commercial profit. However, in priority setting, all concerns must be clearly balanced, respecting ethical aspects but reflecting the actual and potential possibilities of increasing food supplies and alleviating hunger. Robinson, (Robinson, 1999), reported that transgenic crops are about values, which are neither absolute nor universal, and that the controversies to a large extent have polarized society into proponents and opponents, with once seemingly trustworthy and ethically sound scientists being viewed with suspicion by many. He further stated that much of the argument is emotive, with talk of "cashing in on hunger", "Demon seeds", "Terminator Technology" and "Frankenstein Foods", and that it is difficult for anyone to appreciate any underlying truths, should they even exist.

Unarguably farmers have been accidental plant breeders for over ten thousand years, altering the genetic integrity of most crops without any knowledge of heredity. Jones, (Jones, 1994), was of the opinion that farmers were the first genetic engineers, although genetics as such did not come into existence until the work of

Mendel was rediscovered by de Vries et al. (1990). For instance, in traditional breeding techniques, breeders mix thousands of genes when crossing two plants in order to transfer the protein products to enhance one or a few genetic traits. But Charles Darwin once made the following comment on traditional breeding that: "the key is man's power of accumulative selection, nature gives successive variations; man adds them up in certain directions useful to him". Therefore, the odds of naturally occurring toxin being transferred unintentionally are far greater in traditional breeding than in biotechnology. To avoid toxicity, breeders spend many years backcrossing the new plant varieties with distant relatives. Backcrossing involves crossing new plant varieties repeatedly with plants whose genetic components are well known. This slowly dilutes the impact of all those unwanted genetic traits that came along with the few beneficial traits. Plant breeding is generally safe, but biotechnology can make it safer. Biotechnology brings to traditional plant breeding the ability to move single genes instead of having to move thousands and makes possible the identification of these genes and their products that are toxins.

The moral and ethical concerns are important factors in influencing a risk-averse public (Callahan, 1996), and are a pivotal feature of the debate on transgenic crops and their products (Newton et al., 1999). The basic questions which require answers are, do transgenic crops represent the solution to world hunger, and do they pose unacceptable risks to the environment and to human health? Obviously, science has had an enormous impact on human existence, providing numerous innovations which have improved the lives of many, and scientists have been regarded as trustworthy and ethically sound, and agricultural research and its role in food production as being intrinsically good (Hardon, 1997). This view has been altered somewhat by the advent of genetic engineering (GE), although it is generally appreciated that new technologies by their very nature represents a challenge to existing values and systems, and stimulate change in traditional concepts of nature and human identity (Carr and Levidow, 1997).

The effects of transgene escape on the environment are uncertain, but modern technology could limit such "genetic pollution" through, in some cases, engineering sterility into the transgenics to ensure vastly reduced gene flow into the farming and natural environments. Crops do not generally survive outside the farming environment and transgenic crops would probably be out-competed should they spread off farm. According to Holmes, (1997), there is already evidence that many targeted pest species have developed resistance genes. This is a demonstration that nature fights back against the genetic engineer in much the same way as it fights back against the conventional plant breeder and many solutions to pest and disease problems represented by GE are likely to be short-lived. Furthermore, agrochemical control of crop pests is however inefficient and environmentally and ethically unsound and GE could offer a remedy, allowing more precise targeting of pest management (Pimentel, 1995). Proponents of biotechnology are of the opinion, that if genes from any of the most common allergens (milk, eggs, wheat, fish, shellfish, tree nuts (that is, walnuts), and legumes (that is, soybeans) are added to a food via biotechnology, a company has to either demonstrate through scientific data that an allergen is not present in the new food, or label the products to alert allergic consumers. The label must state that a potential allergen has been added to the food, but not that the food was produced via biotechnology. When developing plants through biotechnology, scientists use selectable marker genes to determine whether gene transfer has been successful, and have in the past used antibiotic proteins. The food and Drug Administration (FDA) has reviewed the use of selectable marker genes, and confirmed the safety of antibiotic-resistance marker genes and their rare use in biotechnology. Again in EPA, (1996) approved the use of *Bacillus thuringiensis* (Bt) in corn. Bt is a naturally occurring bacterium present in soil and known for its ability to control pests. Although harmless to most insects, people, birds, and other animals, Bt produces a protein that disrupts the digestive system of target insects.

Weil, (1996) believes that the effects of transgenics on the environment are controversial because of the great difficulty in gauging the associated risks. He argues that all actions are potentially hazardous, and that there have been no problems involving transgenics to compare with those that have been encountered previously as a result of classical plant breeding (e.g. Tcytoplasm and Southern corn leaf blight in the USA in the early 1970's). Moreover, the hazards certainly do not approach the scale of environmental damage wrecked by disasters in traditional industry; the oil spill from the Exxon Valdez (Ubalua and Ezeronye, 2007) and the escape of radioactive fallout from the Chernobyl nuclear power station. Interestingly, while there is a public preoccupation with the potential hazards arising from genetically modified (GM) crops, environmental concerns about conventional crops are few (Annon, 1989a). Concar and Coghlan, (1999)], argued that an oilseed rape variety has been bred in Canada which carries genes for resistance to two herbicides using conventional means rather than through genetic engineering. Arguing further, Ort, (1997), mentioned several crops, including triticale, with its genomes from wheat and rye, which have contained "foreign" genes for a long time without occasioning any public outrage, or indeed causing any environmental damage. It is an immutable fact that human health already suffers as a consequence of agricultural practices. Farmers recognize more than anyone that healthy growing environments define their future.

Thus, they always seek better ways to control weeds with the least toxic herbicides available that do not damage crops. Farmers also would like to reduce their use of insecticides and fungicides, limiting their own exposure to the chemicals. Biotechnology can achieve many of these goals often more efficiently.

The ethical stand of some opponents of biotechnology was reviewed by Robinson, (1999) in which he addressed the position championed by Prince Charles, who claimed that GE takes mankind "into realms that belongs to God and God alone". The implication is that the fate of humankind is in God's hands and that our meddling nature is sinful and goes against his wishes. The question then is: where does divine responsibilities end and that of man begin? The dividing line is not clear, and all human endeavours could be said to interfere with God's will to some extent. Words and terms are probably the root of problems associated with GE. But it should be borne in mind that classification is a man made concept. According to Robinson, (1999), creationist theory views life forms as being fixed and immutable, determined by God, whereas evolutionary theory is based on dynamic concepts and gradualism, whereby small changes (mutations) take place over extended times and the forces of natural selection results in the creation of new species. From all indications, it appears that opinions on transgenic crops are based on value judgments and possible negative health and environmental implications, and such values and attributes are likely to change with time and circumstance and with modifications to conceptual systems.

PROSPECTS

Genetic engineering has opened new avenues to modify crops, and provided new solutions to solve specific needs. The powerful combination of genetic engineering and conventional breeding programmes permits useful traits encoded by transgenes to be introduced into commercial crops within an economically viable time frame. There is great potential for genetic manipulation of crops to enhance productivity through increasing resistance to diseases, pests and environmental stress and by qualitatively changing the seed composition. Plant 'factories' are also being designed for high volume production of pharmaceuticals, nutraceuticals and other beneficial chemicals ((Hansen and Wright, (1999). Transgenic plants might become drug-delivery devices, with both HIV and rabies vaccines being synthesized in plants (Yusibov, 1997), and bananas have been engineered to produce edible vaccines. Moreover, with the establishment and expansion of genomics programmes, a much broader range of genes with potential for crop improvement are being identified and, in some cases, tailored and/or re-designed for further enhancement of their properties within specific crops (Estruch, 1997).

Plant transformation remains an art because of the unique culture conditions required for each crop species. Beyond crop improvement, the ability to engineer transgenic plants is also a powerful and informative means for studying gene function and the regulation of physiological and developmental processes. Transgenic plants are being used as an assay system for the modification of endogenous metabolism or gene inactivation. Advances in tissue culture, combined with improvements in transformation technology, have resulted in increased transformation efficiences. In recent years, many crops, previously classified as recalcitrant because they were stubbornly resistant to the overtures of genetic engineering, have now been transformed. Transformation technologies have advanced to the point of commercialization of transgenic crops. The introduction of transgenic varieties in the market is a multi-step process that begins with registration of the new varieties, followed by field trials and ultimately delivery of the seed to the farmer. Technical improvements that have the greatest opportunities for new approaches, probably in the realm of in planta transformation, will further increase transformation efficiency, extend transformation to elite commercial germplasm and lower transgenic production costs, ultimately leading to lower costs for the consumer (Hansen and Wright, 1999)

Conclusion

Genetic engineering (GE) is the most recent of a variety of 'new' technologies allowing plant breeders to produce plants with new gene combinations while recombinant DNA technology is in many ways an extension of the natural process of adaptation. Transgenic plants can assemble and accumulate many valuable proteins that can be economically extracted or processed. They often show advantages over other expression systems as factories for the production of pharmaceutical and Industrial proteins. Recent results of clinical trials on transgenic plant-derived pharmaceuticals are promising but the possibility of oral tolerance to plant vaccines remains a potential problem in need of further research. The prospects of agricultural biotechnology are obvious notwithstanding the controversies that it is generating especially in Europe. According to Brookes and Barfoot, (2005), there have been substantial economic benefits at the farm level, amounting to a cumulative total of $27 billion. GM technology has also resulted in 172 million kg less pesticide use by growers and a 14% reduction in the environmental footprint associated with pesticide use. GM crops have also made a significant contribution to reducing greenhouse gas emission by over 10 billion kg, equivalent to removing five million cars from the roads for a year (Conner et al., 2003). More work is needed to clarify the relevance of measurements in terms of the environment and soil-microbial, GM interactions.

Opinions are divided on the advantages and disadvantages of transgenic crops. The perceived advantages and disadvantages must be married to each other to provide a crop that is environmentally sound and non-hazardous. Producers of transgenic crops and the agencies that study their effects are aware of this fact. While the debates and campaigns on its adoption or not rages on, one may be inclined to believe that the technology is irreversible considering its spread and benefits so far. Public unease remains strong despite reassurances from scientists, and this may be because GE raises complex philosophical questions about the changing nature of agriculture. Media coverage, and a diminished public trust in regulatory authorities, may explain why GM crops have met rancorous public resistance in Europe. It is only time that will prove us wrong or right. It is obviously too late to keep the genie in its bottle and transgenic crops have been produced in abundance, being that we are in a period of extraordinarily rapid change in modern biology, but modern biotechnology may not diminish the need for conventional agricultural research. However, GE in fairness is not set to replace plant breeding, rather it represents a modern tool for use by the plant breeders, but the truth on the safety of transgenics hinges on reliable communication of information from scientists, policy makers, industry and the press. Whatever the scientific merits of the argument on the safety of GM foods, it is a fact that the "perceived" dangers of trans-genic crops are preventing their widespread adoption, particularly in Europe. Even in countries where the growth of such crops is not seen as a major issue (like the USA), the effect is still felt as export markets for food stuffs derived from genetically engineered crops are affected. Hopefully, with the emergence and future perfection of clean-gene technology, hopes are high that genetically modified crops may become acceptable to all, thereby resting the sceptisms and the fear that have polarized the proponents and the opponents of genetically modified foods.

REFERENCES

Alhassan WS (2002). Agrobiotechnology application in West and Central Africa (Survey outcome). Ibadan: International Institute of Tropical Agriculture. Ibadan, Nigeria.

Annon A (1989). Summaries of Country Reports, May 1989, World Bank-ISNAR-AIDAB-ACIAR, Biotechnology Study Project Papers. ISNAR, The Hague.

Berraquero RF (2006). Microbes and Society", Contributions to Science", Institut d'Estudis Catalans, Barcelona 3(2): 197-202.

Bernstein JA, Bernstein IL, Bucchini L, Goldman LR, Lehrer S, Rubin CH, Sampson HA (2003). Clinical and Laboratory investigation of allergy to genetically modified foods. Environ. Hlth. Perspectives. 111(8): 1114-1121.

Bridges A, Kimberly R, Magin M, Stave JW (2003). Agricultural Biotechnology (GMO). Methods of Analysis, In: Food Analysis. 3rd Edition. KLuvwer Academic/Plenum publishers, New York pp.301-311.

Brookes G, Barfoot P (2005). GM Crops: The global economic and environmental impact-The first nine years, 1996-2004. AgBioForum 8(2&3): 187-196.

Callahan D (1996). Biotechnology and Ethics: A blue print for the future. Keynote: Setting and communicating the limits in biotechnology. http://www.biotech.nwu.edu/nsf/callahan.html.

Carr S, Levidow L (1997). How biotechnology separates ethics from risks, Outlook on Agriculture 26: 145-150.

Concar D, Coghlan A (1999). A question of breeding. New Scientist pp. 4-5.

Conner AJ, Glare TR, Nap J (2003). The release of genetically modified crops into the environment. Part1. Overview of current status and regulations. Plant J. (33)1: 1-18.

CSA (Crops, Soils Agronomy) News (2007). A mixed Outlook for Pharmaceutical crops. www.crops.org

Daniell H, Muthukumar B, Lee SB (2001). Marker free transgenic plants: Engineering the chloroplast genome without the use of antibiotic selection. Curr. Gene. 37: 109-116.

DellaPenna D (2001). Plant metabolic engineering. Plant Physiol. 125: 160-163.

de Vries H (1990). The law of separation of characters in crosses. J. R. Hortic. Soc. 1901(25): 243-248

Dunfield KE, Germida JJ. (2004). Impact of genetically modified crops on soil and plant-associated microbial communities. J. Environ. Qual. 33: 806-815.

Dixon RA (2001). Natural products and plant disease resistance. Nat. 411: 843-847

Eastham K, Sweet J (2002). Genetically modified organisms (GMOs): The significance of gene flow through pollen transfer. Environ. Issue Report. 28 [Online]. Available at http://reports.eea.eu.int/environmental_issue_report_2002_28/en. European Environmental Agency, Copenhagen.

Elbehri A (2005). Biopharming and the food system: Examining the potential benefits and Risks. AgBioForum 8: 18-25.

Environmental Protection Agency (EPA), (1996). http://www.epa.gov.

Estruch JJ (1997). Transgenic plants: an emerging approach to pest control. Nat. Biotechnol. 15: 137-141.

Facchini PJ (2001). Alkaloid biosynthesis in plants: biochemistry, cell biology, molecular regulation, and metabolic engineering applications. Ann. Rev. Plant Physiol. Plant Mol. Biol. 52: 29-66

FAO (2000). Safety aspects of genetically modified food of plant origin. Report of FAO. Expert Consultation on Foods Derived from Biotechnology.

Fischer R, Stoger E, Schillberg S (2004). Plant-based production of Biopharmaceuticals. Current Opinion in Plant Biol. 7: 152-158.

Fraley RT (1991). Genetic Engineering in Crop Agriculture, Commissioned background paper prepared for the office of Technology Assessment.

Gebhard F, Smalla R (1999). Monitoring field releases of genetically modified sugar beets for persistence of transgenic plant DNA and horizontal gene transfer. FEMS Microbiol. Ecol. 28: 261-271.

Giddings G, Allison G, Brooks D (2000). Transgenic plants as factories for Biopharmaceuticals. Nat. Biotechnol. 18: 1151-1155.

Hardon JJ (1997). Ethical issues in plant breeding, biotechnology and conservation. pp. 43-50 In, Ethics and Equity in conservation and use of genetic resources for sustainable food security. Proceedings of a workshop to develop guidelines for the CGIAR, 21-25 April, Foz do Iguacu, Brazil, IPGRI.

Harlander S (1991). Biotechnology in Food Processing in the 1990s, commissioned background paper prepared for the office of Technology Assessment.

Hansen G, Wright MS (1999). Recent advances in the transformation of plants. Trends Plant Sci. 4(6).

Holmes B (1997). Caterpillar's revenge. New Scientist p. 7

Horn EM, Woodward LS, Howard JA (2004). Plant molecular farming. Systems and products. Plant Cell Reproduction. 22: 711-720.

Kass J (2005). Commercialization of transgenic animals: Potential ecological risks. BioSci. 58: 46-58.

Jamie P (2005). Transgenic Animals: How genetics is providing new ways to envision agriculture. Biodiversity-Transgenic animals. http://www.biotech.ubc.ca/biodiversity/transgenicanimals/index.htm.

Jones S (1994). The language of the genes. Flamingo, London, 347p.

LEISA magazine (Magazine of low External Input and Sustainable Agriculture) (2001). GE-not the only option. 17(4): 4.
Ma K-CJ, Drake PMW, Christou P (2003). The production of recombinant pharmaceutical proteins in plants. Nat. Rev. Gene. 4: 794-805.
Ma K-CJ, Barros E, Bock R (2005). Molecular farming for new drugs and Vaccines. EMBO Report 6: 593-599.
Newton P, Brown D, Clover C (1999). Alarm over 'Frankenstein' foods. Electronic Telegraph. (1358) http://www.telegraph.co.uk.
Nielsen KM, Van EJD, Smalla K (2001). Dynamics, horizontal transfer and selection of novel DNA in bacterial populations in the phytosphere of transgenic plants. Ann. Microbiol. 51: 79-94.
Oger P, Petite A, Dessaux Y (1997).Genetically engineered plants producing opines alter their biological environment. Nat. Biotechnol. 15: 369-372.
Ort DR (1997). Pros and cons of foreign genes in crops. Nat. 385: 290
Paget E, Lebrun M, Freyssinet G, Simonet P (1998). The fate of recombinant plant DNA in soil. Eur. J. soil Biol. 34: 81-88.
Pimentel D (1995). Amounts of pesticides reaching target hosts: environmental impacts and ethics. J. Agric. Environ. Ethics 8: 17-29
Riba G, Dattee Y, Couteaudier Y (2000). Les plantes transgeniques et l'environnement. C. R. Acad. Agric. Fr. 86 : 57-65.
Robinson J (1999). Ethics and Transgenic crops: A review. Universidad Catolica de Valparaiso. Electr. J. Biotechnol. Chile. 2(2): 1-16.
Sala F, Rigano MM, Barbante A (2003). Vaccine antigen production in transgenic plants: Strategies, gene constructs and perspectives. Vaccine 21: 803- 808.
Ubalua AO, Ezeronye OU (2007) Determination and Assessment of heavy metal content in fish and shellfish in Aba River, Abia State, Nigeria. KMITL Sci. Technol. J. 7(1): 16-23.
Ubalua AO (2007). Cassava Wastes: Treatment options and value addition alternatives. Afr. J. Biotechnol. 6(18): 2065-2073.
Ubalua AO, Oti E (2008). Evaluation of antimicrobial properties of some medicinal plants for fresh cassava roots preservation. Pakistan J. Nutr. 7(5): 679-681.
Vandekerckhove J (1989). Encephalins produced in transgenic plants using modified 2s seed storage proteins. Biotechnol. 7: 929-936.
Verpoorte R, van der HR, Memelink J, (2000). Engineering the plant cell factory for secondary metabolite production. Transgenic Res. 9: 323-343.
Verpoorte R, Memelink J. (2002). Engineering secondary metabolite production in plants. Curr. Opin. Biotechnol. 13: 181-187.
Weil V (1996). Biotechnology and Ethics: a blue print for the future. Biotechnology: Social impact and quandaries. http://www.biotech.nwu.edu/nsf/weil.html.
Widmer F, Siedler RJ, Donegan KK, Reed GL (1997). Quantification of transgenic plant marker gene persistence in the field. Mole. Ecol. 6: 1-7.
Wolfenbarger LL, Phifer PR (2000). The ecological risks and benefits of genetically engineered plants. Washington DC. Sci. 3: 2088-2093.
Yusibov V (1997). Antigens produced in plants by infection with chimeric plant viruses immunize against rabies virus and HIV-1. Proc. Natl. Acad. Sc. U.S.A 94: 5784-5788.

Transition metal complexes as potential therapeutic agents

Shazia Rafique*, Muhammad Idrees, Anwar Nasim, Haji Akbar and Amin Athar

National Centre of Excellence in Molecular Biology, University of the Punjab, Lahore 53700, Pakistan.

Transition metals have an important place within medicinal biochemistry. Research has shown significant progress in utilization of transition metal complexes as drugs to treat several human diseases like carcinomas, lymphomas, infection control, anti-inflammatory, diabetes, and neurological disorders. Transition metals exhibit different oxidation states and can interact with a number of negatively charged molecules. This activity of transition metals has started the development of metal-based drugs with promising pharmacological application and may offer unique therapeutic opportunities. To provide an update on recent advances in the medicinal use of transition metals, a Medline search was undertaken to identify the recent relevant literature.

Key words: Transition metals, metal complexes, anticancer drugs, metal therapeutics.

INTRODUCTION

Metals have an esteemed place in medicinal chemistry. Transition metals represent the d block element which includes groups 3 - 12 on the periodic table. Their d shells are in process of filling. This property of transition metals resulted in the foundation of coordination complexes. Metal complex or coordination compound is a structure consisting of a central metal atom, bonded to a surrounding array of molecules or anions. Sophus Jorgensen in Denmark synthesized metal conjugates for the first time in the mid 1870's. In 1893 the major breakthrough in this field was occurred when Alfred Werner investigated a series of compounds, which contained cobalt, chlorine and ammonia. He was awarded the Noble Prize in 1913 for his work.

The earliest reports on the therapeutic use of transition metal complexes in cancer and leukemia date from the sixteenth century. In 1960 the anti-tumor activity of an inorganic complex cis-diammine-dichloroplatinum (II) (cisplatin) was discovered. Cisplatin has developed into one of the most frequently used and most effective cytostatic drug for treatment of solid carcinomas. Other metal like gallium, germanium, tin, bismuth, titanium, ruthenium, rhodium, iridium, molybdenum, copper, gold were shown effective against tumors in man and animals.

TRANSITION METAL COMPLEXES AS ANTICANCER AGENTS

Platinum based anticancer drugs

Platinum (II) complexes has been used as anti cancer drugs since long, among them cisplatin has proven to be a highly effective chemotherapeutic agent for treating various types of cancers (Jamieson 1999). Cisplatin moves into the cell through diffusion and active transport. Inside the cell it causes platination of DNA, which involves interstrand and intrastrand cross-linking as well as formation of adducts, usually through guanine, as it is the most electron rich site and hence, easily oxidized. Formation of cisplatin DNA adducts causes distortion and results in inhibition of DNA replication (Lee, 2002). Cisplatin DNA adducts also serve as binding site for cellular proteins such as repair enzymes, histones, transcription factors and HMG-domain proteins (Louie 1999; Volkter 1999; Cohen et al. 2001). The binding of HMG-protein to cisplatin-DNA adduct has been suggested to enhance anticancer effect of the drug (He 2000; Wong 2002). Beside the effectiveness of cisplatin against cancer, it has encountered many side effects.

Drugs like cisplatin does not specifically affect cancer cells but it also effect the rapidly dividing cells of certain normal tissues, such as those found in hair follicles, bone marrow, and the lining of the gastrointestinal tract. Inside

*Corresponding author. E-mail: shaziarafique@gmail.com.

GROUP

PERIOD	IIIB	IVB	VB	VIB	VIIB	VIIIB	VIIIB	VIIIB	IB	IIB
	Sc	Ti	V	Cr	Mn	Fe	Co	Ni	Cu	Zn
	Y	Zr	Nb	Mo	Tc	Ru	Rh	Pd	Ag	Cd
	La	Hf	Ta	W	Re	Os	Ir	Pt	Au	Hg
	Ac	Unq	Unp	Unh						

Figure 1. Transition metals series.

the cell it interacts with a number of other negatively charged bio molecules besides DNA such as proteins, sulphur-containing compounds like metallothioneins and glutathiones that sequester heavy metals like Pt and remove it from the cell. Pt (II) and Pt (IV) complexes are photo reactive. Irradiation of cis-platin modified DNA with UV light can induce cross-links with the protein HMG, which can inhibits RNA transcription (Bartkowiak 2009; Qutob 2004; McKay 2001).

Like many other anticancer drugs, cisplatin (Figure 1) also faces the same problem called "Drug Resistance". It is major complication in cancer chemotherapy (Piulats 2009; Pill 1997), because of decreased intercellular accumulation of cisplatin, it cannot form adduct with DNA (Chu 1994; Stryer 1995). Organometallic compound like Iron (III)-salophene with selective cytotoxic and anti-proliferative properties have been used in platinum-resistant ovarian cancer cells (Lange et al., 2008).

Different strategies have been used to improve efficacy of cisplatin like use of nanotechnology to specify the effect of the drug in the target cells (Liang 2010). Pt (II) complexes have been conjugated to molecules like porphyrin ring. The porphyrin enhances the tumor tissue specificity of the Pt (II) complexes. Porphyrins are used as photodynamic therapeutic agents and can offer additional antitumor activity by photo-induced mechanism (Lottner, 2002).

The clinical use of cisplatin is limited because of the toxicity to the normal cells and drug resistance, therefore, new platinum based anticancer drugs has also been synthesized such as carboplatin, oxaplatin, nedaplatin etc. Other drugs are being developed that have slower hydrolysis rate than cisplatin, are longer acting and require more infrequent doses. One such drug is 2-pincoline Pt complex, which is active by injection and oral administration.

Platinum complexes with distinctively different DNA binding modes from cisplatin may provide higher antitumor activity against cisplatin resistant cells. The trans isomer of diamminedichloro Pt (II) has also been studied, trans DDP offers a different attachment mode with DNA and is used as anticancer drug for cisplatin resistant cancer cells. A series of trans piperazine compounds were reported to have significant cytotoxicity against cisplatin resistant cells (Albertella, 2004; Najajreh, 2002). Pt (II) complexes with thiourea have

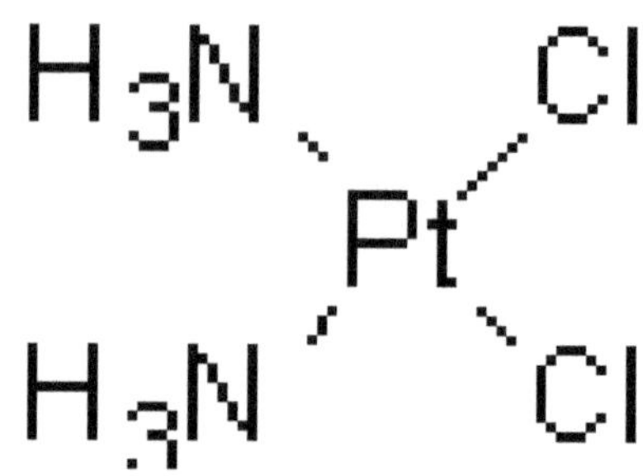

Figure 2. Cisplatin.

showed anti-cancer activity against leukemia cell lines (Martins, 2001; Brow, 2002). Pt (II) has also been complexes with estrogen hormone and used as anticancer agent for the treatment of hormone dependent cancer like breast cancer (Jackson, 2001).

Non-platinum anticancer agents

Platinum is not the only transition metal used in the treatment of cancer, various other transition metals (Figure 2) have been used in anticancer drugs (Chen et al., 2009). Titanium complexes such as Titanocene dichloride had been recognized as active anticancer drug against breast and gastrointestinal carcinomas (Colley et al., 1991).

Gold complexes also show anti-cancer activity, these complexes act through a different mechanism as compared to cisplatin (Au et al., 2008). The target site of Au complexes is mitochondria not DNA. Certain gold complexes with aromatic bipyridyl ligands have shown cytotoxicity against cancer cells (Marcon et al., 2002). The 2-[(dimethylamino) methyl] phenyl gold (III) complex has also proven to be anti tumor agent against human cancers. (Messori et al., 2000). Gold nanoparticles when used in combination with radio therapy or chemotherapy enhance DNA damage and make the treatment target specific (Zheng et al., 2009). Lanthanum has also been used to treat various forms of cancer (Kapoor, 2009). Ansari et al. in 2009 studied some complexes of Mn (III) induce tumor selective apoptosis of human cells.

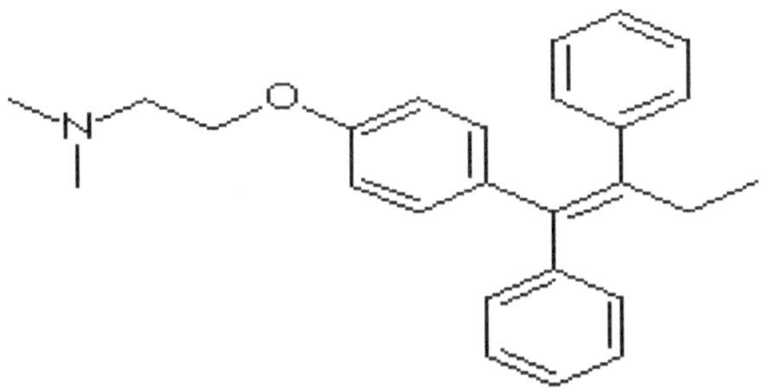

Figure 3. Tamoxifen.

Many ruthenium complexes were studied which showed anti-proliferative effects in human ovarian cancers. Ruthenium complexes with oxidation state +2 or +3 show antitumor activity against metastasis cancers. Ruthenocene derivatives act as anti estrogen (Pascal et al. 2005). The relative binding of ruthenocene derivatives were very high and even better than hydroxyl tamoxifen (Figure 3) which is novel antagonist for estrogen (Clarke, 2003). Ruthenocene complexes with aromatic ligands represent a relatively new group of compounds with antitumor activity. Ru (III) imidazole and Ru (III) indazole exhibit anti cancer properties (Morris et al., 2001).

Galanski and his co-workers studied the anticancer properties of Ru (III) arene complexes (Galanski et al., 2003). Ruthenocene-cymene complexes have shown to damage DNA by forming monofunctional adducts selectively with guanine bases (Allardyce et al., 2001). Ruthenocene complexes show antiproliferative effect on the MCF-7 (ER-positive) breast cancer cell lines. Many of Ru complexes exhibit anti-estrogen properties similar to that observed for novel anti-estrogen Tamoxifen (Anne, 2005).

Complexes of transition metal like Iron have shown remarkable anti-proliferative properties (Lange et al., 2008; Ray et al., 2007; Sun et al., 2007). Ferrocifenes exhibit anticancer activity against hormone dependent and hormone independent breast cancers (Top et al., 2003). The ferrocene derivatives having hydroxyl group in phenyl ring and have high affinity for estrogen receptor.

Many organometallic analogues of tamoxifen used as a vehicle for introducing other cytotoxic agents to the cancer cells (Kiat et al., 2006).

Normally, cancers are diagnosed at a stage of the disease when some anatomical changes occur in the body in the form of well defined tumors. These masses can be removed by surgery however this therapy is not suitable for treatment of small or hidden tumors. Nanotechnology offers potential solutions to this problem for the treatment of various types of cancers. Hirsch et al. (2003) used silica gold nanoshells technology for thermal ablative therapy of cancer. Metal nanoshells are a class of nanoparticles with tunable optical resonances that has highly favorable optical and chemical properties for biomedical imaging and therapeutic applications. Nanoshells absorb light in the near infrared which can be used to deliver a therapeutic dose of heat by using moderately low exposures of extra corporeally applied near-infrared (NIR) light. It has been reported by Asharani et al (2009) silver nanoparticles exhibit anti proliferative activity.

Loo et al. 2004 described several examples of nanoshell-based diagnostic and therapeutic approaches including the development of nanoshell bioconjugates for molecular imaging. Mercaptopurines are well known ant leukemic drugs but their use has been hampered by their short half life. This problem has been overcome by the use of gold nanoparticles in combination with mercaptopurines.

Conjugation therapy of 6-mercaptopurine with gold nano-particles not only enhanced its anti-leukemic and anti-inflammatory activity but also reduces the quantity of dose and side effects of the drug (Podsiadlo et al., 2008).

TRANSITION METAL COMPLEXES AS ANTI INFECTIVE AGENTS

Transition metals like silver have been used for years as anti microbial agents. Silver has low toxicity as compared to other transition metals. Silver nitrate is still given to the infants to prevent the development of opthalmia neonatorum. One of the most commonly used compounds of silver is silver (I) sulfazine; it is used to treat severe burns to prevent them from bacterial infections (Clarke et al. 1975). Chlorohexidine- Silver Sulfadiazine is an anti infective metal complex against catheter infections *in vivo* (Bassetti et al., 2001). Organometallic complexes of Pt, Rh, Ir, Pd, and Os with active organic molecules have been reported to exhibit trypanocidal activity (Lorisean et al., 1992). An increasing amount of data showing the beneficial use of zinc (Zn) in treating diarrhea continues to emerge from epidemiological and clinical trials (Hoque et al., 2009). Fukushima and collegues also studied the role of zinc in maintenance hemodialysis patients (2009). Nitrogen containing macrocyclic complexes of Manganese (II) have shown anti microbial activity. An octahedral geometry for these complexes has been confirmed by spectroscopic analysis. Many manganese complexes have been screened against a number of pathogenic fungi and bacteria to evaluate their growth and potential (Singh and Chaudhary, 2004).

Metal complexes of Pt (II) and Ru (II) with o-vanillin- (4-methyl thiosemicarbazone), and o-vinillin- (4-phenyl thiosemicarbazone) have been prepared, characterized by chemical methods and studied for antibacterial, antifungal, and antiamoebidal activity and have been proven more efficient anti infective agents (Offing et al., 1996).

Transition metals have also been proved useful in the treatment of malaria. One strain plasmodium falciparum has become resistant to major antimalarial drugs such as quinolines. Metal complexes of Ga (III), Al (III), and Fe

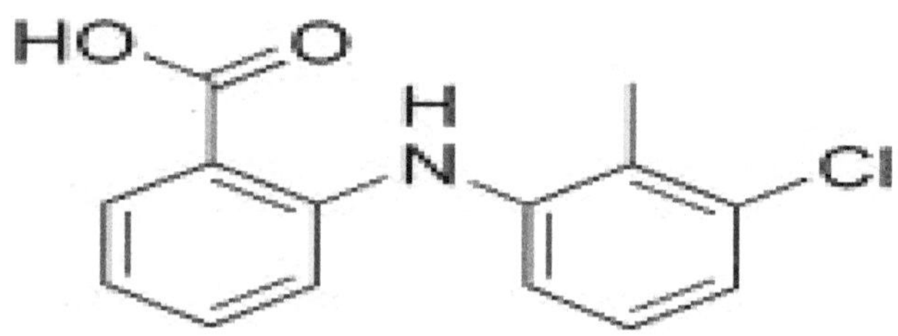

Figure 4. Tolfenamic acid.

(III) were found to be active against malaria. Metal complexes of o-vanillin- (4-methyl thiosemicarbazone), and o-vanillin- (4-phenylthiosemicarbazone) exhibit anti-malarial activity in mice infected with plasmodium berghei indicating that cures were attainable at dose levels upto 60 mg/kg but with toxic death prevalence at higher doses (Singh and Chaudhary, 2004).

TRANSITION METAL COMPLEXES AS ANTI-INFLAMMATORY AGENTS AND FREE RADICAL QUENCHERS

Transition metals have also been used as anti-inflammatory and anti-arthritic agents (Arayne et al., 2009). Several inject able transition gold complexes like sodium aurothiomalate, aurothioglucose and sodium aurothiopropanol are used clinically in the treatment of severe cases of rheumatoid arthritis. Gold and silver nanoparticles conjugated with heparin derivative possess anti-angiogenesis properties (Kemp et al., 2009). Silver nanoparticles are used in the development of an antimicrobial gel formulation for topical use (Jain et al., 2009). Gold has been used in the treatment of peripheral psoriatic arthropathy (Nash and Clegg, 2005). As a product of oxygen metabolism, superoxide anion can trigger oxidative injury to tissues. This activity has been suggested to be associated with riper fusion and inflammatory diseases as well as neurological disorders such as Parkinson's disease and Alzheimer's disease.

In living systems, a natural defense system against superoxide mediated oxidative damages involves SODs, enzymes that catalysis the conversion of superoxide into oxygen and hydrogen peroxide. Metallic gold treatment reduces proliferation of inflammatory cells in brain injury (Pedersen et al., 2009). There are some side effects of anti-arthritic therapy using Au (I), mostly when it is oxidized to Au (III) by some of the potentially strong oxidants such as H_2O_2 available in inflammatory situations. Excessive use of gold complexes for the treatment of juvenile rheumatoid arthritis and osteoarthritis causes pain and fever. Among cutaneous symptoms intolerance was measured at low frequency, wider use of gold salts like gold salicylates and gold pyrazolone derivatives cause urticaria and angioedema (Rethy et al., 2004). Tolfenamic acid (Figure 4) and its metal complexes has been studied as anti-inflammatory, anti proliferative, and radical-scavenging agents by (Kovala et al., 2009).

Among transition metals complexes of Cu and Fe are capable of catalyzing dismutation of the superoxide anion. In addition, Mn complexes dose not bind to NO and react slowly with H_2O_2, demonstrating specificity towards superoxide anion. Interaction of SOD mimics with NO and H_2O_2 levels, both of which can cause high blood pressure and weaken the immune system. NO are an excellent ligand for transition metal ions and these metal nitrosyls having therapeutic values. Sodium nitropruside is used clinically to treat cardiovascular diseases by releasing NO but CN^- toxicity limited its application, however, discovery of new ruthenium nitrosyl complexes offer promising biological applications (Cameron et al., 2003). Over production of NO contributes to various diseases like sepsis, arthiritis, diabetes and epilepsy. Ruthinium polyaminocarboxylate complexes are efficient NO scavengers (Mosi et al., 2002; Spasojevic et al., 2003; Clark et al., 2003).

Many human diseases are associated with the over-production of oxygen free radicals that inflict cell damage. A manganese (II) complex with bis (cyclohexylpyridine)-substituted macrocyclic ligand has designed as a functional mimic of the superoxide dismutase (SOD) enzymes that normally remove these radicals (Li et al., 2002). Mangnese complexes have also been used to treat cell and tissue oxidative injuries by acting as superoxide anion scavenger (Failli et al., 2009).

Magnesium is used for the treatment of asthema in children (Bichara and Goldman, 2009). Some Cu complexes are also active against inflammation but their use is limited (Angelusiu et al., 2009). Cu (II) complexes tend to dissociate and bind to natural ligands such as albumins (Halova et al., 2006; Ward et al., 2005). Zinc has been proved to be involved in the inhibition of pro inflammatory cytokines (Haddad, 2009).

TRANSITION METALS AS ANTI-DIABETIC AGENTS

More than 2 - 8% of world's population is suffering from diabetes (Wild et al., 2004). It is a condition in which body do not produce a hormone called insulin which is necessary for the absorption of glucose in cells (Rother, 2007). Scientists are looking for alternative approaches for the treatment of diabetes (Nahas et al., 2009). Control of the glucose level in the blood plasma has been achieved by administration of vanadium and zinc in form of inorganic salts. It has been shown that elements are poorly absorbed in their inorganic forms and required high doses, which have been associated with undesirable side effects.

Vanadium complexes with organic ligands have proved to be less toxic, with improved solubility and lipophilicity. There are a number of vanadium complexes that have been developed, all of which have insulin-mimetic properties [Badmaey et al., 1999]. The molecular mechanism responsible for the insulin-like effects of vanadium com-

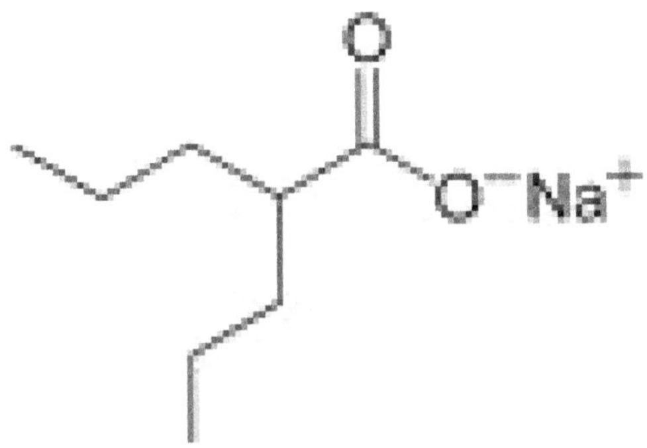

Figure 5. Sodium Valproate.

pounds have been shown to involve the activation of several key components of insulin-signaling pathways. It is interesting that the vanadium effect on these signaling systems is independent of insulin receptor protein tyrosine kinase activity, but it is associated with enhanced tyrosine phosphorylation of insulin receptor substrate (Mehdi et al., 2006). Chromium supplementation significantly improved glycemia among patients with diabetes but do not show any significant effect on glucose metabolism in healthy individuals (Balk et al., 2007). Higher zinc intake has also been associated with a slightly lower risk of type 2 diabetes in women (Sun et al., 2009).

TRANSITION METAL AS NEUROLOGICAL DRUGS

Transition metal complexes are also used in the treatment of neurological disorders (Hashimoto et al., 2003). One example of this is lithium. Lithium has been used to cure many neurological disorders like Huntington's chorea, tardive dyskinesia, spasmodic torticollis, Tourette's syndrome, L-dopa induced hyperkinesias, Parkinsonism, organic brain disorders, drug induced delusional disorders, migraine and cluster headache, periodic hypersomnolence, epilepsy, meniere's disease and periodic hypokalemic paralysis.

The mechanism of action involves Li inhibiting the scavenging pathway for capturing inositol in the resynthesis of polyphosphoinositides in the brain (Camins, 2009). Valproate treatment in combination with lithium delays disease onset, reduces neurological deficits and prolongs survival in an amyotrophic lateral sclerosis mouse model (Feng, 2008) (Figure 5). Other transition metals like zinc are involved as a transmitter in the neuronal signaling pathways. Neuronal Zn (II) serves as an important, highly regulated signaling component responsible for the initiation of a neuroprotective pathway (Aras, 2009).

TRANSITION METALS AS DELIVERY PROBES AND DIAGNOSTIC TOOLS

Metal complexes may be used as research probes of biological function and as potential diagnostic and therapeutic agents (Khan et al. 2009). Unique properties of transition metals like redox activity, lewis acidity, electrophilicity, valency, geometry magnetic spectroscopic and radiochemical properties can be used to measure cellular functions.(Ali et al. 1999; Wenyu et al. 2004; Radha et al., 2004).

Success of selective drug therapies depends on the drug's depot time in the target to treat. Depot time is currently being prolonged, using drug-eluting beads or microspheres for selective internal radiation therapy. The use of transition metals as a tool for radiopharmaceutical tracking of particles injected into tumors is increasing day by day (Luboldt et al., 2009). The use of gold nanorods for photoacoustic molecular imaging with simultaneous multiple targeting is reported.

The potential in improving cancer diagnosis is demonstrated. Gold nanoparticles have been synthesized inside ethosomes, vesicles composed of phospholipids, ethanol, and water, which could be very efficient not only in delivery probes to the skin but also as diagnostic and therapeutic multimodal agents (De la Presa et al., 2009). Being a stable transition metal makes them less reactive and less toxic. Their easy circulated and distribution in the body makes them vehicle of choice for drug delivery (Terentyuk et al., 2009). The nanoparticles injected into the tumor cells increases their ability to absorb radiation of specific wavelength. This property has been used in Lymphotropic nanoparticle-enhanced magnetic resonance imaging of prostate cancer (Ross et al., 2009). Antibody-conjugated gold nanoparticles have been used to deliver immunoglobulins to the retina in the rabbit eye and also for the *in vivo* photoacoustic molecular imaging (Li et al., 2008; Hayashi et al., 2009). Ultrasmall particles of iron oxide exhibits super-paramagnetic properties and are used as negative contrast agent in magnetic resonance imaging (MRI) which significantly improved the sensitivity of detection of inflamed tissues (Radermacher et al., 2009).

CONCLUSION

With the advancement in the field of inorganic chemistry the role of transition metal complexes as therapeutic compounds is becoming increasingly important. Recent advances in inorganic chemistry have made possible formation of number of transition metal complexes with organic ligand of interest, which can be used as therapeutic agent. Significant progress in the synthesis of platinum based anti-cancer drugs like cisplatin has been made. Although, these drugs have proven to be highly effective chemotherapeutic agent for treating various types of cancers, they have encountered the same drawbacks like other anti-cancer drugs are:

(i) Resistance to these drugs occurs when cells once destroyed by a particular drug no longer respond to the treatment with the same drug. When cells become resistant to drug, doses must be increased, a large dose escalation can lead to severe multi-organ toxicities, such as failure of kidney and bone marrow.

(ii) Drug toxicity is not limited; these drugs not specifically destroy cancer cells but also destroy rapidly and dividing normal cells like hair follicles.

Several new derivatives of cisplatin such as carboplatin, nedaplatin, oxaplatin have been made which are less toxic and more effective than cisplatin. Other transition metals like Ru, Ti, Os, Au, Fe, Co, etc have also been used as anticancer therapeutics.

The use of transition metal complexes as therapeutic compounds has become more and more pronounced. These complexes offer a great diversity in their action; they do not only have anti-cancer properties but have also been used as anti-inflammatory, anti-infective and anti diabetic compounds. Development of transition metal complexes as drugs is not an easy task; considerable effort is required to get a compound of interest. Beside all these limitations and side effects transition metal complexes are still the most widely used chemotherapeutic agents and make a large contribution to medicinal therapeutics in a way that is, unimaginable in few years back.

Future direction

Therapeutic applications of transition metal complexes is an under developed area of research and is full of opportunities for further progress. Basic principles to guide the synthesis and development of transition metal based pharmaceuticals are lacking. Development of new methodologies such as combinatorial chemistry will be helpful in the synthesis of inorganic compounds as therapeutic agents. Similarly, the action of metal complexes in the whole living organisms are expected to differ in general from the action of non-metal containing agents and may offer unique research, diagnostic, or therapeutic opportunities.

REFERENCES

Ali H, van Lier JE (1999). Metal complexes as photo and radoisensitizers. Chem Rev. 99: 2379-2450.

Allardyce CS, Dyson PJ, Ellis DJ, Health SL (2001). Ru (p-cymene) Cl2 (pta). A water soluble compound that exhibits pH dependent DNA binding providing selectivity for diseased cells. Chem commun. pp1396-1397.

Angelusiu MV, Almajan GL, Rosu T, Negoiu M, Almajan ER, Roy J (2009). Copper (II) and uranyl(II) complexes with acylthiosemicarbazide: synthesis, characterization, antibacterial activity and effects on the growth of promyelocytic leukemia cells HL-60. Eur. J. Med. Chem. 44(8): 3323-3329.

Anne V, Micheal H, Elizebeth A. Hillard, Emmanuel S (2005). Selective estrogen receptor modulators in the Ruthenocene series. J. Med. Chem. 48: 2814-2821.

Ansari KI, Grant JD, Kasiri S, Woldemariam G, Shrestha B, Mandal SS (2009). Manganese (III)-salens induce tumor selective apoptosis in human cells. J. Inorg. Biochem. 103(5): 818-26.

Aras MA, Hara H, Hartnett KA, Kandler K, Aizenman E (2009). Protein kinase C regulation of neuronal zinc signaling mediates survival during preconditioning. J. Neurochem. 110(1): 106-107.

Au L, Zheng D, Zhou F, Li ZY, Li X, Xia Y (2008). A quantitative study on the photothermal effect of immuno gold nanocages targeted to breast cancer cells. ACS Nano. 2(8): 1645-1652.

Balk EM, Tatsioni A, Lichtenstein AH, Lau J, Pittas AG (2007). Effect of chromium supplementation on glucose metabolism and lipids. A systematic review of randomized controlled trials.Diabetes Care. 30(8): 2154-2163.

Bartkowiak D, Stempfhuber M, Wiegel T, Bottke D (2009). Radiation- and chemoinduced multidrug resistance in colon carcinoma cells. Strahlenther Onkol. 185(12): 815-820.

Bassetti S, Hu J, Agostino Jr. RB, Sherertz RJ (2001). Prolonged anti microbial activity of catheter containing chlorhexidine silver sulfadiazine extends protection againsts catheter infection in vivo. Antimicrob Agent Chemother. 45(5): 1535.

RD (2009). Magnesium for treatment of asthma in children Review. Can Fam Physician. 55(9): 887-889.

Brow JM, Pleatman CR, Bierbach U (2002). Cytotoxic acridinylthiourea and its platinum conjugate produce enzyme- mediated DNA strand breaks. Bioorg. Med. Chem. Lett. 12: 2953 -2955.

Cameron BR, Darkes MC, Yee H, Olsen P (2003). Ruthenium (III)polyaminocarboxylate complexes: efficient and effective nitric oxide scavenger. Inorg. Chem. 42: 1868-1876.

Camins A, Crespo-Biel N, Junyent F, Verdaguer E, Canudas AM, Pallàs M (2009). Calpains as a target for therapy of neurodegenerative diseases: putative role of lithium (Review). Curr. Drug Metab. 10(5): 433-47.

Chen D, Milacic V, Frezza M, Dou QP (2009). Metal complexes, their cellular targets and potential for cancer therapy.Curr. Pharm. Des. 15(7): 777-791.

Chu, G (1994). Journal of Biological Chemistry. 269:787-790.

Clarke MJ (2003) Ruthenium metallopharmaceuticals. Coord. Chem. 236: 299-233.

Cohen SM, Lippard SJ (2001). Cisplatin : from DNA damage to cancer chemotherapy. Prog. Nucleic Acid Res. Mol. Biol. 67: 93-130.

De la Presa P, Rueda T, del Puerto Morales M, Javier Chichón F, Arranz R, Valpuesta JM, Hernando A (2009). Gold nanoparticles generated in ethosome bilayers, as revealed by cryo-electrontomography. J. Phys. Chem. B.12, 113(10): 3051-3057.

Failli P, Bani D, Bencini A, Cantore M, Di C, Mannelli L, Ghelardini C, Giorgi C, Innocenti M, Rugi F, Spepi A, Udisti R, Valtancoli B (2009). A novel manganese complex effective as superoxide anion scavenger and therapeutic agent against cell and tissue oxidative injury. J. Med. Chem. 52(22): 7273-83.

Feng HL, Leng Y, Ma CH, Zhang J, Ren M, Chuang DM (2008). Combined lithium and valproate treatment delays disease onset, reduces neurological deficits and prolongs survival in an amyotrophic lateral sclerosis mouse model. Neurosci. 155(3): 567-572.

Fukushima T, Horike H, Fujiki S, Kitada S, Sasaki T, Kashihara N (2009). Zinc deficiency anemia and effects of zinc therapy in maintenance hemodialysis patients. Ther. Apher. Dial. 3: 213-219.

Galanski M, Arion VB, Jakupec MA, Keppler BK (2003). Recent development in the field of tumor inhibiting metal complexes. Curr. Pharm. 9: 2078-2080.

Haddad JJ (2009). Thymulin and zinc (Zn2+)-mediated inhibition of endotoxin-induced production of proinflammatory cytokines and NF-kappaB nuclear translocation and activation in the alveolar epithetlium: unraveling the molecular immunomodulatory, anti- inflammatory effect of thymulin/Zn2+ in vitro. Mol. Immunol. 47(2-3): 205-214.

Halova-Lajoie B, Brumas V, Fiallo MM, Berthon G (2006). Copper(II) interactions with non-steroidal anti-inflammatory agents. III--3-Methoxyanthranilic acid as a potential *OH-inactivating ligand: a quantitative investigation of its copper handling role in vivo. J. Inorg .Biochem. 100(3): 362-373.

Hashimoto R, Fujimaki K, Jeong MR, Senatorov VV, Christ L, Leeds P, Chuang DM, Takeda M (2003). Neuroprotective actions of lithium. Seishin Shinkeigaku Zasshi. 105(1): 81-86.

Hayashi A, Naseri A, Pennesi ME, de Juan E Jr (2009). Subretinal delivery of immunoglobulin G with gold nanoparticles in the rabbit eye. Jpn J. Ophthalmol. 53(3): 249-256.

He Q, Liang CH Lippard SJ (2000). Steroid hormone induce HMG 1 over expression and sensitized breast cancer cells to cisplatin and carboplatin. Proc. Natl. Acad. Sci. USA 97: 5768-5772.
Hirsch LR, Jackson JB, Lee A, Halas NJ, West JL (2003). A whole blood immunoassay using gold nanoshells. Anal. Chem. 75: 2377-2381.
Jackson A, Davis J, Pither RJ, Rodger A, Hannon MJ (2001). Estrogen-derived steroidal metal complexes: agents for cellular delivery of metal centers to estrogen receptor-positive cells. Inorg. Chem. 40: 3964-3973.
Jain J, Arora S, Rajwade JM, Omray P, Khandelwal S, Paknikar KM (2009). Silver nanoparticles in therapeutics: development of an antimicrobial gel formulation for topical use. Mol. Pharm. 6(5):1388-1401.
Jamieson ER, Lippard SJ (1999). Structure, Recognition and processing of cisplatin DNA adducts. Chem. Rev. 99: 2467-2498.
Kapoor S (2009). Lanthanum and its rapidly emerging role as an anti-carcinogenic agent. J. Cell Biochem.106(2): 193.
Kemp MM, Kumar A, Mousa S, Dyskin E, Yalcin M, Ajayan P, Linhardt RJ, Mousa SA (2009). Gold and silver nanoparticles conjugated with heparin derivative possess anti-angiogenesis properties. Nanotechnol. 20(45): 455104.
Khan G, Merajver S (2009). Copper chelation in cancer therapy using tetrathiomolybdate: an evolving paradigm. Expert Opin Investig Drugs. 4:541-548.
Kiat HC, Weng KL, Gerard J, Lawrence L, Siden T, Anne V (2006). Organometallic cluster analogues of tamoxifen. J. Organomet. Chem. 69: 9-19.
Kovala-Demertzi D, Hadjipavlou-Litina D, Primikiri A, Staninska M, Kotoglou C, Demertzis MA (2009). Anti-inflammatory, antiproliferative, and radical-scavenging activities of tolfenamic acid and its metal complexes. Chem. Biodivers. 6: 948-960.
Lange TS, Kim KK, Singh RK, Strongin RM, McCourt CK, Brard L (2008). Iron (III)-salophene: an organometallic compound with selective cytotoxic and anti-proliferative properties in platinum-resistant ovarian cancer cells. PLoS One. 3(5): e2303.
Lee K-B, Wang D, Lippard SJ, Sharp PA(2002). Transcription-coupled and DNA damage-dependent ubiqitinition of RNA polymerase II in vitro. Proc. Natl. Acad. Sci. USA 99: 4239-4244.
Li H, Qian ZM (2002). Transferrin/transferring receptor-mediated drug delivery. Med. Res. Rev. 22: 225-250.
Li PC, Wang CR, Shieh DB, Wei CW, Liao CK, Poe C, Jhan S, Ding AA, Wu YN (2008). In vivo photoacoustic molecular imaging with simultaneous multiple selective targeting using antibody-conjugated gold nanorods. Opt Express. 16(23): 18605-18615.
Liang XJ, Chen C, Zhao Y, Wang PC (2010). Circumventing tumor resistance to chemotherapy by nanotechnology Methods. Mol. Biol. 596: 467-488.
Loo C, Lin A, Hirsch L, Lee MH, Barton J, Halas N, West J, Drezek R (2004). Nanoshell-enabled photonics-based imaging and therapy of cancer. Technol. Cancer Res. Treat. 3(1): 33-40.
Lorisean PM, Carciunescu DG, Doadrio Villarejo JC, Certad Fombona G, Gayyal P (1992). Pharmacomodulation of new organometallic complexes of Ir, Pt, Os, Rh, Pd: *in vivo* and *in vitro* trypanocidal study against trypanosoma brucci brucci. Trop. Med. Parasito. 43(2): 110-114.
Lottner C, Bart KC, Bernhart G, Brunner H (2002). Hematoporphyrin-derived soluble porphyrin-platinum conjugates with combined cytotoxic and phototoxic antitumour activity. J. Med. Chem. 45: 2064-2078.
Louie AY, Meade TJ (1999). Metal complexes as enzymetic inhibitors. Chem. Rev. 99: 2711-2734.
Luboldt W, Pinkert J, Matzky C, Wunderlich G, Kotzerke J (2009). Radiopharmaceutical tracking of particles injected into tumors: a model to study clearance kinetics. Curr. Drug Deliv. 6(3): 255-260.
Marcon G, Carotti S, Coronnello M, Messori L, Mini E, Orioli P, Mazzie T, Cinellu MA, Minghetti G (2002). Gold (III) complexes with bipyridyl ligands: solution chemistry, cytotoxicity , and DNA binding properties. J. Med. Chem. 45: 1672-1677.
Martins ET, Baruah H, Kramarczyk J, Saluta G, Day CS, Kucera GL, Bierbach U (2001). Design, synthesis, and biological activity of novel non-cisplatin type platinum –acridine pharmacophores. J. Med. Chem. 44: 4492-4496.
McKay BC, Becerril C, Ljungman M (2001). P53 plays a protective role against UV- and cisplatin-induced apoptosis in transcription-coupled repair proficient fibroblasts. Oncogene. 20(46): 6805-6808.
Mehdi MZ, Pandey SK, Theberge JF, Srivastava AK (2006). Insulin signal mimicry as a mechanism for the insulin-like effects of vanadium. Cell Biochem. Biophys. 44(1): 73-81.
Messori L, Abbate F, Marcon G, Orioli P, Fontani M, Mini E, Carroti S (2000). Gold (III) complexes as potential anti tumor agents : solution chemistry, cytotoxicity and DNA binding properties. J. Med. Chem. 43: 3541-3548.
Mosi R, Seguin B, Cameron B, Amankawa L (2002). Mechanistic studies on AMD6221: a ruthenium based nitric oxide scavenger. Biochem Biophys. Res. Commum. 292: 519-529.
Nahas R, Moher M (2009). Complementary and alternative medicine for the treatment of type 2 diabetes (Review). Can. Fam. Physician. 55(6): 591-596.
Najajreh Y, Perez JM, Navarroo-Ranninger C, Gibsin D (2002). Novel soluble cationic trans-diaminedichloroplatinum (II) complexes that are active against cisplatin resistant ovarian cancer cell lines. J. Med. Chem. 45: 5189-5195.
Nash P, Clegg DO (2005). Proriatic arthritis therapy: NSAID and traditional DMARDS. Ann. Rheum. Dis. 64: 74-77.
Offing OE, Etok C, Martelli S (1996). Synthesis and biological activity of Pt group metal complexes of o-vanillin thiosemicarbazones. Farmaco. 51(12): 801-808.
Pedersen MO, Larsen A, Pedersen DS, Stoltenberg M, Penkova M (2009). Metallic gold treatment reduces proliferation of inflammatory cells, increases expression of VEGF and FGF, and stimulates cell proliferation in the subventricular zone following experimental traumatic brain injury. Histol Histopathol. 5: 573-586.
Piulats JM, Jiménez L, García del Muro X, Villanueva A, Viñals F, Germà-Lluch JR (2009). Molecular mechanisms behind the resistance of cisplatin in germ cell tumours. Clin. Transl. Oncol. 12: 780-786.
Podsiadlo P, Sinani VA, Bahng JH, Kam NW, Lee J, Kotov NA (2008). Gold nanoparticles enhance the anti-leukemia action of a 6-mercaptopurine chemotherapeutic agent. Langmuir. 24(2): 568-574.
Qutob SS, Multani AS, Pathak S, Feng Y, Kendal WS, Ng CE (2004). Comparison of the X-radiation, drug and ultraviolet-radiation responses of clones isolated from a human colorectal tumor cell line. Radiat. Res. 161(3): 326-334.
Radermacher KA, Beghein N, Boutry S, Laurent S, Vander Elst L, Muller RN, Jordan BF, Gallez B (2009). In vivo detection of inflammation using pegylated iron oxide particles targeted at E-selectin: a multimodal approach using MR imaging and EPR spectroscopy. Invest. Radiol. 44(7): 398-404.
Radha N, Mostafa AE (2004). Changing Catalytic Activity during Colloidal Platinum Nanocatalysis Due to Shape Changes: Electron-Transfer Reaction. J. Am. Chem. Soc. 126(23): 7194-7195.
Ray S, Mohan R, Singh JK, Samantaray MK, Shaikh MM, Panda D, Ghosh P (2007). Anticancer and antimicrobial metallopharmaceutical agents based on palladium, gold, and silver N-heterocyclic carbene complexes. J. Am. Chem. Soc. 129(48): 15042-15053.
Rethy LA, Balo-Banga JM (2004). The allergic and other side effects of non steroid anti-inflammatory drugs and gold salts. Orv. Hetil. 145(38): 1943-1949.
Ross RW, Zietman AL, Xie W, Coen JJ, Dahl DM, Shipley WU, Kaufman DS, Islam T, Guimaraes AR, Weissleder R, Harisinghani M (2009). Lymphotropic nanoparticle-enhanced magnetic resonance imaging (LNMRI) identifies occult lymph node metastases in prostate cancer patients prior to salvage radiation therapy. Clin. Imaging. 4: 301-305.
Rother KI (2007). "Diabetes treatment—bridging the divide". The New Engl. J. Med. 356(15): 1499-1501.
Singh RV, Chaudhary A (2004). Biologically relevant tetra azamacrocyclic complexes of manganese. Spectral, antimicrobial , antifertility, and anti-inflammatory approach. J. Inorg Biochem. 98(11): 1712-1721.
Spasojevic I, Batinic-Harberle I, Reboculus (2003). Electrostatic contribution in the catalysis of O2- dismutation by superoxide dimutase mimics. J. Med. Chem. 278: 6831-6837.
Stryer L (1995). Biochemistry (4th ed.) W. H. Freeman and Company:

New York.

Sun Q, van Dam RM, Willett WC, Hu FB (2009). Prospective study of zinc intake and risk of type 2 diabetes in women. Diabetes Care. 32(4): 629-634.

Sun RW, Ma DL, Wong EL, Che CM (2007). Some uses of transition metal complexeszas anti-cancer and anti-HIV agents. Dalton Trans. 43: 4884-4892.

Terentyuk GS, Maslyakova GN, Suleymanova LV, Khlebtsov BN, Kogan BY, Akchurin GG, Shantrocha AV, Maksimova IL, Khlebtsov NG, Tuchin VV (2009). Circulation and distribution of gold nanoparticles and induced alterations of tissue morphology at intravenous particle delivery. J. Biophotonics. 5: 292-302.

Volkter WA, Hoffman TJ (1999). Therapeutic radiopharmaceuticals. Chem. Rev. 99: 2269-2292.

Ward PP, Paz E, Conneely OM (2005). Multifunctional roles of lactoferrin: a critical overview. Cell Mol. Life Sci. 62(22): 2540-2548.

Wild S, Roglic G, Green A, Sicree R, King H (2004). "Global prevalence of diabetes: estimates for the year 2000 and projections for 2003". Diabetes Care 27(5): 1047-1053.

Wong B, Masse JE, Yen Y-M, Giannikoupolous P , Feigon J, Johnson RC (2002). Binding of cisplatin modified DNA by saccharomyces cerevisiae HMGB Protien Nhp6A. Biochem. 41: 5404-5414.

Zheng Y, Sanche L (2009). Gold nanoparticles enhance DNA damage induced by anti-cancer drugs and radiation. Radiat. Res. 172(1): 114-119.

Starch phosphorylase: Biochemical and biotechnological perspectives

Rachana Tiwari and Anil Kumar*

School of Biotechnology, Devi Ahilya University, Khandwa Rd., Indore-452001, India.

A dynamic mediatory role between starch synthesis and degradation has been ascribed to starch phosphorylase. However, plant starch phosphorylase is largely considered to be involved in phosphorolytic degradation of starch. It reversibly catalyzes the transfer of glucosyl units from glucose-1-phosphate to the non-reducing end of glucan chain with the release of inorganic phosphate. It is widely distributed in plant kingdom. Enzyme multiplicity is also common in starch phosphorylase and different multiple forms have been predicted to have different roles in starch metabolism. Here, various biochemical properties have been reviewed. Its regulation by aromatic amino acids has also been discussed. Importance of plastidial and cytoplasmic starch phosphorylase has also been discussed. Various biotechnological aspects have been discussed. Its exploitation in production of glucose-1-phosphate, a cytostatic compound has been discussed in the present review.

Key words: Starch phosphorylase, glucose-1-phosphate, amylase, biosensor, aromatic amino acids.

INTRODUCTION

In plant carbohydrate metabolism, a dynamic mediatory role between starch synthesis and degradation has been ascribed to starch phosphorylase (Schneider et al., 1981). Besides, other glycosyl transferring phosphorylases are sucrose phosphorylase (EC. 2.4.1.7); maltose phosphorylase (EC. 2.4.1.8); cellobiose phosphorylase (EC. 2.4.1.20.); 1,3-β-oligoglucan phosphorylase (EC. 2.4.1.30); laminaribiose phosphorylase (EC. 2.4.1.31); cellodextrin phosphorylase (EC. 2.4.1.49); and α-α-trehalose phosphorylase (EC. 2.4.1.64). The catalytic activity of starch phosphorylase is restricted to the non–reducing end of starch. Starch phosphorylase is a form similar to glycogen phosphorylase, except that it acts upon starch instead of glycogen. The plant α- glucan phosphorylase commonly called as starch phosphorylase (EC 2.4.1.1) is largely known for phosphorolytic degradation of starch. It catalyzes the reversible conversion of α-1,4-glucan and inorganic phosphate into glucose-1-phosphate, and plays an important role in starch metabolism.

$$(1{,}4\text{-}\alpha\text{-D-glucosyl})_n + \text{orthophosphate } (P_i) \underset{\text{starch synthesis}}{\overset{\text{starch degradation}}{\rightleftharpoons}} (1{,}4\text{-}\alpha\text{-D-glucosyl})_{n-1} + \alpha\text{-D-glucose-1-phosphate}$$

It has recently been reported that reversible phos-phorylation of starch granule is essential for its subsequent degradation (Fettke et al., 2007). Within starch granules, glucan chains of amylopectin fraction form double helices and get packed to form crystalline lamellae and double helices at the granule surface and get disrupted by the

*Corresponding author. E-mail: ak_sbt@yahoo.com.

Abbreviations: **Pi,** orthophosphate; **mRNA,** messenger ribonucleic acid; **Pho1,** phosphorylase 1; **Pho2,** phosphorylase 2; **KDa,** kilodalton; **Da,** Dalton; **P1,** phosphorylase 1; **P2,** phosphorylase 2; **P3,** phosphorylase 3; **L-form,** light form; **H-form,** heavy form; **ADPglucose,** adenosine diphosphate glucose; **L78,** 78 amino acids long segment; **Km,** Michaelis constant; **AMP,** adenosine monophosphate; **cAMP,** 3',5'-cyclic adenosine monophosphate; **ATP,** adenosine triphosphate; **UDPglucose,** uridine diphosphate glucose; **GDPglucose,** guanidine diphosphate glucose; **NADP$^+$,** nicotinamide adenine dinucleotide; **NADPH,** reduced nicotinamide adenine dinucleotide; **GT,** glycosyl transferase; **PEST-region,** proline, glutamic acid, serine, threonine enriched region.

addition of phosphate groups and individual glucosyl residues. Phosphate addition is mediated by two enzymes, glucan water di kinase (EC. 2.7.9.4) and phosphoglucan water di kinase (EC. 2.7.9.5), respectively. Evidence for the importance of starch phosphorylation *in vivo* comes from the analysis of these enzymes of mutant plants. These plants exhibit reduced growth and accumulate nearly phosphate-free starch to a high level (Zeeman et al., 2007). The glucan water dikinase phosphorylates starch at C6 whereas phosphoglucan water dikinase phosphorylates starch at C3 position has been predicted by molecular modeling (Kossmann and Lloyd, 2000; Blennow et al., 2002; Wischmann et al, 1999; Rathore et al., 2009).

Distribution

Starch phosphorylase is widely distributed in plant kingdom (Fukui, 1982; Steup, 1988; Kumar, 1989). Starch phosphorylase has been described in many plants viz. potato tuber (Kamogawa et al., 1968; Tsay and Kuo, 1980; Schneider et al., 1981; Mohabir and John, 1988; Brisson et al., 1989); sweet potato tuber (Ariki and Fukui, 1975; Chang et al., 1987; Chern et al., 1990; Chiang et al., 1991; Lin et al., 1991); tapioca tuber (Vishwanathan and Krishnan, 1965, 1966; Kumar and Sanwal, 1984), developing maize seeds (Tsai and Nelson, 1969, Tsai et al., 1970); maize kernels (Ozbun et al.., 1973), maize endosperm (Tsai and Nelson, 1968; Chao and Seandalios, 1969); developing barley seeds (Batra and Mehta, 1981); pea seeds (Matheson and Richardson, 1976, 1978; Yang and Steup, 1990); germinating pea seeds (Juliano and Varner, 1969); mung bean and broad bean (Richardson and Matheson, 1977); germinating wheat seeds (Abbott and Matheson, 1972); spinach leaves (Preiss et al., 1980); *Zea mays* leaves (John et al., 1973a; Mateyka and Schnarren-berger, 1984, 1988; Spilatro and Preiss, 1987); *Vitis vinifera* leaves (John et al., 1973b; Downton and Hawker, 1973); loranthus leaves (Khanna et al., 1971); tobacco leaves (Vishwanathan and Shrivastava, 1964); *Aspergillus orisous* (Vishwanathan and Shrivastava, 1964); tapioca leaves (Kumar and Sanwal., 1982b); pea leaves (Steup and Latzko, 1979; Conrads et al., 1986; Yang and Steup, 1990); young banana leaves (Kumar and Sanwal, 1982a), mature banana leaves (Kumar and Sanwal, 1981a, 1983a, b); cabbage leaves (Kumar, 1984; Garg and Kumar, 2008); Indian millet (Garg and Kumar, 2007); *Arabidopsis thaliana* (Caspar et al., 1991); yan (Hamdon and Diapoh, 1991); *Vicia faba* leaves (Ghiena et al., 1993); *Saffron crocus* (Chrungoo and Farooq, 1993), and in developing soyabean seeds (Michiko et al., 1991). Singh et al. (1967) reported the presence of starch phosphorylase activity in the leaves of *Cajanas cajan, Coccinia indica, Impatiens balsamina, Lantana camara, Luffa acutangula, Momordica charantia, Murrava koeniggi, Nerium indicum, Ocimum americanum, Pitheullobium dulce, Santalum album and Tecoma stans.* John et al. (1973b) showed the presence of starch phosphorylase in *Vitis vinifera* berries. Starch phosphorylase has also been reported in banana fruits (Singh and Sanwal, 1975), blue green algae (Fredrick, 1967), *Chlorella vulgaris* (Nakamura and Imamura, 1983; Miyachi and Miyachi, 1987); red sea weed (Yu and Pedersen, 1991, 1991a) red algae (Yu, 1992) and *Escherichia coli* (Yu et al., 1988).

Enzyme multiplicity

Multiple forms of starch phosphorylase have been demonstrated in a number of tissues. The abundance of multiplicity has been suggested as a biochemical response to the heterogeneity of starch molecules and starch granules structure (Beck and Ziegler, 1989). It can result from transcriptional level due to different genes, post transcriptional level, due to splicing of mRNA and post transcriptional level due to glycosylation, proteolytic modifications and aggregation etc (Yu and Pedersen, 1991a). The significance of multiple forms of starch phosphorylase in the regulation of starch metabolism has been suggested. In general, the multiplicity range of starch phosphorylase varies from 2 to 5 (Nighojkar and Kumar, 1997).

Multiple forms of starch phosphorylase have been reported in blue green algae (Fredrick, 1967); spinach (Okita et al., 1979; Steup and Latzko, 1979; Steup et al., 1980, 1980a; Preiss et al., 1980; Shimomura et al., 1982; Schachtele and Steup, 1986), *Vicia faba* leaves (Gerbrandy and Verleur, 1971), *Hordeum distichum* (Baxter and Duffus, 1973), maize leaves (Mateyka and Schnarrenberger, 1984; Spilatro and Preiss, 1987; Mateyka and Schnarrenberger, 1988), pea leaves (Richardson and Matheson, 1977; Levi and Preiss, 1978; Steup and Latzko, 1979; Conrads et al., 1986), red sea weeds (Yu and Pederson, 1991, 1991a), banana leaves (Kumar and Sanwal, 1977), marigold, turnip, cabbage, brinjal, arhardal (*Cajanus*), potato leaves (Kumar and Sanwal, 1982c), potato tuber (Slabnick and Frydman, 1970; Gerbrandy and Verleur, 1971; Gerbrandy and Doorgeest, 1972; Shivaram, 1976), loranthus leaves (Khanna et al., 1971), *Allium cepa* and *Phaseolus vulgaris* (Gerbrandy and Verleur, 1971); sweet corn (Lee, 1972; Lee and Broun, 1973), young and mature banana fruits (Singh and Sanwal, 1975) germinating pea seeds (Matheson and Richardson 1976; Van berkel et al., 1991; Myers and Matheson, 1991), corms of saffron crocus (Chrungoo and Farooq, 1993).

Steup and Latzko (1979) reported two multiple forms in spinach leaves named as Pho1 and Pho2. The Pho1 has low affinity for glycogen and is located in the chloroplast, and Pho2 has high affinity for glycogen and is located in the cytosol (Steup and Latzko, 1979). The subunit molecular weight of Pho2 is found to be about 92 KDa whereas native molecular weight was 194 KDa. The subunit molecular weight of Pho1 is found to be about 104 KDa whereas native molecular weight is reported to be 204 KDa (Preiss et al., 1980).

Three multiple forms of starch phosphorylase are reported in wheat (*Triticum aestivum* L. var. *star*) as resolved by non-denaturing polyacrylamide gel affinity electrophoresis and named as P1, P2 and P3. Compartment analysis of young leaf tissues showed that P3 is plastidic whereas P1 and P2 are cytosolic (Schupp and Zeigler, 2004).

The multiplicity of starch phosphorylase has been shown in different developmental stages with respect to starch synthesis or degradation in maize endosperm (Tsai and Nelson, 1968, 1969), banana fruit pulps (Singh and Sanwal, 1975), growing as well as sprouting tubers and other plant propagules (Gerbrandy and Verleur, 1971).

This multiplicity does not coincide with any functional redundancies, as each form appears to have acquired a distinct and conserved role in starch metabolism (Ball and Morell, 2003). The multiple forms have been readily differentiated by their kinetic behavior or compartmentation. In potato (*Solenum tuberosum*) tubers and leaves, two forms L- and H-types, located in plastids and cytosol have been described corresponding to the low and high affinity for glucans (Mori et al., 1993a). In banana plant, two forms have been isolated in the leaves of *Musa paradisiaca* (type-A and -B) and fruits of *Musa acuminata* (II and I) (Kumar and Sanwal, 1977; Kumar and Sanwal, 1982a; da Mota et al., 2002). Similar pairs of phosphorylases have been found in *Arabidopsis* (AtPHS1 and AtPHS2) (Zeeman et al., 2004b), sweet potato (*Ipomoea batatas)* (Chen et al., 2002) and other higher plants. In general, most of the higher plants possess two types of starch phosphorylases, termed as Pho1 or L-form (plastidic) and Pho2 or H-form (cytosolic). The plastidial L-form has monomeric size of approximately 105,000 Da and has low affinity to branched glucans, glycogen and starch. On the other hand, the H-form has monomeric size of approximately 90,000 Da and has very high affinity to linear and branched glucans. Recent studies indicated that H-form also acts on heteroglycans (Fettke et al., 2004, 2005a, b). A great deal of confusion exists with regard to the nomenclature of these enzymes as in many cases it is not clear if they are isozyme (genetically different).

Importance of starch phosphorylase

Starch and sucrose are the primary products of photosynthesis in the leaves of plants. Starch exhibits major plant storage carbohydrate which gives energy during the times of heterotrophic growth. Starch metabolism has been studied extensively leading to a good knowledge of numerous enzymes involved. On the contrary, understanding of the regulation of starch metabolism is fragmentary (Kötting et al., 2010).

The physiological substrate and biological accurate functioning of cytoplasmic phosphorylase are not yet known in higher plants even after working nearly seventy years. However, there is some information available about the regulation of plant phosphorylase. It is well known that starch phosphorylase is a starch degradative enzyme. Starch biosynthesis takes place by ADPglucose pyrophosphorylase and starch synthase. However, Steup (1988) has shown the role of starch phosphorylase in photosynthetic carbon metabolism.

There are a number of reports for the homogeneous purification and biochemical characterization of starch phosphorylase (Kumar, 1989; Nighojkar and Kumar, 1997). The occurrence of multiple forms and their distinct intra-cellular localization are characteristics of starch phosphorylase (Nighojkar and Kumar, 1997). The heterogeneity of enzymes catalyzing apparently the same reaction is one of the characteristic features of almost all the starch degrading enzyme (Smith et al., 2003; Li et al., 2007; Rathore et al., 2009). Some multiple forms may favor starch biosynthesis over degradation (Gerbrandy and Doorgeest, 1972). Some multiple forms of starch phosphorylase favor starch biosynthesis and the starch thus synthesized may act as a primer for ADP glucose starch synthase (Frydman and Slabnik, 1973).

There are evidences that starch phosphorylase probably contributes to starch synthesis as expression and activity of plastidial starch phosphorylase that strongly correlates with starch synthesis (Brisson et al., 1989; Sonnewald et al., 1995; Yu et al., 2001; Tetlow et al., 2004b; Hannah and James, 2008). Starch phosphorylase in potato (Albrecht et al., 2001) and pho-A and pho-B in *Chlamydomonas reinhardtii* (Dauvillee, 2006) have been shown to involve in starch synthesis. Satoh et al. (2008) have shown that starch phosphorylase plays a crucial role in starch biosynthesis in rice endosperm, and appreciably affects the starch structure. According to the speculative glucan trimming model (Ball et al., 1996; Ball and Morell, 2003), during discontinuous synthesis of starch granules, the short glucan chain released from pre-amylopectin by the action of debranching enzymes are converted to longer glucan chains. The long glucan chains then act as substrate for starch phosphorylase. Recently, a new hypothesis has emerged that proposes multi-protein complex machinery for starch metabolism. It has been shown that phosphorylation of starch phosphorylase and branching enzymes promotes selective protein-protein interaction leading to multi-protein complex and assembly is essential for starch synthesis and degradation (Tetlow et al., 2004a, 2004b; Hennen-Bierwegan et al., 2009).

Chang et al. (1987) reported that the low molecular weight forms of starch phosphorylase stored in roots of sweet potato originate by proteolysis from high molecular weight phosphorylases present in the growing and developing tubers. The accumulation of starch takes place both in the chloroplasts and amyloplasts whereas starch phosphorylase is found in the cytosol and plastids (Schneider et al., 1981; Kumar, 1989).

Importance of plastidial starch phosphorylase in plants

The precise *in vivo* role of plastidic phosphorylases is not yet clear. On the basis of many indirect evidences, it is concluded that starch phosphorylase is primarily involved in phosphorolytic starch degradation. However, the enzyme cannot directly act on intact starch granules. The low affinity of starch phosphorylases for large and branched substrates supports its inability to act on intact starch granules (Zeeman et al., 2004a). The phosphorolytic action of starch phosphorylase on starch is expected across α-1,6-branch point, beyond which the enzyme is unable to act. The starch granules are degraded into branched and linear polyglucans. The branched polymers are reduced to linear glucans by debranching enzymes such as isoamylase (E.C. 3.2.1.68) and β-limit dextrinase (E.C. 3.2.1.142) which specifically cleave α-1, 6-glycosidic bond.

The linear glucans are further degraded by β-amylase and starch phosphorylase into neutral sugar (Smith et al., 2005). However, as the major pathway of starch degradation comes from hydrolysis, β-amylase is considered to be an important enzyme in starch degradation. Therefore, absence of plastidic phosphorylase has no effect on starch degradation. Rather, it has been observed that starch phosphorylase plays an important role in stress tolerance pathway. The exposure to transient dehydration stress leads to the activation of starch phosphorylase that in turn triggers starch degradation. It has been argued that during stress, phosphorolytic starch degradation serves a swift response to ease stress by routing the substrate into oxidative hexose monophosphate pathway (Zeeman et al., 2004b; Lloyd et al., 2005; Rathore et al., 2009).

There are some evidences to show that plastidial starch phosphorylase probably contributes in starch synthesis since expression and activity of plastidial starch phosphorylase favors its role in starch synthesis (Brisson et al., 1989; Sonnewald et al., 1995; Yu et al., 2001; Tetlow et al., 2004b; Hannah and James, 2008; Rathore et al., 2009).

Importance of cytosolic phosphorylase in plants

The physiological substrate and biological functions of cytosolic starch phosphorylase are yet not exactly clear. It is very unlikely that cytosolic starch phosphorylase acts on intact starch granules which are synthesized inside the chloroplast since chloroplastic membrane is impermeable to the starch as well as the enzyme. The mutant studies with potato showed that cytosolic starch phosphorylase had no effect on starch degradation. Studies using antisense RNA for cytosolic phosphorylase indicated that it had effect on sprout growth and flowering (Duwenig et al., 1997; Rathore et al., 2009). The cytosolic phosphorylase (Pho2) lacks L78 insert, which sterically hinders the substrate binding and therefore is more effective in degradation of branched large glucans and therefore may act on intact starch granules (Steup et al., 1983). It is considered that this enzyme may be involved in the degradation of reserved starch where starch containing cells lack compartmental intactness (Buchner et al., 1996; Schupp and Ziegler, 2004).

Biochemical aspects

Kumar (1989) reviewed purification and immobilization of starch phosphorylase from various higher plant tissues. The enzyme has been purified from pea leaves, (Kruger and ap Rees, 1983; Conrads et al., 1986), *Zea mays* (Mateyka and Schnarrenberger, 1988), potato (Brisson et al., 1989), sweet potato (Chang et al., 1987), pea cotyledons (Myers and Matheson, 1991; Van Berkel et al.,1991), yam (Hamdan and Diopoh, 1991), red sea weeds (Yu and Pederson, 1991) rice seedling and red algae (Hsu et al., 2004: Yu et al., 1993).

The enzyme is generally inactivated rapidly above 55°C. The activity of potato phosphorylase has been shown to increase during low temperature storage (Kennedy and Isherwood, 1975). A number of regulatory mechanisms exist for the regulation of starch phosphorylase which include control by pH, temperature, redox potential, oligosaccharides level, allosteric changes and covalent modification (Smith et al., 2005; Orzechowski, 2008). Intracellular compartmentation is also thought to be indirect control on regulation (Brisson et al., 1989) as well as transcriptional level control is also suggested (St Pierre et al., 1996). Since all the metabolites of starch degradation are eventually exported to cytosol, it can be considered that starch degradation is regulated and feed back inhibition is exerted at the cellular level rather than at the individual level (Settke et al., 2007).

Stability of potato phosphorylase is reported at 30°C between pH 6 and 8 but is rapidly inactivated at pH 4.9 (Lee, 1960).

Optimum pH 6 has been reported for several multiple forms of starch phosphorylase from germinating pea seeds (Matheson and Richardson, 1976), tapioca tubers, tapioca leaves and banana leaves (Kumar and Sanwal, 1984, 1982a, b). Preiss et al. (1980), in case of spinach leaves, have reported optimum pH 6 to 7 for the cytoplasmic enzyme and pH 7.5 for the chloroplast starch phosphorylase. The optimum pH value of 6 to 7 has been reported for the enzyme (Conrads et al., 1986; Kamogawa et al., 1968; Tsai and Nelson, 1969; Ariki and Fukui, 1975; Kumar, 1989). The optimum pH 6.5 has been reported for phosphorylase from corms of saffron crocus (Chrungoo and Farooq, 1993) and optimum pH 7 has been reported for both the multiple forms of maize starch phosphorylase during low temperature storage. Starch phosphorylases are generally inactivated rapidly above 55°C (Lee, 1960; Kamogawa et al., 1968).

Starch phosphorylase primarily acts as degrading enzyme for polyglucans via phosphorolysis and is mostly active in the homodimeric form with vitamin cofactor pyridoxal-5'-phosphate (Newgard et al., 1989; Hudson et al., 1993; Buchbinder et al., 2001). Although triose is observed to be the smallest primer required for its catalytic reaction for some of the starch phosphorylases, but in general, larger oligosaccharides have been found to be more efficient (Kumar 1989; Suganuma et al., 1991). There are reports of its primer-less activity, a characteristic that makes it one of the unique proteins. The primer less activity has been attributed to contamination of reagent with traces of oligosaccharides, the presence of glucosyl moieties in the enzyme, or due to some form of metabolic regulation (Venkaiah et al., 1991; Moreno and Tandecarz, 1996; Kossmann and Lloyd, 2000; Young et al., 2006).

Heavy metals such as Hg^{2+}, Ag^{+}, Cu^{2+}, Fe^{2+} and Zn^{2+} inhibit the activity of starch phosphorylase (Kumar, 1989; Hamdan and Diopoh, 1991). However, Mg^{2+} and Ca^{2+} act as effectors (Kastenschmidt et al., 1968). p-chloro-mereuribenzoate is inhibitory for potato tuber, banana leaves, tapioca leaves and tapioca tuber phosphorylases (Kumar, 1989).

Multiple forms of banana fruit phosphorylase are not inhibited by p-chloromercuribenzoate (Singh and Sanwal, 1976). However, ethylene diamine tetra acetic acid (EDTA) suppresses the activity of starch phosphorylase only at high concentration (Hamdan and Dipoh, 1991). Lee (1960) reported that potato phosphorylase does not contain serine phosphate and is unaffected by incubation with mammalian protein kinase, phosphate or AMP. In addition, neither ATP, DTT as reported by Gold et al. (1971) nor glucose-6-phosphate, fructose-6-phosphate, fructose-1, 6-bisphosphate, ribose-5-phosphate or fructose-1-phosphate activates the enzyme (Kamogawa et al., 1968). Lee and Braun (1973) also showed that sweet corn phosphorylase is unaffected by ATP or Mg^{2+}, but gets inhibited by ADP-glucose. Both the multiple forms of phosphorylase from *E. coli* are inhibited by sugar nucleotides, but not by AMP (Steup, 1988a). cAMP, AMP, glucose-1-phosphate, fructose-1, 6-bisphosphate, ADP-glucose, UDP-glucose and GDP-glucose have been shown to have no effect on *Klebsiella pneumoniae* malto-dextrin phosphorylase (Steup, 1988a).

However, granulose phosphorylase of *Clostridium pasteuranium* is inhibited by ADP-glucose, GDP-glucose, UDP-glucose and to a lesser extent by ATP, $NADP^{+}$, and NADPH, and AMP is found to be slightly stimulatory (Steup, 1988a). One form of loranthus leaf phosphorylase has been reported to be activated by ATP and AMP both (Khanna et al., 1971). On the other hand, banana leaf and fruit phosphorylases are not influenced by AMP but get inhibited by ATP (Kumar and Sanwal., 1981b, 1983a, b; Singh and Sanwal, 1976).

The P1 exhibited a strong binding affinity to immobilized glycogen upon electrophoresis whereas P2 and P3 did not exhibit strong binding affinity for immobilized glycogen. The cytosolic leaf phosphorylase has been purified to homogeneity by affinity chromatography. The single polypeptide product constituted both P1 and P2 activity forms. Probes for the detection of phosphorylase transcripts have been derived from cDNA sequences of cytosolic and plastidic phosphorylases, and these together with activity assays and a cytosolic phosphorylase specific antiserum have been used to monitor phosphorylase expression in leaves and seeds (Schupp and Zeigler, 2004).

The activity of rabbit glycogen phosphorylase has been reported to be regulated by a lipoprotein (Rutherford et al., 1986).

The phosphorylases from *Chlorella* have been reported to be activated by ammonium ions (Miyachi and Miyachi, 1987; Pan et al., 1988). It is also reported that β-amylase in sweet potato is an inhibitor of starch phosphorylase. Besides, aromatic amino acids viz. phenylalanine, tyrosine and tryptophan have been shown to be potent inhibitors of banana leaves (Kumar and Sanwal, 1983a), tapioca leaves (Kumar and Sanwal, 1982b), tapioca tubers (Kumar and Sanwal, 1984), one form of banana fruit phosphorylase (Singh and Sanwal, 1976). Since aromatic amino acids are originally derived from starch, it is believed that inhibition may exert a control via negative feedback mechanism ((Kumar and Sanwal, 1988).

Unlike animal phosphorylases, serine phosphate could not be detected in plant phosphorylases (Kumar, 1989). However, role of serine phosphate in animal phosphorylase is yet to be explained.

All the α-glucan phosphorylases either from animals, plants or microorganisms, contain pyridoxal phosphate as a prosthetic group (Graves and Wang, 1972; Iwata and Fukui, 1973; Steup, 1988; Kumar, 1989). On the other hand, Kumar and Sanwal (1982a, b) reported no pyridoxal phosphate in banana leaf and tapioca leaf starch phosphorylases. It has been explained that pyridoxal phosphate is involved neither in direct catalysis nor in regulation, but is present as a structural determinant essential for the enzyme activity (Kumar, 1989).

Starch phosphorylase and glycogen phosphorylase have a kinetic mechanism which involves random order of substrate with a rapid attainment of equilibrium and with inter-conversion of ternary complexes at rate limiting step (Steup and Schachtele, 1981; Kruger and ap Rees, 1983).

The Km value for the substrate depends not only on the temperature and pH but on the concentration of other

Table 1. General biochemical characteristics of the enzyme.

S. No.	Biochemical characteristics	Value
1	pH of the enzyme assay in the direction of starch (polysaccharide) synthesis	Around pH 6
2	pH of the enzyme assay in the direction of starch (polysaccharide) degradation	Around pH 7 to 7.5
3	pH stability	pH 6 to 8
4.	Temperature stability	Around 30°C, upper limit 55°C
4	Sub-unit molecular weight	Around 90,000 to 110,000 Da
5	Km for glucose -1 phosphate	0.6 to 5 mM
6	Km for inorganic phosphate	1 to 15 mM

substrate also. The values of Km for inorganic phosphate are normally in the range of 1 to 15 mM, and for glucose-1-phosphate normally in the range of 0.6 to 5 mM. The Km value of plastidic starch phosphorylase for glycogen and amylopectin in the direction of glucan synthesis are much higher than those of the cytoplasmic enzymes. Similar case is in the direction of phosphorolysis. However, Km values for the inorganic phosphate with both the enzymes are of the similar order. General biochemical characteristics of the enzyme are mentioned in Table 1.

Generally, all plant phosphorylases exhibit classical Michaelis kinetics but banana leaf phosphorylases showed biphasic kinetics indicating mixed positive-negative co-operativity and which is more prominent to the presence of aromatic amino acids. Aromatic amino acids are found to be inhibitors of banana leaf phosphorylases (Kumar and Sanwal., 1988). The multiple forms of phosphorylases are well differentiated by their kinetic behavior (Matheson and Rechardson, 1978; Steup et al., 1980a; Preiss et al., 1980; Steup and Schachtele, 1981; Shimomura et al., 1982; Kruger and ap Rees, 1983; Mateyka and Schnarrenberger, 1988; Myers and Matheson, 1991; Chrungoo and Farooq 1993).

Animal glycogen phosphorylases are regulated allosterically upon binding an effector or by covalent modulation at the N-terminal Ser14 residue by specific protein kinases. Two conformational states namely T (tight) and R (relaxed) are characterized by their inactive and active state, respectively (Johnson, 1992). Earlier, plant starch phosphorylases were being considered to be always active species only without any regulatory mechanism. Afterwards, this view of non-regulatory mechanism of starch phosphorylases, and in general for other starch-degrading enzymes has been found to be no longer true. A number of regulatory mechanisms for starch degrading enzymes have been shown like control by pH, redox potential, oligosaccharide level, allosteric control and covalent modulations (Smith et al., 2005; Orzechowski, 2008; Rathore et al., 2009). Studies indicated regulatory mechanisms such as intracellular compartmentation as an indirect control (Brisson et al., 1989) or at the transcriptional level (St Pierre et al., 1996) for starch phosphorylase. Fettke et al. (2007) argued that the starch degradation is regulated and feed back inhibition is exerted at the cellular level rather than at the individual organelles since all the metabolites of starch degradation are eventually exported to cytosol. Studies have shown that starch phosphorylase gets inhibited by aromatic amino acids (Kumar and Sanwal, 1983a, b, 1988; Venkaiah and Kumar, 1996). Inhibition of starch phosphorylase by aromatic amino acids has been correlated as feedback inhibition mechanism since basically aromatic amino acids are synthesized from starch (Kumar, 1988). Allosteric regulation of starch phosphorylases by L-tyrosine and ATP has also been reported (Singh and Sanwal, 1973; Kumar and Sanwal, 1983b, 1988).

Protein phosphorylation by kinase and dephosphorylation by phosphatase represents a ubiquitous mechanism of regulation for a variety of cellular processes. Tetlow et al. (2004a, b) demonstrated regulation by phosphorylation in starch branching enzyme and *Triticum aestivum* starch phosphorylase. There are reports that multiple forms of starch branching enzyme and Pho 1 are phosphorylated at one or more serine residues by plastidial protein kinase(s) in amyloplasts. Attempts are also done to search starch phosphorylase specific protein kinases. Young et al. (2006) reported a novel protein kinase that phosphorylates the L78 peptide of Pho 1 in sweet potato roots. However, this protein kinase does not phosphorylate Pho 2 and proteolytically modified Pho 1 lacking L78 indicating that this kinase is specific for L78 peptide of Pho 1 (Rathore et al., 2009). The molecular weight of this protein kinase has been reported to be nearly 338,000 (Young et al., 2006). This kinase phosphorylates only one seryl residue in L78 in spite of the fact that computational analyses showed many putative phosphorylation sites (Young et al., 2006; Rathore et al., 2009).

STRUCTURE AND FUNCTION

The animal counterpart of starch phosphorylase, glycogen phosphorylase is comparatively a more studied enzyme. Glycogen phosphorylase from rabbit muscle is especially most widely studied enzyme among all glycogen phosphorylases. Over 100 structures of glycogen phosphorylases from several different sources, their mutants and complexes have been reported in Protein Data Bank (PDB, available at http://www.rcsb.org). It has been shown that temporarily blood glucose level in diabetic patients may be controlled by inhibiting glycogen phosphorylase enzyme activity. Although, it is a temporary solution but it has the advantage of no risk of hypoglycemia which is considered to be main side effect of several antidiabetic agents. Therefore, many studies are in progress searching for structure based drug discovery (Buchbinder et al., 2001; Oikonomakos, 2002, Rathore et al., 2009).

Amino acid composition of many animal phosphorylases and potato tuber phosphorylase has been reported (Kamogawa et al., 1968; Cohen et al., 1973). The amino acid composition of phosphorylases isolated from young and mature banana leaves is found to be quite same. However, it is not same in amino acid composition from rabbit muscle and potato phosphorylases (Kumar and Sanwal, 1982d).

Gene of *E. coli* α-glucan phosphorylase has been cloned and sequenced (Yu et al., 1988). Lin et al. (1991) has reported the primary structure of sweet potato starch phosphorylase deduced from its cDNA sequence. Mori et al. (1991) reported cloning and nucleotide sequencing of the gene coding for a potato tuber phosphorylase isozyme.

A few attempts have been made to clone the genes encoding plastidic and cytosolic phosphorylase (Kossmann and Frohberg, 2004; Schupp and Zeigner, 2004, Rathore et al., 2009). The plastidial phosphorylases are characterized by the presence of approximately 80 residue long insertion region in the centre known as L78 insertion region which was first identified in *Solanum tuberosum* (Nakano and Fukui, 1986). It is suggested that L78 segment might be derived from an intron ancestor (Chen et al., 2002). The oligomeric state of starch phosphorylase is homodimeric in its native state (Buchbinder et al., 2001), although heterodimeric or tetrameric forms have also been observed. In potato, two pho1 enzymes namely pho 1a (1.1) and pho 1b (1.2) exist (Sonnewald et al., 1995; Albrecht et al., 1998, 2001) and occur both in the homodimeric (pho1a) and heterodimeric (pho1a-pho1b) state (Albrecht et al., 1998). These enzymes are α-β proteins and belong to characteristic GT-B fold of glycosyl transferase enzyme (Breton et al., 2006). The monomeric unit contains two conserved N-terminal and C-terminal domains (Buchbinder et al., 2001). These two domains are thought to have arisen by the fusion of two ancestral genes each coding for a protein of approximately 50 KDa (Nakano and Fukui, 1986). Non vascular organisms such as unicellular green algae lack the insertion region in the pho1 (Dauvillee et al., 2006).

The L78 insert region is rich in ionic residues (Nakano and Fukui, 1986).

The sequences of amino acids and comparison of their primary structures have been reported (Palm et al., 1985; Nakano and Fukui, 1986; Yu et al., 1988, 1993; Mori et al., 1991). These studies indicated that α-1-4-glucan phosphorylase is an evolutionary well conserved protein. Kumar (1990) has predicted the secondary structure of maltodextrin phosphorylase from *E. coli* using the Chou Fasman model.

α - Glucan phosphorylases from plants, animals and prokaryotes occur as dimers and tetramers composed of apparently identical monomeric subunits (Fukui, 1982).

The apparent molecular weight of the monomers has been reported in the range of 100,000 Da (Graves and Wang, 1972). White potato tuber phosphorylase has been reported to have a native molecular weight 207,000 Da, as calculated from the sedimentation constant (Lee, 1960). There are two multiple forms in sweet potato having native molecular weight of 180,000 to 210,000 Da, respectively (Ariki and Fukui, 1975). The molecular weight of different multiple forms of potato phosphorylase as determined by gel filtration chromatography are 180,000, 320,000, 520,000 and 620,000 Da (Gerbrandy and Doorgeest, 1972). However, a molecular weight of 220,000 Da as calculated from the sedimentation coefficient has been reported for potato phosphorylase (Kamogawa et al., 1968; Iwata and Fukui, 1973).

Sweet corn phosphorylase has been reported to have the molecular weight of 315000 Da (Lee and Braun 1973). However, two phosphorylases purified from hybrid maize have been reported to have the molecular weight 160000 and 200000 Da, respectively (Burr and Nelson, 1975). Richardson and Metheson (1977) reported two multiple forms of phosphorylase along with their molecular weights in alaska pea cotyledons, banana fruit, broad bean, mung bean, tobacco leaf, pea cotyledons and potato tuber. One of the multiple forms has been reported to have the molecular weight in the range of 170,000 to 200,000 Da, whereas the molecular weight of other form is reported in the range of 330,000 to 490,000 Da.

Spinach leaf cytoplasmic and chloroplastic phosphorylases have been reported to have the molecular weight, 194,000 and 203,000 Da, respectively (Preiss et al., 1980). Kumar and Sanwal (1977) reported the molecular weights of 450000 and 220000 Da, respectively for two multiple forms present in banana leaves.

Hamdon and Diopah (1991) reported the molecular weight of yam starch phosphorylase to be 200000 Da. In *Gracilaria sordida*, molecular weight has been reported to be 243000 and 337000 Da, respectively (Yu and Pedersen, 1991a). Insect flight muscle phosphorylase multiple forms have molecular weights 320,000, 140,000 and 100000, respectively (Childress and Secktor, 1970). The molecular weight of pea leaf cytoplasmic phosphorylase is 160000 Da, the subunit molecular weight being 88,700 Da (Conrads et al., 1986). The cytoplasmic starch phosphorylase of maize has molecular weight of 195,000 Da for the native enzyme and 87,000 and

57,000 Da for the SDS treated enzyme protein.

This implies a dimeric structure for cytoplasmic starch phosphorylase and a tetrameric structure for the chloroplastic starch phosphorylase (Mateyka and Schnarrenberger 1988), Kumar and Sanwal (1981a, b) have reported subunit molecular weight of 55000 ± 3000 in case of mature banana leaf phosphorylase. It was shown that the enzyme is made up of identical polypeptide chains of molecular weight 110000 and on heating, there is splitting of the chain from the middle, similarly as has been assumed in the case of sweet potato phosphorylase (Ariki and Fukui, 1975). Pho 1 is susceptible to degradation either by proteolysis or by heating. The monomeric unit of mature banana (*Musa paradisiaca*) and tapioca (*Manihot utilissima)* leaf phosphorylases and L-form of starch phosphorylase from potato and sweet potato upon proteolysis dissociate into two subunits of 55,000 size, corresponding to two N- and C-terminal domains (Kumar and Sanwal, 1981b, 1982b; Brisson et al., 1989; Chen et al., 2002). It has been reported that starch phosphorylase retains its quaternary structure as well as full catalytic activity indicating the possible mechanism of proteolytically driven regulation of its activity (Chen et al., 2002; Young et al., 2006). The characteristic degradation susceptibility of Pho 1 as against Pho 2 suggested that L-78 peptide is the primary target of proteolytic degradation. The sequence analysis of this L78 peptide showed the occurrence of PEST region (rich in proline, glutamic acid, serine and threonine). The PEST regions are shown to occur in proteins that are degraded rapidly (Rogers et al., 1986). These are generally flanked by many positively charged amino acids clusters (Chen et al., 2002; Rathore et al., 2009).

As mentioned earlier, comparatively plant starch phosphorylases have been lesser studied. To the best of our knowledge, no crystal structure of either L- or H- form has been reported yet. Many people including our laboratory are engaged in studies related to determi-nation of crystal structures of starch phosphorylases aiming to elucidate the structure and intriguing role of L78 region. Computational modeling studies are also in progress. The computational modeling work has the advantage of probing the dynamics of this region. However, it has been mentioned that the scope of such studies is limited due to many factors. The L78 loop requires *ab inito* modeling - a challenging task given the size of the fragment. The details of computationally modeled structure vary depending on the choice of the template – unregulated or regulated structures and the latter in active or inactive states (Rathore et al., 2009). Besides, orientation of momoners in the dimeric structure, and the details of dimeric interface also vary among phosphorylases (Watson et al., 1997).

There is a report of crystal structure in PDB carrying the name of starch phosphorylase (referred as StP; PDB code: '24 cm') from soil bacterium *Corynebacterium callunae*. It is a crystallographic tetramer. This starch phosphorylase has been reported to belong to non-regulatory class of α-glucan phosphorylase and has preference for starch over glycogen and has also been mentioned as the first bacterial starch phosphorylase (Weinhäusel et al., 1997). This bacterial starch phosphorylase like other prokaryotic phosphorylases does not have L78 region in its structure. This bacterial starch phosphorylase may be considered to have important biotechnological application since its stability may be increased using conformational engineering approach (Nidetzky et al., 2003). It has been shown that upon phosphate ion binding with the enzyme protein, there is much stabilization and therefore this property can be exploited for studies on manipulation via conformational engineering. Griessler et al. (2004) probed the forces which stabilizes its quaternary structure.

It has been shown that all the α-glucan phosphorylases have conserved catalytic and structural features, and pyridoxal phosphate is an essential cofactor for catalytic activity. It is also shown that 5'-phosphate moiety in pyridoxal phosphate is the main catalytic domain (Palm et al., 1990; Buchbinder et al., 2001). However, Kumar and Sanwal (1982a, b) in case of banana and tapioca leaves starch phosphorylases reported absence of pyridoxal phosphate and still are catalytically active. It has been shown that pyridoxal phosphate is attached to conserved lysyl residue side chain in the active site through schiff base in between its 4'-aldehyde group and ε- amino group of lysine. It is also shown that functional oligomeric state of starch phosphorylase is homodimeric state (Buchbinder et al., 2001). Of course, tetrameric and heterodimeric states have also been observed. In potato, two Pho1 enzymic forms [Pho1a (L1) and Pho1b (L2)] have been observed (Rathore et al., 2009; Sonnewald et al., 1995; Albrecht et al., 2001). It is also shown that Pho1 occurs both in homodimeric and heterodimeric form (Albrecht et al., 1998). Breton et al. (2006) reported that these proteins are α/β proteins and belong to the GT-B fold of glycosyl transferase enzymes. The monomeric subunit has two conserved N-terminal and C-terminal domains (Buchbinder et al., 2001). It is also shown that the N-terminal domain contains most of the ligand binding residues and therefore is called as regulatory domain, and C-terminal domain contains most of the active site residues and pyridoxal phosphate binding residues, and therefore is called as catalytic domain. It is also predicted that the active site is located in between the N-terminal and C-terminal domains and regulatory domain has residues involved in dimeric contacts and are responsible for transmitting conformational changes across subunits upon ligand binding in regulated phosphorylases. These two domains are thought to have possibly evolved by the fusion of two ancestral genes each coding for a protein of nearly 50,000 molecular weight (Nakano and Fukui, 1986). Hudson et al. (1993) reported that the residues in the functional and dimeric interface sites in starch phosphorylase and all other glucan phosphorylases are highly conserved. In non-vascular organisms viz. unicellular green algae, *Chlamydomonas reinhardtii*, Dauvillee et al. (2006) reported absence of insertion region of L78.

It has also been shown that Pho1 like other plastidial starch metabolizing enzymes also possess N-terminal transit peptide, in the precursor protein which gets cleaved upon translocation. The L78 insert region is rich in ionic residues (Nakano and Fukui, 1986). The transit peptide, L78 and its flanking proline-rich C-peptides are highly diverse. Excluding these regions, Pho1 enzymes share high sequence identity (~80%). Molecular modeling has suggested that the adjacent L78 loop might sterically block the active site resulting in low affinity for the substrates (Nakano and Fukui, 1986). The functional role of L78 insert has been studied by constructing chimeric enzymes (Mori et al., 1993a, b). Removal of L78 from L-type makes it completely inactive. However, the replacement of L78 region and flanking residues by the corresponding residues of Pho2 or the rabbit muscle phosphorylase containing glycogen storage site, showed its significantly high affinity for branched glucans.

Possible role of plastidic phosphorylases

The exact *in vivo* roles of plastidial phosphorylases are not yet confirmed. On the basis of indirect evidences, it has been said that plastidial phosphorylases are involved in phosphorolytic starch degradation. However, still many questions are unanswered. Sonnewald et al. (1995) demonstrated that the antisense repression of starch phosphorylase activity indicated no significant influence on starch accumulation in the leaves of transgenic potato plants. The T-DNA insertion in Pho1 gene of *Arabidopsis* suggested for an alternative route of starch phosphorylase (Zeeman et al., 2004b). The loss of plastidial starch phosphorylase does not cause a significant change in the level of starch accumulation in the day time or its mobilization at night. These results indicated that the absence of plastidial starch phosphorylase has no effect on starch degradation. However, it has been demonstrated that starch phosphorylase plays a role in stress tolerance pathway. The exposure to transient dehydration stress leads to the activation of starch phosphorylase that in turn triggers starch degradation. It has been argued that during stress, phosphorolytic starch degradation serves as a swift response to alleviate stress by feeding substrates into oxidative pentose phosphate pathway (Zeeman et al., 2004b; Lloyd et al., 2005).

On the other hand, there are evidences that plastidial phosphorylase (Pho 1) contributes in starch synthesis as expression and activity of plastidial starch phosphorylase correlates with starch synthesis (Brisson et al., 1989; Sonnewald et al., 1995; Yu et al., 2001; Tetlow et al., 2004b; Hannah and James, 2008). Albrecht et al. (2001) showed involvement of Pho 1a in starch synthesis in *Solanum tuberosum*. Dauvillee et al. (2006) showed involvement of plastidial phosphorylase Pho A and Pho B in starch synthesis.

d'Hulst et al. (2007) demonstrated that the switching off a gene of starch phosphorylase in many transgenic plants increases starch grain size and the amount of starch. Satoh et al. (2008) showed that Pho1 plays a crucial role in starch biosynthesis in rice endosperm, and it appreciably affects the starch structure. It remains unsolved, what mechanism exactly underlies the starch phosphorylase driven starch biosynthesis (Rathore et al., 2009). Other groups showed that disproportionating enzymes (D-enzymes; EC 2.4.1.25), in association with starch phosphorylase may contribute to overall starch synthesis by recycling the intermediate glucans, released in the process (Takaha et al., 1998a; Tetlow et al., 2004b). Colleoni et al. (1999) demonstrated that D-enzymes stimulate the phosphorolytic reaction of phosphorylase. According to glucan trimming model, it has been speculated that during discontinuous synthesis of starch granules, the short glucan chains released from pre-amylopectin by the action of debranching enzymes are converted to longer glucan chains by D-enzymes (Ball et al., 1996; Ball and Morell, 2003). The long glucan chains are then substrates for starch phosphorylase. The released glucose-1-phosphate is used up by ADP-glucose pyrophosphorylase (EC 2.7.7.27) for starch synthesis. Therefore, it may be said that reaction of starch phosphorylase in the direction of starch degradation ultimately contributes to its overall synthesis of starch. Reconciliation of the multiple lines of evidences on specific role of starch phosphorylase in starch metabolism requires new thinking and experiments. Experiments on the predicted role of proteolytically driven L78, acting as switch for synthesis or degradation are important (Chen et al., 2002). Others have proposed multi-protein complex machinery for starch metabolism. Studies have shown that the phosphorylation of starch phosphorylase and branching enzymes promote selective protein-protein interactions leading to the multi-protein complex, and this assembly is essential for starch synthesis and degradation (Tetlow et al., 2004a, b; Hennen-Bierwagen et al., 2009).

Possible roles of cytosolic phosphorylases

The physiological substrate and the biological function(s) of cytosolic phosphorylase have remained unsolved for a long time. The controversy surrounded over the exact role of cytoplasmic starch phosphorylase since starch granules are found inside the chloroplasts or amyloplasts and are impermeable to the surrounding membranes. The elimination studies in *Solenum tuberosum* indicated that it had no effect on starch degradation. However, studies on anti-sense inhibition of cytosolic phosphorylase indicated that it has effect on sprout growth and flowering (Duwenig et al., 1997). Steup et al. (1983) showed that the cytosolic Pho2 lacks Pho1 type of L78 insert, which sterically hinders the substrate binding and therefore they are more effective in degrading large branched glucans, and may even attack starch granules. Therefore, it has also been predicted that this enzyme is involved in starch degradation of reserve starch in plant organs where the starch containing cells have lost their compartmental integrity, such as in the cotyledons of germinating legumes (Buchner et al., 1996; Schupp and Ziegler, 2004). On the other hand, in cells with intact plastids, cytosolic phosphorylase may be involved in metabolism of products of starch degradation, exported from the chloroplast and the regulation of cytosolic glucose-1-phosphate level.

Steup group demonstrated that the cytosolic phosphorylase functions in a novel glycan metabolism. The experiments on pea (*Pisum sativam*) and *Arabidopsis* indicated that soluble heteroglycans (SHG) with a complex pattern of glycosidic linkages are the substrates for the cytosolic phosphorylase (Fettke et al., 2004, 2005b; Lu et al., 2006). Fettke et al. (2005a) showed that in transgenic potato plants, anti-sense inhibition of Pho2 resulted in increased glucosyl and rhamnosyl contents of the glycans, whereas its over-expression decreased their contents. This is perhaps the first *in vivo* indication for the physiological substrate of the cytosolic phosphorylase. Lu et al. (2006) demonstrated that cytosolic AtPHS2 (Pho2) and disproportionating enzymes (DPE2) in *Arabidopsis* act on maltose and SHGs, akin to amylomaltase and maltodextrin phosphorylase orthologs in *E. coli*. The T-DNA insertion line of the AtPHS2 gene increased the night time maltose level. Chia et al. (2004) showed that in DPE2 deficient mutants of *Arabidopsis*, the activity of the cytosolic phosphorylase increased, indicating the role of the enzyme as a buffer between starch degradation and cytosolic metabolism.

Biotechnological aspects/ industrial applications

The α- glucan phosphorylase can be used for the production of glucose-1-phosphate, to synthesize amylose and a variety of novel glucans for designing starch and development of transgenic varieties with modified starches (Fujii et al., 2003, 2007; Kossmann and Frohberg, 2004; Morell and Myers, 2005; Yanase et al., 2006a, 2007; d'Hulst et al., 2007; van der Vlist et al., 2008; Rathore et al., 2009). Glucose-1-phosphate is an important metabolite and has been used in cardiotherapy as a cytostatic compound (Venkaiah and Kumar, 1994). Glucose-1-phosphate has also been used as an antibacterial, anti-inflammatory and anti-tumor agent (Shin et al., 2000; Sakata et al., 2007; Rathore et al., 2009). It has also been used for the chemical synthesis of glucuronic acid and trehalose production (Takahashi et al., 1996; Rathore et al., 2009).

Amylose is a functional biomaterial and is also used to produce cycloamylose/ cyclodextrin and other glucose polymers. Amylose is not available as an industrial raw material, since its separation from amylopectin in starch is difficult and has a low yield. As the molecular size of amylose can be controlled by glucose-1-phosphate / primer molar ratio, the starch phosphorylase may be a potentially good system for its manufacturing. Since starch phosphorylase can synthesize only amylose, desired characteristic ratio of amylose to amylopectin in industrial starch can be manipulated. Amylose can also be used as inclusion compounds and has applications in drug delivery and nanotechnology (Ohdan et al., 2006, 2007a). The synthetic amylose can be designed with more mechanical strength and better moldability by controlling molecular size and branching and therefore it can be used as biodegradable plastics (Masao et al., 2005). Starch phosphorylase has been used for increased storage life of potato as the reduction of phosphorylase content in potato tubers leads to a substantial decrease in the accumulation of sugars (Kawchuk et al., 1999). On inactivation of starch phosphorylase, it has been observed that there is increase in the starch grain size in transgenic plant. Therefore, it is useful for increasing yields in the extraction and purification of starch on an industrial scale since smaller starch grains are generally lost during washing (d'Hulst et al., 2007). Starch phosphorylase is also used as a biosensor to detect amount of inorganic phosphate pollution in the environment and in the micro-estimation of inorganic phosphate under pathological conditions (Hüwell et al., 1997).

The glucan production in an cheaper way has been made possible by the joint use of α-glucan phosphorylase and other glucan enzymes. Its biochemical production using glucose-1-phosphate at the industrial scale is costly (Rathore et al., 2009). This has been accomplished in two different manners viz. by phosphorolysis of sucrose or cellobiose by sucrose phosphorylase (EC 2.4.1.7) or cellobiose phosphorylase (EC 2.4.1.20) respectively in the presence of inorganic phosphate. The joint system of sucrose phosphorylase, glucan phosphorylase and cellobiose phosphorylase as well as glucan phosphorylase has been used for enzymatic industrial production of amylose (Ohdan et al., 2006, 2007a, b; Yanase et al., 2006a; Fujii et al., 2007; Rathore et al., 2009).

The immobilized starch phosphorylase can be used for the synthesis of glucose-1-phosphate. It can also be used in microestimation of inorganic phosphate under pathological conditions and also to monitor the environmental pollution. Starch phosphorylase for the first time was immobilized on polyacrylamide by Kumar and Sanwal in 1981. In the past few years, immobilized starch phosphorylase or starch phosphorylase co-immobilized with other enzymes has been exploited for many purposes. Sucrose production from starch and fructose has been reported using co-immobilized starch phosphorylase, amylosucrase and sucrose phosphorylase enzyme bed reactor (Daurat et al., 1982). Szulczynski (1986) reported immobilization of starch phosphorylase on chitin derivatives.

Venkaiah and Kumar (1994, 1995) immobilized starch phosphorylase from sorghum leaves and potato waste water and developed bioreactors for the production of glucose-1-phosphate. They used various solid supports viz. egg shell, gelatin, alumina. Srivastava et al. (1996) immobilized starch phosphorylase from *Cuscuta reflexa* filaments using egg shell as solid support. Upadhye and Kumar (1996) immobilized starch phosphorylase from bengal gram seeds using alginate beads and DEAE cellulose as solid support.

The immobilization of starch phosphorylase from Indian millet has been reported for production of glucose-1-phosphate (Garg and Kumar, 2007). They used brick dust as solid support Garg and Kumar (2008) also reported immobilization of starch phosphorylase from cabbage leaves on egg shell and showed these immobilized enzymes as very good for production of glucose-1-phosphate.

CONCLUSION

Starch phosphorylase plays a dynamic role between starch synthesis and degradation. However, plant starch phosphorylase is largely considered to be involved in phosphorolytic degradation of starch. It reversibly catalyzes the transfer of glucosyl units from glucose-1-phosphate to the non-reducing end of glucan chain with the release of inorganic phosphate. It is widely distributed in plant kingdom. Enzyme multiplicity is also common in starch phosphorylase and different multiple forms have been predicted to have different roles in starch metabolism. Various biochemical properties have been reviewed. Its regulation by aromatic amino acids has also been discussed. Importance of plastidial and cytoplasmic starch phosphorylase has also been discussed. Various biotechnological aspects and industrial applications have been reviewed. Its exploitation in production of glucose-1-phosphate, a cytostatic compound has been discussed in the present review. Amylose is a functional biomaterial and is also used to produce cycloamylose/cyclodextrin and other glucose polymers. Amylose is not available as an industrial raw material, since its separation from amylopectin in starch is difficult and has a low yield. As the molecular size of amylose can be controlled by glucose-1-phosphate/primer molar ratio, starch phosphorylase may be a potentially good system for its manufacturing. Since starch phosphorylase can synthesize only amylose, desired characteristic ratio of amylose to amylopectin in industrial starch can be manipulated. Amylose can also be used as inclusion compounds and has applications in drug delivery and nanotechnology. The synthetic amylose can be designed with more mechanical strength and better moldability by controlling molecular size and branching, and therefore it can be used as biodegradable plastics. Starch phosphorylase is also used as a biosensor to detect amount of inorganic phosphate pollution in the environment and in the micro-estimation of inorganic phosphate under pathological conditions.

ACKNOWLEDGEMENTS

The authors acknowledge the facilities of the Department of Biotechnology, Ministry of Science and Technology, Government of India, New Delhi (DBT) under the Bioinformatics sub centre used in the preparation of the manuscript. The authors also acknowledge UGC, New Delhi for partial financial support.

REFERENCES

Abbott IR, Matheson NK (1972). Starch depletion in germinating wheat, wrinkled seeded peas and senescing tobacco leaves. Phytochemistry 11:1261-1272.

Albrecht T, Greve B, Pusch K, Kossmann J, Buchner P, Wobus U, Steup M (1998). Homodimers and heterodimers of Pho1-type phosphorylase isoforms in *Solanum tuberosum L.* as revealed by sequence-specific antibodies. Eur. J. Biochem. 251:343-352.

Albrecht T, Koch A, Lode A, Greve B, Schneider-Mergener J, Steup M (2001). Plastidic (Pho1-type) phosphorylase isoforms in potato (*Solanum tuberosum L.*) plants: expression analysis and immunochemical. characterization. Planta 213:602-613.

Ariki M, Fukui T (1975). α-Glucan phosphorylase from sweet potato: Isolation and properties of the partially degraded enzyme. Biochim. Biophys. Acta 386:301-308.

Ball S, Guan HP, James M, Myers A, Keeling P, Mouille G, Buléon A, Colonna P, Preiss J (1996). From glycogen to amylopectin: a model for the biogenesis of the plant starch granule. Cell 86:349-352.

Ball SG, Morell MK (2003). From bacterial. glycogen to starch: understanding the biogenesis of the plant starch granule. Annu. Rev. Plant Biol. 54:207-233.

Batra VIP, Mehta SL (1981). Enzymes of starch metabolism in developing grains of high lysine barley mutant. Phytochemistry 20:635-640.

Baxter ED, Diffus CM (1973). Enzymes of carbohydrate metabolism in developing Hordeum distichum grain, Phytochemistry 12:1923-1928.

Beck E, Ziegler P (1989). Biosynthesis and degradation of starch in higher plants. Annu. Rev. Plant Physiol. Plant Mol. Biol. 40:95-117.

Blennow A, Engelsen SB, Nielsen TH, Baunsgaard L, Mikkelsen R (2002). Starch phosphorylation: A new front line in starch research. Trends Plant Sci. 7:445-450.

Breton C, Šnajdrová L, Jeanneau C, Koča J, Imberty A (2006). Structures and mechanisms of glycosyltransferases. Glycobiology 16:29R-37R.

Brisson N, Giroux H, Zollinger M, Camirand A, Simard C (1989). Maturation and subcellular compartmentation of potato starchphosphorylase. Plant Cell 1:559-566.

Buchbinder JL, Rath VL, Fletterick RJ (2001). Structural relationships among regulated and unregulated phosphorylases. Annu. Rev. Biophys. Biomol. Struct. 30:191-209.

Buchner P, Borisjuk L, Wobus U (1996). Glucan phosphorylases in *Vicia faba L.*: cloning, structural analysis and expression patterns of cytosolic and plastidic forms in relation to starch. Planta 199:64-73.

Burr B, Nelson OE (1975). Maize α- glucan phosphorylase. Eur. J. Biochem. 56:539-546.

Caspar T, Lin TP, Kakefuda G, Benhow L, Preiss J, Sommerville C (1991). Mutants of *Arabidopsis* with altered regulation of starch degradation. Plant Physiol. 95:1181-1188.

Chang TC, Lee SC, Su JC (1987). Sweet potato starch phosphorylase-purification and characterization. Agr. Biol. Chem. 51:187-195.

Chao SK, Scandalios JG (1969). Identification and genetic control of starch degrading enzymes in maize endosperm. Biochem. Genet. 3:537-547.

Chen HM, Chang SC, Wu CC, Cuo TS, Wu JS, Juang RH (2002). Regulation of the catalytic behavior of L-form starch phosphorylase from sweet potato roots by proteolysis. Physiol. Plant. 114:506-515.

Chern MS, Mo YC, Juang RH, Su JC (1990). Probing the protein structure of sweet potato starch phosphorylase with monoclonal. antibodies. J. Chin. Biochem. Soc. 19:55-64.

Chia T, Thorneycroft D, Chapple A, Messerli G, Chen J, Zeeman SC, Smith SM, Smith AM (2004). A cytosolic glucosyltransferase is required for conversion of starch to sucrose in *Arabidopsis* leaves at night. Plant J. 37:853-863.

Chiang CL, Lu YL, Juang RH, Lee PD, Su JC (1991). Native and degraded forms of sweet potato starch phosphorylase. Agric. Biol. Chem. 55:641-646.

Childress CC, Secktor B (1970). Regulation of glycogen metabolism in insect flight muscle: Purification and properties of phosphorylases *in vitro* and *in vivo*. J. Biol. Chem. 245:2927-2936.

Chrungoo NK, Farooq S (1993). Partial. purification and physico-chemical properties of starch phosphorylase from corms of Saffron crocus (*Crocus sativus* L.). Plant Physiol. Biochem. 20:19-23.

Cohen P, Saari JC, Fischer EH (1973). Comparative study of dog fish and rabbit muscle phosphorylases. Biochemistry 12:5233-5241.

Colleoni C, Dauvillée D, Mouille G, Morell M, Samuel M, Slomiany MC, Liénard L, Wattebled F, d'Hulst C, Ball S (1999). Biochemical. characterization of the *Chlamydomonas reinhardtii* α-1,4 glucanotransferase supports a direct function in amylopectin biosynthesis. Plant Physiol. 120:1005-1014.

Conrads J, van Berkel J, Schachtele C, Steup M (1986). Non-chloroplastic α-1,4- glucan phosphorylase from pea leaves: Characterization and *in situ* local.ization by indirect immunofluorescence. Biochim. Biophys. Acta 882:452-463.

Da Mota RV, Cordenunsi BR, do Nascimento JR, Purgatto E, Rosseto MR, Lajolo F M. (2002) Activity and expression of banana starch phosphorylases during fruit development and ripening. Planta 216:325-333.

Daurat RL, Lena H, Whelan WJ (1982). Enzyme reactors for the continuous synthesis of sucrose from starch. J. Appl. Biochem. 4:133-152.

Dauvillée D, Chochois V, Steup M, Haebel S, Eckermann N, Ritte G, Ral. JP, Colleoni C, Hicks G, Wattebled F, Deschamps P, d'Hulst C, Liénard L, Cournac L, Putaux JL, Dupeyre D, Ball SG (2006). Plastidial. phosphorylase is required for normal starch synthesis in *Chlamydomonas reinhardtii*. Plant J. 48:274-285.

d'Hulst C, Planchot V, Chatterjee M (2007). Method for improving plants. US patent application 20070209088.

Downton WJS, Hawker JS (1973). Enzymes of starch metabolism in leaves and berries of Vitis vinifera. Phytochemistry 12:1557-1563.

Duwenig E, Steup M, Willmitzer L, Kossmann J (1997). Antisense inhibition of cytosolic phosphorylase in potato plants (*Solanum tuberosum L.*) affects tuber sprouting and flower formation with only little impact on carbohydrate metabolism. Plant J. 12: 323-333.

Fettke J, Eckermann N, Kötting O, Ritte G, Steup M (2007). Novel starch-related enzymes and carbohydrates. Cell. Mol. Biol. 52:OL883-OL904.

Fettke J, Eckermann N, Poeste S, Pauly M, Steup M (2004). The glucan substrate of the cytosolic (Pho2) phosphorylase isozyme from *Pisum sativum L.*: identification, linkage and subcellular localization. Plant J. 39:933-946.

Fettke J, Eckermann N, Tiessen A, Geigenberger P, Steup M (2005b). Identification, subcellular localization and biochemical. characterization of water-soluble heteroglycans (SHG) in leaves of *Arabidopsis thaliana L.*: distinct SHG reside in the cytosol and in the apoplast. Plant J. 43:568-586.

Fettke J, Poeste S, Eckermann N, Tiessen A, Pauly M, Geigenberger P, Steup M (2005a). Analysis of cytosolic heteroglycans from leaves of transgenic potato (*Solanum tuberosum L.*) plants that under- or overexpress the Pho 2 phosphorylase isozyme. Plant Cell Physiol. 46:1987-2004.

Fredrick JF (1967). Glucosyl transferase isozymes in algae. Phytochemistry 6:1041-1046.

Frydman RB, Slabnik E (1973). The role of phosphorylase in starch biosynthesis. Ann. New York Acad. Sci. 210:153-169.

Fujii K, Takata H, Yanase M, Terada Y, Ohdan K, Takaha T, Okada S, Kuriki T (2003). Bioengineering and application of novel glucose polymers. Biocatal. Biotransform. 21: 167-172.

Fujii K, Terada Y, Yanase M, Ohdan K, Takata H, Takaha T, Kuriki T, Okada S (2007). Production method and preparation method of glucans. US patent, pp. 7, 229, 801.

Fukui T (1983). Plant Phosphorylases: Structure and Function. In: Akazawa T, Imaseki H (eds.), Martinus Nijhuff/ Dr. W. Junk Publishers, The Hague. N. Front. Plant Biochem. pp. 71-82.

Fukui T, Shimomura S, Nakano K (1982). Potato and rabbit muscle phosphorylases: comparative studies on the structure, function and regulation of regulatory and nonregulatory enzymes. Mol. Cell Biochem. 42:129-144.

Garg N, Kumar A (2008). Immobilization of starch phosphorylase from cabbage leaves: Production of glucose-1-phosphate. Brazilian J. Chem. Eng. 25:229-235.

Garg S, Kumar A (2007). Immobilization of starch phosphorylase from seeds of Indian millet (*Pennisetum typhoides*) variety KB560. Afr. J. Biotechnol. 6:2715-2720.

Gerbrandy SJ, Verleur JD (1971). Phosphorylase isoenzymes: Localization and occurrence in different plant organs in relation to starch metabolism. Phytochemistry 10:261-266.

Gerbrandy SjJ, Doorgeest A (1972). Potato phosphorylase isoenzymes. Phytochemistry 11:2403-2407.

Ghiena C, Schulz M, Schnabl H (1993). Starch phosphorylation and distribution of the starch degrading enzymes in *Vicia faba* leaves. Plant Physiol. 101:73-79.

Gold AM, Johnson RM, Sanchez GR (1971). Kinetic mechanism of potato starch phosphorylase. J. Biol. Chem. 246:3444-3450.

Graves JP, Wang J (1972). α- Glucan phosphorylase, chemical and physical basis of catalysis and regulation. In Boyer PD (ed.), Academic Press, New York. Enzymes pp.435-482,

Griessler R, Psik B, Schwarz A, Nidetzky B (2004). Relationships between structure, function and stability for pyridoxal. 5'-phosphate-dependent starch phosphorylase from *Corynebacterium callunae* as revealed by reversible cofactor dissociation studies. Eur. J. Biochem. 271:3319-3329.

Hamdan I, Diopoh J (1991). Purification and physico-chemical. properties of starch phosphorylase from yam (*Dioscorea cayenensis*). Plant Sci. 76:1-7.

Hannah LC, James M (2008). The complexities of starch biosynthesis in cereal. endosperms Curr. Opin. Biotechnol. 19:160-165.

Hennen-Bierwagen TA, Lin Q, Grimaud F, Planchot V, Keeling P L, James MG, Myers AM (2009). Proteins from multiple metabolic pathways associate with starch biosynthetic enzymes in high molecular weight complexes: a model for regulation of carbon allocation in maize amyloplasts. Plant Physiol. 149:1541-1559.

Hsu JH, Yang CC, Su JC, Lee PD (2004). Purification and characterization of a cytosolic starch phosphorylase from etiolated rice seedlings. Bot. Bull. Acad. Sin. 45:187-196.

Hudson JW, Golding GB, Crerar MM (1993). Evolution of allosteric control in glycogen phosphorylase. J. Mol. Biol. 234:700-721.

Hüwell S, Haalck L, Conrath H, Spencer F (1997). Maltose phosphorylase from *Lactobacillus brevis*: purification, characterization, and application in a biosensor for orthophosphate. Enzyme Microb. Technol. 21:413-420.

Iwata S, Fukui T (1973). The subunit structure of α- glucan phosphorylase from potato. FEBS Lett. 36:222-226.

John W, Downton S, Hawker JS (1973a). Enzymes of starch and sucrose metabolism in zeamays leaves. Phytochemistry 12:1551-1556.

John W, Downton S, Hawker JS (1973b). Enzymes of starch metabolism in leaves and berries of *Vitis vinifera*. Phytochemistry 12:1557-1562.

Johnson LN (1992). Glycogen phosphorylase: control by

phosphorylation and allosteric effectors. FASEB J. 6:2274-2282.
Juliano BO, Varner JE (1969). Enzymic degradation of starch granules in the cotyledons of germinating peas. Plant Physiol. 44:886-892.
Kamogawa A, Fukui T, Nikuni A (1968). Potato α-glucan phosphorylase: crystallization, amino acid composition and enzymatic reaction in the absence of primer; (Tokyo). J. Biochem. 63:361-369.
Kastenschmidt LL, Kastenschmidt J, Melmreich E (1968). Subunit interaction and their relationship to the allosteric properties of rabbit skeletal. muscle phosphorylase b. Biochemistry 7:3590-3608.
Kawchuk LM, Armstrong JD, Lynch D R, Knowles NR (1999). Potatoes having improved quality characteristics and methods for their production. US patent pp. 5, 998, 701.
Kennedy MGH, Isherwood FA (1975). Activity of phosphorylase in *Solanum tuberosum* during low temperature storage. Phytochemistry 14:667-670.
Khanna SK, Krishnan PS, Sanwal GG (1971). Glucan phosphorylases in the leaves of *Dendrophthoe falcata*: Purification and characterization of enzyme. Phytochemistry 10:551-559.
Kossmann J, Frohberg C (2004). Nucleic acid molecules encoding starch phosphorylase from maize. US Patent pp. 6, 686, 514.
Kossmann J, Lloyd J (2000). Understanding and influencing starch biochemistry. Crit. Rev. Plant Sci. 19:171-226.
Kötting O, Pusch K, Tiessen A, Geigenberger P, Steup M, Ritte G (2005). Identification of a novel enzyme required for starch metabolism in *Arabidopsis* leaves. The phosphoglucan, water dikinase. Plant Physiol. 137:242-252.
Kruger NJ, ap Rees T (1983). Properties of α- glucan phosphorylase from pea chloroplasts. Phytochemistry 22:1891-1898.
Kumar A (1984). Partial purification and characterization of starch phosphorylase from cabbage (*Elephantopus acaber*) leaves. Indian J. Plant Physiol. 27:209-213.
Kumar A (1989). Starch phosphorylase in plants. J. Scientific Industrial. Res. 48:568-576.
Kumar A (1990). Predicted secondary structure of maltodextrin phosphorylase from *Escherichia coli* as deduced using Chou-Fasman model. J. Biosci. 15:53-58.
Kumar A, Sanwal GG (1977). Multiple forms of starch phosphorylase from banana leaves. Phytochemistry 16:327-328
Kumar A, Sanwal GG (1982b). Starch phosphorylase from tapioca leaves: absence of pyridoxal. phosphate. Arch. Biochem. Biophys. 217:341-350.
Kumar A, Sanwal GG (1982c). Subcellular localization of multiple forms of starch phosphorylase in plant leaces during development. Ind. J. Plant Physiol. 25:317-323.
Kumar A, Sanwal GG (1982d). Amino acid composition of starch phosphorylase from banana (*Musa paradisiaca*) leaves. Ind. J. Biochem. Biophys. 19:62-63.
Kumar A, Sanwal GG (1983a). Starch phosphorylase from mature banana (*Musa paradisiaca*) leaves: Part I - kinetics and inhibition studies at optimum pH. Indian J. Biochem. Biophys. 20:280-284.
Kumar A, Sanwal GG (1983b). Starch phosphorylase from mature banana (*Musa paradisiaca*) leaves: Part II - Aromatic amino acid inhibition at pH 7.0. Ind. J. Biochem. Biophys. 20:285-289.
Kumar A, Sanwal GG (1984). Starch phosphorylase from tapioca (Manihot utilissima) tuber. Isolation and some physico-chemical. properties. Indian J. Biochem. Biophys. 21:241-247.
Kumar A, Sanwal GG (1981b) Characterization of purified starch phosphorylase from mature banana (*Musa paradisiaca*) leaves. Indian J. Biochem. Biophys. 18:421-424.
Kumar A, Sanwal GG (1981a). Immobilization of starch phosphorylase from mature banana (*Musa paradisiaca*) leaves. Indian J. Biochem. Biophys. 18:114-119.
Kumar A, Sanwal GG (1988). Kinetics of starch phosphorylase from young banana leaves. Phytochemistry 27:983-988.
Kumar A, Sanwal GG (1982a). Purification and physicochemical. properties of starch phosphorylase from young banana leaves. Biochemistry 21:4152-4159.
Lee EYC (1972). Multiple forms of α- glucan phosphorylase in sweet corn. FEBS Lett. 27:341-344.
Lee EYC, Braun JJ (1973). Sweet corn phosphorylase: Purification and properties. Arch. Biochem. Biophys. 156:276-286.
Lee YP (1960). Potato phosphorylase. I. Purification, physico-chemical. properties and catalytic properties. Biochim. Biophys. Acta 43:18-24.
Levi C, Preiss J (1978). Amylopectin degradation in pea chloroplast extracts. Plant Physiol. 61:218-220.
Li Z, Morell MK, Rahman S (2007). Rice and products thereof having starch with an increased proportion of amylase. US patent application 20070300319.
Lin CT, Lin MT, Chou HY, Lee PD, Su JC (1995). The gene structure of starch phosphorylase from sweet potato. Plant Physiol. 107:277-278.
Lin TP, Yeh KW, Lee PD, Su JC (1991). Primary structure of sweet potato starch phosphorylase deduced from its cDNA sequence. Plant Physiol. 95:1250-1253.
Lloyd JR, Kossmann J, Ritte G (2005). Leaf starch degradation comes out of the shadows. Trends Plant Sci. 10:130-137.
Lu Y, Steichen JM, Yao J, Sharkey TD (2006). The role of cytosolic α-glucan phosphorylase in mal.tose metabolism and the comparison of amylomaltase in *Arabidopsis* and *Escherichia coli*. Plant Physiol. 142:878-889.
Masao T, Junichi T, Yasushi M (2005). Enzymatic synthesis of amylose and its medical Application. Biol. Ind. 22:58-66.
Mateyka C, Schnarrenberger C (1984). Starch phosphorylase isoenzymes from mesophyll and bundle sheath cells of the corn leaves. Plant Sci. Lett. 36:119-123.
Mateyka C, Schnarrenberger C (1988). Purification and properties of mesophyll and bundle sheath cell α-glucan phosphorylase from *Zea mays* L. Plant Physiol. 86:417-422.
Matheson NK, Richardson RH (1976). Starch phosphorylase enzymes in developing and germinating pea seeds. Phytochemistry 15:887-892.
Matheson NK, Richardson RH (1978). Kinetic properties of two starch phosphorylases from pea seeds. Phytochemistry 17:195-200.
Michiko M, Sugimoto T, Mitsura M (1991). Starch breakdown in developing soybean seeds (*Glycine max* cv. Enrei). Agric. Biol. Chem. 55:67-71.
Miyachi S, Miyachi S (1987). Some biochemical. changes related to starch breakdown induced by blue light illumination and by addition of ammonia to chlorella cells. Plant Cell Physiol. 28:309-314.
Mohabir G, John P (1988). Effect of temperature on starch synthesis in potato tuber tissue and in amyloplasts. Plant Physiol. 88:1222-1228.
Morell MK, Myers AM (2005). Towards the rational design of cereal. starches Curr. Opin. Plant Biol. 8:204-210.
Moreno S, Tandecarz JS (1996). Analysis of primer independent phosphorylase activity in potato plants: high levels of activity in sink organs and sucrose-dependent activity in cultured stem explants. Cell Mol. Biol. 42:637-643.
Mori H, Tanizawa K, Fukui T (1991). Potato tuber type H phosphorylase isozyme. J. Biol. Chem. 266:18446-18453.
Mori H, Tanizawa K, Fukui T (1993a). A chimeric α-glucan phosphorylase of plant type L and H isozymes. Functional. role of 78-residue insertion in type L isozyme. J. Biol. Chem. 268:5574-5581.
Mori H, Tanizawa K, Fukui T (1993b). Engineered plant phosphorylase showing extraordinarily high affinity for various α-glucan molecules. Protein Sci. 2:1621-1629.
Myers DK, Matheson NK (1991). Affinity purification and kinetics of pea cotyledon starch phosphorylase. Phytochemistry 30:1079-1087.
Nakamura Y, Imamura M (1983). Characteristics of α- glucan phosphorylase from *Chlorella vulgaris*. Phytochemistry 22:835-840.
Nakano K, Fukui T (1986). The complete amino acid sequence of potato α -glucan phosphorylase. J. Biol. Chem. 261:8230-8236.
Newgard CB, Hwang PK, Fletterick RJ (1989). The family of glycogen phosphorylases: Structure and function. Crit. Rev, Biochem. Mol. Biol. 24:69-99.
Nidetzky B, Griessler R, Pierfederici FM, Psik B, Sciré A, Tanfani F (2003). Mutagenesis of the dimer interface region of *Corynebacterium cal.lunae* starch phosphorylase perturbs the phosphate-dependent conformational. relay that enhances oligomeric stability of the enzyme. (Tokyo). J. Biochem. 134:599-606.
Nighojkar SA, Kumar A (1997). Starch phosphorylase: Biochemical, molecular and biotechnological aspects. Genet. Eng. Biotechnol. 17:189-202.
Ohdan K, Fujii K, Yanase M, Takaha T, Kuriki T (2006). Enzymatic synthesis of amylose. Biocatal. Biotransform. 24:77-81.
Ohdan K, Fujii K, Yanase M, Takaha T, Kuriki T (2007a).

Phosphorylase coupling as a tool to convert cellobiose into amylose. J. Biotechnol. 127:496-502.
Ohdan K, Takaha T, Kuriki T, Kudo K, Wada M, Sunako M, Takahara J (2007b). Method of converting β-1,4-glucan to α-glucan. US patent application 20070092949.
Oikonomakos NG (2002). Glycogen phosphorylase as a molecular target for type 2 diabetes therapy. Curr. Protein Peptide Sci. 3:561-586.
Okita THW, Greenberg E, Kuhn DN, Preiss J (1979). Subcellular localization of starch degradative and biosynthetic enzymes of spinach leaves. Plant Physiol. 64:187-192.
Orzechowski S (2008). Starch metabolism in leaves. Acta Biochim. Polonica 55:435-445.
Ozbun JL, Hawker JS, Greenberg E, Lammel C, Preiss J, Lee EYC (1973). Starch synthetase, phosphorylase, ADPglucose pyrophosphorylase and UDPglucose pyrophosphorylase in developing maize kernels. Plant Physiol. 51:1-5.
Palm D, Goerl TC, Juang RH, Su JC (1985). Evolution of catalytic and regulatory sites in phosphorylases. Nature 313: 500-502.
Palm D, Klein HW, Schinzel R, Buehner M, Helmreich EJ (1990). The role of pyridoxal. 5'-phosphate in glycogen phosphorylase catalysis. Biochemistry 29:1099-1107.
Pan SM, Chang TC, Juang RR, Su JC (1988). Starch phosphorylase inhibitor is β-amylase. Plant Physiol. 88:1154-1156.
Preiss J, Okita TW, Greenberg E (1980). Characterization of the spinach leaf phosphorylases. Plant Physiol. 66:864-869.
Rathore RS, Garg N, Garg S, Kumar A (2009). Starch phosphorylase: Role in starch metabolism and biotechnological applications. Crit. Rev. Biotechnol. 29:214-224.
Richardson RH, Matheson NK (1977). Some molecular properties of plant starch phosphorylases and the levels of activity of the multiple forms. Phytochemistry 16:1875-1879.
Rutherford WM, Jacob PV, McCracken DA, Mathew NJ (1986). Lipoprotein as a regulator of starch synthesizing enzymes activity. Biochem. Biophys. Res. Commun. 135:701-707.
Sakata M, Kawai T, Kayane S, Ooshima H (2007). Kinetic study of phosphorolysis of dextrin by potato phosphorylase. J. Chem. Eng. Jpn. 40:441-446.
Satoh H, Shibahara K, Tokunaga T, Nishi A, Tasaki M, Hwang SK, Okita TW, Kaneko N, Fujita N, Yoshida M, Hosaka Y, Sato A, Utsumi Y, Ohdan T, Nakamura Y (2008). Mutation of the plastidial. α -glucan phosphorylase gene in rice affects the synthesis and structure of starch in the endosperm. Plant Cell 20:1833-1849.
Schachtele C, Steup M (1986). α-1,4-Glucan phosphorylase from leaves of spinach (*Spinacia oleracea* L.) I. *In situ* localization by indirect immunofluorescence. Planta 167:444-451.
Schneider EM, Becker JU, Volkmann D (1981). Biochemical. properties of potato phosphorylase change with its intracellular local.ization as revealed by immunological. methods. Planta 151:124-134.
Schupp N, Ziegler P (2004). The Relation of starch phosphorylases to starch metabolism in wheat. Plant Cell Physiol. 45:1471-1484.
Shimomura S, Nagai M, Fukui T (1982). Comparative glucan specificities of two types of spinach leaf phosphorylase. (Tokyo). J. Biochem., 91:703-717.
Shin HJ, Shin Y, Lee DS (2000). Formation of α -D-glucose-1-phosphate by thermophilic α -1,4-D-glucan phosphorylase. J. Ind. Microbiol. Biotechnol. 24:89-93.
Shivaram KN (1976). Purification and properties of potato phosphorylase isoenzymes. Z. Naturforsch. 316:424-432.
Singh M, Beg MU, Singh DV, Tewari KK, Krishnan PS (1967). Biochemical. aspects of parasitism by angiosperm parasites. Part II-Phosphatase and phosphorylase activity in dodder and infected hosts. Ind. J. Biochem. 4:146-151.
Singh S, Sanwal GG (1973). An allosteric α -glucan phosphorylase from banana fruits. Biochim. Biophys. Acta 309:280-288.
Singh S, Sanwal GG (1975). Characterization of multiple forms of α-glucan phosphorylase from *Musa paradisiaca* fruits. Phytochemistry 14:113-118.
Singh S, Sanwal GG (1976). Multiple forms of α-glucan phosphorylase in banana fruits. Phytochemistry 15:1447-1451.
Slabnik E, Frydman RB (1970). A phosphorylase involved in starch biosynthesis. Biochem. Biophys. Res. Commun. 38:709-714.
Smith AM, Zeeman SC, Smith SM (2005). Starch degradation. Annu. Rev. Plant Biol. 56:73-98.
Smith AM, Zeeman SC, Thorneycroft D, Smith SM (2003). Starch mobilization in leaves. J. Exp. Bot. 54:577-583.
Sonnewald U, Basner A, Greve B, Steup M (1995). A second L-type isozyme of potato glucan phosphorylase: cloning, antisense inhibition and expression analysis. Plant Mol. Biol. 27:567-576.
Spilatro SR, Preiss J (1987). Regulation of starch synthesis in the bundle sheath and mesophyll of Zeamays L. intracellular compartmentation of enzymes of starch metabolism and the properties of the ADPglucose pyrophosphorylase. Plant Physiol. 83: 621-627.
Srivastava S, Nighojkar A, Kumar A (1996). Immobilization of *Cuscuta reflexa* starch phosphorylase: Production of glucose-1-phosphate using bioreactors. J. Fermen. Bioeng. 81:355-357.
Steup M (1988). Starch degradation. In Preiss J (ed.), The Biochemistry of Plants. Academic Press, San Diego, 14:255-266.
Steup M (1988a). Role of phosphorylases in photosynthetic carbon metabolism In Singh R, Sawhney RK (eds.), Advances in frontier Areas of Biochemistry Prentice, Hall of India Pvt. Ltd. New Delhi pp.127-146,
Steup M, Latzko E (1979). Intracellular localization of phosphorylases in spinach and pea leaves. Planta 145:69-75.
Steup M, Robenek H, Melkonian M (1983). In-vitro degradation of starch granules isolated from spinach chloroplasts. Planta 158:428-436.
Steup M, Schachtele C (1981). Mode of glucan degradation by purified phosphorylase forms from spinach leaves. Planta 153:351-361.
Steup M, Schachtele C, Latzko E (1980). Purification of a non-chloroplastic α- glucan phosphorylase from spinach leaves. Planta, 148: 168-173.
Steup M, Schachtele C, Latzko E (1980a). Separation and partial. characterization of chloroplast and non-chloroplast α- glucan phosphorylase from spinach leaves. Z. Pflanzenphysiol. 96:365-374.
St-Pierre B, Bertrand C, Camirand A, Cappadocia M, Brisson N (1996). The starch phosphorylase gene is subjected to different modes of regulation in starch-containing tissues of potato. Plant Mol. Biol. 30:1087-1098.
Suganuma T, Kitazono JI, Yoshinaga K, Fujimoto S, Nagahama T (1991). Determination of kinetic parameters for maltotriose and higher malto-oligosaccharides in the reactions catalyzed by α $_D$-glucan phosphorylase from potato. Carbohydr. Res. 217:213-220.
Szulczynski M (1986). Studies on the potato phosphorylase immobilization. Biol. Chem. Hoppe-Seyler 367:209-213.
Takaha T, Critchley J, Okada S, Smith SM (1998a). Normal starch content and composition in tubers of antisense potato plants lacking D-enzyme (4- α -glucanotransferase). Planta 205:445-451.
Takahashi E, Wada T, Konai Y (1996). Process for producing trehalose. US patent, pp. 5, 565, 341.
Tetlow IJ, Morell MK, Emes MJ (2004b). Recent developments in understanding the regulation of starch metabolism in higher plants. J. Exp. Bot. 55:2131-2145.
Tetlow IJ, Wait R, Lu Z, Akkasaeng R, Bowsher CG, Esposito S, Kosar-Hashemi B, Morell MK, Emes MJ (2004a). Protein phosphorylation in amyloplasts regulates starch branching enzyme activity and protein-protein interactions. Plant Cell 16:694-708.
Tsai CY, Nelson OE (1968). Phosphorylase I and II of maize endosperm. Plant Physiol. 43:103-112.
Tsai CY, Nelson OE (1969). Two additional. phosphorylases in developing maize seeds. Plant Physiol. 44:159-167.
Tsai CY, Salamani F, Nelson OE (1970). The Enzymes of carbohydrate metabolism in developing endosperm of maize. Plant Physiol. 46:299-306.
Tsay JS, Kuo CG (1980). Enzymatic activities of starch synthesis in potato tubers of different sizes. Physiol. Plant. 48:460-462.
Upadhye SP, Kumar A (1996). Immobilisation of starch phosphorylase from bengal gram seeds: Production of glucose-1-phosphate. Genet. Eng. Biotechnol. 16:145-151.
Van Berkel J, Conrads JS, Steup M (1991). Glucan phosphorylase forms in cotyledons of *Pisum sativum* L.: Localization, developmental changes, *in vitro* translation and processing. Planta 188:432-439.
Van der Vlist J, Reixach MP, van der Maarel M, Dijkhuizen L, Schouten

AJ, Katja Loos K (2008). Synthesis of branched polyglucans by the tandem action of potato phosphorylase and *deinococcus geothermalis* glycogen branching enzyme. Macromol. Rapid Commun. 29:1293-1297.

Venkaiah B, Kumar A (1994). Egg shell bound starch phosphorylase packed bed reactor for the continuous production of glucose-1-phosphate. J. Biotechnol. 36:11-17.

Venkaiah B, Kumar A (1995). Process for recovery and immobilization of starch phosphorylase from starch based industrial waste water. Biotechnol. Appl. Biochem. 21:77-85.

Venkaiah B, Kumar A (1996) Multiple forms of starch phosphorylase from *Sorghum* leaves. Phytochemistry 41:713-717

Venkaiah B, Srivastava S, Kumar A (1991). Starch phosphorylase from Banana (*Musa paradisiaca*) leaves. Plant Physiol. Biochem. 18:54-56.

Vishwanathan PN, Krishnan PS (1965). Metabolic activity of starch granules from the tapioca plants. Part I- UDPG starch glucosyl transferase. Ind. J. Biochem. 2:16-21.

Vishwanathan PN, Krishnan PS (1966). Metabolic activity of starch granules from the tapioca (*Manihot utilissima*) plants. Part III- starch granules as multienzyme units. Ind. J. Biochem. 3:228-234.

Vishwanathan PN, Shrivastava LN (1964). Search for uridine diphosphate glucose starch synthetase and phosphorylase activity in polyfructosan bearing tissues. Ind. J. Biochem. 1:133-136.

Watson KA, Schinzel R, Palm D, Johnson LN (1997). The crystal structure of *Escherichia coli* maltodextrin phosphorylase provides an explanation for the activity without control in this basic archetype of a phosphorylase. EMBO J. 16:1-14.

Weinhäusel A, Griessler R, Krebs A, Zipper P, Haltrich D, Kulbe KD, Nidetzky B (1997). α -1,4-D-glucan phosphorylase of gram-positive *Corynebacterium callunae*: isolation, biochemical properties and molecular shape of the enzyme from solution X-ray scattering. Biochem. J. 326:773-783.

Wischmann B, Nielsen TH, Møller BL (1999). *In vitro* biosynthesis of phosphorylated starch in intact potato amyloplasts. Plant Physiol. 119:455-462.

Yanase M, Takaha T, Kuriki T (2006a). α -Glucan phosphorylase and its use in carbohydrate engineering. J. Sci. Food Agric. 86:1631-1635.

Yanase M, Takaha T, Kuriki T (2007). Developing and engineering enzymes for manufacturing amylose. J. Appl. Glycosci. 54:125-131.

Yang Y, Steup M (1990). Polysaccharide fraction from higher plants which strongly interacts with cytosolic phosphorylase isozyme. Plant Physiol. 94:960-969.

Young GH, Chen HM, Lin CT, Tseng KC, Wu JS, Juang RH (2006). Site-specific phosphorylation of L-form starch phosphorylase by the protein kinase activity from sweet potato roots. Planta 223:468-478.

Yu F, Jen Y, Takeuchi E, Inouye M, Nakayama H, Tagaya M, Fukui T (1988). α -Glucan phosphorylase from *Escherichia coli*: Cloning of the gene and purification and characterization of the protein. J. Biol. Chem. 263:13706-13711.

Yu S (1992). Enzymes of floridean starch and floridoside degradation in red algae: Purification, characterization and physiological studies. Acta Univ. Ups 372:1-47.

Yu S, Pederson M (1991). Purification and properties of α-1,4- glucan phosphorylase from the red sea weed (*Gracilaria sordida* Harvey). Physiol. Plant. 81:149-155.

Yu S, Pederson M (1991a). One step purification to homogeneity and isoforms of α -1,4- glucan phosphorylase of red sea weeds. Plant Physiol. Biochem. 29:9-16.

Yu S, Pinchetti JLG, Ek B, Reina GG, Pedersen M (1993). Subcellular immuno-localization, amino acid composition and partial amino acid sequences of α-1,4-glucan phosphorylase of *Gracilaria* spp. (Rhodophyta). Physiol. Plant. 89:11-20.

Yu Y, Mu HH, Wasserman BP, Carman GM (2001). Identification of the maize amyloplast stromal. 112-kD protein as a plastidic starch phosphorylase. Plant Physiol. 125:351-359.

Zeeman SC, Delatte T, Messerli G, Umhang M, Stettler M, Mettler T, Streb S, Reinhold H, Kötting OK (2007). Starch breakdown: recent discoveries suggest distinct pathways and novel mechanisms. Funct. Plant Biol. 34:465-473.

Zeeman SC, Smith SM, Smith AM (2004a). The breakdown of starch in leaves. New Phytol. 163:247-261.

Zeeman SC, Thorneycroft D, Schupp N, Chapple A, Weck M, Dunstan H, Haldimann P, Bechtold N, Smith AM, Smith SM (2004b). Plastidial α -glucan phosphorylase is not required for starch degradation in *Arabidopsis* leaves but has a role in the tolerance of abiotic stress. Plant Physiol. 135:849-858.

Biotechnology and food security in developing countries

Nyerhovwo J. Tonukari[1]* and Douglason G. Omotor[2]

[1]Department of Biochemistry, Delta State University, Abraka, Nigeria.
[2]Department of Economics, Delta State University, Abraka, Delta State, Nigeria.

Food security is a necessity for every individual, home, community and nation. In developing countries, food security could be substantially improved by increased investment and policy reforms. Biotechnology's ability to eliminate malnutrition and hunger through production of crops resistant to pests and diseases, having longer shelf-lives, refined textures and flavours, higher yields per units of land and time, tolerant to adverse weather and soil conditions, and generate employment, cannot be over-emphasized. This technology can be applied to improve agriculture in order to improve food production for the human population in an environmentally sustainable manner. However, there is need for government and public-private collaborations to invest in agricultural biotechnology-based companies, researches, or initiatives, in order to make the gene revolution beneficial to developing countries.

Key words: Biotechnology, food security, developing countries, agriculture, public investment, policy reform.

INTRODUCTION

The increasing world population has led to increased demand for food and reduced per capita availability of arable land and irrigation water. Compounding this problem is the fact that most farmers in the developing world own only small plots of land that have the potential to feed one family and generate income. Low soil fertility and crop losses from pests and droughts have reduced harvests to below subsistence levels (Vasil, 1998; Conway and Toenniessen, 2003). This situation has, undeniably, led to serious food insecurity.

Availability of food, access to food, and risks related to either availability or access are the essential determinants of food security (von Braun et al., 1992). National food security implies that within a country the amount of food available, if evenly distributed, is enough to meet people's food needs. At the household level, "a household is food secure when it has access to the food needed for a healthy life for all its members (adequate in terms of quality, quantity, safety, and culturally acceptable), and when it is not at undue risk of losing such access" (UN ACC/SCN, 1991).

Both rural and urban poor people suffer from food insecurity and poor nutrition, caused in large measure by poverty and lack of nutritional balance in the diet they can afford. About 1.2 billion people, or one of every five humans, live in a state of absolute poverty, on the equivalent of US$1/day or less (World Bank, 1999). About 800 million people are food insecure (FAO, 1999), and 160 million preschool children suffer from energy-protein malnutrition, which results in the death of over 5 million children under the age of five each year (ACC/SCN and IFPRI, 1999). A much larger number of people suffer from deficiencies of micronutrients such as iron and vitamin A. For example, 2 billion people (one of every three) are anaemic, basically as a result of iron deficiency. Food insecurity and malnutrition result in serious public health problems and loss of human potential in developing countries (Pinstrup-Andersen and Cohen, 2000).

Because land and water for agriculture are diminishing resources, there is no option but to produce more food and other agricultural commodities from less arable land and irrigation water. Thus, the need for more food has to be met through higher yields per units of land, water, energy and time. There is need, therefore, to examine how science can be mobilized to raise further the biological productivity ceiling without associated ecological harm (Swaminathan, 2000).

*Corresponding author. E-mail: tonukari@gmail.com.

Biotechnology may help achieve the productivity gains needed to feed a growing global population, introduce resistance to pests and diseases without costly purchased inputs, heighten crops' tolerance to adverse weather and soil conditions, improve the nutritional value of some foods, and enhance the durability of products during harvesting or shipping. New crop varieties and biocontrol agents may reduce reliance on pesticides, thereby reducing farmers' crop protection costs and benefiting both the environment and public health (Pinstrup-Andersen and Cohen, 2000). Biotechnology-based agriculture combines elements of ecological agriculture with crop varieties designed to perform well under low-input and stress conditions, uses inorganic inputs very judiciously, and engages farmers themselves in analyzing their needs and adapting new varieties and agronomic practices to their own conditions. Greater commitments and new partnerships are needed to sustain and expand this revolution in agriculture to small-scale farming families across all Africa (Conway and Toenniessen, 2003).

Biotechnology research could aid the development of drought-tolerant and insect-resistant crops, to the benefit of small farmers and poor consumers. Research on genetic modification to achieve appropriate weed control can increase farm incomes and reduce the time women farmers spend weeding, allowing more time for the child care that is essential for good nutrition. This technology may also offer cost-effective solutions to micronutrient malnutrition, such as vitamin A- and iron-rich crops (Pinstrup-Andersen and Cohen, 2000).

CAUSES OF FOOD INSECURITY

Food is the most basic of human needs. Despite the "green revolution" between 1970 and 1990 almost half of the world's less developed countries suffer a decline in aggregate food supply, and more than a quarter suffer an increase in hunger. Malnutrition is a major barrier to economic and social development, leaving populations unable to maintain normal lives and to be economically and socially less productive (Jenkins and Scanlan, 2001; Conway, 1999).

Population growth and urbanization

Within the next 20 years, more poor and under-nourished people in developing countries will live in the cities than in rural areas. High rates of urbanization mean that urban food insecurity and malnutrition are concerns even for regions like Africa and Asia, where current levels of urbanization are relatively low (Ruel et al., 1998). By the year 2030, the rural population would have grown by more than 235 million, but the urban population would have grown by 2.4 billion (United Nations, 1998). The number of people living in cities in Africa will be more than triple, from 251 million to 864 million.

Poverty

Food availability means that the overall supply should potentially cover overall nutritional needs in terms of quantity (energy) and quality (providing all essential nutrients); furthermore, it should be safe (free of toxic factors and contaminants) and of good food quality (taste, texture, and so on). Also, the types of foodstuffs commonly available (nationally, in local markets, and eventually at the household level) should be culturally acceptable (Oshaug, 1994). Food expenditures can make up as much as 60 to 80 percent of total income among low-income urban households (Tabatabai, 1993; Maxwell et al., 1998).

The importance of being able to earn cash income also means that the ability to stay healthy, to get a good job (and therefore the ability to acquire good education and training), and to have access credit to smooth consumption, or expand, or start up businesses, are all critical to urban food and nutrition security. With enough income, prices can rise and families can still buy enough to eat. Millions of urban poor, however, are vulnerable to price rises or sharp declines in income, say due to illness or loss of job by the principal income-earner. The majority of the urban labour force works in sectors like petty trade and services where wages are low and job tenure uncertain. In urban Nigeria and most of sub-Saharan Africa, employment in sectors that pay regular wages, such as manufacturing and industry, accounts for less than 10 percent of total employment (Rondinelli and Kasarda, 1993). Urban poverty thus is not primarily the result of lack of work but the lack of well-paying, steady jobs (Ruel et al., 1998).

Health

Health is determined by a series of factors that act at three levels. At the community level, factors such as the quality of the overall environment (biological pathogens and chemical pollutants in air, food, and water), and the availability, cost, and quality of services such as water, electricity, sewage, refuse disposal, and health services are important health determinants. At the household level, the most important factors include the general conditions of the household, including the type of housing, the availability and cost of water and hygienic facilities, and the number of rooms per household member (an indicator of crowding); the availability of food; and household caring behaviours related to the use of preventive and curative health services, the use of water and hygienic facilities to provide a healthy, hygienic and safe environment, and food-related behaviours such as the acquisition of food, the intra-household allocation of resources, feeding practices (including breast-feeding), and food preparation methods. At the individual level, the determinants of health relate to the interactive mechanisms among an individual's food and nutrient intake, nutritional status, and health status (Ruel et al., 1998).

High rates of malnutrition and escalating rates of diet-related diseases such as diabetes and hypertension in developing countries are attributable to various concerns: inadequate sources of dietary protein, foods with high levels of anti-nutritional components and toxicants, a disproportionate amount of highly digestible, high glycaemic index carbohydrates which constitute the staple foods such as yams, maize and rice, and limited alternatives. In addition, socio-economic factors such as urbanization and migration to urban areas have led to changes in lifestyles to include imported highly processed foods and modification in eating patterns and food habits. In many rural areas, there are challenges of protein malnutrition as well as inadequate vitamin and mineral intake. Furthermore, these problems are compounded by inadequate prenatal nutrition which leads to deficiencies in both mothers and children. Nutritional needs of postpartum nursing mothers are often not met, subsequently leading to inadequate nutrition in children. The quality of staple foods and other foods that are commonly consumed in most of these areas could be optimized to improve their quality (Niba, 2003).

In developing countries where mechanization is still limited, human labour provides much of the power for productive economic work. For the poor, labour is their most important asset. Work capacity, performance, and productivity of workers are therefore generally important for income (Pryer and Crook, 1988). This can lead to a situation where poor, malnourished workers perform poorly because of decreased working capacity, and therefore do not get access to better-paying jobs. Various studies support this assumption, showing positive associations between wage achievement and nutritional status as proxied by either body weight, body-mass index, stature, or caloric and protein intake (Haddad and Bouis, 1991; Strauss and Thomas, 1996; Satyanarayana et al., 1980; Pryer and Crook, 1988). Furthermore, the poor sanitary conditions of many developing-countries are notorious, and this combined with poor hygiene practices, create a situation in which food safety is severely compromised (Ruel et al., 1998).

Outdated farming system

Small-scale farmers in developing countries are faced with many problems and constraints. Pre- and post-harvest crop losses due to insects, diseases, weeds, and droughts result in low and fluctuating yields, as well as risks and fluctuations in incomes and food availability. Low soil fertility and lack of access to reasonably priced plant nutrients, along with acid, salinated, and water-logged soils and other abiotic factors, contribute to low yields, production risks, and degradation of natural resources as poor farmers try to eke out a living. They are often forced to clear forest or farm ever more marginal land to cultivate crops. Poor infrastructure and poorly functioning markets for inputs and outputs together with lack of access to credit and technical assistance add to the impediments facing these farmers (Pinstrup - Andersen and Cohen, 2000).

Politics

Eight hundred million people on earth are poor and malnourished. They live on less than a dollar a day and cannot be sure that their fields will yield enough food or that they will earn enough money to buy food. Forty thousand people die each day of malnutrition, one-half of them children. The increase in food production enabled by the Green Revolution unfortunately did not solve the problems of malnutrition and hunger. There were about a billion hungry people some 40 years ago, and population projections show that there may still be 600 million poor people by 2025, when the earth's population would have grown to 8 billion. The Green Revolution did many things, but it did not wipe out poverty. Not enough jobs were created in either the rural areas or the cities to generate the purchasing power that provides farmers with the incentive to grow more food. It is ironic that hunger persists while the prices for agricultural commodities are at an all-time low (Chrispeels, 2000). Several countries that have millions of hungry people are exporting food and other agricultural products to countries where people are already well fed. Yet, most of these countries that are poor, with so many hungry people, seem to be able to grow food quite abundantly (Clover, 2003).

COMBATING FOOD INSECURITY

To eliminate malnutrition and hunger, food production and purchasing power both need to increase in developing countries. In addition, food production needs to increase in developed countries as well so that grain can be exported at a price the poor can afford. Agricultural biotechnology as the solution to the problem of global food insecurity has been reviewed by Soetan (2008). Since land and water are the most limiting resources for food production, there is only one option: to increase yields on the available land. Indeed, there is very little extra land that can be put to the plow. By 2020, the world's farmers will have to produce 40% more grain (200 million extra tons in the developed countries and 500 million extra tons in the developing countries).

According to the forecasts of the International Food Policy Research Institute in Washington, DC

(Pinstrup-Anderson et al., 1999), the less-developed countries will double their grain imports (mostly maize and wheat) by 2020. The reason is that the projected production increase of 500 million tons in those less-developed countries will still not satisfy demand. The imported grains will come from North America, Australia, the European Union, and the former Soviet Union. Thus trade will increase (assuming that prices remain stable and low), but redistribution is not the answer to the problem of hunger because there is not enough production capacity in the developed countries to satisfy the expected world demand.

The answer to the problems of the poor, according to a number of organizations that oppose genetically modified (GM) crops, is more organic, regenerative agriculture. There is certainly the need for more sustainable regenerative agricultural practices (Pretty, 1995), but "organic" farming is the type of agriculture already practiced by the poor, primarily because they do not have the means to buy fertilizers, pesticides, and irrigation equipment. According to Dyson (1999), sub-Saharan Africa, where most food crop production is "organic," is unlikely to see much improvement in its already dismal food situation. Exhaustion of the soil caused by the lack of fertilizers is depressing yields and pushing agriculture onto more erodable soils. Organic agriculture is nearly always nitrogen starved unless land is set aside for the sole purpose of producing green manures, a luxury the poor can ill afford. Agriculture as it is practiced now in much of sub-Saharan Africa is environmentally unsustainable and a new approach that will require considerable investment in agricultural research is needed. This new approach must be research-driven and will most certainly include GM crops (Chrispeels, 2000).

Agriculture must figure prominently in poverty alleviation strategies of developing countries. Accelerated public investments are needed to facilitate agricultural and rural growth through:

- Yield-increasing crop varieties, including those that are drought and salt tolerant and pest resistant, and producing improved livestock.
- Yield-increasing and environmentally friendly production technology.
- Reliable, timely, and reasonably priced access to appropriate inputs such as tools, fertilizer, and, when needed, pesticides, as well as the credit often needed to purchase them.
- Strong extension services and technical assistance to communicate timely information and developments in technology and sustainable resource management to farmers and to relay farmer concerns to researchers.
- Improved rural infrastructure and effective markets.
- Particular attention to the needs of women farmers, who grow much of the locally produced food in many developing countries.
- Primary education and health care, clean water, safe sanitation, and good nutrition for all.

These investments need to be supported by good governance and an enabling policy environment, including trade, macroeconomic, and sectoral policies that do not discriminate against agriculture, and policies that provide appropriate incentives for the sustainable management of natural resources, such as secure property rights for small farmers. Development efforts must engage poor farmers and other low-income people as active participants, not passive recipients; unless the affected people have a sense of ownership, development schemes have little likelihood of success (Pinstrup-Andersen and Cohen, 2000).

There are many aspects of providing food for the poor that are well beyond the control of either laboratory scientists or agricultural advisors in the field. The government must realize that agriculture can be an important engine of economic growth and therefore must invest more in agricultural research. Agricultural development and creation of rural infrastructure that will permit crop surpluses to be marketed should be encouraged. Cheap food policies that favour the urban poor are attractive to city dwellers but discourage development of food production capacity in the countryside. Such policies amount to a transfer of wealth from the agricultural sector to the industrial sector. Developing countries need to examine whether these policies would benefit only the big farmers who rely primarily on purchased outside inputs, or also the smaller farmers who might be engaged in more sustainable practices. If the entire framework for supporting agricultural development is put into place, then biotechnology can also play a role. "Biotechnology is only one tool, but a potentially important one, in the struggle to reduce poverty, improve food security (Rosegrant and Cline, 2003; Serageldin, 1999), reduce malnutrition, and improve the livelihoods of the rural and the urban poor" (Persley, 1999).

BIOTECHNOLOGY AND FOOD PRODUCTION

Scientific innovation and its derivative benefits have had profound implications to humanity within the last century. The exciting discipline of biotechnology has drawn the interests of traditional biologists, biochemists, microbiologists, medical and agricultural scientists into applying mathematical and engineering models to understanding biology. Furthermore, several scientists in the exact sciences of mathematics, physics, and chemistry have begun to use system approaches to unravel the mystery and complexity of biology. And from the side, diagnostic, biopharmaceutical, biochemical and agriculcultural industries are rapidly drawing from and applying the research results of biotechnology. Biotechnology is experiencing a revolution like none before in the life sciences and is affecting every facet of our lives, from crop improvement to commerce, and drugs to sustainable development (Tonukari et al., 2003; Soetan and Abatan, 2008). This technology has the potential to address problems not solved by conventional research. At the same time, biotechnology may speed up research process and increase research precision. GM crops have been developed and rapidly disseminated since the early 1990s (Huang et al., 2002).

At present, there is very little commercial utilization of results from modern biotechnology research in developing countries. As a result, the potential contributions of biotechnology to poverty alleviation, enhanced food security and nutrition in developing countries have received little attention, beyond blanket statements of support or opposition (Pinstrup-Andersen and Cohen, 2000).

The gene revolution

Mendel's laws of genetics were rediscovered in 1900. Mendel had published his work on inheritance patterns in pea in 1865, but it took 35 years for others to grasp their significance. Since 1900, there has been steady progress in our understanding of the genetic makeup of all living organisms ranging from microbes to man. A major step in human control over genetic traits was taken in the 1920s when Muller and Stadler discovered that radiation can induce mutations in animals and plants. In the 1930s and 1940s, several new methods of chromosome and gene manipulation were discovered, such as the use of colchicine to achieve a doubling in chromosome number, commercial exploitation of hybrid vigour in maize and other crops, use of chemicals such as nitrogen mustard and ethyl methane sulphonate to induce mutations and techniques like tissue culture and embryo rescue to get viable hybrids from distantly related species. The double helix structure of DNA (deoxyribonucleic acid), the chemical substance of heredity, was discovered in 1953 by James Watson and Francis Crick. This triggered explosive progress in every field of genetics.

Just as it took 35 years for biologists to understand the significance of Mendel's work, it may take a couple of decades more to understand fully the benefits and risks associated with genetically improved foods. It would be prudent to apply scientific and precautionary principles in areas of human health and environmental safety. The 1990s have seen dramatic advances in our understanding of how biological organisms function at the molecular level, as well as in our ability to analyze, understand, and manipulate DNA molecules, the biological material from which the genes in all organisms are made. The entire process has been accelerated by the Human Genome Project, which has poured substantial resources into the development of new technologies to work with human genes. The same technologies are directly applicable to all other organisms, including plants. Thus, the new scientific discipline of *genomics* has arisen, which has contributed to powerful new approaches in agriculture and medicine, and has helped to promote the biotechnology industry (Swaminathan, 2000).

The quality of food and food plants can be modified and optimized to meet the nutritional and health needs of at-risk and compromised populations prevalent in most of the developing countries. High rates of malnutrition, infectious disease as well as diet-related diseases such as diabetes and hypertension are prevalent in many developing countries. These are as a result of compromised immune function, inadequate sources of nutritious and quality foods and limited access to healthy and suitable foods. Biotechnology and genetic modification techniques have been proposed and applied for the improvement of the quality of various food crops. These have typically been geared towards increasing yields and pest resistance of cash crops. Furthermore, the application of biotechnology techniques for the development of functional food plants with higher levels of bioactive components or increased availability of nutrients would greatly benefit most populations in developing countries and improve the health and nutritional status overall (Niba, 2003).

In certain areas, biotechnology and genetic modification techniques are being optimized for the production and development of healthy foods, and improvement in the levels and activity of biologically active components in food plants (phytochemicals). Biotechnology techniques in developing countries however have mostly been targeted at increasing yields of cash crops. Food crops or the improvement of food quality and functional foods have garnered much less attention.

Techniques applied in genetic modification include mutation breeding, improved conventional breeding, transgenic modifications, DNA insertion, gene transfer and somatic hybridization (Bouis et al., 2003; Christou, 1997; Mazur, 2001; Mackay, 1991; Yan and Kerr, 2002).

While there has been some hesitation with regard to the acceptability and adoption of biotechnology products in certain developing countries, achievements such as the development of high-vitamin A rice have greatly increased the acceptability of biotechnology for human food applications among hitherto skeptical consumers. Furthermore, it provides insights into the potential for application of biotechnology in developing improved quality and functional foods for human nutrition and health (Soetan and Abatan, 2008), rather than simply for use in agricultural technology for improved yields and pest resistance. The production of increased levels of beta-carotene (the precursor to vitamin A) in plants is especially important, as its precursor, lycopene has been shown to have physiological chemo-preventive effects with regard to various cancers (Yan and Kerr, 2002). Furthermore, lycopene, commonly found in various carotenoid containing plants such as tomatoes and carrots, is an essential ingredient in maintaining eye health and vision.

Modifications that have been targeted and developed by various biotechnology companies include improvement in the oil content and composition of oil seeds such as legumes (Mazur, 2001; Mazur et al., 1999; Uzogara, 2000). Improvement in soybean oil quality includes stabilization of the unsaturated fatty acids by increasing levels of the antioxidant, vitamin E (Yan and Kerr, 2002). These successes indicate a relevant and important role for biotechnology in improving food quality and developing functional foods, particularly those targeted for needy populations in developing countries, such as children and pre-natal women (Niba, 2003).

Genetically modified (GM) crops

Modern agricultural biotechnology is still in an early phase, and the focus is overwhelmingly on production in industrial country farms and for industrial country markets. In 1998, 85% of the land planted with genetically improved (GI) crops was in just five developed countries (Australia, Canada, France, Spain, and the United States), with the United States alone accounting for about 75% of the area. Argentina, China, Mexico, and South Africa cultivated the remaining 15%, and the countries other than China include a substantial number of large-scale, capital-intensive farms that produce primarily for industrial country markets. Among the crops produced in these four developing countries are insect-resistant cotton and maize, herbicide-resistant soybean, and tomatoes with a long shelf life. Globally, herbicide-resistant soybean, insect-resistant maize, and genetically improved cotton (containing insect resistance and/or herbicide tolerance genes) account for 85 percent of all plantings.

The unprecedented rapid adoption of transgenic crops during the initial five-year period (1996 - 2000) when genetically modified (GM) crops were first adopted, reflects the significant multiple benefits realized by large and small farmers in industrial and developing countries that have grown transgenic crops commercially. Between 1996 and 2000, a total of fifteen countries - 10 industrial and 5 developing - contributed to more than a twenty five fold increase in the global area of transgenic crops from 1.7 million hectares in 1996 to 44.2 million hectares in 2000. The accumulated area of transgenic crops planted in the five-year period 1996 - 2000 total 125 million hectares, equivalent to more than 300 million acres.

Adoption rates for transgenic crops are unprecedented and are the highest for any new technology by agricultural industry standards. High adoption rates reflect growing satisfaction with products that offer significant benefits ranging from more convenient and flexible crop management, higher productivity and/or net returns per hectare, health benefits and a safer environment through decreased use of conventional pesticides, which collectively contribute not only to improved weed and insect pest control (attainable with transgenic herbicide-tolerant and insect-resistant *Bt* crops) but also benefits of lower input and production costs; genetically modified crops offer significant economic advantages to farmers compared with corresponding conventional crops. The severity of weed and insect pests varies from year to year and hence this will have a direct impact on pest control costs and the consequent economic advantage (James, 2003).

Despite the on-going debate on GM crops, particularly in countries of the European Union, millions of large and small farmers in both industrial and developing countries continue to increase their plantings of GM crops because of the significant multiple benefits they offer. This high adoption rate is a strong vote of confidence in GM crops, reflecting grower's satisfaction. Many recent studies have confirmed that farmers planting herbicide-tolerant and insect-resistant *Bt* crops (*Bacillus thuringiensis* toxin- or Bt toxin-producing crops) are more efficient in managing their weed and insect pests. An estimated 3.5 million farmers grew transgenic crops to health and economic advantages.

In coming years, the number of farmers planting GM crops is expected to grow substantially and the global area of GM crops is expected to continue to grow. Global population would exceed 6 billion by 2050, when approximately 90% of the global population will reside in Asia, Africa and Latin America. Today, 815 million people in the developing countries suffer from malnutrition and 1.3 billion are afflicted by poverty. Transgenic crops, often referred to as GM crops, represent promising technologies that can make a vital contribution to global food, feed and fibre security (James, 2003).

During the six-year period 1996 - 2001, the global area of transgenic crops increased by more than 30-fold, from 1.7 million hectares in 1996 to 52.6 million hectares in 2001. This high rate of adoption reflects the growing acceptance of transgenic crops by farmers using GM technology in both industrial and developing countries (Table 1). During the six-year period 1996 - 2001, the number of countries growing transgenic crops more than doubled, increasing from 6 in 1996 to 9 in 1998, to 12 countries in 1999 and 13 in 2000 and 2001 (James, 2003).

The experience of the first six years, 1996 to 2001, during which a cumulative total of over 175 million hectares (almost 440 million acres) of transgenic crops were planted globally in 16 countries, 10 industrialized and 6 developing, has vindicated the vision of the pioneers of crop biotechnology who have seen their early promises of transgenic crops fulfilled - GM crops have met the expectations of large and small farmers planting transgenic crops in both industrial and developing countries.

The unprecedented rapid adoption of transgenic crops during the first six-year period, 1996 - 2001, reflects the significant multiple benefits realized by large and small

Table 1. Global area of transgenic crops in 2000 and 2001: by countries (million hectares).

	2000	%	2000	%	+/-	%
USA	30.3	68	35.7	68	+5.4	+18
Argentina	10.0	23	11.8	22	+1.8	+18
Canada	3.0	7	3.2	6	+0.2	+6
China	0.5	1	1.5	3	+1.0	+200
South Africa	0.2	<1	0.2	<1	<0.1	+33
Australia	0.2	<1	0.2	<1	<0.1	+37
Mexico	<0.1	<1	<0.1	<1	<0.1	-
Bulgaria	<0.1	<1	<0.1	<1	<0.1	-
Uruguay	<0.1	<1	<0.1	<1	<0.1	-
Romania	<0.1	<1	<0.1	<1	<0.1	-
Spain	<0.1	<1	<0.1	<1	<0.1	-
Indonesia	-	-	<0.1	<1	<0.1	-
Germany	<0.1		<0.1	<1	<0.1	-
France	<0.1	<1	-	-	-	-
Total	44.2	<1	52.6	100	+8.4	+19%

Source: James (2003).

farmers. There is a growing body of evidence that clearly demonstrates the improved weed and insect pest control attainable with transgenic herbicide-tolerant and insect resistant *Bt* crops that also benefit from lower input and production costs. Despite the ongoing debate on GM crops, particularly in countries of the European Union, millions of large and small farmers in both industrial and developing countries continue to increase their plantings of GM crops because of the significant multiple benefits they offer. More specifically, the use of transgenic crops results in:

- More sustainable and resource-efficient crop management practices that require less energy and fuel and conserves natural resources.
- More effective control of insect pests and weeds.
- A reduction in the overall amount of pesticides used in crop production, which impacts positively on biodiversity, protects predators and non-target organisms and contributes to a safer environment.
- Less dependency on conventional pesticides that can be a health hazard to producers and consumers; the potential health benefits associated with fewer pesticide poisonings from *Bt* cotton in China is an important finding, with significant implications for other developing
- Countries where small farmers may be at similar risk from heavy and over-use of conventional pesticides.
- *Bt* maize, which has reduced levels of the fumonisin mycotoxin provides safer and healthier food and feed products.
- Greater operational flexibility in timing of herbicide and insecticide applications.
- Conservation of soil moisture, structure, nutrients and control of soil erosion through no or low-tillage practices as well as improved quality of ground and surface water with less pesticide residues.

There is now considerable evidence that transgenic crops are delivering significant economic benefits. The global economic advantage to farmers deploying herbicide tolerant (HT) soybean, HT canola and *Bt* corn was estimated to be of the order of $ 700 million in 1999, equally shared between developing and industrial countries. In addition to these direct economic advantages that farmers derive from transgenic crops, there are also significant additional indirect benefits to others in society. For crops such as herbicide-tolerant soybean, these indirect benefits to consumers globally can be of the same order of magnitude as the direct economic advantages to farmers. Thus, the global direct and indirect economic advantage of GM crop in 1999 was of the order of $ 1 billion or more.

2001 is the initial year of the second quinquennium (2001 - 2005) during which GM crops is being comercialized. In 2001, coincidental with increased political, policy and institutional support for GM crops due to their acknowledged contribution to global food security, the global area of transgenic crops benefited from renewed growth which resulted in a 19% increase in global GM crop area in 2001 - almost twice the growth rate in 2000. The number of farmers that benefited from GM crops increased from 3.5 million farmers in 2000 to an estimated 5.5 million in 2001. More than three-quarters of the GM farmers that benefited from GM crops in 2001 were resource-poor farmers planting *Bt* cotton, mainly in eight provinces in China and also in the Makathini Flats in KwaZulu Natal province in South Africa. The well documented experience of China with *Bt* cotton presents a remarkable case study where 5 million small resource-poor farmers in 2001 are already benefiting from significant agronomic, environmental, health and economic advantages - this is a unique example of how biotechnology can impact on poverty alleviation, as advocated in the 2001 UNDP Human Development Report. The China experience with *Bt* cotton lends itself for introduction and replication to carefully selected developing countries in Asia, Latin America and Africa where resource-poor farmers can learn, share and benefit from the rich experience of China – the majority of the hectare of global cotton is grown in developing countries. Following a successful launch of *Bt* cotton in 2001, Indonesia is also expected to expand its *Bt* cotton in 2002, and India, the largest cotton growing country in the world, has approved commercial cultivation of *Bt* cotton in 2002.

There is cautious optimism that global area and the number of farmers planting GM crops will continue to grow in 2002 in the six principal countries already growing GM crops - USA, Argentina, Canada, China, South Africa and Australia. The other seven countries growing transgenic crops in 2001 are expected to report modest growth in GM crop area in 2002. The commercialization of herbicide-tolerant soybean in Brazil will be dependent on resolving the outstanding regulatory issues between the Ministries of Agriculture, Environment and Justice. The commercialization of GM crops in India and Brazil would represent a watershed for GM crops in developing countries in that the three most populous countries in Asia - China, India and Indonesia with 2.5 billion people, as well as the three major economies of Latin America - Argentina, Brazil and Mexico, plus South Africa would then all be commercializing and benefiting from transgenic crops.

As new and novel products with input and output traits will become available for commercialization in the near future, it is critical that these products be deployed in an integrated strategy in which both conventional and biotechnology applications are applied to attain the challenging goal of global food security. Adoption of such a strategy will allow society to continue to benefit from the vital contributions that both conventional and modern plant breeding offer. Technology is only one of several elements that can contribute to an integrated global food security strategy where population control and improved food distribution systems are also essential elements. Biotechnology can play its appropriate and essential role in achieving food security in the developing world. The experience of China, Argentina and South Africa, that are already deriving significant benefits from GM crops, can be shared with other developing countries in the three continents of the South which face similar challenges (James, 2003).

Adapting biotechnology for increased crop production

What is needed in developing countries is an understanding that goes beyond conventional, orthodox wisdom to work more strategically in developing and implementing effective, international, national and regional policies (Clover, 2003). Funding for agricultural research has declined 50% on a worldwide basis. Funding for research in general is at a dismal low in developing countries. The intrusion of intellectual property rights into the arena of crop improvement, while beneficial to the economies of the developed world is making the lives of many researchers even more difficult. Private industry has dominated research in biotechnology (there are a few exceptions: for example, Rockefeller Foundation support for research on rice, USDA's role in developing the terminator technology, and modest programs at IARCs. Consolidation of the industry has proceeded rapidly since 1996, with more than 25 major acquisitions and alliances worth US$15 billion. Little private-sector agricultural biotechnology research so far has focused on developing country food crops other than maize (Pinstrup-Andersen and Cohen, 2000).

Production of genetically modified crops is not a complex technology and is clearly within the capabilities of national research institutes in many developing countries. Genetic modification of crops using recombinant DNA technology is also within reach of the institutes of the Consultative Group on International Agricultural Research (CGIAR), including Centro Internacional de Mejoramiento de Maiz y Trigo (CIMMYT) in Mexico, International Rice Research Institute (IRRI) in The Philippines, and the International Institute for Tropical Agriculture (IITA) in Nigeria. Furthermore, these institutes have already assumed responsibility for biotechnological research and a number of crop improvement projects are under way. These institutes see biotechnology as a tool and not as an end in itself. Crop improvement through biotechnology need not be equated with transgenic plants. For example, marker-assisted breeding is a powerful biotechnology that can find widespread application with the crops of the poor. Detailed linkage maps of these crops will be tremendously useful. As these CGIAR institutes focus on their needs, they will want and need to reach out to public institutions in developed countries. Alliances such as the Cassava Biotechnology Network that bring together researchers from many countries are an effective way to create synergy toward a common goal.

The private sector leads in every aspect of the agricultural biotechnology revolution and activities in the public sector will have to marshal the strength of the private sector through public-private partnerships. Such partnerships must be based upon mutual trust and common goals. The private sector can work with the CGIAR institutes and with national research institutions (the foreign equivalents of the U.S. Department of Agriculture) of developing countries to transfer technologies, train scientists, provide hands-on experience in intellectual property management, and facilitate the no-cost or low-cost licensing of inventions. Since the cost of these inventions is being charged to the consumers in developed countries, such approaches amount to a transfer of wealth by large corporations from the developed world to the developing world.

The application of biotechnology to the problems of the poor will not be straightforward and the models we have from developed countries will probably not be applicable. Agricultural research for the crops and problems of the poor has to proceed from the bottom up, not from the top down. Crops have to be created that fit not only in the agro-ecology of the poorest regions often characterized by marginal and heterogeneous environments, but the crops must also fit into the social and economic systems. Agricultural research has to start with studying farming practices (so called "on-farm research"), asking the farmers - men and women - what they want, allowing the farmers to make choices between often conflicting objectives such as higher yield versus yield stability, and examining the possibility of marketing the excess production. Researchers have to begin by soliciting the help of the farmers to describe farming practices and analyze these practices to pinpoint problem areas and opportunities. Together, the researchers and farmers have to generate a range of choices that the farmers could implement. The major objective of this approach is not the transfer of technology, but empowerment of the farmer to improve production (Chrispeels, 2000).

PRESPECTIVES

Inadequate nutrition, sub-par quality foods and limited food processing capabilities have led to compromised, sub-par health status and a prevalence of diet-related diseases in many developing countries, most especially among children, prenatal and post-partum women. There is clear potential for the application of biotechnology and genetic modification as tools to combat these challenges and improve the situation of at-risk populations. While there is concerted and dedicated focus on combating more apparent challenges such as infectious disease and producing enough food to feed the growing populations, the quality of the foods produced ought to be an important consideration. The feasibility and cost of applying modern and emerging genetic modification technologies in these areas is certainly a daunting task, as sometimes the primary need is often just survival. However, as scientists and policy makers make progress towards alleviating these debilitating conditions, consideration should also be given to the possibility and potential for augmenting food quality and developing functional foods for improved nutritional and health status (Niba, 2003).

In spite of an impressive stockpile of scientific discoveries and technological innovations, poverty and social and gender inequities are increasing. According to the World Bank, 1.3 billion people lived on less than US$1 per day and another 3 billion lived on less than US$2 per day in 1993. Illiteracy, particularly among women, is still high in many developing countries. It is not only in opportunities for education that children of many developing countries remain handicapped, but even more alarming, in opportunities for the full expression of their innate genetic potential for physical and mental development.

There are now uncommon opportunities to harness the power of such synergy to address contemporary development issues such as the growing rich-poor divide, feminization of poverty, famine of jobs, human numbers exceeding the population-supporting capacity of ecosystems, climate change, and loss of forests and biodiversity. Whether in economics or in ecology, experience has shown that a trickle-down approach does not work. Fortunately, modern information technology provides opportunities to reach the unreached. Virtual colleges (computer-aided and internet-connected) linking scientists and women and men living in poverty can be established at local, national, and global levels to launch a knowledge and skill revolution. This will help to create better awareness of the benefits and risks associated with genetically improved organisms, so that both farmers and consumers will get better insights into the processes leading to the creation of novel genetic combinations.

The future of small scale farm families will depend on precision agriculture, which involves the use of the right inputs at the right time and in the right way. Biotechnology will play an important role in the major components of precision farming: integrated gene management, soil health care, efficient water management, integrated pest management, integrated nutrient supply, and efficient post-harvest management. Ecotechnology-based precision farming can help to cut costs, enhance marketable surplus, and eliminate ecological risks. This is the pathway to an ever-green revolution in small-farm agriculture (Swaminathan, 2000).

The ethical considerations of genetic engineering of crops pale in comparisons to the ethical considerations of not improving the lives of the poor (Chrispeels, 2000). Those who oppose GM crops are also quick to point out that this technology primarily benefits the multinational corporations that sell the seeds, and that these corporations are more interested in their own bottom line (always referred to as "corporate greed") than in "feeding the poor." True enough, the big corporations are not working on the crops of the poor, such as cassava, millets, sorghum, sweet potatoes, yams, and legumes (other than soybeans). Furthermore, they are not giving away their technology to poor countries because they want to recover the costs of their investments in biotechnology. The poor will not have the resources to purchase transgenic seeds from multinationals. Research on these crops in the public sector is also unfortunately quite limited. Rice, an important crop of the poor, is an exception, with some research in the corporate sector and considerable research in the public sector taking place, primarily as a result of the Rockefeller Foundation's initiatives (Chrispeels, 2000).

An important feature of the Green Revolution is that the research was carried out in the public domain, and that the genetically improved crop varieties were given away free to the farmers without concerns for the intellectual property rights of those who produced them. Food production was raised substantially in large areas of the developing world, but other areas, especially Africa, were bypassed. It is unfortunate that "public-sector support for agricultural development has collapsed across the board" according to Robert Paarlberg, with a 57% drop in foreign aid to agriculture in poor countries between 1988 and 1996 and a 47% decrease in lending by the World Bank for agriculture and rural development between 1986 and 1998 (Paarlberg, 2000).

Research focused on how to reduce the need for inputs and increase the efficiency of input could lead to the development of crops that use water more efficiently and extract phosphate from the soil more effectively. The development of cereal plants capable of capturing nitrogen from the air could contribute greatly to plant nutrition, helping poor farmers who often cannot afford fertilizers. By raising productivity in food production, agricultural biotechnology could help further reduce the need to cultivate new lands and help conserve biodiversity and protect fragile ecosystems (Pinstrup-Andersen and Cohen, 2000).

Research and technology alone will not drive agricultural growth. The full and beneficial effects of agricultural research and technological change will materialize only if government policies are conducive to and supportive of poverty alleviation and sustainable management of natural resources. Public investment in agricultural research is of particular importance for achieving food security in developing countries (Pinstrup-Andersen and Cohen, 2000). Expanded enlightened adaptive research on agricultural biotechnology can contribute to food security in developing countries, provided that it focuses on the needs of poor farmers and consumers in those countries, identified in consultation with poor people themselves.

REFERENCES

ACC/SCN and IFPRI United Nations Administrative Committee on Coordination/Subcommittee on Nutrition (ACC/SCN) and International Food Policy Research Institute (1999). Fourth Report on the World Nutrition Situation, Draft 1. Geneva: ACC/SCN and Washington: IFPRI.

Bouis HE, Chassy BM, Ochanda JO (2003). Genetically modified food crops and their contribution to human nutrition and food quality. Trends Food. Sci. Tech. 14: 191-209.

Braun JV, Bouis H, Kumar S, Pandya-Lorch R (1992). Improving food security of the poor: Concept, policy, and programs. Washington, D.C.: International Food Policy Research Institute.

Chrispeels MJ (2000). Biotechnology and the Poor. Plant Physiol. 124: 3-6.

Christou P (1997). Biotechnology applied to grain legumes. Field Crops Res. 53: 83-97.

Clover J (2003). Food security in sub-Saharan Africa. Afr. Sec. Rev. 12(1): 5-15.

Conway G (1999). The Doubly Green Revolution. Comstock Publishing Associates, Ithaca, NY.

Conway G, Toenniessen G (2003). Science for African Food Security. Sci. 299(5610): 1187-1188.

Dyson T (1999). World food trends and prospects to 2025. Proc. Nat. Acad. Sci. USA 96: 5929-5936.

FAO (1999). The State of Food Insecurity in the World 1999. Rome: FAO.

Haddad LJ, Bouis HE (1991). The impact of nutritional status on agricultural productivity: Wage evidence from the Philippines. Oxf. Bull. Econ. Stat. 53(1): 45-68.

Huang J, Hu R, Wang O, Eeley J, Zepeda JF (2002). Agricultural biotechnology development, policy and impact in China. Economic and Political Weekly [India] 37 (27) (Review of Science Studies), 6-12 July, p. 2756-2761 (http://www.ids.ac.uk/ids/env/PDFs/China%20PaperEPW.pdf).

James C (2003). Global review of commercialized transgenic crops. Curr. Sci. 84(3): 303-309.

Jenkins JC, Scanlan SJ (2001). Food Security in Less Developed Countries, 1970 to 1990. Am. Sociol. Rev. 66: 718-744.

Mackay GR (1991). Genetic Manipulation of the potato for improved processing. J. Sci. Food. Agric. 57: 449-458.

Maxwell D, Levin C, Armar-Klemesu M, Ahiadeke C, Ruel M, Morris S (1998). Urban livelihoods, food and nutrition security in greater Accra. Research Report. Washington, D.C.: International Food Policy Research Institute.

Mazur B, Krebbers E, Tingey S (1999). Gene discovery and product development for grain quality traits. Sci. 285: 372-377.

Mazur BJ (2001). Developing transgenic grains with improved oils, proteins and carbohydrates. Novartis Found. Symp. 236: 233-239.

Niba LL (2003). The relevance of biotechnology in the development of functional foods for improved nutritional and health quality in developing countries. Afr. J. Biotechnol. 2(12): 631-635.

Oshaug O (1994). Nutrition security in Norway? A situation analysis. 1994. Scand. J. Food Nutr. 38(28): 1-68.

Paarlberg R (2000). The global food fight. Foreign Aff. 79(3): 24-38.

Persley GJ (1999). Agricultural biotechnology and the poor: Promethean science. (http://www.cgiar.org/biotech/rep0100/persley.pdf).

Pinstrup-Andersen P, Cohen MJ (2000). Modern Biotechnology for Food and Agriculture: Risks and Opportunities for the Poor. In: Agricultural biotechnology and the poor. Persley GJ and Lantin MM (eds). Proceedings of an international conference, The World Bank. Washington DC.

Pinstrup-Anderson P, Pandya-Lorch R, Rosegrant MW (1999). World food prospects: Critical issues for the early twenty-first century. International food policy research institute, Washington, DC.

Pretty JN (1995). Regenerating agriculture. Joseph Henry Press, Washington, DC.

Pryer J, Crook N (1988). Cities of hunger. Urban malnutrition in developing countries. Oxford: OXFAM.

Rondinelli D, Kasarda JD (1993). Job creation needs in Third World cities. In: Third World cities: problems, policies, and prospects. Kasarda JD and Parnell AM (eds). Sage Publications.

Rosegrant MW, Cline SA (2003). Global Food Security: Challenges and Policies. Sci. 302(5652): 1917-1919.

Ruel MT, Garrett JL, Morris SS, Maxwell D, Oshaug A, Engle P, Menon P, Slack A, Haddad L (1998). Urban challenges to food and nutrition security: A review of food security, health, and care giving in the cities. FCND Discussion Paper No. 51.

Satyanarayana K, Nadamuni NA, Narasinga RBS (1980). Agricultural employment, wage earnings and nutritional status of teenage rural Hyderabad boys. Ind. J. Nutr. Diet 17: 281-286.

Serageldin I (1999). Biotechnology and food security in the 21st century. Sci. 285: 387-389.

Soetan KO (2008). The Dynamic and Ubiquitous nature of Biotechnology-A Review.Afr. J. Biotechnol. 7(16): 2768-2772.

Soetan KO, Abatan MO (2008). Biotechnology A key tool to Breakthrough in Medical and Veterinary Research- A Review. Biotechnol. Mole. Biol. Rev. 3(4): 088-094.

Strauss J, Thomas D (1996). Wages, schooling, and background: Investments in men and women in urban Brazil. In Opportunity foregone: Education in Brazil. Washington, D.C.: Inter-American Development Bank.

Swaminathan MS (2000). Genetic Engineering and Food Security: Ecological and Livelihood Issues. In: Agricultural Biotechnology and the Poor. Persley GJ and Lantin MM (eds). Proceedings of an International Conference. The World Bank. Washington DC.

Tabatabai H (1993). Poverty and food consumption in urban Zaire. Working Paper 47. Ithaca, N.Y., U.S.A.: Cornell Food and Nutrition Policy Program.

Tonukari NJ, Ikea JK and Ude G (2003). Biotechnology research. Afr. J. Biotechnol. 2(12): 477.

UN ACC/SCN (United Nations Administrative Committee on Coordination-Subcommittee on Nutrition) (1991). Brief on policies to alleviate underconsumption and malnutrition in deprived areas. Draft. Revised 12 February. Geneva.

United Nations (1998). World urbanization prospects: The 1996 revision. New York.

Uzogara SG (2000). The impact of genetic modification of human foods in the 21st century: A review. Biotechnol. Adv. 18: 179-206.

Vasil IK (1998). Biotechnology and food security for the 21st century: A real-world perspective. Nat. Biotechnol. 16(5): 399.

World Bank (1999). "Poverty Data: Income Poverty." http://www.worldbank.org/poverty/data/trends/income.htm.

Yan L, Kerr PS (2002). Genetically engineered crops: their potential use for improvement of human nutrition. Nutr. Rev. 60: 135-141.

21

Computational aspects of thematics: application to protein tyrosine phosphatase role in diabetes mellitus

Allam Appa Rao

Center for Biotechnology, Andhra University College of Engineering, Visakhapatnam-530003, India. E-mail: principal@aucevizag.ac.in.

Computation plays an important role in functional genomics and proteomics. Theoretical Microscopic Titration Curves (Thematics) are being employed to predict active binding sites of enzymes. The principal reasons are that the pace of discovery of new proteins is increasing, outpacing the ability to characterize them in conventional biochemical and structural techniques; in addition, advances in computational, structural and force data are used in an interative manner to improve accuracy of active site prediction. From methods using amino acid and nucleotide sequences evidence is available that residues in the enzyme core are selected for stability while those at the surface, which are sites of protein interaction, trade off stability for ligand interactivity. Thematics is a computational method that predicts chemical and electrostatic properties of residues in enzymes and utilizes information contained in those predictions to identify active sites. Discussion of the chemical basis for the predictive powers of Thematics is featured in this paper, and the role of protein tyrosine phosphatase in type 2 diabetes discussed briefly.

Key words: Computational aspects, thematics, protein tyrosine phosphatase, diabetes mellitus.

TABLE OF CONTENT

1. Introduction
2.'Lock and key' Concept and computational prediction
3. Molecular docking
4. Theoretical microscopic titration curves)
5. Prediction of interface residues and sites
6. Correlated mutations and its detection
7. Method of Raj Chakrabarti
8. Protein topology and stability
9. Protein tyrosine phosphatase and its relationship to diabetes
10.Conclusion
11. Acknowledgment
12. References

INTRODUCTION

Target identification might seem to be the sole domain of bioinformatics methods. Computational science or computer science has been playing a major role in these methods. Bioinformatics methods have to find the genes that are up or down regulated in certain diseases and then point out which proteins are expressed by these genes. These proteins might either be enzymes catalyzing certain metabolic reactions or be proteins involved in signaling pathways. Manual analysis of such a gigantic data is impossible, computer analysis became a necessity. Access to powerful methods for prediction of binding constants for ligands to target receptors is highly desirable in the drug discovery process. Computational methods for the estimation of binding constants (or, equivalently, free energies of binding) can be particularly fruitful when the structure of the receptor is known. Predictions of the affinity of a given inhibitor for different enzyme variants and mutants may be equally valuable.

Enzymes are protein molecules, which perform a variety of cellular and regulatory functions by interacting with ligands at specific locations called 'active sites.' The active site depends on a variety of factors from the arrangement of protein chain in space and time (secondary, tertiary and quaternary structures). The active site is dynamic with interactions at a variety of physical and chemical factors in the surroundings. With more data flooding can be annotated, computational methods are employed in this area (Terribilini, 2006). Predicting the active site leads to better understanding of a proteins function and therefore can be useful to design drugs rationally that can modulate interaction of proteins by modifying their structure as well as to establish interactions among networks of proteins (Liang, 2006; Halperin, 2003).

The motivation for 'reverse engineering' of protein structure and function is the assumption that sequences of naturally occurring proteins are defined by selective pressure evolutionarily; the pressure is exerted by a balance of function and stability (Koehl, 2002). Studies comparing stability and solubility of proteins from their sequences have shown that core residues in proteins appear to follow the need for stability (Koehl, 2002).

Efforts were made to use structure of known proteins and predict the amino acid sequences so that the information can be utilized to predict structure of n Sequence of amino acids (primary structure) to can be utilized to predict structure of new proteins whose amino acid sequences alone are known (Raja, 2000). Many protein sequence design methods have been employed such as homology modeling in which attempts are made to match unknown to known, and then thread the sequence to a known backbone based on energy expression. Energy expression is in turn determined by search algorithms such as stochastic and deterministic.

'Lock and key' Concept and computational prediction

Emil Fischer, a chemist, summarized the action of enzymes as follows: 'In order to be able to act chemically on one another, an enzyme and its substrate must fit together like a lock and key.' The original 'lock and key' concept has been revised and refined to describe the site where substrate and enzyme interact. The region of the enzyme containing the binding and catalytic sites is called the active site (Palmar, 2004). It usually occurs near or on the surface of the protein as a cleft for the substrate to fit. The active site is located in three dimensions, with amino acids or cofactors comprising binding and active sites. These arrangements are further affected by amino acids in the microenvironment of the active site.

Identification of binding and catalytic sites is a complex process, involving a variety of processes including the use of substrate analogues, modification of amino acid side chains of enzymes, generation of fusion protein and by site-directed mutagenesis (Palmar, 2004; Kennelly, 2006) In essence, to understand enzyme action at the active site requires not only the identity of amino acids, their spatial arrangement. Information about enzyme structure has been traditionally obtained by x-ray crystallography and NMR spectrometry. Considering the complexity of the physical and chemical properties in which the interactions occur, computational prediction requires, besides, the order of amino acids, the various physical and chemical interactions of atoms lining the active site gorge. Once the raw data is available, computational, mathematical and statistical techniques are employed to predict the active site.

Various computational methods for estimating ligand-binding affinity have been devised, each representing a different choice of computational demand versus accuracy (Ajay, 1995). Free energy perturbation (FEP) theory (Kolllman, 1993; Jorgensen, 1989; Beveridge, 1989) combined with conformational sampling by molecular dynamics (MD) or Monte Carlo (MC) simulation can be used to calculate free energy changes upon small modifications of the ligand or receptor (Singh, 1988a, b). However, FEP calculations become quite complicated and computationally expensive for structurally dissimilar inhibitors and for calculation of absolute free energies of binding. In the latter case, precautions must also be taken to ensure a correct standard state when molecules are annihilated or created (Cammon, 1997). Scoring functions applied to single conformations of the docked complex is a more empirical approach to affinity prediction. They are generally based on identifying individual points of intermolecular interaction such as hydrogen bonds, ionic interactions and hydrophobic interactions, as well as entropy estimates, in a given conformation of the receptor-ligand complex and as signing a binding energy score to each contributing factor (Bohm, 1994; Wallqvist, 1995; Smyithe, 1996; Jain, 1996). These methods include estimates of the effect of solvation, either implicitly by their parameterization to fit a set of experimental values or explicitly by measuring the change in the solvent-accessible surface area upon binding. There are also examples of utilizing scaled molecular mechanics energies for minimized structures to obtain binding energy estimates (Holloway, 1995).

Yuko Tsuchiya et al (Yuko, 2004) analyzed 63 protein-DNA complexes by focusing their attention on the shape of the molecular surface of the protein and DNA, along with the electrostatic potential on the surface, and constructed a new statistical evaluation function to make predictions of DNA interaction sites on protein molecular surfaces. Recently Kengo et al. (2007) have developed a method to predict ligand-binding sites in a new protein structure by searching for similar binding sites in the Protein Data Bank (PDB).

The similarities are measured according to the shapes of the molecular surfaces and their electrostatic potentials. They provided new web server, to interface the search method (URL: http://eF-site.hgc.jp/eF-seek). It has been observed that the density of conserved residue positions is higher at the interface regions of interacting protein surfaces, except for antibody–antigen complexes, where a very low number of conserved positions is observed at the interface regions. Yuhua et al. (2005) identified putative interacting regions on the surface of interacting partners. Using the free energy along with conservation information and other descriptors used in the literature for ranking docking solutions, such as shape complementarity and pair potentials, they developed a global ranking procedure that significantly improves the docking results by giving top ranks to near-native complex structures

Molecular docking

Assigning function to uncharacterized enzymes discovered through genome projects has provided a great challenge to the fields of informatics, enzymology and structural biology. Docking potential ligands into flexible models of protein structures and docking potential high-energy intermediates, rather than substrates, into known structures are two new computational methods that have provided a much-needed boost to the field (Karen, 2007). The use of structure based docking correctly predicts the function of BC0371 enzyme of enolase family (Ling, 2007) and S-adenosylhomocysteine deaminase of an enzyme from the amidohydrolase family (Jhoannes, 2007). In this work the authors approached the problem with the rational that because proper interaction, and thus corrected computational scoring are the reflection of fit of enzyme to ligand, the use of more complementary ligand will produce superior results. Since the 1990s, hundreds of genomes have been sequenced, providing the great promise of new drug targets, a comprehensive understanding of cell metabolism and a deep well of knowledge for protein engineering (Marvin, 2003). Reliable molecular docking and estimation of binding affinities remain an important goal in structure based drug design applications (Blaney, 1993; Jones, 1995; Lynbard, 1995; Rosenfeld, 1995; Gschwend, 1996; Lengauer, 1996). Given a novel inhibitor against an enzyme of known structure, it is important to predict how the ligand is going to bind. Successful prediction allows one to prioritize synthetic resource into producing compounds, which are more likely to be useful and biologically active. Docking studies are also important when designing new ligands in order to confirm the binding mode and to form the basis for an energetic assessment of their binding affinity. Docking methodology can be used to screen compound databases (Kuntz, 1982; Miller, 1994; Welch, 1996) provided the docking is fast enough and there is a method to provide a reasonably reliable ranking order for the biological activity of the ligands. Kmuncek et al. (????) explained the applicability of evaluate the applicability of automated molecular docking techniques and quantum mechanical calculations to the construction of a set of structures of enzyme-substrate complexes for use in Comparative binding energy analysis to obtain 3D structure-activity relationships (Jan, 2003).

Thematics (Theoretical microscopic titration curves)

Thematics is based on finite difference Poisson-Boltzmann (FDPB) methods for calculating the electrical potential function for a complex array of charges, coupled with a Monte-Carlo procedure for determining the average charge as a function of pH for the ionizable residues in that protein's calculated potential. This method identifies reactive sites, including residues involved in catalysis (Ondrechen, 2001). This technique utilizes Deep View-Swiss Pdb Viewer method to obtain the electrical potential function for the protein, followed by calculation of the predicted titration curves for all of the ionizable residues in the protein structure. The novel feature of Thematics is that it extracts information from the shapes of the theoretical titration curves for the ionizable residues in the protein structure, as derived from a FDPB calculation. In a protein the residues generally considered to be ionizable are all of the Arg, Asp, Glu, His, Lys, and Tyr residues, all Cys residues that are not involved in disulfide bridges, plus the N- and C- termini. A typical equation for the pH as a function of the concentrations of an acid HA and its conjugate base as reported (Jaeju, 2005).

In an enzyme, the protein environment influences the chemical and electrostatic properties of active site residues in order to facilitate catalysis and recognition. Any active-site residue that acts as a catalytic Brønsted acid (/base) must then act as a Brønsted base (/acid) in order to restore itself to its initial state for the next turnover cycle. A perturbed titration curve increases the range of conditions over which a residue may act as both an acid and a base, as is required by the definition of a catalyst. Similarly, such unusual titration behavior expands the pH range over which a residue may exist in both charged and neutral forms, thus assisting reversible recognition. Finding these perturbed titration properties not only identifies the active site but may also give clues about the type of chemistry that is catalyzed by that active site.

In proteins, active sites usually comprise hydrophobic pockets that involve side-chain atoms. Methods which use interaction energies between the receptor and different probes to locate energetically favorable sites are complex. They need the assignment of proton locations and partial charges to the receptor atoms. While van der Waals energies can indicate sterically available regions, the long-range nature of electrostatic potentials make the interpretation of energy levels difficult. The purpose of the Site Finder application in MOE is to calculate possible active sites in a receptor from the 3D atomic coordinates of the receptor. The underlying concept is that the pH dependent electrical potential caused by ionizable residue, perturbs the substrate and the catalytically active residues, so that efficient proton transfer is enabled between them over the desired pH range (Fariselli, 2002; Yan, 2006)

Prediction of interface residues and sites

Identification of protein-protein interaction sites (Rao, 2006) and detection of specific amino acid residues that contribute to the specificity and strength of protein interactions is an important problem with applications ranging from rational drug design to analysis of metabolic and signal transduction networks. Computational methods, mostly sequence based, were developed to predict interface sites at differing resolution: at the entire domain, a sequence or at the level of each amino acid (Halperin, 2003). In general, interface residues are predicted by their known characteristics, evolutionary conservation of residues and by residue energy distribution on protein surfaces (Liang, 2006). Known characteristics include the fact that protein interfaces occur at the planar and accessible paths, their surface is less flexible. Shehadi et al. (Ihsan, 2005) suggests that the Thematics can be applicable to a much larger set of proteins for which an experimentally determined structure is unavailable. With a few exceptions, the predicted active sites in the comparative model structures are similar to that of the corresponding template structure.

Correlated mutations and its detection

It has long been suggested that, in the course of evolution, residue substitutions, which tend to destabilize a particular structure must be compensated by other substitutions that confer greater stability on that structure (Tuffery, 1999). Altschuh et al. (Alstchuh, 1987) analyzed the amino acid substitutions in the coat protein structure of tobacco mosaic virus and seven related viruses and showed that some pairs of positions with identical patterns of amino acid substitutions are close together using correlated mutations, residues close to interaction site are expected to mutate simultaneously during evolution, in contrast to convolution approach where simultaneously mutations are looked at two interacting partners rather than a single protein (Halperin, 2003) Sequence-based methods depend on the genomic contex, that is, the primary structure, on gene order conservation and on domain homologues (Halperin, 2003). It should be emphasized that such predictive methods need an iterative experimental characterization and production, even though it assures a 'comprehensive measure of the compatibility of a sequence with a structure' (Raha, 2000). In addition, a variety of statistical predictive methods are employed using neural networks and machine learning techniques (Raha, 2000). Often a combination of methods are used, involving neural networks, docking and superimposition (Li, 2005). Such predictive methods sometimes use well known phage display libraries (Site Light method) (Halperin, 2003).

The remarkable conservation of protein structure, compared to that of sequences, suggests that, in the course of evolution, residue substitutions which tend to destabilize a particular structure must be compensated by other substitutions that confer greater stability on that structure. Given the compactness of proteins, spatially close residues are expected to undergo the compensatory process. Surprisingly, approaches designed to detect such correlated changes have led, until now, only to limited success in detecting pairs of residues adjacent in the three-dimensional structures (Friesner, 2004). Thematics achieves a high success rate of interaction site prediction; about 86% correct or partially correct using CatRes/CSA annotations only and about 93% with an expanded reference set (Ying, 2007). Success rates for catalytic residue prediction are similar to those of other structure-based methods, but with substantially better precision and lower false positive rates. The method requires only the structure of the query protein as input. Thematics predictions may be obtained via the web from structures in PDB format at: http://pfweb.chem.neu.edu/thematics/submit.html

Method of Raj Chakrabarti

Given the fact that most of the amino acid sequences in a protein are optimized for structural stability and are often buried within the structure, attempts were made to perform a more comprehensive physico-chemical and electrical interaction map of amino acids at the surface of the protein which are likely to be sites of interactions with other proteins and chemicals (Li, 2005).

In their original study, Chakrabarti et al chose a set of 10 amino acid residues that form essential contacts to the ligand for sequence optimization. The essential contacts were identified by residues necessary for hydrogen-bonds, salt bridges, van der Waals forces or hydrophobic contacts. The latter was obtained using data and by multiple sequence alignments (Chakrabarti, 2005). The three principal steps in the procedure included determining the lowest-energy protein structure for each residue, following sampling of side-chain conformation and calculation of ligand-binding affinity for each. In this way it was observed that most binding-site amino acids were 'optimized for simple scoring functions based on ligand-binding affinity, under the constraint that residues involveed in catalysis are restricted to catalytically favorable conformations (Chakrabarti, 2005).

Protein topology and stability

The sequences of naturally occurring proteins are defined by evolutionary selective pressure, which is controlled by a fine balance of function, stability, and kinetics. Although most random mutations of sequences are unlikely to enhance stability or function, they can be accepted by natural selection as long as they are neutral (or near neutral). As a consequence, the size of the sequence space

space compatible with a given protein fold is very large (although small compared with the full space a protein sequence can explore, whose size is 20^N, where N is the number of residues of the protein). The number of compatible amino acids at a given position in a protein is structure-dependent: some local structures such as tight turns have energetic constraints that can be satisfied only by small amino acids such as glycine, alanine, or proline. In an exploration of sequence space that was associated with a given protein attribute, all- atom models were used along with a physical energy function (Chakrabarti, 2005). It was found that the volume of sequence space was defined by protein length, protein topology and stability of protein fold. Folding free energy function selects naturally available protein sequences in the core but not on the surface (Koehl, 2002). This is utilized to search sequence space for sequences that satisfy stability constraints for known protein structures, which are then compared with naturally occurring counterparts. Information of native sequence appears to be 'encoded' in the backbone structure of protein (Jarmillo, 2002). Core amino acid sequences were on an average 50% identical to their native counterparts. Surface region identity scores were much lower than core amino acid sequences (Heinz, 1992). Newer methods such as Hidden Markov models are being explored to improve performance of identifying appropriate sequences. It has been suggested that sequences the surface of enzymes may have been selected at least in part, for mediating interactions at the expense of protein stability. Amino acids at the surface were therefore optimized not for protein stability, but for functional reasons (Heinz, 1992).

Many proteins maintain their structure while undergoing extensive mutations. For example, alanine substitution of 10 consecutive residues in bacteriophage T4 lysozyme leads to only minor structural differences (Cordes, 1999). On the other hand, a single double mutation can generate a dramatic structural change, as observed in the Arc repressor for which the interchange of the sequence position of residues 11 and 12 leads to a new structure in which each β-strand is replaced by an α-helix (Bogarad, 1999). These seemingly conflicting results have lead to a complicated picture of protein sequence evolution: it is not clear whether a protein fold can evolve into a new fold by accumulation of simple point mutations (Drexler, 1981). As a first step toward a better understanding of evolution, studies have focused on characterizing the protein sequence space compatible with a given protein structure, the so-called inverse folding problem (Pabo, 1983; Arnold,1998). A large range of methods, including *in vitro* experiments mimicking evolution (Arnold, 1998 a, b; Desjarlais, 1995) and fully automated computer protein design (Dahiyat, 1997; Hellinga, 1998; Muegge, 2001), has been proposed for searching sequences that would stabilize a given protein structure with improved stability or with a new activity.

In summary there has been synergistic use of data and methods obtained from different sources to give meaningful biological solutions to the structure and func-tion of proteins, that are available in geometric progression, faster than can be characterized by hitherto tradi-tional methods. Computational methods offer simulation and analytical opportunities to characterize, identify and modify proteins with appropriate characteristics to catalyze the specific biochemical reaction that is required.

The accurate prediction of binding affinity is one of the most important tasks in de novo ligand design to direct the synthesis of potential modeled ligands bound to receptors. The continuously growing number of protein structures and protein structural models, resulting from the knowledge of genomes, represents enormous promise for development of drugs and other bioactive compounds. Full utilization of the potential requires an armory of computational methods for prediction of ligand–protein-binding affinities. Structure-based prediction methods with various levels of accuracy and speed are available today: fast docking and scoring approaches (Leach, 2006) for reduction of vast chemical libraries to several hundreds of probable binders, the second-pass methods selecting a dozen or so compounds with a high probability of success, and the most sophisticated methods like Free Energy Perturbation (Akash, 2007) and Thermodynamic Integration (King, 1998) which can e.g., discern the affinity of closely related analogs.

Protein tyrosine phosphatase and its relationship to diabetes

The global prevalence of diabetes mellitus has increased continuously and it has been predicted that the number of adult diabetics will double within 30 years. Almost 250 million people, nearly 6% of adults in the world, have diabetes (Mayor, 2006). The problem with diabetes is that, even if the exact same mutation caused it in everyone, it would look different from person to person and family to family, depending on environmental influences, the genetic background it's laid upon, and modifier genes (Martin, 1992 a, b). Its expression would be variable. Furthermore, studies have shown that diabetes is not simple; it's genetically complex, involving multiple genes, and multiple gene-environment interactions

Type 1 diabetes mellitus (T1D) results from an autoimmune destruction of beta-cells in genetically predisposed individuals (Castano, 1990). In prediction of T1D the presence of circulating autoantibodies to insulin (IAA), glutamic acid decarboxylase (GADA) and the protein tyrosine phosphatase like IA-2 (IA-2A) are important and multiple autoantibodies confer a higher risk of developing T1D (Bingley, 1994), especially when combined with the risk genotype (Kimpimaki, 2002).

Type 2 diabetes, formerly known as non-insulin-dependent diabetes (NIDDM), accounts for most cases of diabetes mellitus worldwide. It is estimated that in 2005 there were approximately 160 million individuals with the

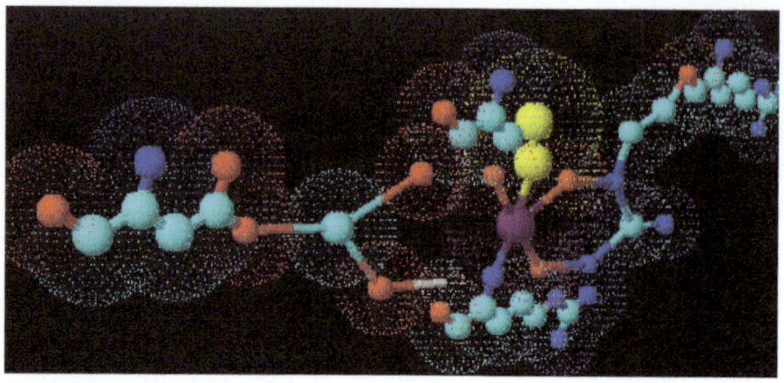

Figure.1. 3D structure of PTP 1B enzyme phosphate formation.

disease and that this number is likely to double by 2025 (Swan, 1982). Swan GW first applied the engineering optimal control theory to equations describing insulin and glucose interactions (Jannik, 2003). T2DM result from complex interaction between genetic and environmental factors. The high incidence of T2DM in certain populations and among first-degree relatives of T2DM patients, as well as the high concordance in identical twins, provides strong evidence that genetic factors underlie susceptibility to T2DM (Elebein, 2002).

Protein tyrosine phosphatase 1B (PTP1B) (Figure 1) has been shown to be involved in the negative regulation of both insulin and leptin action at the *in vitro, ex vivo* and *in vivo* levels. A growing body of human genetic data also support the hypothesis that PTP1B has an important role in insulin signaling and possibly in obesity in humans. The protein tyrosine phosphatases (PTP) constitute a family of closely related key regulatory enzymes that dephosphorylate phosphotyrosine residue in their protein substrates.

Malfunctions of PTP activity are linked to various diseases, ranging from cancer to neurological disorders and diabetes. The family of protein tyrosine phosphatases, which is encoded by 100 genes in humans, plays a critical in the regulation of signal transduction. Recently a variety of links between aberrant PTP function and human diseases has been defined (Liu, 2003) Protein tyrosine phosphatase 1B (PTP1B) has been implicated as one of the key regulators of insulin and leptin signal transduction pathways. Inhibiting PTP1B action using antisense oligonucletides and small molecule inhibitors represents novel therapeutic approach for the treatment of insulin resistance, type II diabetes, and obesity. Liu Gang reviewed the recentaction, particularly small molecules PTP1B inhibitors, and inhibitors with drug like properties (Zhong, 2002). Zhong (2002) summarized the correlation of PTP summarized the correlation of PTP structure and the correlation of PTP structure and function from mutagenesis experiments, in this the molecular basis for PTP1B substrate reorgani zation is discussed, a powerful strategy is presented for creating specific and high affinity bidentate PTP inhibitors that simultaneously bind both the active site and unique adjacent site (Ebina, 1985).

The major sites of autophosphorylation within the pro-tein tyrosine kinase (PTK) domain of the β subunit are Tyr-1158, Tyr-1162, and Tyr-1163 (using the numbering of (White, 1988) which are contained within the PTK activation segment.

Autophosphorylation of all three sites is required for maximal activation (Patti, 1998). Following activation, additional autophosphorylation reactions occur in the juxtamembrane segment and the C-terminal tail of the β subunit. IRK-catalyzed phosphorylation of proteins, such as IRS-1 and IRS-2, creates high-affinity binding sites for proteins that contain Src homology 2 (SH2) domains, such as phosphatidylinositol 3-kinase (PI3-K), resulting in the assembly of the multiprotein signaling complexes that mediate the effects of insulin on cellular metabolism, growth, and glucose homeostasis (Ahmad, 1997).

In the context of insulin signaling, protein phosphorylation is a reversible process, regulated by the coordinated action of protein kinases and phosphatases. Therefore, dephosphorylation of the insulin receptor by protein tyrosine phosphatases (PTPs) is also a critical component of the control of insulin signaling.

Numerous studies have demonstrated that in humans, and in animal mo-dels, insulin resistance in type 2 diabetes and obesity is accompanied by increases in PTP activity and increases in the level of expression of defined members of the PTP family.

In skeletal muscle and adipose tissue from insulin-resistant obese and diabetic subjects, the increased PTP activity has been linked primarily to changes in the expression of the receptor-like PTP LAR (leukocyte common Antigen related) and the cytoplasmic enzyme PTP1B (Ahmad, 1995).

Studies in cell culture models have indicated a functional link between LAR and dephosphorylation of the insulin receptor (Kennedy, 1999).

However, disruption of the *LAR* gene in mice yields a complex phenotype consistent with a postreceptor defect in insulin signaling but associated with impaired activation of downstream signals, such as PI 3-kinase (Desmarais, 1999). Therefore, attention has focused on PTP1B as a major regulator of insulin signaling.

The observation that peptides with tandem pTyr residues bind with high affinity to PTP1B is in agreement with a recent study by Ramachandran and colleagues who observed that pep-tides containing two adjacent nonhydrolyzable analogs of pTyr (difluorophosphonomethyl-Phe) were potent inhibitors of the challenges associated with developing PTP1B advances in various approaches for attenuating PTP1B (Alastair, 2006).

One of the main challenges for the design of potent inhibitors is to overcome the small size and polar nature of the active-site pocket. In the case of PTP1B, bidentate inhibitors have been designed that target two surface pockets. This concept could also be applicable to other family members. The polarity of the active site and difficulties in identifying potent pTyr mimetic compounds resulted in phosphatase inhibitors with unfavorable pharmacological properties. To generate inhibitors that can be used *in vivo*, these properties must be improved.

Recently Alastair J. Barr and Stefan Knapp constructed phylogenetic tree for PTP 1D which will be useful for development of new PTP inhibitors (Kennedy, 1999). Kennedy et al reviewed its role in diabetes and obesity (Kenn, 2000; Burke, 1998)

Design of small molecule compounds inhibiting the enzymatic function of PTP1B is of great medicinal interest. It is actively being pursued in many academic (Moller, 2000) and industrial (Kole, 1996) organizations. It is appreciated that PTP1B is a negative regulator of insulin receptor signaling and a suitable drug target for treatment of insulin resistance associated with diabetes and obesity (Bandyopadhya, 1997). Clinical studies have found a correlation between insulin resistance states and levels of PTP1B expression in muscle and adipose tissues (Ahmad, 1997; Elchelby, 1999). Recent PTP1B knockout studies revealed that mice lacking functional PTP1B exhibit increased sensitivity toward insulin and are resistant to obesity (Chernoff, 1990).

PTP1B is a prototypical intracellular protein tyrosine phosphatase found in a wide variety of human tissues (Salmeen, 2000). The exact roles of PTP1B in relation to insulin resistance are not fully understood. It has been demonstrated that the interaction of insulin with its receptor leads to the phosphorylation of certain tyrosine residues (1158, 1162 and 1163) within the receptor protein, thus activating the receptor kinase (Ahmad, 1995). PTP1B, probably together with other phosphatases (LAR, PTPα and SH-PTP2) (Tonks, 2001), dephosphorylate the activated insulin receptor. Recent studies with LAR knocked out mice have not shown altered glucose homeostasis, however (Malamas, 2000). The dephosphorylated insulin receptor loses the tyrosine kinase activity that is required for further down-stream signaling (Sarmiento, 2000). Inhibition of PTP1B prevents, at least to some extent, the activated insulin receptor from being inactivated. In addition to the above function, PTP1B dephosphorylates a number of other receptor tyrosine kinases, including the EGF receptor and the PDGF receptor (Tuffery, 1999). Protein- tyrosine phosphatases (PTPs), together with protein –tyrosine kinases, control the tyrosine phosphorylation state in the cell (Figure 1), which is important for cellular activities such as growth , differentiation ,motility ,cell-cell interact-tions ,metabolism, gene transcription, and the immune response(Hunter, 1995; Tonks, 1996). PTPs constitute a growing family of transmembrane (receptor- like) and intracellular enzymes (cytoplasmic) that rival the protein tyrosine kinase in terms of structural diversity and complexity. Extensive biochemical and structural studies of PTPs during the past ten years have led to a detailed understanding of the mechanism by which PTPs catalyze phosphate monoester hydrolysis. PTP1B can effectively accommodate peptides of varying sequence, in a catalytically viable fashion that differs from that observed in the PTP1B.DADEpYL structure .considering that the region surrounding the active site is covered with many charged cervices and protrusions, compensatory interact-tions may be available. A further understanding of the specific functional role that PTPs play in cellular signaling and how they are mechanically able to carry out these roles requires a detailed description of the structural features that control PTP substrate specificity. Inspite of the remarkable progress in the identification and characterrization of new PTPs distinguish and in the understanding of the PTP catalysis, the molecular basis by which PTPs distinguish and recognize the diverse substrates that they encounter in the cell has still not been firmly established. This is partly because the physiological substrates for most PTPs remain unknown and it remains a major obstacle to obtain quantities of specifically and stoichiometrically tyrosine phosphory-lated proteins required for detailed enzymological studies. So the application of Thematics for the study of this type of complex proteins will be useful for *in silico* drug discovery studies. (Figure 2) represents the proposed comparison view of active site prediction / drug target identification using proteomic and thematic approaches.

Conclusion

PTP1B is tractable to structure-based drug design, and the crystal structure is well known. Recently, the crystal structure of the closely related T-cell protein tyrosine phosphatase (TCPTP) has become available, raising the possibility for designing selective inhibitors (Johnson, 2002). It is concluded that PTP1B is playing an important role not only in type 2 diabetes but in other diseases like cardiovascular and alzheimer's. As previously reported by us for mathematical, bioinformatic and structural analysis of complex proteins in diabetes and its complications is useful to understand the pathophysiology of diseases (Bhremeramba, 2007; Rao, 2006; Rao, 2006; Sridhar, 2006; Sridhar, 2005). Thematics are useful for further development of specific inhibitors to proteins (PTP1B), potential anti-diabetic drugs, chemical probes for clarification of the roles of PTP1B in normal cellular processes as well as in pathogenic pathways. It is proved beyond doubt that the computational analysis do necessary for any of the protein like PTP which a complex protein structurally.

ACKNOWLEDGMENT

Author is thankful to Srinubabu Gedela for his help during manuscript preparation.

REFERENCES

Ahmad F, Azevedo JL, Cortright R, Dohm GL, Goldstein BJ (1997). Alterations in Skeletal Muscle Protein-Tyrosine Phosphatase Activity

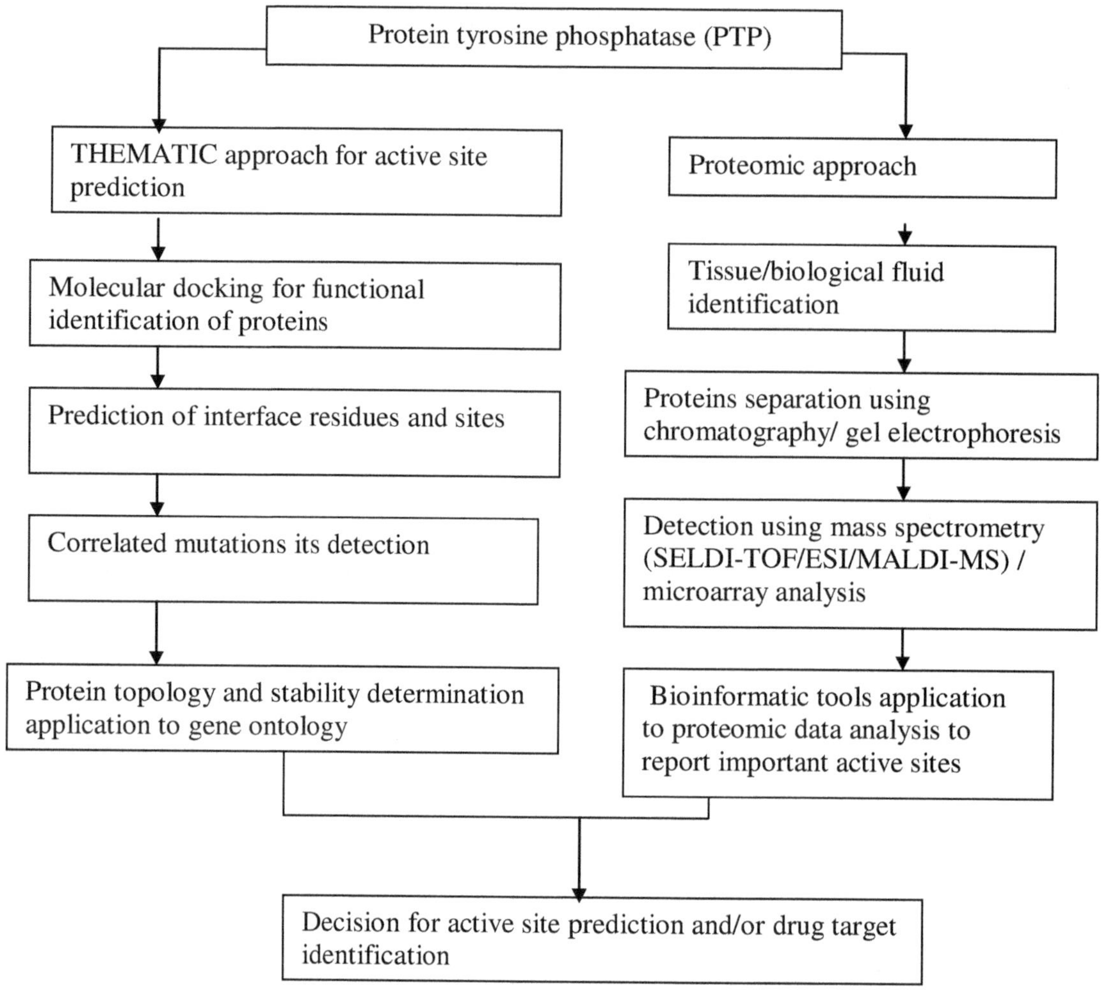

Figure 2. Proposed comparison view of proteomic and thematics approaches for active site prediction; SELDI-TOF: Surface enhanced laser desorption ionization time of flight mass spectrometry; ESI: electro spray ionization mass spectrometry; MALDI: Matrix assisted laser desorption ionization mass spectrometry.

and Expression in Insulin-resistant Human Obesity and Diabetes. Metabolism 46 : 1140-1145

Ahmad F, Azevedo JL, Cortright R, Dohm GL, Goldstein BJ (1997). Alterations in Skeletal Muscle Protein-Tyrosine Phosphatase Activity and Expression in Insulin-resistant Human Obesity and Diabetes J. Clin. Invest. 100: 449–458.Ahmad F, Considine RV, Bauer TL, Ohannesian JP, Marco CC, Goldstein BJ (1997). Improved sensitivity to insulin in obese subjects following weight loss is accompanied by reduced protein-tyrosine phosphatases in adipose tissue .Metabolism. 46:1140-1145.

Ahmad F, Goldstein BJ (1995). Protein-Tyrosine phosphatases:Emerging Targets for Therapeutic invention ofinType2 Diabetes and related states of Insulin Resistance. Biochim. Biophys. Acta. 1248: 57-63.

Ahmad F, Goldstein BJ, (1995). J. Biol. Chem. 272: 448–453.

Ajay, Murcko MA (1995). Computational methods. to. predict. binding. free-energy. in. ligand-. receptor complexes. J. Med. Chem. 38: 4953-4967

Akash KÆ, Stefan B (2007). Improved estimation of ligand–macromolecule binding affinities by linear response approach using a combination of multi-mode MD simulation and QM/MM methods. J. Comput. Aided. Mol. Des. 21: 131–137.

Alastair J, Barr K, Stefan K, T (2006), MAPK-specific tyrosine phosphatases: new targets for drug discovery?. Pharmacol. Sci. 27: 525-530.

Altschuh D, Lesk AM, Bloomer AC, Klug A (1987). Normal modes of symmetric protein assemblies. Application to the tobacco mosaic virus protein disk. J. Mol. Biol. 193: 693-698.

Arnold FH (1998). When blind is better: Protein design by evolution. Nat. Biotechnol. 16: 617-622.

Arnold FH (1998). Design by directed evolution. Acc. Chem. Res. 31: 125-129.

Arnold FH (1998). Enzyme engineering reaches the boiling point Proc. Natl. Acad. Sci. USA. 95: 2035-2039.

Bandyopadhyay D, Kusari A, Kenner KA, Liu F, Chernoff J, Gustafson TA, Kusari J (1997). Protein tyrosine phosphatase role in type 1diabetes J. Biol. Chem 272:1639-1643.

Beveridge DL, DiCapua FM (1989). Free energy via molecular simulations: application to chemical and biochemical system .Annu. Rev. Biophys. Chem. 18: 439-445.

Bhremeramba, Sridhar GR, Rao AA (2007). 'Diabetes Molecular Genetics, Signaling Pathways and Integrated Physiology (J1), sponsored by Biovitrum AB, January 14-19, Keystone Resort, Keystone, Colorado, USA. p. 87.

Bingley PJ, Christie MR, Bonifacio E, Bonfanti R, Shattock M, Fonte MT (1994). Combined analysis of autoantibodies improves prediction of IDDM in islet cell antibody-positive relatives. Diabetes. 43:1304–1310.

Blaney JM, Dixon JS, (1993). Perspect. Drug Discov. Design. 1: 301-306

Bogarad L, Deem M (1999). Random multi-recombinant PCR for the construction of combinatorial protein libraries. Proc. Natl. Acad. Sci. USA. 96: 2591-2597.

Böhm HJ (1994). Protein-protein interactions as targets for antiviral chemotherapy J. Comput.-Aided Mol. Design. 8:243-247.

Burke Jr, TR, Zhang ZY (1998). Protein-tyrosine phosphatases: Structure, mechanism, and inhibitor discovery. Biopolymers. 47: 225-229.

Castano L, Eisenbarth GS (1990). Type-I diabetes: a chronic autoimmune disease of human mouse and rat. Annu. Rev. Immunol. 8: 647–679.

Chakrabarti R, Kilbanov AM, Friesner RA (2005). Computational prediction of native protein ligand-binding and enzyme active site sequences. Proc. Natl. Acad. Sci. USA. 102: 10153-10158.

Chen H, Wertheimer SJ, Lin CH, Amrein KE, Burn P, Quon MJ (1997). protein tyrosine phosphatase inhibition. J. Biol. Chem. 272: 8026-8031.

Chernoff J, Schievella AR, Jost CA, Erikson RL, Neel BG (1990). Cloning of a cDNA for a Major Human Protein-Tyrosine-Phosphatase. Proc. Natl. Acad. Sci. USA. 87: 2735-2739.

Cordes M, Walsh N, McKnight C , Sauer R (1999). Sequence determinants of a conformational switch in a protein structure Sci. 284: 325-329

Dahiyat BI, Mayo SL (1997). De Novo Protein Design: Fully Automated Sequence Selection. Sci. 278: 82-86.

Desjarlais J, Handel T (1995) . strategies in protein design Protein Sci. 4: 2006-2009.

Desmarais S, Friesen RW, Zamboni R, Ramachandran C., Difluro(phosphono) methyl]phenylalanine-containing peptide inhibitors of protein tyrosine phosphatases., Biochem J (1999). 337: 219–223.

Drexler KE (1981). Concepts and schemes for the re-engineering of physical protein modules: generating nanodevices via targeted replacements with constrained amino acids. Proc. Natl. Acad. Sci. USA 78: 5275-5279.

Ebina Y, Ellis L, Jarnagin K, Edery M, Graf L, Clauser E, Ou JH , Masiarz F, Kan YW, Goldfine ID (1985). The human insulin receptor cDNA: The structural basis for hormone-activated transmembrane signalling. Cell. 40: 747–758.

Elbein SC (2002). Perspective: The Search for Genes for Type 2 Diabetes in the Post-Genome Era. Endocrinology. 143: 2012–2018.

Elchelby M, Payette P, Michaliszyn E, Cromlish W, Collins S, Lee LA, Normandin D, Cheng A, Himms HJ, Chan CC, Ramachandran C, Gresser MJ, Tremblay ML, Kennedy BP (1999).enzyme inhibition prediction. Sci. 283:1544.

Fariselli P, Pazos F, Valencia A, Casadio R (2002). Prediction of protein-protein interaction sites in hetero-complexes withneural networks Eur. J. Biochem. 269:1356-1361

Friesner RA, Banks JL, Murphy RB, Halgren TA, Klicic JJ, Mainz DT, Repasky MP, Knoll EH, Shelley M, Perry JK, Shaw DE, Francis P,Shenkin PS (2004). A new approach for rapid, accurate docking and scoring,1. Method and assessment of Docking. J. Med. Chem. 47:1739-49.

Gschwend DA, Good AC, Kuntz ID (1996) Molecular interaction sites of interleukin-2 RNA and methods of modulating the same . J. Mol. Recognition. 9: 175-179.

Lengauer T, Rarey M (1996). Transcription Factors and Neoplasia Vistas in Novel Drug Design Curr. Opin. Struct. Biol. 6: 402-407.

Halperin I, Wolfson H, Nussinov R (2003). Epitope Mapping: The First Step in Developing Epitope-Based Vaccines. Drug. Dev. Protein Sci.12:1344-59.

Head RD, Smythe ML, Oprea TL, Waller CL, Greene S, Marshall GR (1996). Validate - A New Method for the Receptor-Based Prediction of Binding Affinities of Novel Ligands. J.Am.Chem. Soc.118: 3959-3962.

Heinz, D, Baase W, Matthews B (1992). Folding and function of a T4 lysozyme containing 10 consecutive alanines illustrate the redundancy of information in an amino acid sequence. Proc. Natl. Acad. Sci. USA . 89: 3751-3755.

Hellinga HW (1998). Computational protein engineering. Nat. Struct. Biol. 5:525-529.

Holloway MK, Wai JM, Halgren TA, Fitzgerald PM, Vacca JP, Dorsey BD, Levin RB, Thompson Wj, Chen LJ, deSolms SJ, Gaffin N, Ghosh AK, Giuliani EA, Graham SL, Guare JP, Hungate RW, Lyle TA, Sanders WM, Tucker TJ, Wiggins M, Wiscount CM, Woltersdorf OW, Young SD, Darke PL, Zugay JA (1995) Molecular mechanics calculations on protein-ligand complexes. Med. Chem. 38: 305-309.

Hunter T (1995). Protein kinases and phosphatases: the yin and yang of protein phosphorylation and signaling. Cell. 80: 225-236.

Ihsan AS, Alexej A, Alper U, Ying W, Leonel F, Murga M 02115, Usa, Valentin I, Mary JO (2005) J. Bioinformatics. Comput. Biol. 3: 127-143.

J. Clin. Invest. Proteinin tyrosine phosphatase in obesity. 100: 449-453.

Jaeju KO, Leonel F, Murga, Ying W, Mary JO (2005), Prediction of active sites for protein structures from computed chemical properties Bioinformatics. 21: i258-i265.

Jain AN (1996). Scoring noncovalent protein-ligand interactions:A continuous differentiable function tuned to compute binding affinities.J. Comput.-Aided Mol. Design.10:427-432.

Jan K, Michal B, Santos L, Federico G, Rebecca C. Wade JD (2003). J. Comput-Aided Mol. Design. 17: 299–311.

Jannik N, Andersen Nicholas K, tonks (2003). protein tyrosine phosphatase based therapeutics: lessons from PTP1B, Protein phosphatases, Springer Berlin publisher.

Jarmillo A, Wernisch L, Hery S, Woodak SJ (2002). High-Resolution Protein Design with Backbone Freedom. Proc. Natl. Acad. Sci. USA. 99:13554-13559.

Johannes CH, Ricardo MA, Alexander A, Fedorov E, Steven C, Almo B K, Shoichet, Frank M (2007). Nature published online 1 July 2007 (doi: 10.1038/nature 05981).

Johnson TO, Ermolieff J, Michael RJ (2002). Probing acid replacements of thiophene PTP1B inhibitors . Nat. Rev.Drug. Disc.1: 696-699.

Jones G, Willett P (1995). Dose-dependent Activation of Antiapoptotic and Proapoptotic Pathways by Ethanol Treatment in Human Vascular Endothelial Cells. Curr. Opin. Biotech. 6: 652-658.

Jorgensen WL (1989). Free Energy calculations: A Breakthrough for Modeling Organic Chemistry in Solution. Chem. Res. 22: 184-189.

Karen NA (2007). PTP in diabetes. Nat. Chem. Biol. 3: 452-453.

Kenn BP, Kennedy BP, Ramachandran C (2000). Biochemical Pharmacol. 60: 877-883.

Kennedy BP (1999). Role of protein tyrosine phosphatase-1B in diabetes and obesity. Biomed Pharmacother. 53: 466-470.

Kennelly PJ, Rodwell VW (2006). In: Murray RK, Granner DK, Rodwell VW Harper's Illustrated Biochemistry. McGraw Hill, Boston. 27 ed, pp. 49-60.

Kimpimaki T, Kulmala P, Savola K, Kupila A, Korhonen S, Simell T (2002) . Natural History of ß-Cell Autoimmunity in Young Children with Increased Genetic Susceptibility to Type 1 Diabetes Recruited from the General Population. J. Clin. Endocrinol. Metab. 87: 4572–4579.

King H, Aubert RE, Herman WH (1998). Global burden of diabetes. 1995–2025: prevalence numerical estimates and projections. Diabetes Care. 21: 1414–1431.

Koehl P, Levit M (2002). Roles of mutation and recombination in the evolution of protein thermodynamics. Proc. Natl. Acad. Sci. USA. 99: 120-125

Koehl P, Levitt M (2002). Structural Determinants of the Rate of Protein Evolution in Yeast. Proc Natl Acad. Sci. USA. 99:1280-5.

Kole HK, Garant MJ, Kole S, Bernier M (1996). Protein tyrosine phosphatase role in type 2 diabetes. J. Biol. Chem. 271:14302-14307.

Kollman P (1993). Computational and Experimental tools for Studying Protein Structure and Stability. Chem. Rev. 93: 2395-2417.

Kuhlman B, Baker D (2000). Structural Bioinformatics: Methods, Concepts and Application to Blood Coagulation Proteins.Proc. Natl. Acad. Sci. USA. 97:10383-8.

Kuntz ID, Blaney JM, Oatley SJ, Langridge R, Ferrin TE (1982). Molecular speleology: the exploration of crevices in proteins for prediction of binding sites, design of drugs and analysis of surface recognition. J. Mol. Biol. 161: 269-276.

Leach AR, Shoichet BK, Peishoff CE (2006). Prediction of protein-ligand interactions. Docking and scoring: successes and gaps. J. Med. Chem. 49:5851-5856.

Li B, Gallin WJ (2005). A7DB: a relational database for mutational phy-

siological and pharmacological data related to the α7 nicotinic acetylcholine receptor. BMC Structur. Biol. 5: 16-21.
Liang S, Zhand C, Liu S, Zhou Y (2006). Protein binding site prediction using an empirical scoring function. Nucleic. Acids. Res.34:3698-3707. Ling S, Chakrapani K, Alexander A, Fedorov EV, Margaret E, Shoshana B, Heidi J, Patricia C, Steven CA, Matthew PJ, John AG (2007). Roles of galectins in chronic inflammatory microenvironments. Nat. Chem. Biol. 3:486-491.
Liu G (2003). Protein Tyrosine Phosphatase 1B Inhibition: Opportunities and Challenges. Curr. Med. Chem. 10: 1407-1421.
Lybrand TP (1995). Ligand-protein docking and rational drug design. urr. Opin. Struct. Biol. 5: 224-229.
Malamas MS, Sredy J, Gunawan I, Mihan B, Sawicki DR, Seestaller L, Sullivan D, Flam BR (2000). α-Ketocarboxylic acid-based inhibitors of protein tyrosine phosphatases J. Med. Chem. 43: 995-1002.
Martin BC, Warram JH, Krolewski AS, Bergman RN, Soeldner JS, Kahn CR (1992) . Role of glucose and insulin resistance in development of type 2 diabetes mellitus: results of a 25-year follow-up study. Lancet. 340: 925–929.
Martin BC, Warram JH, Rosner B, Rich SS, Soeldner JS ,Krolewski AS (1992). Familial clustering of insulin sensitivity. Diabetes. 41: 850–854.
Marvin EF, Gary MJ, David G, Thomassen C, EO, Aristides P (2004) Harnessing the power of genomics for energy and the environment Sci. 300: 290 – 293.
Mayor S (2006). News Severe adverse reactions prompt call for trial design changes. BMJ. 9: 191-1194.
Miller MD, Kearsley SK, Underwood DJ Sheridan RP (1994). Structure activity relationship approaches and applications J. Comput.-Aided Mol. Design. 8: 153-159.
Møller NPH, Iversen LF, Andersen HS, McCormack JG (2000). PTP1B inhibited drugs.,Curr. Opin. Drug Dis. Dev. 3: 527-533.
Muegge I, Rarey M (2001). Proteins in computer sci. Rev. Comput. Chem. 17:1-8.
Ondrechen MJ, Clifton JG, Ringe D (2001). Protein structure to function: insights from computation Proc Natl Acad Sci USA. 98: 12473-12478.
Pabo C (1983). Molecular technology: Designing proteins and peptide. Nature (London). 301: 200-205.
Palmer T (2004) Enzymes. Affiliated East West Press, New Delhi. pp. 175-909.
Palmer T (2004). Enzymes. Affiliated East West Press, New Delhi. pp. 67-75.
Patti ME, Kahn CR (1998). The insulin receptor--a critical link in glucose homeostasis and insulin action. J. Basic Clin. Physiol. Pharmacol. 9: 89–109.
Raha K, Wollacott AM, Italia MJ, Desjarlais JR (2000).). Prediction of aminoacid sequence from structure. Protein Sci. 9:1106-19.
Rao AA, Bhaskar D (2006). Integration specificity of phage c31 integrase in Human genome Int. J. Bioinformatics. 1: 1-5
Rao AA, Bhremeramba, Sridhar GR (2006). Proceedings of 18th IEEE International Conference on Tools with Artificial Intelligence (ICTAI'06). pp. 810-821.
Rao AA, Sridhar GR (2006). Butylcholinesterase connection between type 2 diabetes and Alzheimer's disease. Lipids in Health and Disease. 5: 28-32.
Rao AA., Proceedings of 18th IEEE International Conference on Tools with Artificial Intelligence (ICTAI'06). pp. 12-13.
Ren JM, Li PM, Zhang WR, Sweet LJ, Cline G, Shulman GI, Livingston JN, Goldstein BJ, . Transgenic mice deficient in the LAR protein-tyrosine phosphatase exhibit profound defects in glucose homeostasis. Diabetes 1998). 47: 493–497.
Rosenfeld R, Vajda S, Delisi C (1995). Ann. Rev. Biophys. Biomol. Struct. 24: 677-683.
Salmeen A, Andersen JN, Myers MP, Tonks NK, Barford D (2000). Molecular basis for recognition and dephosphorylation of the activation segment of the insulin Cell. 6: 1401-1406.
Sarmiento M, Puius YA, Vetter SW, Keng YF, Wu L, Zhao Y, Lawrence DS, Almo SC, Zhang ZY (2000). Structural Basis of Plasticity in Protein Tyrosine Phosphatase 1B Substrate Recognition Biochem. 39: 8171-8176.
Singh UC (1988). Probing the Salt Bridge in the Dihydrofolate Reductase-Methotrexate Complex by Using the Coordinate-Coupled Free-Energy Perturbation Method. Proc. Natl. Acad. Sci. USA. 85: 4280-4585.
Singh UC, Benkovic S, McCammon JA, Gilson K, Given JA, Bush BL (1997). Computation of affinity and selectivity: Binding of 2,4-diaminopteridine and 2,4-diaminoquinazoline inhibitors to dihydrofolate reductases Biophys. J. 72: 1047-1053.
Sridhar GR, Nirmala G, Rao AA, Madhavi AS, Sreelatha S, Sudha Rani J, Vijayalakshmi P (2005). Serum butyrylcholinesterase in type 2 diabetes mellitus: a biochemical and bioinformatics approach. Lip. Health and Dis. 4: 18-23.
Sridhar GR, PV Lakshmi, Allam Appa Rao (2006). Phylogenetic tee construction. J. Assoc. Physicians.India. 54:122-128.
Swan GW (1982). An optimal control model of diabetes mellitus, Bull. Mathematical. boil. 44 : 793-808.
Terribilini M, Lee JH, Yan C, Jernigan RL, Honavar V, Dobbs D (2006). Prediction of RNA binding sites in proteins from amino acid sequence. RNA.12:1450-1462.
Tonks NK, Neel BG (1996). Protein Tyrosine Phosphatases in the Vessel Wall. 87: 365-368.
Tonks NK, Neel BG (2001). The T-cell protein tyrosine phosphatase is phosphorylated on Ser-304 by cyclin-dependent protein kinases in mitosis. Curr. Opin. Cell Biol 13: 182-187.
Tuffery P, Durand M, Darlu P (1999). How possible is the detection of correlated mutations?, Theor. Chem. Acc. 101: 9-14.
Wallqvist A, Jernigan RL, Covell DG (1995). A preference-based free-energy parameterization of enzyme-inhibitor binding. Applications to HIV-1-protease inhibitor design Protein Sci. 4:1881-1903
Welch W, Ruppert J , Jain AN (1996). Evidence for the Evolution of Bdelloid Rotifers Without Sexual Reproduction or Genetic Exchange. Chem. Biol. 3: 449-455.
White MF, Shoelson SE, Keutmann H, Kahn CR, (1988). A cascade of tyrosine autophosphorylation in the beta-subunit activates the phosphotransferase of the insulin receptor. J. Biol. Chem. 263: 2969–2980.
Yan C, Terribilini M, Wu F, Jernigan RL, Dobbs D, Honavar V (2006). Alternative Splicing: New Insights from Global Analyses BMC. Bioinformatics. 7: 262-268.
Ying W, Jaeju K, Leonel F, Murga, Mary JO (2007). Active site prediction fo rcomparative model structures with Thematics. BMC Bioinformatics. 2007. 8: 119-125.
Yuhua D, Boojala VB, Reddy, Yiannis N, Kaznessis (2005).Physicochemical and residue conservation calculations to improve the ranking of protein–protein docking solutions Protein Sci. 14:316-328.
Yuko T, Kengo K , Haruki N (2004). Protein structure function bioinformatics. 55: 885-894.Kengo K, Yoichi M, Haruki N (2007). Identification of protein biochemical functions by similarity search using the molecular surface database eF-site. Nucleic Acids Res. 35: W398–W402.
Zhong YZ (2002). Annual. Rev. Pharmcol. Toxicol. 42: 209-234

Permissions

All chapters in this book were first published in BMBR, by Academic Journals; hereby published with permission under the Creative Commons Attribution License or equivalent. Every chapter published in this book has been scrutinized by our experts. Their significance has been extensively debated. The topics covered herein carry significant findings which will fuel the growth of the discipline. They may even be implemented as practical applications or may be referred to as a beginning point for another development.

The contributors of this book come from diverse backgrounds, making this book a truly international effort. This book will bring forth new frontiers with its revolutionizing research information and detailed analysis of the nascent developments around the world.

We would like to thank all the contributing authors for lending their expertise to make the book truly unique. They have played a crucial role in the development of this book. Without their invaluable contributions this book wouldn't have been possible. They have made vital efforts to compile up to date information on the varied aspects of this subject to make this book a valuable addition to the collection of many professionals and students.

This book was conceptualized with the vision of imparting up-to-date information and advanced data in this field. To ensure the same, a matchless editorial board was set up. Every individual on the board went through rigorous rounds of assessment to prove their worth. After which they invested a large part of their time researching and compiling the most relevant data for our readers.

The editorial board has been involved in producing this book since its inception. They have spent rigorous hours researching and exploring the diverse topics which have resulted in the successful publishing of this book. They have passed on their knowledge of decades through this book. To expedite this challenging task, the publisher supported the team at every step. A small team of assistant editors was also appointed to further simplify the editing procedure and attain best results for the readers.

Apart from the editorial board, the designing team has also invested a significant amount of their time in understanding the subject and creating the most relevant covers. They scrutinized every image to scout for the most suitable representation of the subject and create an appropriate cover for the book.

The publishing team has been an ardent support to the editorial, designing and production team. Their endless efforts to recruit the best for this project, has resulted in the accomplishment of this book. They are a veteran in the field of academics and their pool of knowledge is as vast as their experience in printing. Their expertise and guidance has proved useful at every step. Their uncompromising quality standards have made this book an exceptional effort. Their encouragement from time to time has been an inspiration for everyone.

The publisher and the editorial board hope that this book will prove to be a valuable piece of knowledge for researchers, students, practitioners and scholars across the globe.

List of Contributors

Anuj Kumar Chandel
Department of Microbiology, Osmania University, Hyderabad-500 007, India
Department of Biotechnology, Jawaharlal Nehru Technological University, Hyderabad-500 007, India

Chan ES
School of Engineering and Information Technology, Universiti Malaysia Sabah,Kotakinabalu, 88999, Malaysia

Ravinder Rudravaram
Department of Microbiology, Osmania University, Hyderabad-500 007, India

M. Lakshmi Narasu
Department of Biotechnology, Jawaharlal Nehru Technological University, Hyderabad-500 007, India

L. Venkateswar Rao
Department of Microbiology, Osmania University, Hyderabad-500 007, India

Pogaku Ravindra
School of Engineering and Information Technology, Universiti Malaysia Sabah,Kotakinabalu, 88999, Malaysia

Duduku Krishnaiah
Department of Chemical Engineering, School of Engineering and Information Technology University Malaysia Sabah, 88999, Kota Kinabalu, Malaysia

Rosalam Sarbatly
Department of Chemical Engineering, School of Engineering and Information Technology University Malaysia Sabah, 88999, Kota Kinabalu, Malaysia

Awang Bono
Department of Chemical Engineering, School of Engineering and Information Technology University Malaysia Sabah, 88999, Kota Kinabalu, Malaysia

K Sreenivasulu
Department of Biotechnology, KL University, Vaddeswaram 522502, India

M Vijaya Lakshmi
Department of Botany, Acharya Nagarjuna University (ANU), Guntur, India

Huma Habib
Department of Biotechnology, The University of Kashmir, P/O Naseembagh, Hazratbal, Srinagar -190006, Jammu and Kashmir, India

Khalid Majid Fazili
Department of Biotechnology, The University of Kashmir, P/O Naseembagh, Hazratbal, Srinagar -190006, Jammu and Kashmir, India

Anthony I Okoh
Department of Biochemistry and Microbiology, University of Fort Hare, Private Bag X1314, Alice 5700, South Africa

Sarder Nasir Uddin
Biotechnology and Genetic Engineering Disciplne, Khulna University, Khulna-9208, Bangladesh

Godliving Y. S. Mtui
Department of Molecular Biology and Biotechnology, University of Dar es Salaam, P. O. Box 35179, Dar es Salaam, Tanzania

Bishnu Joshi
Central Department of Biotechnology, Tribhuvan University, Kirtipur, Nepal

Megh Raj Bhatt
Central Department of Biotechnology, Tribhuvan University, Kirtipur, Nepal

Dinita Sharma
Central Department of Biotechnology, Tribhuvan University, Kirtipur, Nepal

Jarina Joshi
Central Department of Biotechnology, Tribhuvan University, Kirtipur, Nepal

Rajani Malla
Central Department of Biotechnology, Tribhuvan University, Kirtipur, Nepal

Lakshmaiah Sreerama
Department of Chemistry, St Cloud State University, St Cloud, MN 56301, USA

L. Sudarsana Reddy

V. Sarojamma
Department of Microbiology, Kurnool Medical College, Kurnool - 518 001, INDIA

V. Ramakrishna
Department of Biotechnology, Sri Krishnadevaraya University, Anantapur - 515

Mamoudou H. Dicko
Laboratoire de Biochimie, UFR-SVT, CRSBAN, Université de Ouagadougou, 03 BP. 7021, Ouagadougou 03, Burkina Faso
Laboratory of Food Chemistry, Department of Agrotechnology and Food Sciences, Wageningen University, PO Box 8129, 6700 EV Wageningen, The Netherlands
Laboratory of Biochemistry, Department of Agrotechnology and Food Sciences, Wageningen University, PO Box 8128, 6700 ET Wageningen, The Netherlands

Harry Gruppen
Laboratory of Food Chemistry, Department of Agrotechnology and Food Sciences, Wageningen University, PO Box 8129, 6700 EV Wageningen, The Netherlands

Alfred S. Traoré
Laboratoire de Biochimie, UFR-SVT, CRSBAN, Université de Ouagadougou, 03 BP. 7021, Ouagadougou 03, Burkina Faso,

Alphons G. J. Voragen
Laboratory of Food Chemistry, Department of Agrotechnology and Food Sciences, Wageningen University, PO Box 8129, 6700 EV Wageningen, The Netherlands

Willem J. H. van Berkel
Laboratory of Biochemistry, Department of Agrotechnology and Food Sciences, Wageningen University, PO Box 8128, 6700 ET Wageningen, The Netherlands

Anuj Kumar Chandel
Centre for Biotechnology, Jawaharlal Nehru Technological University, Hyderabad-500 085, India
Department of Biotechnology, School of Engineering of Lorena, University of Sao Paulo, Lorena- 12.602.810, Brazil

Gajula Chandrasekhar
Centre for Biotechnology, Jawaharlal Nehru Technological University, Hyderabad-500 085, India

Konakalla Radhika
Centre for Biotechnology, Jawaharlal Nehru Technological University, Hyderabad-500 085, India

Rudravaram Ravinder
Celestial Labs Limited, Banjara Hills, Hyderabad -500 034 (A.P) India

Pogaku Ravindra
School of Engineering and Information Technology, University Malaysia Sabah, Kotakinabalu, 88999 Sabah Malaysia

H. I. Kankia
Department of Biochemistry, Umaru Musa Yar'adua University, Katsina, Katsina State, Nigeria

Gowhar Shafi
Department of Genomics, Stem cells and Translational Medicine Indo-American Cancer Institute and Research Centre Road # 14, Banjara Hills, Hyderabad-500034, A.P. India

Kaiser Jamil
Department of Genomics, Stem cells and Translational Medicine Indo-American Cancer Institute and Research Centre Road # 14, Banjara Hills, Hyderabad-500034, A.P. India

Atya Kapley
National Environmental Engineering Research Institute, Nehru Marg, Nagpur

Hemant J. Purohit
National Environmental Engineering Research Institute, Nehru Marg, Nagpur

Mohana Ch Vamsy
Department of Genomics, Stem cells and Translational Medicine Indo-American Cancer Institute and Research Centre Road # 14, Banjara Hills, Hyderabad-500034, A.P. India

Ahmed E Fazary
Department of Chemical Engineering, National Taiwan University of Science and Technology, Taipei 106-07, Taiwan, ROC

Yi-Hsu Ju
Department of Chemical Engineering, National Taiwan University of Science and Technology, Taipei 106-07, Taiwan, ROC

I. O. Okonko
Department of Biochemistry and Microbiology, Lead City University, Ibadan, Nigeria

M. O. Ojezele
Department of Nursing Science and Public Health, Lead City University, Ibadan, Nigeria

E. T. Babalola
Department of Microbiology, Crawford University, Igbesa, Ogun State, Nigeria

J. C. Nwanze
Department of Pharmacology and Therapeutics, Igbinedion University, Okada, Edo State, Nigeria

O. K. Mejeha
Department of Microbiology, Federal University of Technology, Owerri, Imo State, Nigeria

T. A. Amusan
Medical Laboratory Unit, Department of Health Services, University of Agriculture, Abeokuta, Ogun State, Nigeria

D. Balachandar
Department of Agricultural Microbiology, Tamil Nadu Agricultural University, Coimbatore - 641 003, India

P. Raja
Department of Agricultural Microbiology, Tamil Nadu Agricultural University, Coimbatore - 641 003, India

K. Kumar
Department of Agricultural Microbiology, Tamil Nadu Agricultural University, Coimbatore - 641 003, India

SP. Sundaram
Department of Agricultural Microbiology, Tamil Nadu Agricultural University, Coimbatore - 641 003, India

Alfred O. Ubalua
Microbiology/Tissue Culture laboratories, National Root Crops Research Institute (NRCRI), Umudike, P. M. B 7006 Umuahia, Abia State, Nigeria

Shazia Rafique
National Centre of Excellence in Molecular Biology, University of the Punjab, Lahore 53700, Pakistan

Muhammad Idrees
National Centre of Excellence in Molecular Biology, University of the Punjab, Lahore 53700, Pakistan

Anwar Nasim
National Centre of Excellence in Molecular Biology, University of the Punjab, Lahore 53700, Pakistan

Haji Akbar
National Centre of Excellence in Molecular Biology, University of the Punjab, Lahore 53700, Pakistan

Amin Athar
National Centre of Excellence in Molecular Biology, University of the Punjab, Lahore 53700, Pakistan

Rachana Tiwari
School of Biotechnology, Devi Ahilya University, Khandwa Rd., Indore-452001, India

Anil Kumar
School of Biotechnology, Devi Ahilya University, Khandwa Rd., Indore-452001, India

Nyerhovwo J. Tonukari
Department of Biochemistry, Delta State University, Abraka, Nigeria

Douglason G. Omotor
Department of Economics, Delta State University, Abraka, Delta State, Nigeria

Allam Appa Rao
Center for Biotechnology, Andhra University College of Engineering, Visakhapatnam-530003